D1234037

KUIPERS and NIEDERREITER—Uniform Distribution of Sequences

LOVELOCK and RUND—Tensors, Differential Forms, and Variational Principles

MAGNUS, KARRASS, and SOLITAR—Combinatorial Group Theory

MARTIN—Nonlinear Operators and Differential Equations in Banach Spaces

MELZAK—Companion to Concrete Mathematics

MELZAK—Mathematical Ideas Modeling and Applications (Volume II of Companion to Concrete Mathematics)

MEYER—Introduction to Mathematical Fluid Dynamics

MORSE—Variational Analysis: Critical Extremals and Sturmian Extensions

NAYFEH—Perturbation Methods

ODEN and REDDY—An Introduction to the Mathematical Theory of Finite Elements

PRENTER—Splines and Variational Methods

RIBENBOIM—Algebraic Numbers

RIBENBOIM—Rings and Modules

RICHTMYER and MORTON—Difference Methods for Initial-Value Problems, 2nd Edition

RIVLIN—The Chebyshev Polynomials

RUDIN—Fourier Analysis on Groups

SAMELSON—An Introduction to Linear Algebra

SIEGEL—Topics in Complex Function Theory

Volume 1—Elliptic Functions and Uniformization Theory
Volume 2—Automorphic Functions and Abelian Integrals
Volume 3—Abelian Functions and Modular Functions of Several Variables

STOKER—Differential Geometry

STOKER—Nonlinear Vibrations in Mechanical and Electrical Systems

STOKER—Water Waves

TRICOMI—Integral Equations

WHITHAM—Linear and Nonlinear Waves

YOSIDA—Lectures on Differential and Integral Equations

AN INTRODUCTION TO THE MATHEMATICAL THEORY OF FINITE ELEMENTS

AN INTRODUCTION TO THE MATHEMATICAL THEORY OF FINITE ELEMENTS

J. T. ODEN

The University of Texas, Austin

J. N. REDDY

The University of Oklahoma, Norman

A WILEY-INTERSCIENCE PUBLICATION

JOHN WILEY & SONS, **New York • London • Sydney • Toronto**

Library of Congress Cataloging in Publication Data

Oden, John Tinsley, 1936–
An introduction to the mathematical theory of finite elements.

(Pure and applied mathematics)
"A Wiley-Interscience publication."
Includes bibliographical references and index.
1. Boundary value problems—Numerical solutions. 2. Differential equations, Elliptic—Numerical solutions. 3. Approximation theory. 4. Finite element method. I. Reddy, Junuthula Narasimha, 1945– joint author. II. Title.

QA379.03 515′.353 76-6953
ISBN 0-471-65261-X

Printed in the United States of America

10 9 8 7 6 5 4 3 2

To the memory of

JOHN JAMES ODEN

PREFACE

During a remarkably short span of years the subject of finite elements has expanded from a collection of effective techniques for solving practical problems in engineering and science to a rich and exciting branch of applied mathematics. The aim of this book is to present the student of engineering science or applied mathematics an introductory account of this mathematical theory.

The book has developed as a result of seminars and courses on finite-element theory taught by the authors at five universities in recent years to students with diverse backgrounds and often modest mathematical preparation. For such an audience, we have found it effective to begin the study with basic mathematical concepts and to systematically build on these the elements of approximation theory, Hilbert spaces, and partial differential equations essential to an understanding of the most important aspects of linear finite-element theory. This book essentially follows this plan. To keep the size and scope of the work within reasonable limits, it has been necessary to omit several important topics in favor of more basic ones. For example, we have not included material on nonlinear problems, integral equations, or eigenvalue problems. However, some of these subjects should be easily mastered by the reader of this book; other subjects must await study in future works.

We owe a great deal to those who developed the mathematical theory of finite elements in recent years. We have been particularly influenced by the work of Ivo Babuška and J. P. Aubin, and we have profited not only in writing this book but also in our own research, from the writings of Philippe Ciarlet, P. A. Raviart, J. L. Lions, and others, and from numerous discussions with our colleague, Ralph Showalter. The first author registers a special note of thanks to the Finite-Element Circus and to certain members of the Circus who have patiently discussed the subject with him; particularly Ivo Babuška, Jim Douglas, Ridgway Scott, Gilbert Strang, Al Schatz, Bruce Kellogg, Mary Wheeler, and James Bramble. We are also thankful for the advice we have received from several colleagues who read an early draft of the manuscript. In particular, we have benefited from the suggestions of John Cannon and Linda Hayes, who read the entire manuscript, and from the comments of Philippe Ciarlet. We also express thanks to M. G. Sheu, N. Kikuchi, and C. T. Reddy who helped with the proofreading. Much of our work on finite-element methods has been supported through the Air Force Office of Scientific Research and the U. S. National Science Foundation. We express our sincere gratitude for this support.

J. T. Oden
J. N. Reddy

Austin, Texas
Norman, Oklahoma
January 1976

CONTENTS

PART II THE THEORY OF FINITE ELEMENTS

AN INTRODUCTION TO THE MATHEMATICAL THEORY OF FINITE ELEMENTS

1

INTRODUCTION

1.1 THE FINITE-ELEMENT METHOD

The finite-element method emerged from the engineering literature of the 1950's as one of the most powerful methods ever devised for the approximate solution of boundary-value problems. A relatively complete historical account is given in [1.1]. It is a variational method of approximation, making use of global or variational statements of physical problems and employing the Rayleigh–Ritz–Galerkin philosophy of constructing coordinate functions whose linear combinations represent the unknown solutions. The key to the success of the method is the unique way in which these coordinate functions are constructed: A given domain is represented as a collection of a number of geometrically simple subdomains (finite elements) connected together at certain nodal points. The variational problem is formulated approximately, for arbitrary boundary conditions, over each subdomain, necessitating only a choice of simple local coordinate functions for each element. These are generally polynomials. The global model of the problem is then obtained by simply fitting the elements together to depict a given domain, and summing, so to speak, the local contributions furnished by each element. We describe these ideas more precisely in Chapter 8.

The importance of this process of building a global approximation from a number of local ones is that it embodies a systematic procedure for constructing coordinate functions for arbitrary domains—thereby finally freeing the analyst of traditional difficulties due to irregular geometries and boundary conditions. Moreover, the coordinate functions generated in this way are usually piecewise polynomials with local compact support; i.e., they assume nonzero values only in a relatively small neighborhood of certain nodal points. This property normally leads to stable, well-conditioned equations for the properly posed boundary-value problem. Indeed, for linear elliptic problems, the method leads to a banded system of linear algebraic equations which can be solved using any of a variety of well-known techniques. All these features combine to make a quite general and effective method which has found fruitful application in practically every area of mathematical physics.

1.2 THE MATHEMATICS OF FINITE ELEMENTS

To the structural engineer, the idea of building up a structure by fitting a number of structural elements together is quite a natural one. That is precisely how he visualizes a structure as being built, and engineers have developed and used stress analysis techniques based on this simple fact for many decades. The translation of these ideas into concrete mathematical statements, however, is not a trivial task and has come about only in relatively recent times.

It is not uncommon in the history of applied mathematics for a mathematical concept or method to be used successfully for years before its mathematical basis is completely understood. Expanding technology demands solutions to physical problems, and the competent technician must develop methods to solve them. When his methods fail or are too fragile for wide application, they are abandoned and forgotten. When they succeed and are sufficiently broad, they eventually attract the attention of mathematicians who are equipped to dissect them and unravel their intrinsic properties. Such was the case with complex variable theory, operational calculus and the theory of distributions, much of probability theory, and numerous other branches of mathematics.

The mathematical theory of finite elements was no exception. By 1965, hundreds of papers on the method had appeared in the engineering literature; it was widely used in industrial applications, was the subject of courses at most technical universities, and was firmly established as a general and powerful method of analysis. However, only a handful of mathematically oriented papers on the method appeared in the engineering literature of the 1960's, and only one or two purely mathematical papers on the subject appeared in 1968 and 1969. A complete mathematical theory began to be pieced together in the 1970's. It was about this time that the strength and elegance of the method and its relation to contemporary research in interpolation theory, splines, and differential equations began to be appreciated by the mathematical world. In a remarkably short time, a virtual flood of mathematical literature appeared on the subject, and a variety of important aspects of the method was quickly put on a sound mathematical footing.

Today the theory has reached a fairly high degree of development, at least as it applies to linear elliptic boundary-value problems, and its foundations are now recognized to be a natural union of spline theory and the modern theory of partial differential equations. In addition, finite-element methods occupy an increasingly important place in modern numerical analysis, and their implementation continues to prompt developments in computational methods and computer software.

1.3 THE PRESENT STUDY

As the mathematical theory of finite elements continues to be developed in the framework described above, it becomes more inaccessible to the very practitioners who developed it, not to mention the beginning student who wishes to learn the underlying mathematical features. This book was written with these readers in mind. It represents a systematic introduction to the mathematical theory of finite elements, starting with relatively elementary mathematical concepts and proceeding to an exposé of basic mathematical properties of finite-element approximations of linear boundary-value problems.

The study is divided into two parts. Part I involves mathematical foundations and contains Chapters 2 through 5. Here all the groundwork for a theory of finite elements is laid. It begins with an elementary introduction to distribution theory, generalized derivatives, and mollifiers, and proceeds to an introductory account of the theory of Sobolev spaces. This is followed by a chapter on interpolation spaces and trace theorems, and a chapter on the theory of elliptic equations. Part II of the study is devoted to the theory of finite elements. It begins with Chapter 6, which is a chapter on finite-element interpolation in which many of the basic properties of finite-element models are described. Here we develop the special local and global properties of finite-element representations that deal with the connectivity of collections of elements and the decomposition of a given domain into a collection of elements. We also cite examples of several important families of finite elements, and we present a collection of theorems that establish the error of finite-element interpolations of smooth functions and of functions in certain Sobolev spaces. These theorems, which involve a priori error estimates for finite-element interpolations, are fundamental to the approximation theory taken up in Chapters 7 through 9. Chapter 7 contains an introduction to the theory of elliptic variational boundary-value problems, with some emphasis on properties of coercive bilinear forms and the equivalence of certain variational problems to classical ones. We also establish an extended version of the Lax–Milgram theorem, following Babuška, which establishes conditions for the existence of unique solutions to an abstract class of elliptic variational problems. We then show that practically all linear elliptic problems of interest fall within those covered by the theorem. In Chapter 8, all the theory developed in previous chapters is brought to bear on the central problem of the book: the approximation of linear elliptic boundary-value problems using finite elements. The basic elements of an approximation theory for finite elements is established, including criteria for stability, consistency, and convergence, as well as a priori error estimates. In Chapter 8 we also consider several special topics in finite-element theory such as boundary errors, L_∞- and L_2-error estimates, and quadrature errors, and we present a brief account of the theory of mixed and hybrid finite-element models. Finally, in Chapter 9 we present a summary account of finite-element methods applied to time-dependent problems wherein we confine ourselves to the simplest linear parabolic and hyperbolic problems.

1.4 NOTATIONS AND PRELIMINARIES

As a prerequisite to reading this book, the reader should be equipped with an introductory course in functional analysis of a level roughly comparable to that covered in the texts of Kolmogorov and Fomin [1.2, 1.3]. However, much of what we cover in the earlier chapters assumes no more background than calculus and real analysis.

We employ standard notation throughout. For instance, $\in$ denotes membership, $\not\in$ negates $\in$, $\forall$ is read "for every" or "every," a set A of elements a having property P is described as $A=\{a\colon a \text{ has property } P\}$, $\subset$, $\cap$, $\cup$, denote the usual set operations, etc. A function is an injection if it is one-to-one, a surjection if it is onto, and a bijection if it is both an injection and a surjection. The space of real numbers (the real line) is denoted R, the complex numbers C, and $\mathsf{R}^n=\prod^n\mathsf{R}=\mathsf{R}\times\mathsf{R}\times\cdots\times\mathsf{R}$ (n times) denotes n-dimensional euclidean space. We use the symbol ■ to denote the conclusion of a logical unit, such as the completion of an example or the proof of a theorem.

As usual, $L_p(\Omega)$, $1\leqslant p\leqslant\infty$, denotes the space of all equivalence classes of real-valued (or complex-valued) Lebesgue-measurable functions u, defined almost everywhere on an open, bounded domain $\Omega\subset\mathsf{R}^n$, whose pth powers $|u|^p$ are Lebesgue-integrable over Ω. The norm on $L_p(\Omega)$ is given by

$$\|u\|_{L_p(\Omega)}=\left(\int_\Omega |u|^p\,dx\right)^{1/p} \tag{1.1}$$

$dx=mes(dx_1\,dx_2\cdots dx_n)$. The dual of $L_p(\Omega)$ is $L_q(\Omega)$, where $1/p+1/q=1$, and if $u\in L_p(\Omega)$ and $v\in L_q(\Omega)$, they satisfy Hölder's inequality

$$\int_\Omega uv\,dx\leqslant \|u\|_{L_p(\Omega)}\|v\|_{L_q(\Omega)} \tag{1.2}$$

There are several basic theorems concerning properties of normed linear spaces, Hilbert spaces, and linear operators that play an important role in some of the developments to be presented later. For example, we mention

1. The Hahn–Banach theorem
2. The Riesz representation theorem
3. The projection theorem for Hilbert spaces
4. The closed graph theorem and the Banach theorem for linear operators on Banach spaces
5. The spectral theorem for linear operators.

Statements and proofs of these theorems can be found in standard texts on functional analysis (see, e.g., [1.2–1.7]). However, we record statements of some of these theorems in appropriate places in the chapters to follow. In the sequel, other notations and conventions are defined where they first appear.

REFERENCES

1.1. Oden, J. T., *Finite Elements of Nonlinear Continua*, McGraw-Hill, New York, 1972.

1.2. Kolmogorov, A. N. and Fomin, S. V., *Elements of the Theory of Functions and Functional Analysis, Vol. I: Metric and Normed Spaces*, Translated from the 1954 Russian edition by L. F. Boron, Graylock Press, Rochester, N.Y., 1957.

1.3. Kolmogorov, A. N. and Fomin, S. V., *Elements of the Theory of Functions and Functional Analysis, Vol. II: Measure. The Lebesque Integral. Hilbert Space*, Translated from the 1960 Russian edition by H. Kamel and H. Kromm, Graylock Press, Albany, N.Y., 1960.

1.4. Naylor, A. W. and Sell, G. R., *Linear Operator Theory in Engineering and Science*, Holt, N.Y., 1971.

1.5. Taylor, A. E., *Introduction to Functional Analysis*, John Wiley, N.Y., 1958.

1.6. Dunford, N. and Schwartz, J. T., *Linear Operators, Part I: General Theory*, Interscience, N.Y., 1958.

1.7. Yosida, K., *Functional Analysis*, 3rd ed., Springer-Verlag, New York, 1971.

PART I
MATHEMATICAL FOUNDATIONS

2

DISTRIBUTIONS, MOLLIFIERS, AND MEAN FUNCTIONS

2.1 INTRODUCTION

In many problems in mathematical physics we encounter relations that can be defined precisely only in terms of an integral—or, more generally, in terms of the action of some operator on functions of a given class. For example, the Dirac delta function $\delta(x)$ has the property

$$\int_0^\infty f(x)\delta(x-a)\,dx = f(a) \tag{2.1}$$

where f is, say, some continuous function on $[0,\infty)$ and $a \in [0,\infty)$. Attempts at actually defining $\delta(x)$ as a *function* of x are at best only symbolic; $\delta(x)$ is simply not a function in the usual sense. Its nonzero values do not exist for any x, and we can interpret it only in terms of the numbers $f(a)$ it produces when it operates, via (2.1), on functions f of a given class.

The need for handling such mathematical objects as the Dirac delta almost always arises in problems with discontinuous data and especially point sources, e.g., points of concentrated nonhomogeneous data, discontinuous coefficients or, more importantly, in attempts to *differentiate* discontinuous functions. Concentrated forces on elastic bodies, impulsive forces on bodies in dynamics, and point electric charges in electrostatics are examples of point sources encountered in physical situations that have mathematical descriptions akin to (2.1), and all of which lead to mathematical difficulties if they are interpreted merely as functions in the classical sense. All these difficulties in handling point sources are resolved by defining a given source by its *action* on a special class of test functions. The "source" $\delta(x-a)$ in (2.1), for example, is defined by $f(a)$; the symbolism $\delta(x-a)$ really signifies that we have some sort of operation whose effect on a member f of a certain class of functions $\mathcal{F}$ is the real number $f(a)$. Effectively, $\delta(x-a)$ symbolizes a functional on $\mathcal{F}$, since it associates with each $f \in \mathcal{F}$ a real number $f(a)$. We refer to functionals of this type as *distributions*, a term we define more precisely below. For convenience in discussing properties of distributions, we often assign to each distribution, such as the Dirac delta $\delta: \mathcal{F} \rightarrow \mathbf{R}$, a special symbolism that makes it appear to be a function, e.g., $\delta(x)$; the symbolic $\delta(x)$ is called a *generalized function*. We now set out to generalize and make more precise all these concepts.

2.2 FUNCTIONALS AND TEST FUNCTIONS ON ONE-DIMENSIONAL DOMAINS

We first need to reinforce our concept of a functional, i.e., a mapping of a linear space into the real numbers. For simplicity, let us begin our study by considering the special but illuminating case in which the domain of the functions considered is the real line $\mathbf{R}=(-\infty,\infty)$. Let $C(\mathbf{R})$ denote the space of continuous functions on $\mathbf{R}$ and recall that it is customary to regard a function $f \in C(\mathbf{R})$ as a mapping f: $\mathbf{R} \rightarrow \mathbf{R}$ defined by some sort of rule, $y=f(x)$. Here y [or $f(x)$] is the *value* of f at x or the image of x under the mapping f, and the rule or law of correspondence of y and x is established by a specific formula for $f(x)$. Geometrically, we can plot a curve showing the variation of y with x by preparing a table of values such as

x	x_1	x_2	x_3	$\cdots$	x_n	$\cdots$
y	$f(x_1)$	$f(x_2)$	$f(x_3)$	$\cdots$	$f(x_n)$	$\cdots$

Then, with each point x_i that we choose in $\mathbf{R}$, a definite value $f(x_i)$ of y exists, and f is viewed as a collection of pairs of points in the xy plane.

Now the function $f(x)$ can also be characterized by a different table of weighted values. Suppose that $\mathcal{D}(\mathbf{R})$ is a class of "smooth" accessory functions $\phi(x)$ defined on $\mathbf{R}$, the exact character of which will be described later, and consider the scalar product s: $f(\mathbf{R}) \times \mathcal{D}(\mathbf{R}) \rightarrow \mathbf{R}$ defined by the Lebesgue integral,

$$\langle f,\phi\rangle = \int_{\mathbf{R}} f\phi\,dx = \int_{-\infty}^{\infty} f\phi\,dx \tag{2.2}$$

If $\mathcal{D}(\mathbf{R})$ is well defined, we can construct a new representation of f in terms of its action on $\mathfrak{D}(\mathbf{R})$ via a table of the type

ϕ	$\phi_1(x)$	$\phi_2(x)$	$\phi_3(x)$	$\cdots$	$\phi_n(x)$	$\cdots$
f	$\langle f,\phi_1\rangle$	$\langle f,\phi_2\rangle$	$\langle f,\phi_3\rangle$	$\cdots$	$\langle f,\phi_n\rangle$	$\cdots$

In other words, f is now described by a collection of *weighted averages* over $\mathbf{R}$, where the weight functions $\phi(x)$ are members of a certain class $\mathfrak{D}(\mathbf{R})$.

Now we recall that a mapping q that assigns to every element ϕ of a linear vector space V a real number $q(\phi)$ is called a *functional* on V. Since $\mathfrak{D}(\mathbf{R})$ can be easily endowed with the structure of a linear space, it is now clear that the second tabular representation of f given above describes f as a functional on the space $\mathfrak{D}(\mathbf{R})$ of accessory functions $\phi(x)$. This latter visualization of f is considerably more general than the pointwise function representation described earlier and, in fact, it allows us to extend the usual idea of a function in an important way. Roughly speaking, such functionals are distributions, but we continue to postpone a fuller definition until a few additional properties of $\phi(x)$ have been established.

Obviously, if the functional representation of f is to make sense, the functions $\phi(x)$ must be chosen so as to have several desirable properties. We would expect, for example, that $\phi(x)$ must be chosen so that two different continuous functions $f_1(x)$ and $f_2(x)$ generate different functionals. We obtain this and other desirable properties by requiring the weight functions $\phi(x)$ to belong to a class of functions called *test functions*. To describe this class more precisely, we first introduce the following definition.

Support of a Function

The *support* of a function f on $\mathbf{R}$ is the closure $\bar{I}$ of the set of points $I\subset\mathbf{R}$ on which $f(x)\neq 0$; we use the notation $\bar{I}=\operatorname{supp} f(x)$.

We refer to the subset of $C^\infty(\mathbf{R})$ consisting of functions with *compact support* as $C_0^\infty(\mathbf{R})$. That is, the set

$$C_0^\infty(\mathbf{R})\subset C^\infty(\mathbf{R}) \tag{2.3}$$

contains infinitely differentiable functions which, together with their derivatives of all orders, assume nonzero values only on some compact set $K\subset\mathbf{R}$.

Test Functions

We now introduce a special class of real-valued functions defined on the real line $\mathbf{R}=(-\infty,\infty)$ called the class of *test functions* on $\mathbf{R}$ and denoted

$\mathfrak{D}(\mathbf{R})$, which has the following properties:

(i) Each $\phi(x) \in \mathfrak{D}(\mathbf{R})$ is in $C_0^\infty(\mathbf{R})$.

(ii) There is a special topology associated with $\mathfrak{D}(\mathbf{R})$ in which the convergence of sequences of test functions is defined in the following way. A sequence $\{\phi_n(x)\}_{n=1}^\infty \in \mathfrak{D}(\mathbf{R})$ *converges* to a test function $\phi(x) \in \mathfrak{D}(\mathbf{R})$ if

(ii.1) The supports of all $\phi_n(x)$ are contained in a fixed compact set $K \subset \mathbf{R}$.

(ii.2) The derivative of *any* given order r of the $\phi_n(x)$ converge uniformly, as $n \to \infty$, to the corresponding derivative of order r of $\phi(x)$.

It is of course possible to define test functions on $\mathbf{R}^n = \mathbf{R} \times \mathbf{R} \times \cdots \times \mathbf{R}$ and, in fact, on bounded domains Ω. We then use the notation $\mathfrak{D}(\mathbf{R}^n)$ or $\mathfrak{D}(\Omega)$ to describe these classes.

REMARKS. 1. Property (i) of test functions merely means that each $\phi(x)$ in $\mathfrak{D}(\mathbf{R})$ is infinitely differentiable and the support of each $\phi(x)$ is contained in a bounded interval of $\mathbf{R}$.

2. Convergence in $\mathfrak{D}(\mathbf{R})$ is a fairly delicate property designed (as we soon see) to provide a broad generalization of the concept of derivatives and differentiation. To better appreciate property (ii) consider what is meant by a *null sequence* of test functions $\{\phi_n(x)\}$. Every $\phi_n(x)$ must vanish outside some common interval of $\mathbf{R}$ [i.e., the intersection of the supports of $\phi_n(x)$ is nonempty], and $\phi_n(x)$ and *all* its derivatives with respect to x must approach zero uniformly as $n \to \infty$:

$$\lim_{n\to\infty} \left| \frac{d^r \phi_n(x)}{dx^r} \right| = 0 \qquad \forall x \in \mathbf{R}; \quad r \geq 0 \tag{2.4}$$

3. The set $\mathfrak{D}(\mathbf{R})$ of all test functions on $\mathbf{R}$ is a linear vector space. Indeed, if α and β are scalars (i.e., real or complex numbers) and $\phi_1(x)$ and $\phi_2(x)$ are two test functions, $\alpha\phi_1(x) + \beta\phi_2(x)$ is obviously infinitely differentiable and has finite support. Hence $\alpha\phi_1(x) + \beta\phi_2(x) \in C_0^\infty(\mathbf{R})$.

4. The spaces $\mathfrak{D}(\mathbf{R})$, $\mathfrak{D}(\mathbf{R}^n)$, and $\mathfrak{D}(\Omega)$ are in fact locally convex linear topological vector spaces. For example, suppose Ω is a bounded interval in $\mathbf{R}$ and let $\mathfrak{D}(K)$ denote the functions in $C_0^\infty(\Omega)$ whose supports are contained in a compact set $K \subset \Omega$. The functions

$$N_{K,m}(\phi) = \sup_{\substack{0 \leq k \leq m \\ x \in K}} \left| \frac{d^k \phi(x)}{dx^k} \right| \qquad m < \infty \tag{2.5}$$

define a family of seminorms on $\mathfrak{D}(K)$. Now if we regard Ω as the union of a sequence of compact sets, $\cup_{i=1}^{\infty} K_i$, $K_1 \subseteq K_2 \subseteq \cdots \subseteq K_i \subseteq \cdots$, the space $\mathfrak{D}(\Omega)$ is obtained as the inductive limit of the $\mathfrak{D}(K)$'s as K ranges over all the compact sets $K_i \subset \Omega$. When topologized in this way, $C_0^{\infty}(\Omega)$ becomes $\mathfrak{D}(\Omega)$. The space $\mathfrak{D}(\Omega)$ [or $\mathfrak{D}(\mathbf{R})$], however, is not metrizable.

Fortunately, we do not need to call on this rather complicated topological structure in any of the developments that follow. Our needs are fully served by an *operational* concept of test functions and distributions which we continue to develop below. For additional details on the topological structures involved, see Schwartz [2.1] or, for a summary account, Yosida [2.2].

5. Clearly, derivatives of test functions of any order are also test functions.

6. If $g(x)$ is any infinitely differentiable function, its product $g(x)\phi(x)$ with any test function $\phi(x)$ obviously has finite support, because $\phi(x)$ does. Hence $g\phi \in \mathfrak{D}(\mathbf{R})$, i.e., $g(x)\phi(x)$ is also a test function. For additional properties, see [2.1]. ■

EXAMPLE 2.1. In discussions of test functions, a standard example is the function

$$\phi(x) = \begin{cases} 0 & |x| \geqslant a \\ \exp \dfrac{1}{x^2 - a^2} & |x| < a \end{cases}$$

This function obviously has finite support, $|x| \leqslant a$. Is it infinitely differentiable? Clearly, for $x \in (-a, a)$,

$$\phi = e^{u(x)} \qquad \frac{d\phi}{dx} = e^u \frac{du}{dx} \qquad \frac{d^2\phi}{dx^2} = e^u \frac{d^2u}{dx^2} + e^u \left(\frac{du}{dx}\right)^2 \qquad \cdots$$

where

$$u(x) = \frac{1}{x^2 - a^2} \qquad \frac{du}{dx} = \frac{-2x}{(x^2 - a^2)^2} \qquad \frac{d^2u}{dx^2} = \frac{8x^2}{(x^2 - a^2)^3} - \frac{2}{(x^2 - a^2)^2} \qquad \cdots$$

Thus $\phi(x)$ is clearly differentiable infinitely many times in the open interval $(-a, a)$. At an end point a, we must investigate the behavior of terms like $\exp[1/(x^2 - a^2)]/(x^2 - a^2)^n$. It can be shown that

$$\lim_{x \to a} \frac{1}{(x^2 - a^2)^n} \exp \frac{1}{x^2 - a^2} = 0$$

This means that both the left derivatives of ϕ at $x=a$ and the right derivatives of ϕ at $x=a$ vanish. Consequently, all derivatives of ϕ vanish at $x=a$. By symmetry, the same is also true at $x=-a$. Hence $\phi(x)$ is a test function. Notice that the sequence $\{(1/n)\phi(x)\}$ and all its derivatives converge uniformly to zero. ■

EXAMPLE 2.2. Let $\phi(x)$ be the exponential function defined in the previous example. Then the translation

$$\psi(x)=\phi\left(\frac{x-x_0}{\mu}\right)$$

is also a test function. The function $\psi(x)$ takes on nonzero (positive) values in the finite interval $x_0-\mu a<x<x_0+\mu a$. ■

2.3 DISTRIBUTIONS

With the proper properties of the accessory (test) functions ϕ now established, we can return to the functional representation of the function $f(x)$ described in tabular form earlier. We recall that by a *functional* on the space of test functions $\mathfrak{D}(\mathbf{R})$ we mean a mapping q that assigns to each $\phi\in\mathfrak{D}(\mathbf{R})$ a real number $q(\phi)$. A functional q is linear if it is a linear mapping, i.e., if

$$q(\alpha\phi_1+\beta\phi_2)=\alpha q(\phi_1)+\beta q(\phi_2)$$

for every $\phi_1, \phi_2\in\mathfrak{D}(\mathbf{R})$. A functional q: $\mathfrak{D}(\mathbf{R})\to\mathbf{R}$ is continuous if it is a continuous operator; i.e., if the inverse image under q of open sets is open.

Distribution

A linear functional q on the space $\mathfrak{D}(\mathbf{R})$ is called a distribution or a generalized function if and only if for every compact set $K\subset\mathbf{R}$ there is a constant $C_K>0$ and a nonnegative integer m such that

$$|q(\phi)|\leqslant C_K\sup_{\substack{k\leqslant m\\ x\in K}}\left|\frac{d^k\phi(x)}{dx^k}\right| \qquad \forall\phi\in\mathfrak{D}(K) \tag{2.6}$$

where $\mathfrak{D}(K)$ is the space of $C_0^\infty(\mathbf{R})$ functions whose supports are contained in int$K$.

The Dual Space $\mathfrak{D}(\mathbf{R})'$

It is not difficult to show that a linear functional q on the space of test functions $\mathfrak{D}(\mathbf{R})$ is a distribution if and only if it is continuous (see, e.g., Yosida [2.2, p. 47]), and therefore the definition of a distribution as any continuous linear functional on $\mathfrak{D}(\mathbf{R})$ is often used as an alternative to that given above.

Now we recall that, if $\mathfrak{U}$ and $\mathfrak{V}$ are two linear topological spaces defined on the field $\mathbf{R}$ of real numbers, the totality of all continuous linear operators on $\mathfrak{U}$ into $\mathfrak{V}$ is also a linear space and is denoted $\mathfrak{L}(\mathfrak{U},\mathfrak{V})$. In particular, if $\mathfrak{V}$ is itself $\mathbf{R}$, then the space $\mathfrak{L}(\mathfrak{U},\mathbf{R})$ of continuous linear functionals on $\mathfrak{U}$ is called the *dual space* of $\mathfrak{U}$ and is denoted $\mathfrak{U}'$:

$$\mathfrak{U}' = \mathfrak{L}(\mathfrak{U},\mathbf{R}) \tag{2.7}$$

We gather from all these observations that the dual space $\mathfrak{D}(\mathbf{R})'$ of the space of test functions is the *space of distributions*. We can therefore immediately list several basic properties of $\mathfrak{D}(\mathbf{R})'$:

(i) If $q \in \mathfrak{D}(\mathbf{R})'$, we write, as before,

$$q(\phi) \equiv \langle q, \phi \rangle \qquad \phi \in \mathfrak{D}(\mathbf{R}) \tag{2.8}$$

where $\langle \cdot, \cdot \rangle$ is now seen as a *duality pairing*; i.e., a bilinear map of $\mathfrak{D}(\mathbf{R})' \times \mathfrak{D}(\mathbf{R})$ into $\mathbf{R}$.

(ii) $\mathfrak{D}(\mathbf{R})'$ is a linear space, for if q_1 and q_2 are two distributions and α and β are scalars, $\alpha q_1 + \beta q_2$ is also a distribution:

$$\langle \alpha q_1, \phi \rangle + \langle \beta q_2, \phi \rangle = \langle \alpha q_1 + \beta q_2, \phi \rangle \tag{2.9}$$

(iii) The space $\mathfrak{D}(\mathbf{R})'$ is endowed with the so-called weak star (weak *) topology, in which a sequence of distributions $\{q_n\}$ converges to a distribution q if, for any $\phi \in \mathfrak{D}(\mathbf{R})$,

$$\lim_{n\to\infty} \langle q_n, \phi \rangle = \langle q, \phi \rangle \tag{2.10}$$

We can likewise speak of a series of distributions $q_1 + q_2 + \cdots$ which converges to a distribution q if the nth partial sum $s_n = \Sigma_{k=1}^{n} q_k$ converges to q in the sense of (2.10).

EXAMPLE 2.3. Consider the following sequences:

$$s_n(x)=\frac{1}{\pi}\frac{n}{1+n^2x^2}$$

$$b_n(x)=\frac{n}{\sqrt{\pi}}\exp(-n^2x^2)$$

$$d_n(x)=\begin{cases} 0 & |x|>1/2n \\ 4n^2x+2n & -1/2n\leqslant x\leqslant 0 \\ -4n^2x+2n & 0\leqslant x\leqslant 1/2n \end{cases}$$

$$e_n(x)=\begin{cases} -n & |x|<1/2n \\ 2n & 1/2n\leqslant x\leqslant 3/2n \\ 0 & \text{otherwise} \end{cases}$$

The sequences $s_n(x)$, $b_n(x)$, $d_n(x)$ and $e_n(x)$ have the property that, if ϕ is any test function,

$$\lim_{n\to\infty}\int_{-\infty}^{\infty} s_n(x)\phi(x)\,dx=\lim_{n\to\infty}\int_{-\infty}^{\infty} b_n(x)\phi(x)\,dx=\phi(0)$$

$$\lim_{n\to\infty}\int_{-\infty}^{\infty} d_n(x)\phi(x)\,dx=\lim_{n\to\infty}\int_{-\infty}^{\infty} e_n(x)\phi(x)\,dx=\phi(0)$$

Many other sequences have this same property. Obviously, $\{s_n\}$, $\{b_n\}$, $\{d_n\}$, and $\{e_n\}$ are sequences of distributions that converge to the functional $q(\phi)=\phi(0)$.

The generalized function (i.e., the distribution) associated with this particular functional q is called the *Dirac delta function* and is denoted $\delta(x)$. All sequences $\{q_n\}$, such as those above, that produce in the limit the action of the Dirac delta; i.e., symbolically,

$$\lim_{n\to\infty}\langle q_n,\phi\rangle=\int_{-\infty}^{\infty}\delta(x)\phi(x)\,dx=\phi(0)$$

are called *delta sequences*. We observe that pointwise,

$$\lim_{n\to\infty} s_n(x)=\begin{cases} 0 & x\neq 0 \\ \infty & x=0 \end{cases}$$

but

$$\lim_{n\to\infty} e_n(x)=\begin{cases} 0 & x\neq 0 \\ -\infty & x=0 \end{cases}$$

Thus it is impossible to interpret $\delta(x)$ as a function in the usual way (at least, it cannot be interpreted pointwise in a unique way). In fact, the observation that $\delta(x)$ is defined only by its action on a typical test function $\phi(x)$ is precisely a motivation for introducing the concept of distributions in the first place. ■

EXAMPLE 2.4. Let $c_n(x) = 1/2 + \arctan nx/\pi$. Then, if $\{s_n(x)\}$ is the sequence described in the previous example, it can be shown by elementary calculations that

$$c_n(x) = \int_{-\infty}^{x} s_n(t)\,dt \qquad \lim_{n\to\infty} c_n(x) = \begin{cases} 0 & x<0 \\ \frac{1}{2} & x=0 \\ 1 & x>0 \end{cases}$$

The discontinuous *Heaviside step function*

$$H(x) = \begin{cases} 0 & x<0 \\ \frac{1}{2} & x=0 \\ 1 & x>0 \end{cases}$$

can thus be defined pointwise and as the limit of certain sequences in $\mathcal{D}(\mathbf{R})'$; as another example,

$$H(x) = \lim_{n\to\infty} \frac{1}{\pi} \int_{\infty}^{x} \frac{\sin n\xi}{\xi}\,d\xi$$

Sequences converging to $H(x)$ are called *H sequences*.

Clearly, H "generates" a distribution with the property

$$\langle H, \phi \rangle = \int_{-\infty}^{\infty} H(x)\phi(x)\,dx = \int_{0}^{\infty} \phi(x)\,dx \quad ■$$

It is clear from the above examples that in some cases generalized functions can be interpreted pointwise as limits of sequences of integrable functions, whereas in other cases this appears to be impossible. We examine such questions more closely in the next section.

2.4 LOCALLY INTEGRABLE GENERATORS, REGULAR AND SINGULAR DISTRIBUTIONS

The simplest type of distribution is that generated by a locally integrable function.

Locally Integrable Function

A function f is said to be *locally integrable* if the Lebesgue integral

$$\int_a^b |f(x)|\,dx$$

exists for every compact interval $[a,b]$. Here $dx = d\mu(x)$ is the Lebesgue measure.

Let K be any compact set in R, f be locally integrable, and $\operatorname{supp}\phi \subset K$. Then we may introduce the distribution F defined by

$$F(\phi) = \int_{-\infty}^{\infty} f\phi\,dx \leqslant \sup_{x\in K} |\phi(x)| \int_K |f|\,dx$$

Since the last integral is finite, we see from (2.6) (with $m=0$) that F is a distribution. Thus, *every locally integrable function f defines a distribution.* We call such functions *locally integrable generators*, and we say that f *generates* the distribution F.

In such cases, we do not distinguish between f and F, and henceforth we shall write

$$F(\phi) = \langle f,\phi\rangle = \int_{-\infty}^{\infty} f(x)\phi(x)\,dx \tag{2.11}$$

Regular and Singular Distributions

A distribution q that can be generated from a locally integrable function is called a *regular distribution*. If a distribution is not regular, it is said to be *singular*.

EXAMPLE 2.5. The Heaviside step function $H(x)$ is locally integrable. Hence, it generates the distribution $\langle H,\phi\rangle = \int_0^\infty \phi\,dx$, and $\langle H,\phi\rangle$ is regular. ■

EXAMPLE 2.6. The Dirac delta function $\langle \delta,\phi\rangle$ is a singular distribution. To see this, suppose that $\phi_a(x)$ are test functions similar to those described earlier; i.e.,

$$\phi_a(x) = \begin{cases} \exp\left(\dfrac{a^2}{x^2 - a^2}\right) & |x| < a \\ 0 & |x| \geqslant a \end{cases}$$

Hence, $\phi_a(0)=1/e$, and $\phi_a(x)\leqslant 1/e$. Now assume that $\delta(x)$ is a locally integrable function and observe that

$$\left|\int_{-\infty}^{\infty}\delta(x)\phi_a(x)\,dx\right|\leqslant\frac{1}{e}\int_{-a}^{a}\delta(x)\,dx$$

If $\delta(x)$ were locally integrable, then $\lim_{a\to 0}\int_{-a}^{a}\delta(x)\,dx=0$, but this cannot be so, because $\langle\delta(x),\phi_a(x)\rangle=1/e\neq 0$. Hence there does not exist a locally integrable function f such that $\int_{-\infty}^{\infty}f\phi\,dx=\phi(0)$. ■

We recall that two locally integrable functions f_1 and f_2 are said to be equal *almost everywhere* if they differ only on *a set of measure zero*. Then

$$\langle f_1,\phi\rangle=\langle f_2,\phi\rangle$$

for every $\phi\in\mathfrak{D}(\mathbf{R})$. Obviously, two locally integrable functions that are equal almost everywhere generate the same distribution; if they are not equal almost everywhere, they generate different distributions.

EXAMPLE 2.7. The functions

$$f_1(x)=\begin{cases}3 & x \text{ is rational;}\\ 2 & x \text{ is irrational;}\end{cases}\qquad f_2(x)=2\qquad\text{for all } x$$

are equal *almost everywhere* on $[0,1]$:

$$\int_0^1 f_1(x)\,dx=\int_0^1 f_2(x)\,dx=2$$

This is true because the rationals are countable and thus have zero measure. ■

It is of fundamental importance in many problems to understand the difference between locally integrable functions and functions that are defined pointwise. In particular, it is illuminating to observe the following theorem.

THEOREM 2.1. The only continuous function f for which $\langle f,\phi\rangle=0$, $\forall\phi\in\mathfrak{D}(\mathbf{R})$, is the zero function $f(x)=0$.

PROOF. This can be seen by assuming, to the contrary, that $f(x_0)>0$ for some $x_0\in\mathbf{R}$. Since $f(x)$ is, by hypothesis, continuous, there exists a neighborhood $(x_0-\varepsilon,x_0+\varepsilon)$ of x_0 wherein $f(x)>0$. Now pick a test function ϕ such that $\operatorname{supp}\phi\subset(x_0-\varepsilon,x_0+\varepsilon)\neq\phi$; i.e., ϕ is nonzero in the interval $|x-x_0|<\varepsilon$. Suppose $\phi(x)>0$ in this interval. Then $\langle f,\phi\rangle>0$, which is a contradiction. A similar conclusion is reached if we assume that $f(x)<0$. Thus the conjecture is proved. ■

Let $C(a,b)$ denote the space of continuous functions on the interval (a,b), and $L_1(a,b)$ the space of Lebesque-integrable functions on (a,b). Then, according to what has been just shown, particularly the above theorem, we have

$$\langle f,\phi\rangle=0 \qquad \forall\phi\in\mathfrak{D}(\mathbf{R}) \text{ then } f\in C(a,b)\Rightarrow f(x)=0$$

$$\langle f,\phi\rangle=0 \qquad \forall\phi\in\mathfrak{D}(\mathbf{R}) \text{ then } f\in L_1(a,b)\Rightarrow f(x)=0 \text{ almost everywhere} \tag{2.12}$$

Properties (2.12) and Theorem 2.1 are related to (and sometimes referred to as) the fundamental theorem of the calculus of variations.

2.5 SOME PROPERTIES OF DISTRIBUTIONS

With the basic structure of a distribution now in hand, we can uncover a variety of their intrinsic properties. We list several of these below.

Translation

If f is locally integrable, it generates a distribution $\langle f,\phi\rangle$. Its translation $f(x-a)$ generates the translated distribution

$$\langle f(x-a),\phi(x)\rangle=\langle f(x),\phi(x+a)\rangle \tag{2.13}$$

Similarly, for an arbitrary generalized function q, we define

$$\langle q(x-a),\phi(x)\rangle=\langle q(x),\phi(x+a)\rangle \tag{2.14}$$

From this general rule it follows that

$$\langle\delta(x-a),\phi(x)\rangle=\langle\delta(x),\phi(x+a)\rangle=\phi(a)$$

We often use the notation

$$\delta_a(\phi)=\phi(a)$$

Multiplication of a Distribution by an Infinitely Differentiable Function

If f is infinitely differentiable and q is a distribution we define

$$\langle fq,\phi\rangle=\langle q,\bar{f}\phi\rangle \tag{2.15}$$

If $f \not\in C^\infty(\mathbf{R}^n)$, $f\phi$ is not necessarily a test function, but in some cases (2.15) may still make sense. For example, if f is continuous only at $x=0$,

$$\langle f(x)\delta(x), \phi(x)\rangle = \langle \delta(x), f(x)\phi(x)\rangle = f(0)\phi(0)$$

Then $f(x)\delta(x)$ is the generalized function $f(0)\delta(x)$.

Note that in general the product of two distributions *cannot* be defined. Hence $\delta^2(x) = \delta(x)\delta(x)$ is not defined.

Change in Scale

If f is locally integrable, and $\alpha \neq 0$,

$$\langle f(\alpha x), \phi(x)\rangle = \frac{1}{|\alpha|}\left\langle f(x), \phi\left(\frac{x}{\alpha}\right)\right\rangle \tag{2.16}$$

We define the change in scale of a distribution in a like manner.

Derivatives of Distributions

Suppose that $f(x)$ has the property that its derivative $f'(x) = df(x)/dx$ is locally integrable. Then $f'(x)$ generates the distribution $\langle f', \phi\rangle$. Integrating by parts, and recalling that the test functions ϕ have finite support [which implies that $\phi(-\infty) = \phi(\infty) = 0$], we have

$$\begin{aligned}\langle f', \phi\rangle &= \int_{-\infty}^{\infty} f'(x)\phi(x)\,dx = -\int_{-\infty}^{\infty} f(x)\phi'(x)\,dx + f\phi\Big|_{-\infty}^{\infty}\\ &= -\langle f, \phi'\rangle\end{aligned}$$

In other words, the action of f' on ϕ equals that of f on $-\phi'$. We now use this result as a basis for *defining* the derivative q' of *any* distribution q.

Let q be an arbitrary distribution. The functional p defined by

$$\langle p, \phi\rangle = -\langle q, \phi'\rangle \qquad \forall \phi \in \mathcal{D}(\mathbf{R}) \tag{2.17}$$

is called the *distributional derivative* of q, and we use the notation

$$p = q' \tag{2.18}$$

that is, (2.17) is made to appear as an integration by parts:

$$\langle q', \phi\rangle = -\langle q, \phi'\rangle \qquad \forall \phi \in \mathcal{D}(\mathbf{R}) \tag{2.19}$$

The quantity q' is also a distribution. Since q is a distribution, it satisfies (2.6). However, $|q'(\phi)|$ must then be bounded by $C_K \sup_{k+1 \leqslant m}\{|d^{k+1}\phi/dx^{k+1}|, x \in K\}$. Since q' is also linear, we have established that q' is a distribution. Moreover, higher derivatives of distributions are defined in a similar manner. For example, the kth derivative $q^{(k)}$ of a distribution q is the distribution $q^{(k)}$ defined by

$$\langle q^{(k)}, \phi \rangle = (-1)^k \left\langle q, \frac{d^k\phi}{dx^k} \right\rangle \qquad \forall \phi \in \mathfrak{D}(\mathbf{R}) \tag{2.20}$$

Since we associate symbolically with the distribution q a *generalized function* q, the distribution $q^{(k)}$ is some times assigned the *symbolism*

$$q^{(k)} = \frac{d^k q}{dx^k} \tag{2.21}$$

and $d^k q/dx^k$ is referred to as the kth *generalized derivative* of q.

Convergence of Sequences of Distributions

We now return to the question of convergence of sequences of distributions. First, it is important to observe that *every convergent sequence* $\{q_n\}$ *of distributions can be differentiated* (distributionally) *term by term and as often as is desired. Indeed, if* $\{q_n\} \to q$, $\{q_n'\} \to q'$, *because* $\langle q_n', \phi \rangle = -\langle q_n, \phi' \rangle$ *and* $\langle q_n, \phi' \rangle \to \langle q, \phi' \rangle = -\langle q', \phi \rangle$.

This property can be used to describe convergence of sequences $\{f_n(x)\}$ of functions which may otherwise fail to converge pointwise or uniformly. What is usually needed is a distributional interpretation of the sequence.

Suppose $\{f_n(x)\}$ is a pointwise *divergent* sequence which is equal to the kth derivative of a sequence of functions $\{g_n(x)\}$ that converges uniformly to $g(x)$. This means that

$$f_n(x) = \frac{d^k g_n(x)}{dx^k} \qquad \text{and} \qquad \lim_{n\to\infty} g_n(x) = g(x)$$

Now every uniformly convergent series $\{g_n(x)\}$ is also distributionally convergent. Therefore, as n tends to ∞,

$$\langle g_n, \phi \rangle \to \langle g, \phi \rangle \qquad \text{and} \qquad \left\langle \frac{d^k g_n}{dx^k}, \phi \right\rangle = (-1)^k \left\langle g_n, \frac{d^k\phi}{dx^k} \right\rangle$$

which means that our previously divergent series now converges distributionally to $d^k g/dx^k$:

$$\lim_{n\to\infty} \langle f_n(x), \phi(x)\rangle = \left\langle \frac{d^k g(x)}{dx^k}, \phi(x)\right\rangle = (-1)^k \left\langle g, \frac{d^k \phi}{dx^k}\right\rangle$$

EXAMPLE 2.8. Consider the piecewise continuous function f shown in Fig. 2.1. The derivative of f exists in open segments between a finite set of points $x_1, x_2, \ldots, x_n$, and at these points $f(x)$ suffers jumps $[f(x_i)]$. While

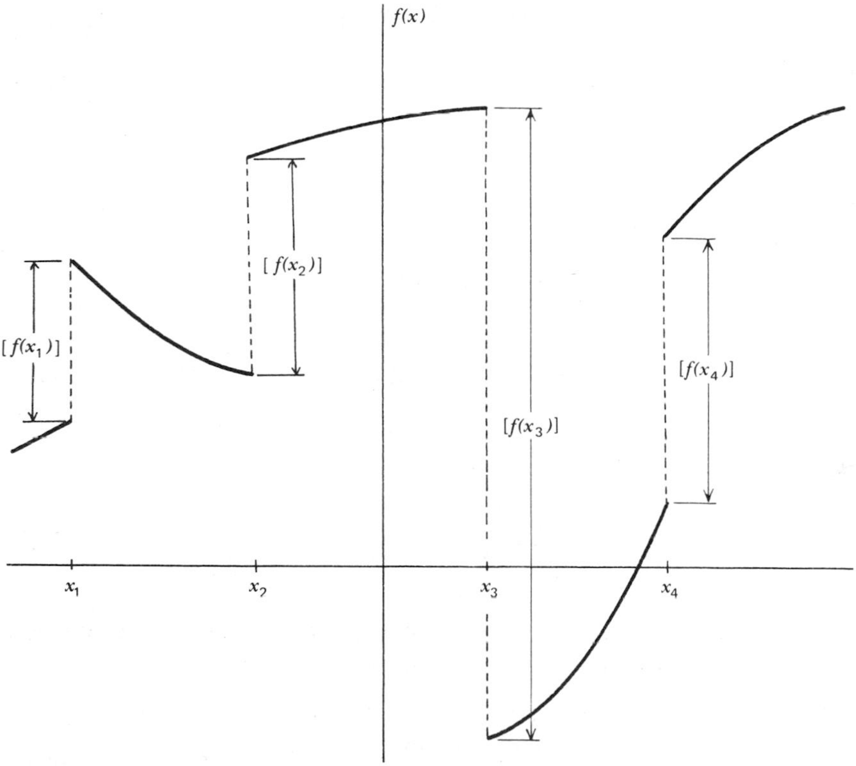

Figure 2.1. A function suffering jumps at a finite number of points in **R**.

$f(x)$ does not possess an ordinary derivative everywhere, it does possess a generalized derivative:

$$\langle f', \phi\rangle = -\langle f, \phi'\rangle$$

or, integrating by parts,

$$-\langle f,\phi'\rangle=\int_{-\infty}^{x_1}\frac{df}{dx}\phi\,dx-f(x)\phi(x)\Big|_{-\infty}^{x_1}+\int_{x_1}^{x_2}\frac{df}{dx}\phi\,dx-f(x)\phi(x)\Big|_{x_1}^{x_2}$$

$$-\cdots+\int_{x_n}^{\infty}\frac{df}{dx}\phi\,dx-f(x)\phi(x)\Big|_{x_n}^{\infty}$$

$$=\int_{-\infty}^{\infty}\frac{df}{dx}\phi\,dx+\phi(x_1)[f(x_1+)-f(x_1-)]$$

$$+\phi(x_2)[f(x_2+)-f(x_2-)]+\cdots+\phi(x_n)[f(x_n+)-f(x_n-)]$$

where df/dx is the usual derivative of $f(x)$. Denoting the jumps in $f(x)$ by

$$[f(x_i)]=f(x_i+)-f(x_i-)$$

we have

$$\langle f',\phi\rangle=\left\langle\frac{df}{dx},\phi\right\rangle+\sum_{i=1}^{n}\phi(x_i)[f(x_i)]$$

$$=\left\langle\frac{df}{dx},\phi\right\rangle+\sum_{i=1}^{n}[f(x_i)]\langle\delta(x-x_i),\phi(x)\rangle \tag{2.22}$$

Symbolically,

$$f'=\frac{df}{dx}+\sum_{i=1}^{n}[f(x_i)]\delta(x-x_i) \tag{2.23}$$

Evidently, the generalized derivative of a piecewise continuous function $f(x)$ involves delta functions corresponding to each jump. ■

EXAMPLE 2.9. The derivative of a delta function, denoted $\delta'(x)$, is called the *dipole distribution* and is defined by

$$\langle\delta',\phi\rangle=-\langle\delta,\phi'\rangle=-\phi'(0) \tag{2.24}$$

Interestingly enough, δ' can also be defined as the limit of the difference quotient $\lim_{\alpha\to 0}\langle(1/\alpha)[\delta(x+\alpha)-\delta(x)],\phi\rangle=\lim_{\alpha\to 0}(1/\alpha)[\phi(-\alpha)-\phi(0)]$ $=-\phi'(0)$. ■

The singular distribution $\delta(x)$ is referred to as a *first-order singularity* or *first-order distribution.* All regular distributions are of zero order, while the dipole δ' is of second order. Similarly, the nth generalized derivative $\delta^{(n)}$ of δ is of order $n+1$. Roughly speaking, a singular distribution is of order n if n antidifferentiations are required to make it a regular distribution.

EXAMPLE 2.10. The generalized derivative of the Heaviside step function $H(x)$ is the Dirac delta. To see this, note that

$$\langle H',\phi\rangle = -\langle H,\phi'\rangle = -\int_0^\infty \frac{d\phi}{dx}\,dx = -\phi\Big|_0^\infty = \phi(0) = \langle\delta,\phi\rangle \qquad (2.25)$$

In other words, symbolically,

$$\frac{dH(x)}{dx} = \delta(x) \qquad (2.26)$$

■

EXAMPLE 2.11. Consider the sequence

$$\{f_n(x)\} = \{\sin nx\}$$

This sequence is divergent, in the usual sense of the term (except at points x, which are multiples of π, in which case $\{f_n\}\to 0$). It so happens that $f_n(x) = dg_n(x)/dx$, where $g_n(x) = (-1/n)\cos nx$ and $g_n(x)\to 0$ uniformly in $(-\infty,\infty)$. Therefore $g_n'(x)\to 0$ distributionally:

$$\lim_{n\to\infty}\langle f_n,\phi\rangle = \lim_{n\to\infty}\langle g_n',\phi\rangle = 0 \quad ■$$

EXAMPLE 2.12. Consider the sequence of functions

$$f_n(x) = \sum_{k=1}^{n} \frac{\sin kx}{k} \qquad (2.27)$$

which converges to

$$f(x) = \tfrac{1}{2}(\pi - x) \qquad 0 < x < 2\pi$$

and $f(x)$ is periodic with period 2π. Derivatives of $f_n(x)$ converge distributionally; in fact, it follows from (2.23) that, symbolically,

$$\left.\begin{aligned}\sum_{k=1}^{\infty}\cos kx &= -\tfrac{1}{2}+\pi\sum_{k=-\infty}^{\infty}\delta(x-2\pi k)\\ \sum_{k=1}^{\infty}k\sin kx &= -\pi\sum_{k=-\infty}^{\infty}\delta'(x-2\pi k)\\ \sum_{k=1}^{\infty}k^2\cos kx &= -\pi\sum_{k=-\infty}^{\infty}\delta''(x-2\pi k)\end{aligned}\right\}\qquad(2.28)$$

etc. ■

EXAMPLE 2.13. There are numerous examples of series that are pointwise divergent but weakly (or "distributionally") convergent. For example, the series

$$f(x)=2\sum_{n=1}^{\infty}\cos nx$$

is divergent in a traditional sense but convergent to $-1+2\pi\Sigma_{-\infty}^{\infty}\delta(x-2\pi n)$ in the sense of distrubutions. Also, the complex Fourier series $\Sigma_{-\infty}^{\infty}c_n\exp(inx)$ is uniformly convergent if for large $|n|$, $|c_n|\leqslant M/n^2$, but possibly pointwise divergent if $|c_n|\leqslant M/n^\alpha$, M being a number >0 and α an integer. Since the series $\Sigma_{-\infty}^{\infty}(in)^{-5}c_n\exp(inx)$ then converges uniformly to some $g(x)$, the original series converges weakly to $c_0+g^{(5)}(x)$, where $g^{(5)}(x)$ is the fifth generalized derivative of $g(x)$. ■

2.6 DISTRIBUTIONAL DIFFERENTIAL EQUATIONS

Since it is possible to define derivatives of distributions of all orders, it is natural to consider the problem of determining a distribution once combinations of its generalized derivatives are specified, i.e., the problem of solving differential equations in distributions. For example, if f is a given distribution, the problem of finding q such that

$$\langle q',\phi\rangle=\langle f,\phi\rangle \qquad \forall\phi\in\mathfrak{D}(\mathbf{R})$$

is a first-order distributional differential equation in q. Equivalently,

$$-\langle q,\phi'\rangle=\langle f,\phi\rangle \qquad \forall\phi\in\mathfrak{D}(\mathbf{R}) \qquad (2.29)$$

In general, we can consider the linear operator

$$A = a_0(x)\frac{d^n}{dx^n} + a_1(x)\frac{d^{n-1}}{dx^{n-1}} + \cdots + a_n(x) \tag{2.30}$$

where the coefficients $a_0(x)$, ..., $a_n(x)$ are assumed to be infinitely differentiable. By partial integrations, we find that, if u is any n-times differentiable function,

$$\langle Au, \phi\rangle = \langle u, A^*\phi\rangle \qquad \forall \phi \in \mathcal{D}(\mathbf{R}) \tag{2.31}$$

where A^* is clearly the formal adjoint of A:

$$A^*\phi = (-1)^n \frac{d^n\,(a_0(x)\phi)}{dx^n} + \cdots + a_n(x)\phi = \sum_{k=0}^{n} (-1)^k \frac{d^k\,(a_{n-k}\phi)}{dx^k} \tag{2.32}$$

Following previous patterns, we can use relations such as (2.31) as means for dealing with distributional equations. Indeed, the operator A can generate distributional equations of the type

$$\langle Aq, \phi\rangle = \langle f, \phi\rangle \qquad \forall \phi \in \mathcal{D}(\mathbf{R}) \tag{2.33}$$

In fact, A and q are interpreted in the sense of distributions if by (2.33) we actually mean: Find the distribution q such that

$$\langle q, A^*\phi\rangle = \langle f, \phi\rangle \tag{2.34}$$

for every $\phi \in \mathcal{D}(\mathbf{R})$. The nonhomogeneous part of (2.34) is defined by f. If f is continuous on $\mathbf{R}$, q may be a function with n continuous derivatives. If f is only locally integrable, or is a singular distribution, the differentiability required of q must be weakened or defined distributionally. We classify these cases as follows:

1. If $q(x)$ is a function, sufficiently differentiable in a classical sense, that satisfies the differential equation

$$A(x)q(x) = f(x) \tag{2.35}$$

at every point x, $q(x)$ is called a *classical* (or *strict*) solution of (2.35).

2. If q is a locally integrable function satisfying (2.33) [but not necessarily of sufficient differentiability to satisfy (2.35)], q is called a *strong solution*.

3. If q is a (e.g., singular) distribution satisfying (2.34), q is called a *weak* or *generalized solution*.

It is easily seen that every classical solution is also a strong and a weak solution; likewise, every strong solution is also a weak solution. However, there may be weak solutions which are neither classical nor strong.

Fundamental Solutions

Any generalized solution q of the equation

$$\langle q, A^*\phi\rangle = \phi(\xi) \qquad \forall \phi \in \mathfrak{D}(\mathbf{R}) \tag{2.36}$$

where $\xi \in (-\infty, \infty)$ is called a fundamental solution for the operator A with pole ξ. Equivalently, fundamental solutions are represented symbolically by

$$Aq(x) = \delta(x - \xi) \tag{2.37}$$

Green's Functions

Consider the differential equation (2.37), where A is an nth-order linear differential operator. Suppose that the solution $q(x)$ is further constrained by the imposition of n linearly independent boundary conditions. A distribution $G(x,\xi)$ is called a *Green's function* if and only if $G(x,\xi)$ is a fundamental solution and $G(x,\xi)$ satisfies all n boundary conditions.

In physical problems, $G(x,\xi)$ is also known as an *influence function*, since it gives the "influence" at x (i.e., the solution at x) due to an impulse at ξ. In addition, if the Green's function $G(x,\xi)$ is known for a given operator, we can use it to solve inhomogeneous equations such as

$$\langle q, A^*\phi\rangle = \langle f, \phi\rangle \qquad \forall \phi \in \mathfrak{D}(\mathbf{R})$$

The solution is

$$q(x) = \int_{-\infty}^{\infty} G(x,\xi) f(\xi)\, d\xi \tag{2.38}$$

since

$$\begin{aligned}\int_{-\infty}^{\infty}\int_{-\infty}^{\infty} & G(x,\xi) f(\xi) A^*\phi(x)\, dx\, d\xi \\ &= \int_{-\infty}^{\infty} f(\xi)\left[\int_{-\infty}^{\infty} G(x,\xi) A^*\phi(x)\, dx\right] d\xi \\ &= \int_{-\infty}^{\infty} f(\xi)\phi(\xi)\, d\xi \\ &= \langle f, \phi\rangle\end{aligned}$$

EXAMPLE 2.14. Consider the distributional differential equation

$$\frac{dq}{dx}=0$$

that is, we wish to find a distribution whose distributional derivative is precisely the zero distribution. Since

$$\langle q',\phi\rangle=-\langle q,\phi'\rangle=0$$

a necessary and sufficient condition that a distribution q be a solution of this equation is that

$$\langle q,\psi\rangle=0$$

for every test function ψ that is the derivative of another test function ϕ. This means that

$$\int_{-\infty}^{\infty}\psi\,dx=\phi(\infty)-\phi(-\infty)=0$$

Let $\phi_1(x)$ be a test function with the property that

$$\int_{-\infty}^{\infty}\phi_1(x)\,dx=1$$

and make use of the obvious identity for any test function $\phi(x)\in\mathfrak{D}(\mathbf{R})$,

$$\phi(x)=\phi_1(x)\int_{-\infty}^{\infty}\phi(x)\,dx+\left[\phi(x)-\phi_1(x)\int_{-\infty}^{\infty}\phi(x)\,dx\right]$$

Then the function in brackets is precisely a test function $\psi(x)$ with the desired property $\int_{-\infty}^{\infty}\psi\,dx=0$. Hence, for any $\phi(x)$,

$$\phi(x)=\phi_1(x)\int_{-\infty}^{\infty}\phi(x)\,dx+\psi(x) \tag{2.39}$$

and

$$\langle q,\phi\rangle=\left\langle q,\phi_1\int_{-\infty}^{\infty}\phi\,dx+\psi\right\rangle=\left(\int_{-\infty}^{\infty}\phi\,dx\right)\langle q,\phi_1\rangle$$

But $\langle q,\phi_1\rangle=c=$ a constant, independent of $\phi(x)$. Hence

$$\langle q,\phi\rangle=c\int_{-\infty}^{\infty}\phi\,dx$$

That is, the solution q of the distributional differential equation $q'=0$ is the *constant distribution*, $q=c$. In other words, the only distributions satisfying $q'=0$ are regular distributions generated by constant functions. ■

EXAMPLE 2.15. Now consider the distribution equation

$$\frac{dq}{dx}=f$$

where f is an arbitrary distribution. Equivalently,

$$-\langle q,\phi'\rangle=\langle f,\phi\rangle$$

Using the decomposition (2.39), we have

$$\begin{aligned}\langle q,\phi\rangle&=\left\langle q,\phi_1\int_{-\infty}^{\infty}\phi\,dx\right\rangle+\langle q,\psi\rangle\\&=\langle q,\phi_1\rangle\int_{-\infty}^{\infty}\phi\,dx+\langle q_0,\phi\rangle\end{aligned}$$

Here $\phi_1(x)$ is the "unit" test function described earlier, and q_0 is a specific distribution with the property

$$\langle q_0,\phi\rangle=-\langle f,\tilde{\phi}\rangle \tag{2.40}$$

and $\tilde{\phi}(x)$ is a test function whose derivative is $\psi=\phi(x)-\phi_1(x)\int_{-\infty}^{\infty}\phi(x)\,dx$; e.g.,

$$\tilde{\phi}(x)=\int_{-\infty}^{x}\phi(s)\,ds-\int_{-\infty}^{\infty}\phi(s)\,ds\int_{-\infty}^{x}\phi_1(s)\,ds$$

To obtain (2.40), we used the observation that $\langle q,\phi\rangle=c=\text{constant}$, and $c'=0$. Hence,

$$\langle q,\phi\rangle=\langle c,\phi\rangle+\langle q_0,\phi\rangle$$

where c is the constant distribution described earlier. It follows that the distributional differential equation has solutions of the form

$$q=q_0+c$$

where q_0 is defined by (2.40). ■

EXAMPLE 2.16. Consider the distributional differential equation

$$x\frac{dq(x)}{dx}=0$$

The classical solution of this equation is

$$q(x)=c$$

where c is a constant. However, the weak solution is

$$q(x)=c_1H(x)+c_2$$

where $H(x)$ is the Heaviside step function. Since $H(x)$ is not differentiable at $x=0$ in the classical sense, the weak solution is of course not a classical solution.

The weak solution follows from the solution of the distribution equation $\langle xp,\phi\rangle=0$. It can be shown that $\langle xp,\phi\rangle=\langle p,\eta\rangle$, where $\eta(x)=x\phi(x)$ is a test function if and only if $\eta(0)=0$. Thus pick a specific test function $\tilde{\phi}(x)$ with the property $\tilde{\phi}(0)=1$. Then notice that any $\phi(x)$ is of the form $\phi(x)=\phi(0)\tilde{\phi}(x)+[\phi(x)-\phi(0)\tilde{\phi}(x)]$, where we may take $[\phi(x)-\phi(0)\tilde{\phi}(x)]$ $=\eta(x)$. Thus, if p is such that $\langle p,\eta\rangle=0$, $\langle p,\phi\rangle=\phi(0)\langle p,\tilde{\phi}\rangle=\phi(0)c$, where c is the constant $\langle p,\tilde{\phi}\rangle$. Thus $p=c\delta$, and a simple integration gives $q=c_1H(x)+c_2$. ■

EXAMPLE 2.17. As an example of a distributional solution consider the symbolic equation

$$\frac{d^2q}{dx^2}=\delta^{(2)}(x)=\delta''(x)$$

The solution of this equation is $q(x)=\delta(x)+c_1x+c_2$. ■

2.7 DISTRIBUTIONS AND GENERALIZED FUNCTIONS IN $\mathbf{R}^n$

The extension of all the ideas covered thus far to functions defined on $\mathbf{R}\times\mathbf{R}\times\cdots\times\mathbf{R}=\mathbf{R}^n$ is, with a few adjustments, straightforward. We now briefly examine some of the details.

Multi-index Notation

To be able to work effectively with the calculus of functions of n variables, we need a more compact notation. Toward this end, we interrupt our discussion briefly here to describe the extremely useful conventions of multi-index notation.

Let Z_+^n denote the set of ordered n-tuples of nonnegative integers,

$$\boldsymbol{\alpha}=(\alpha_1,\alpha_2,\ldots,\alpha_n) \qquad \alpha_i \geqslant 0;\ \alpha_i=\text{integer} \tag{2.41}$$

On all operations involving such multi-integers, we adopt the following conventions:

$$|\boldsymbol{\alpha}|\equiv\alpha_1+\alpha_2+\cdots+\alpha_n \qquad \boldsymbol{\alpha}!\equiv\alpha_1!\alpha_2!\cdots\alpha_n!$$
$$\binom{\boldsymbol{\alpha}}{\boldsymbol{\beta}}\equiv\frac{\boldsymbol{\alpha}!}{\boldsymbol{\beta}!(\boldsymbol{\alpha}-\boldsymbol{\beta})!} \qquad \text{for } \alpha_i \geqslant \beta_i \tag{2.42}$$

Moreover, if $u(\mathbf{x})$ is a function of points $\mathbf{x}=(x_1,x_2,\ldots,x_n)\in\mathbb{R}^n$, we denote

$$\mathbf{x}^{\boldsymbol{\alpha}}=x_1^{\alpha_1}x_2^{\alpha_2}\cdots x_n^{\alpha_n} \tag{2.43}$$

and

$$D^{\boldsymbol{\alpha}}u(\mathbf{x})=\frac{\partial^{|\boldsymbol{\alpha}|}u(\mathbf{x})}{\partial x_1^{\alpha_1}\partial x_2^{\alpha_2}\cdots\partial x_n^{\alpha_n}}=\left(\frac{\partial^{\alpha_1}}{\partial x_1^{\alpha_1}}\right)\left(\frac{\partial^{\alpha_2}}{\partial x_2^{\alpha_2}}\right)\cdots\left(\frac{\partial^{\alpha_n}}{\partial x_n^{\alpha_n}}\right)u(\mathbf{x}) \tag{2.44}$$

This notation makes it possible to calculate in $\mathbb{R}^n$ in much the same fashion as in $\mathbb{R}^1$. For example, if Ω is a bounded subset of $\mathbb{R}^n$ in which a given function $u(\mathbf{x})$ has all partial derivatives of order $\leqslant p+1$ continuous in Ω, and if $\mathbf{x}$ and $\mathbf{y}$ are two points in Ω, the extended mean-value theorem allows us to represent $u(\mathbf{x}+\mathbf{y})$ by the expansion

$$u(\mathbf{x}+\mathbf{y})=\sum_{|\boldsymbol{\alpha}|\leqslant p}\frac{\mathbf{y}^{\boldsymbol{\alpha}}}{\boldsymbol{\alpha}!}D^{\boldsymbol{\alpha}}u(\mathbf{x})+R_{p+1}(\mathbf{x},\mathbf{y},\theta) \tag{2.45}$$

Here $R_{p+1}(\mathbf{x},\mathbf{y},\theta)$ is the remainder; e.g.,

$$R_{p+1}(\mathbf{x},\mathbf{y},\theta)=\frac{1}{(p+1)!}\sum_{|\boldsymbol{\alpha}|=p+1}\mathbf{y}^{\boldsymbol{\alpha}}D^{\boldsymbol{\alpha}}u(\boldsymbol{\xi})$$
$$\boldsymbol{\xi}=(1-\theta)\mathbf{x}+\theta\mathbf{y} \qquad 0<\theta<1 \tag{2.46}$$

EXAMPLE 2.18. Suppose $\Omega\subset\mathbb{R}^2$, $x_1=x$, $x_2=y$; and $\mathbf{0}=(0,0)$ is the origin. Consider the function

$$u(\mathbf{x})=\sum_{|\boldsymbol{\alpha}|\leqslant 2}\frac{\mathbf{x}^{\boldsymbol{\alpha}}}{\boldsymbol{\alpha}!}D^{\boldsymbol{\alpha}}u(\mathbf{0})$$

Then

$$u(\mathbf{x})=u(x,y)=\frac{\mathbf{x}^{(0,0)}}{(0,0)!}D^{(0,0)}u(0,0)+\frac{\mathbf{x}^{(1,0)}}{(1,0)!}D^{(1,0)}u(0,0)$$

$$+\frac{\mathbf{x}^{(0,1)}}{(0,1)!}D^{(0,1)}u(0,0)+\frac{\mathbf{x}^{(2,0)}}{(2,0)!}D^{(2,0)}u(0,0)$$

$$+\frac{\mathbf{x}^{(1,1)}}{(1,1)!}D^{(1,1)}u(0,0)+\frac{\mathbf{x}^{(0,2)}}{(0,2)!}D^{(0,2)}u(0,0)$$

According to (2.42) and (2.43),

$$\mathbf{x}^{(0,0)}=x^0y^0=1 \qquad \mathbf{x}^{(1,0)}=x^1y^0=x \qquad \text{etc.}$$

$$(0,0)!=0!0!=1 \qquad (1,0)!=1!0!=1 \qquad \text{etc.}$$

Also, from (2.44),

$$D^{(0,1)}u(0,0)=\frac{\partial^{0+1}}{\partial x^0\,\partial y^1}u(0,0)=\frac{\partial u(0,0)}{\partial y}$$

$$D^{(1,1)}u(0,0)=\frac{\partial^{1+1}}{\partial x^1\,\partial y^1}u(0,0)=\frac{\partial^2 u(0,0)}{\partial x\,\partial y} \qquad \text{etc.}$$

Thus

$$u(x,y)=u(0,0)+x\frac{\partial u(0,0)}{\partial x}+y\frac{\partial u(0,0)}{\partial y}$$

$$+\frac{x^2}{2!}\frac{\partial^2 u(0,0)}{\partial x^2}+xy\frac{\partial^2 u(0,0)}{\partial x\,\partial y}+\frac{y^2}{2!}\frac{\partial^2 u(0,0)}{\partial y^2}$$

$$=\sum_{|\alpha|\leq 2}\frac{\mathbf{x}^\alpha}{\alpha!}D^\alpha u(\mathbf{0}) \qquad \mathbf{x}\in\mathsf{R}^2$$

■

We now return to our discussion of distributions on functions defined in n-dimensional domains. Suppose that $\mathbf{x}=(x_1,x_2,\ldots,x_n)$ is a point in a bounded region $\Omega\subset\mathsf{R}^n$. A function $u(\mathbf{x})$ is referred to as *locally integrable* on R^n if its Lebesgue integral $\int_K|u(\mathbf{x})|\,dx$ is finite for any compact set $K\subset\mathsf{R}^n$, where $dx=dx_1\,dx_2\cdots dx_n$. Likewise, if Γ is some hypersurface in R^n, $u(\mathbf{x})$ is locally integrable on Γ if $\int_{\bar\Gamma}u(\mathbf{x})\,ds$ is finite for every compact subset $\bar\Gamma\subset\Gamma$.

As in the one-dimensional case, every locally integrable function $u(\mathbf{x})$ generates a regular distribution on the space of test functions $\mathfrak{D}(\mathbf{R}^n)$ via the duality pairing

$$\langle u,\phi\rangle = \int_{-\infty}^{\infty}\int_{-\infty}^{\infty}\cdots\int_{-\infty}^{\infty} u(\mathbf{x})\phi(\mathbf{x})\,dx_1\,dx_2\cdots dx_n$$

$$\equiv \int_{\mathbf{R}^n} u\phi\,dx \tag{2.47}$$

Here $\mathfrak{D}(\mathbf{R}^n)$ is the space of test functions on $\mathbf{R}^n$; an example of a test function $\phi(\mathbf{x})\in\mathfrak{D}(\mathbf{R})$, in analogy with Example 2.1, is the function

$$\phi(\mathbf{x}) = \begin{cases} \exp\dfrac{1}{|\mathbf{x}|^2 - a^2} & |\mathbf{x}| < a \\ 0 & |\mathbf{x}| \geqslant a \end{cases} \tag{2.48}$$

where $|\mathbf{x}| = (\Sigma_{i=1}^n |x_i|^2)^{1/2}$. The continuous linear functional p given by

$$\langle p,\phi\rangle = (-1)^{|\alpha|}\langle q, D^{\alpha}\phi\rangle \tag{2.49}$$

$\forall\phi\in\mathfrak{D}(\mathbf{R}^n)$, where q is a distribution, is called the $\boldsymbol{\alpha}$th *generalized partial derivative* of q, and we often use the symbolic notation

$$p = D^{\alpha} q \tag{2.50}$$

In the case of singular distributions for test functions defined on $\mathbf{R}^n$, such as the Dirac delta, we now use the symbolism

$$\delta(\mathbf{x}) = \delta(x_1)\delta(x_2)\cdots\delta(x_n) \tag{2.51}$$

where $\delta(x_i)$ are one-dimensional delta functions. Hence

$$\langle\delta(\mathbf{x}),\phi(\mathbf{x})\rangle = \phi(0,0,\ldots,0) = \phi(\mathbf{0}) \tag{2.52}$$

In fact, if $q_1(x_1), q_2(x_2),\ldots,q_n(x_n)$ are one-dimensional regular distributions, their *product* is defined as an n-dimensional distribution given by

$$\langle q_1 q_2\cdots q_n,\phi\rangle = \int_{\mathbf{R}^n} q_1(x_1)q_2(x_2)\cdots q_n(x_n)\phi(\mathbf{x})\,dx_1\,dx_2\cdots dx_n \tag{2.53}$$

■

Virtually all the other properties developed for distributions in the one-dimensional case carry over to the n-dimensional case. A few n-

dimensional cases that deserve special attention, aside from those already mentioned, are discussed in the following examples.

EXAMPLE 2.19. Consider the function

$$r(\mathbf{x}) = \begin{cases} x_1 x_2 \cdots x_n & \text{if all } x_i \geqslant 0 \\ 0 & \text{if any } x_i < 0 \end{cases} \tag{2.54}$$

Consider the distributional derivative

$$\frac{\partial r(\mathbf{x})}{\partial x_i} \sim (-1) \int_{\mathbb{R}^n} r \frac{\partial \phi}{\partial x_i}\, dx = \int_{\{\mathbf{x}:\mathbf{x}\in\mathbb{R}^n, x_i \geqslant 0\}} x_1 \cdots x_{i-1} x_{i+1} \cdots x_n \phi(\mathbf{x})\, dx$$

Continuing in this way, we have

$$D^{(1,1,1,\ldots,1)} r \sim \int_{\{\mathbf{x}:\ \mathbf{x}\in\mathbb{R}^n, x_j \geqslant 0\}} \phi(\mathbf{x}) dx = \int_{\mathbb{R}^n} H(\mathbf{x}) \phi(\mathbf{x})\, dx$$

where H is the n-dimensional Heaviside step function

$$H(\mathbf{x}) = \begin{cases} 1 & \forall x_i \geqslant 0 \\ 0 & \text{if some } x_i < 0 \end{cases} \tag{2.55}$$

Hence

$$D^{(1,1,\ldots,1)} r(\mathbf{x}) = H(\mathbf{x}) \tag{2.56}$$

■

EXAMPLE 2.20. Using (2.51) and (2.55), we can easily show that

$$D^{(1,1,\ldots,1)} H(\mathbf{x}) = \phi(0,0,\ldots,0) = \delta(\mathbf{x}) \tag{2.57}$$

■

EXAMPLE 2.21. (Generalized Green's Formula). Let Ω be a smooth bounded simply connected domain in $\mathbb{R}^n$ with a smooth boundary $\partial\Omega \subset \mathbb{R}^{n-1}$ having a unit exterior normal $\mathbf{n}$. So-called *surface distributions* can be defined on $\partial\Omega$ by considering restrictions of test functions $\phi \in \mathcal{D}(\mathbb{R}^n)$ to $\partial\Omega$; e.g. if $\boldsymbol{\sigma}$ is a generic point on $\partial\Omega$, we may introduce the functionals

$$\langle u(\boldsymbol{\sigma}) \delta(\mathbf{x} - \boldsymbol{\sigma}), \phi \rangle = \int_{\partial\Omega} u(\boldsymbol{\sigma}) \phi(\boldsymbol{\sigma})\, ds = \langle u, \phi \rangle_{\partial\Omega} \tag{2.58}$$

$$\left\langle w(\boldsymbol{\sigma}) \frac{\partial}{\partial n} \delta(\mathbf{x} - \boldsymbol{\sigma}), \phi \right\rangle = \int_{\partial\Omega} w(\boldsymbol{\sigma}) \frac{\partial \phi(\boldsymbol{\sigma})}{\partial n}\, ds = \left\langle w, \frac{\partial \phi}{\partial n} \right\rangle_{\partial\Omega} \tag{2.59}$$

where $u(\boldsymbol{\sigma})$ and $w(\boldsymbol{\sigma})$ are integrable functions on $\partial\Omega$ and ds is an element of surface area of $\partial\Omega(u(\mathbf{x})=0\ \mathbf{x}\notin\partial\Omega$, etc.).

Now let u be a function which has locally integrable derivatives of second order in Ω, and let $\gamma_0 u = u(\boldsymbol{\sigma})$ and $\gamma_1 u = \partial u(\boldsymbol{\sigma})/\partial n$ be the values of extensions of u and its normal derivative to points $\boldsymbol{\sigma}$ on $\partial\Omega$. Let q be the distribution generated by u. Then the distributional Laplacian of q may be defined, for every $\phi\in\mathfrak{D}(\mathbf{R}^n)$, by

$$\langle\Delta q,\phi\rangle=\langle u,\Delta\phi\rangle-\langle\gamma_1 u,\phi\rangle_{\partial\Omega}+\left\langle\gamma_0 u,\frac{\partial\phi}{\partial n}\right\rangle_{\partial\Omega} \tag{2.60}$$

wherein $\Delta u=\sum_{i=1}^{n}\partial^2 u/\partial x_i^2$. Equation (2.60) is referred to as a *generalized Green's formula* because of its similarity to the classical formula. ■

2.8 FOURIER TRANSFORMS, RAPIDLY DECAYING FUNCTIONS, AND TEMPERED DISTRIBUTIONS

There are several situations in which it is important to extend somewhat the idea of distributions. One that is particularly relevant to our study involves the concept of Fourier transforms—or specifically, extensions of the idea of Fourier transformation to include special types of distributions. This section is devoted to a brief study of this collection of ideas.

We begin by recalling the definition of the Fourier transformation. Let $u(\mathbf{x})$ denote a function in $L_2(\mathbf{R}^n)$. Then the Fourier transform of u in $L_2(\mathbf{R}^n)$ is the function $\hat{u}$ defined by

$$\hat{u}(\mathbf{y})=\mathfrak{F}(u)=\frac{1}{(2\pi)^{n/2}}\int_{\mathbf{R}^n}u(\mathbf{x})\exp(-i\mathbf{x}\mathbf{y})\,dx \tag{2.61}$$

where $\mathbf{x}=(x_1,x_2,\ldots,x_n)$ and $\mathbf{y}=(y_1,y_2,\ldots,y_n)$ are points in $\mathbf{R}^n$, $i=\sqrt{-1}$, $\mathbf{x}\mathbf{y}=x_1y_1+x_2y_2+\cdots+x_ny_n$, $dx=dx_1\,dx_2\cdots dx_n$, and the integral converges in an L_2 sense. The Fourier transform $\mathfrak{F}$ so defined is an isomorphism of $L_2(\mathbf{R}^n)$ onto itself, and the inverse Fourier transform is given almost everywhere by

$$u(\mathbf{x})=\mathfrak{F}^{-1}(\hat{u})=\frac{1}{(2\pi)^{n/2}}\int_{\mathbf{R}^n}\hat{u}(\mathbf{y})\exp(i\mathbf{x}\mathbf{y})\,dy \tag{2.62}$$

Now the class of functions that possess a Fourier transform (2.61) may seem to be rather small. For example, simple functions such as $u(\mathbf{x})$ $=$constant are not in this class, because the integral (2.61) does not converge for every $\mathbf{y}$ when u is constant. We can, however, considerably extend the notion of a Fourier transform, and the class of objects that are transformable, by introducing a special kind of distribution called a *tempered distribution* or a *distribution of slow growth*. We describe them below.

To develop an appropriate definition of the Fourier transform of a distribution q, we are immediately tempted to use the tricks characteristic of the usual distributions described earlier. Operations on distributions are defined by performing the appropriate operation on a typical test function ϕ. Thus why not define

$$\hat{q}(\phi) = q(\hat{\phi})$$

for $\phi \in \mathfrak{D}(\mathsf{R}^n)$? The problem is that the term $\exp(-i\mathbf{xy})$ in the definitions of the Fourier transform is *not* a test function, since it does not have compact support. We overcome this difficulty by simply introducing a new and more appropriate class of test functions and, correspondingly, a new class of distributions. For brevity, we itemize the principal properties as follows (for additional details, see Schwartz [2.1]):

1. The space $\mathfrak{S}(\mathsf{R}^n)$ of *test functions of rapid decay* is defined by

$$\mathfrak{S}(\mathsf{R}^n) = \{\phi\colon \mathbf{x}^{\boldsymbol{\alpha}} D^{\boldsymbol{\beta}}\phi \in L_2(\mathsf{R}^n);\ \forall \boldsymbol{\alpha} \text{ and } \boldsymbol{\beta} \in Z_+^n\} \tag{2.63}$$

Clearly, every $\phi \in \mathfrak{S}(\mathsf{R}^n)$ is (almost everywhere equal to) a function infinitely differentiable on R^n and, since $\mathbf{x}^{\boldsymbol{\alpha}} D^{\boldsymbol{\beta}}\phi \in L_2(\mathsf{R}^n)$ for every $\boldsymbol{\alpha}, \boldsymbol{\beta}$, it can be shown that

$$|\mathbf{x}^{\boldsymbol{\alpha}} D^{\boldsymbol{\beta}}\phi(\mathbf{x})| \to 0 \qquad \text{if } |\mathbf{x}| \to \infty;\ \forall \boldsymbol{\alpha}, \boldsymbol{\beta} \in Z_+^n \tag{2.64}$$

and

$$\operatorname*{ess\,sup}_{\mathbf{x} \in \mathsf{R}^n} |\mathbf{x}^{\boldsymbol{\alpha}} D^{\boldsymbol{\beta}}\phi(\mathbf{x})| < \infty \qquad \forall \boldsymbol{\alpha}, \boldsymbol{\beta} \in Z_+^n \tag{2.65}$$

2. Every "ordinary" test function $\phi \in \mathfrak{D}(\mathsf{R}^n)$ is in $\mathfrak{S}(\mathsf{R}^n)$, but $\mathfrak{S}(\mathsf{R}^n)$ contains elements not in $\mathfrak{D}(\mathsf{R}^n)$; e.g., $\phi(\mathbf{x}) = \exp(-|\mathbf{x}|^2) \in \mathfrak{S}(\mathsf{R}^n)$ but does not have compact support.

3. Property (2.64) suggests an appropriate property of convergence in $\mathfrak{S}(\mathsf{R}^n)$; e.g., a sequence $\{\phi_m\} \in \mathfrak{S}(\mathsf{R}^n)$ is a null sequence if, for every $\boldsymbol{\alpha}$ and $\boldsymbol{\beta}$,

$$\lim_{m \to \infty} \operatorname*{ess\,sup}_{\mathbf{x} \in \mathsf{R}^n} |\mathbf{x}^{\boldsymbol{\alpha}} D^{\boldsymbol{\beta}}\phi_m| = 0 \tag{2.66}$$

The space $\mathcal{S}(\mathbf{R}^n)$ is in fact a locally convex linear topological space with the topology induced by the family of seminorms

$$\hat{N}(\phi) = \sup_{\mathbf{x}\in\mathbf{R}^n} |p(\mathbf{x})D^{\alpha}\phi(\mathbf{x})| \tag{2.67}$$

$p(\mathbf{x})$ being a polynomial, and therefore its dual $\mathcal{S}(\mathbf{R}^n)'$ is also a locally convex linear topological space.

4. *The Fourier transform $\hat{\phi}$ of every $\phi \in \mathcal{S}(\mathbf{R}^n)$ exists and is in $\mathcal{S}(\mathbf{R}^n)$.* To demonstrate this important property, it suffices to consider the one-dimensional case ($\mathbf{R}^n = \mathbf{R}$). Then

$$\hat{\phi}(y) = \frac{1}{\sqrt{2\pi}} \int_{-\infty}^{\infty} \phi(x)\exp(-ixy)\,dx$$

and since $\phi(x)$ decays rapidly at $|x| = \infty$, the integral

$$I_k(\phi) \equiv \frac{1}{\sqrt{2\pi}} \int_{-\infty}^{\infty} (-ix)^k \phi(x)\exp(-ixy)\,dx$$

$k = 0, 1, \ldots$ is convergent. However, we see that

$$\left|\frac{d^k\hat{\phi}}{dy^k}\right| = |I_k(\phi)| \leqslant \frac{1}{\sqrt{2\pi}} \int_{-\infty}^{\infty} |x^k\phi|\,dx$$

and therefore $|d^k\hat{\phi}/dy^k|$ is bounded for all y. Furthermore, if we multiply $d^k\hat{\phi}/dy^k$ by $(iy)^m$, where m is arbitrary, we obtain

$$(iy)^m \frac{d^k\hat{\phi}}{dy^k} = (-1)^m \frac{1}{\sqrt{2\pi}} \int_{-\infty}^{\infty} \exp(-ixy)\frac{d^m}{dx^m}\left[(-ix)^k\phi(x)\right] dx$$

so that $|y^m d^k\hat{\phi}/dy^k|$ is bounded for all y and for arbitrary m and k. It follows that $\hat{\phi}$ is also in $\mathcal{S}(\mathbf{R}^n)$.

We can now generalize the idea of distributions.

5. A continuous linear functional on $\mathcal{S}(\mathbf{R}^n)$ is called a *tempered distribution* or a *distribution of slow growth*. Every tempered distribution therefore belongs to the dual space $\mathcal{S}(\mathbf{R}^n)'$ of $\mathcal{S}(\mathbf{R}^n)$, and we call $\mathcal{S}(\mathbf{R}^n)'$ the space of tempered distributions.

Since $\mathfrak{D}(\mathbf{R}^n)$ is contained (and in fact is dense) in $\mathcal{S}(\mathbf{R}^n)$ and has stronger topology, the restriction of a tempered distribution to $C_0^{\infty}(\mathbf{R}^n)$ is a distribution in the usual sense. Thus

$$\mathfrak{D}(\mathbf{R}^n) \subset \mathcal{S}(\mathbf{R}^n) \subset \mathcal{S}(\mathbf{R}^n)' \subset \mathfrak{D}(\mathbf{R}^n)' \tag{2.68}$$

6. In the same way that locally integrable functions can be used to generate regular distributions, such functions can also generate tempered distributions, provided they do not increase too rapidly with increasing $|\mathbf{x}|$. For this purpose we say that a function $u(\mathbf{x})$ has *slow growth* if it is locally integrable and if, for every $\mu>0$, there are constants C and multi-integers $\boldsymbol{\alpha}$ such that $|u(\mathbf{x})|<C|\mathbf{x}^{\alpha}|$ for $|\mathbf{x}|>\mu$, where $|\mathbf{x}|$ is the euclidean norm of $\mathbf{x}$, $|\mathbf{x}|=(x_1^2+x_2^2+\cdots+x_n^2)^{1/2}$; i.e., $u(\mathbf{x})$ has slow growth if it grows more slowly than some polynomial. It is not difficult to show that *every function* $u(\mathbf{x})$ *having slow growth generates a* (regular) *distribution of slow growth* (i.e., a tempered distribution) by means of the formula

$$u(\phi)=\langle u,\phi\rangle=\int_{\mathbf{R}^n} u(\mathbf{x})\phi(\mathbf{x})\,dx \tag{2.69}$$

$\forall\phi\in\mathcal{S}(\mathbf{R}^n)$.

Finally, we come to a central definition.

7. Let $q\in\mathcal{S}(\mathbf{R}^n)'$ be any tempered distribution. Then the *Fourier* transform $\hat{q}$ of q is a tempered distribution defined by

$$\langle\hat{q},\phi\rangle=\langle q,\hat{\phi}\rangle \qquad \forall\phi\in\mathcal{S}(\mathbf{R}^n) \tag{2.70}$$

It is easy to see that this definition makes sense. Since $\mathcal{F}:\mathcal{S}(\mathbf{R}^n)\to\mathcal{S}(\mathbf{R}^n)$ is linear and continuous, so also is the composition $q(\mathcal{F}(\phi))=q(\hat{\phi})\equiv\hat{q}(\phi)$; hence $\hat{q}\in\mathcal{S}(\mathbf{R}^n)'$.

Our next result establishes the fact that (2.70) is in fact an extension of the usual definition of Fourier transforms.

THEOREM 2.2. Let $u(\mathbf{x})\in L_2(\mathbf{R}^n)$ have a Fourier transform $\hat{u}=\mathcal{F}(u)$. Then the tempered distribution q with the property

$$\langle\hat{q},\phi\rangle=\int_{\mathbf{R}^n} u\hat{\phi}\,dx \qquad \forall\phi\in\mathcal{S}(\mathbf{R}^n) \tag{2.71}$$

coincides with u; i.e.,

$$\hat{q}=\hat{u} \tag{2.72}$$

PROOF. Clearly

$$\begin{aligned}\langle\hat{q},\phi\rangle&=\int_{\mathbf{R}^n} u(\mathbf{x})\frac{1}{(2\pi)^{n/2}}\int_{\mathbf{R}^n}\phi(\mathbf{y})\exp(-i\mathbf{x}\mathbf{y})\,dy\,dx\\&=\int_{\mathbf{R}^n}\phi(\mathbf{y})\frac{1}{(2\pi)^{n/2}}\int_{\mathbf{R}^n}u(\mathbf{x})\exp(-i\mathbf{x}\mathbf{y})\,dx\,dy\\&=\int_{\mathbf{R}^n}\phi(\mathbf{y})\hat{u}(\mathbf{y})\,dy=\langle\hat{u},\phi\rangle\end{aligned}$$

■

In other words, the definition of the Fourier transform of a distribution q coincides with the usual definition of a Fourier transform whenever q is (generated by) a function that has a Fourier transform.

EXAMPLE 2.22. We observed earlier that the simple function $u(x)=1$ $=\text{constant}$, $x\in\mathsf{R}$, does *not* possess a Fourier transform in the classical sense. We now demonstrate that its Fourier transform is well defined in a distributional sense. Using the definition (2.70), we have

$$\begin{aligned}\langle\hat{1},\phi\rangle&=\langle 1,\hat{\phi}\rangle\\&=\int_{-\infty}^{\infty}\hat{\phi}(y)\,dy\\&=\lim_{x\to 0}\int_{-\infty}^{\infty}\hat{\phi}(y)\exp(ixy)\,dy\cdot\frac{\sqrt{2\pi}}{\sqrt{2\pi}}\\&=\sqrt{2\pi}\left[\frac{1}{\sqrt{2\pi}}\int_{-\infty}^{\infty}\hat{\phi}(y)\exp(ixy)\,dy\right]_{x=0}\\&=\sqrt{2\pi}\,\left(\mathcal{F}^{-1}(\hat{\phi})\right)_{x=0}\\&=\sqrt{2\pi}\,\phi(0)\end{aligned}$$

Therefore

$$\hat{1}=\sqrt{2\pi}\,\delta \tag{2.73}$$

where δ is the Dirac delta distribution. We remark that in the n-dimensional case, the same procedure leads us to the conclusion that

$$\hat{1}=(2\pi)^{n/2}\delta \tag{2.74}$$

where δ is now the n-dimensional δ distribution. ■

EXAMPLE 2.23. Using the notion of transforms of distributions, we can also calculate the Fourier transform of δ. Indeed, for $n=1$,

$$\begin{aligned}\langle\hat{\delta},\phi\rangle&=\langle\delta,\hat{\phi}\rangle\\&=\delta\left[\frac{1}{\sqrt{2\pi}}\int_{-\infty}^{\infty}\phi(x)\exp(-ixy)\,dx\right]\\&=\frac{1}{\sqrt{2\pi}}\int_{-\infty}^{\infty}\phi(x)\,dx=\frac{1}{\sqrt{2\pi}}\langle 1,\phi\rangle\end{aligned}$$

Therefore

$$\hat{\delta} = \frac{1}{\sqrt{2\pi}} \tag{2.75}$$

In view of (2.73), we also have

$$(\hat{\hat{\delta}}) = \delta \qquad \text{and} \qquad (\hat{\hat{1}}) = 1 \tag{2.76}$$

■

EXAMPLE 2.24. Now consider $\delta(x-\xi) = \delta_\xi$

$$\begin{aligned}
\langle \hat{\delta}_\xi, \phi \rangle &= \langle \delta_\xi, \hat{\phi} \rangle \\
&= \delta_\xi \left[\frac{1}{\sqrt{2\pi}} \int_{-\infty}^{\infty} \phi(x) \exp(-ixy)\, dx \right] \\
&= \frac{1}{\sqrt{2\pi}} \int_{-\infty}^{\infty} \exp(-i\xi x) \phi(x)\, dx \\
&= \left\langle \frac{1}{\sqrt{2\pi}} \exp(-i\xi x), \phi \right\rangle
\end{aligned}$$

Therefore

$$\hat{\delta}_\xi = \frac{1}{\sqrt{2\pi}} \exp(-i\xi x) \tag{2.77}$$

■

We are now ready to record a number of fundamental identities involving Fourier transforms. We begin by considering transforms of the rapidly decaying test functions ϕ of the space $\mathcal{S}(\mathbf{R}^n)$.

THEOREM 2.3. Let $\phi(\mathbf{x})$ be an arbitrary test function in the space $\mathcal{S}(\mathbf{R}^n)$ defined in (2.63) and let $\mathcal{F}(\phi) = \hat{\phi}$ denote its Fourier transform. Then the following identities hold for every ϕ in $\mathcal{S}(\mathbf{R}^n)$:

(i) $\mathcal{F}(D_{\mathbf{x}}^{\boldsymbol{\alpha}} \phi) = (i\mathbf{y})^{\boldsymbol{\alpha}} \mathcal{F}(\phi) \qquad \forall \boldsymbol{\alpha} \in Z_+^n$ (2.78)

(ii) $D_{\mathbf{y}}^{\boldsymbol{\alpha}} \mathcal{F}(\phi) = \mathcal{F}[(-i\mathbf{x})^{\boldsymbol{\alpha}} \phi(\mathbf{x})] \qquad \forall \boldsymbol{\alpha} \in Z_+^n$ (2.79)

(iii) $\displaystyle\int_{\mathbf{R}^n} \mathcal{F}(\phi) \psi\, dx = \int_{\mathbf{R}^n} \phi \mathcal{F}(\psi)\, dx \qquad \forall \phi, \psi \in \mathcal{S}(\mathbf{R}^n)$ (2.80)

(iv) Moreover, for every $\phi, \psi \in L_2(\mathbf{R}^n)$,

$$(\phi, \psi)_{L_2(\mathbf{R}^n)} = (\mathcal{F}(\phi), \mathcal{F}(\psi))_{L_2(\mathbf{R}^n)} \tag{2.81}$$

where

$$(\phi,\psi)_{L_2(\mathbf{R}^n)} = \int_{\mathbf{R}^n} \phi(\mathbf{x})\,\overline{\psi(\mathbf{x})}\,dx \tag{2.82}$$

and, finally,

(v) (Plancherel's equality) for $\phi \in L_2(\mathbf{R}^n)$,

$$\|\phi\|_{L_2(\mathbf{R}^n)} = \|\mathscr{F}(\phi)\|_{L_2(\mathbf{R}^n)} \tag{2.83}$$

PROOF. To prove (i), we use the definition (2.61):

$$\begin{aligned}
\mathscr{F}(D_{\mathbf{x}}^{\boldsymbol{\alpha}}\phi) &= \frac{1}{(2\pi)^{n/2}} \int_{\mathbf{R}^n} \exp(-i\mathbf{x}\mathbf{y})\, D_{\mathbf{x}}^{\boldsymbol{\alpha}}\phi(\mathbf{x})\,dx \\
&= \frac{1}{(2\pi)^{n/2}} (i\mathbf{y})^{\boldsymbol{\alpha}} \int_{\mathbf{R}^n} \exp(-i\mathbf{x}\mathbf{y})\phi(\mathbf{x})\,dx \\
&= (i\mathbf{y})^{\boldsymbol{\alpha}} \mathscr{F}(\phi)
\end{aligned}$$

where we integrated by parts and used the fact that $\mathbf{x}^{\boldsymbol{\alpha}} D^{\boldsymbol{\beta}} u(\mathbf{x}) \to 0$ as $|\mathbf{x}^{\boldsymbol{\alpha}}| \to \infty$ for every $\boldsymbol{\alpha}$ and $\boldsymbol{\beta}$.

To prove (ii), consider

$$\begin{aligned}
D_{\mathbf{y}}^{\boldsymbol{\alpha}}(\mathscr{F}(\phi)) &= D_{\mathbf{y}}^{\boldsymbol{\alpha}} \left[\frac{1}{(2\pi)^{n/2}} \int_{\mathbf{R}^n} \exp(-i\mathbf{x}\mathbf{y})\phi(\mathbf{x})\,dx \right] \\
&= \frac{1}{(2\pi)^{n/2}} \int_{\mathbf{R}^n} D_{\mathbf{y}}^{\boldsymbol{\alpha}} [\exp(-i\mathbf{x}\mathbf{y})]\phi(\mathbf{x})\,dx \\
&= \frac{1}{(2\pi)^{n/2}} \int_{\mathbf{R}^n} \exp(-i\mathbf{x}\mathbf{y})(-i\mathbf{x})^{\boldsymbol{\alpha}}\phi(\mathbf{x})\,dx \\
&= \mathscr{F}[(-i\mathbf{x})^{\boldsymbol{\alpha}}\phi(\mathbf{x})]
\end{aligned}$$

which is precisely (2.79).

Identity (iii) follows from the fact that

$$\begin{aligned}
\int_{\mathbf{R}^n} \mathscr{F}(\phi)\psi(\mathbf{x})\,dx &= \int_{\mathbf{R}^n} \left[\frac{1}{(2\pi)^{n/2}} \int_{\mathbf{R}^n} \exp(-i\mathbf{x}\mathbf{y})\phi(\mathbf{y})\,dy \right] \psi(\mathbf{x})\,dx \\
&= \int_{\mathbf{R}^n} \left[\frac{1}{(2\pi)^{n/2}} \int_{\mathbf{R}^n} \exp(-i\mathbf{x}\mathbf{y})\psi(\mathbf{x})\,dx \right] \phi(\mathbf{y})\,dy \\
&= \int_{\mathbf{R}^n} \mathscr{F}(\psi)\phi(\mathbf{y})\,dy
\end{aligned}$$

To establish (iv), we note that, for ϕ and ψ in $L_2(\mathbf{R}^n)$,

$$(\phi,\psi)_{L_2(\mathbf{R}^n)}=\int_{\mathbf{R}^n}\left[\frac{1}{(2\pi)^{n/2}}\int_{\mathbf{R}^n}\hat{\phi}(\mathbf{y})\exp(i\mathbf{y}\mathbf{x})\,dy\right]\bar{\psi}(\mathbf{x})\,dx$$

$$=\frac{1}{(2\pi)^{n/2}}\int_{\mathbf{R}^n}\hat{\phi}(\mathbf{y})\left[\int_{\mathbf{R}^n}\exp(i\mathbf{y}\mathbf{x})\bar{\psi}(\mathbf{x})\,dx\right]dy$$

$$=\int_{\mathbf{R}^n}\hat{\phi}(\mathbf{y})\overline{\left[\frac{1}{(2\pi)^{n/2}}\int_{\mathbf{R}^n}\exp(-i\mathbf{y}\mathbf{x})\psi(\mathbf{x})\,dx\right]}\,dy$$

$$=\int_{\mathbf{R}^n}\hat{\phi}(\mathbf{y})\overline{\hat{\psi}(\mathbf{y})}\,dy=(\hat{\phi},\hat{\psi})_{L_2(\mathbf{R}^n)}$$

The Plancherel equality (2.83) follows immediately from (2.81) and the fact that $\|\phi\|^2_{L_2(\mathbf{R}^n)}=(\phi,\phi)_{L_2(\mathbf{R}^n)}$. ■

REMARK. Equation (2.81) establishes that the Fourier transform on $L_2(\mathbf{R}^n)$ is a *unitary* mapping of $L_2(\mathbf{R}^n)$ into itself; i.e., it preserves inner products. From this observation it can be shown to follow that the adjoint of $\mathcal{F}$ is the inverse Fourier transformation $\mathcal{F}^{-1}$. If A is any bounded linear operator on $L_2(\mathbf{R}^n)$, independent of $\mathbf{x}$, any operator of the form $\mathcal{F}A\mathcal{F}^{-1}$ is *unitarily equivalent* to A. ■

A remarkable result, and one that we make good use of in Chapter 4, is that some of the formulas (2.78) through (2.83) also hold when ϕ is replaced by a tempered distribution q. We record this fact in the following theorem.

THEOREM 2.4. Let q be an arbitrary tempered distribution in $\mathcal{S}(\mathbf{R}^n)'$. Then $\forall\alpha\in Z^n_+$,

$$\text{(i)}\qquad \mathcal{F}(D^{\alpha}_{\mathbf{x}}q)=(i\mathbf{y})^{\alpha}\mathcal{F}(q) \tag{2.84}$$

$$\text{(ii)}\qquad D^{\alpha}_{\mathbf{y}}\mathcal{F}(q)=\mathcal{F}((-i\mathbf{x})^{\alpha}q) \tag{2.85}$$

where the Fourier transforms $\mathcal{F}$ and the partial differentiations D^{α} are to be interpreted in the sense of distributions.

PROOF. By $\mathcal{F}(D^{\alpha}_{\mathbf{x}}q)$ we mean the distribution

$$\langle\widehat{D^{\alpha}_{\mathbf{x}}q},\phi\rangle=\langle D^{\alpha}_{\mathbf{x}}q,\hat{\phi}\rangle=(-1)^{|\alpha|}\langle q,D^{\alpha}_{\mathbf{x}}\hat{\phi}\rangle$$

Hence

$$\langle \widehat{D_{\mathbf{x}}^{\alpha} q}, \phi \rangle = (-1)^{|\alpha|} q \left[\frac{1}{(2\pi)^{n/2}} \int_{\mathsf{R}^n} D_{\mathbf{x}}^{\alpha} \phi(\mathbf{y}) \exp(-i\mathbf{x}\mathbf{y})\, dy \right]$$

$$= q \left[\frac{1}{(2\pi)^{n/2}} \int_{\mathsf{R}^n} (i\mathbf{y})^{\alpha} \phi(\mathbf{y}) \exp(-i\mathbf{x}\mathbf{y})\, dy \right]$$

$$= q\left(\widehat{(i\mathbf{y})^{\alpha}\phi(\mathbf{y})}\right)$$

$$= \langle \hat{q}, (i\mathbf{y})^{\alpha} \phi \rangle$$

$$= \langle (i\mathbf{y})^{\alpha} \hat{q}, \phi \rangle$$

Therefore

$$\widehat{D_{\mathbf{x}}^{\alpha} q} = (i\mathbf{y})^{\alpha} \hat{q}$$

which is precisely (2.84).

To prove (ii), we begin with an interpretation of the right side in the sense of distributions:

$$\langle \mathcal{F}((-i\mathbf{x})^{\alpha} q), \phi \rangle = \langle \widehat{(-i\mathbf{x})^{\alpha} q}, \phi \rangle$$

$$= \langle (-i\mathbf{x})^{\alpha} q, \hat{\phi} \rangle$$

$$= (-1)^{|\alpha|} \langle q, (i\mathbf{x})^{\alpha} \hat{\phi} \rangle$$

Since $\hat{\phi} \in \mathcal{S}(\mathsf{R}^n)$, we may use (2.84) to obtain

$$\langle \widehat{(-i\mathbf{x})^{\alpha} q}, \phi \rangle = (-1)^{|\alpha|} \langle q, \widehat{D_{\mathbf{y}}^{\alpha} \phi} \rangle$$

$$= (-1)^{|\alpha|} \langle \hat{q}, D_{\mathbf{y}}^{\alpha} \phi \rangle$$

$$= \langle D_{\mathbf{y}}^{\alpha} \hat{q}, \phi \rangle$$

Hence

$$\widehat{(-i\mathbf{x})^{\alpha} q} = D_{\mathbf{y}}^{\alpha} \hat{q}$$

which is (2.85). ■

REMARK. In view of Theorem 2.4, properties (2.84) and (2.85) also hold for distributions generated by functions in $L_2(\mathsf{R}^n)$. We make use of this fact in Chapter 4. ■

2.9 WEAK AND STRONG DERIVATIVES IN $L_p(\Omega)$

The concepts of weak and strong topologies on Banach spaces play an important role in certain aspects of analysis. Here we wish to use the concepts to distinguish between types of generalized derivatives of functions $u(x_1,x_2,\ldots,x_n)=u(\mathbf{x})$ in the space $L_p(\Omega)$, $p\geqslant 1$, Ω being an open bounded domain in $\mathbf{R}^n$. Recall that a sequence $\{u_k\}$ of functions is *strongly* convergent in $L_p(\Omega)$ to the function $u\in L_p(\Omega)$ if

$$\lim_{k\to\infty}\|u_k-u\|_{L_p(\Omega)}\equiv\lim_{k\to\infty}\left(\int_\Omega |u_k-u|^p\,dx\right)^{1/p}=0 \tag{2.86}$$

Likewise, a sequence $\{u_k\}$ is said to be *weakly* convergent to $u\in L_p(\Omega)$ if, for an arbitrary bounded linear functional $l\in L_q(\Omega)$, $1/p+1/q=1$, ($p<\infty$),

$$\lim_{k\to\infty} l(u_k)=l(u) \tag{2.87}$$

Since any functional $l\in L_q(\Omega)$ can be represented in the form of a scalar product via an integral, $l(u)=\int_\Omega vu\,dx$, $v\in L_q(\Omega)$, (2.87) can also be written

$$\lim_{k\to\infty}\int_\Omega u_k v\,dx=\int_\Omega uv\,dx \tag{2.88}$$

Here v is an arbitrary function in $L_q(\Omega)$. We can now use these definitions to manufacture strong and weak distributional derivatives.

Strong Derivatives in $L_p(\Omega)$

Let $\{\phi_k\}$ denote a bounded sequence of m-times continuously differentiable functions with compact support in $\Omega\subset\mathbf{R}^n$ [i.e., $\phi_k\in C_0^m(\Omega)$] which are strongly convergent in an L_p sense to a function $u\in L_p(\Omega)$:

$$\lim_{k\to\infty}\|\phi_k-u\|_{L_p(\Omega)}=0$$

Then u is said to have *strong L_p derivatives up to order m* if there exist functions $v^{\boldsymbol\alpha}\in L_p(\Omega)$, $\forall|\boldsymbol\alpha|\leqslant m$, such that

$$\lim_{k\to\infty}\|D^{\boldsymbol\alpha}\phi_k-v^{\boldsymbol\alpha}\|_{L_p(\Omega)}=0 \tag{2.89}$$

The function $v^{\boldsymbol\alpha}$ is called the $\boldsymbol\alpha$th strong L_p derivative of u; it is unique in the sense that any two functions satisfying (2.89) for a given $\boldsymbol\alpha$ are equal almost everywhere.

Weak Derivatives

A function u that is locally integrable on $\Omega \subset \mathbb{R}^n$ is said to have a *weak derivative* $w^{\boldsymbol{\alpha}}$ of order $\boldsymbol{\alpha}$ if a locally integrable function $w^{\boldsymbol{\alpha}}$ exists such that

$$\int_\Omega w^{\boldsymbol{\alpha}}\phi\, dx = (-1)^{|\boldsymbol{\alpha}|}\int_\Omega uD^{\boldsymbol{\alpha}}\phi\, dx \tag{2.90}$$

for every test function $\phi \in \mathfrak{D}(\Omega)$. In other words, $w^{\boldsymbol{\alpha}}$ is the $\boldsymbol{\alpha}$th generalized derivative of the regular distribution generated by u. It follows from previous observations that $w^{\boldsymbol{\alpha}}$ is unique (up to a set of measure zero).

THEOREM 2.5. If a function $u \in L_p(\Omega)$ has strong L_p derivatives $v^{\boldsymbol{\alpha}}$ of order $\leqslant m$, u has weak derivatives $w^{\boldsymbol{\alpha}}$ of order $\leqslant m$ and $v^{\boldsymbol{\alpha}} = w^{\boldsymbol{\alpha}}$ for all $\boldsymbol{\alpha}, |\boldsymbol{\alpha}| \leqslant m$.

PROOF. Suppose $\phi_k \to u$ and $D^{\boldsymbol{\alpha}}\phi_k \to v^{\boldsymbol{\alpha}}$ in $L_p(\Omega)$, $|\boldsymbol{\alpha}| \leqslant m$. Let $\phi \in C_0^\infty(\Omega)$. Then

$$\int_\Omega \phi D^{\boldsymbol{\alpha}}\phi_k\, dx = (-1)^{|\boldsymbol{\alpha}|}\int_\Omega \phi_k D^{\boldsymbol{\alpha}}\phi\, dx$$

Letting $k \to \infty$, we have

$$\int_\Omega \phi v^{\boldsymbol{\alpha}}\, dx = (-1)^{|\boldsymbol{\alpha}|}\int_\Omega uD^{\boldsymbol{\alpha}}\phi\, dx$$

Thus u also has weak derivatives $w^{\boldsymbol{\alpha}}$ and, by (2.90), $w^{\boldsymbol{\alpha}} = v^{\boldsymbol{\alpha}}$. ∎

It is easily seen that, whenever the classical derivative $D^{\boldsymbol{\alpha}}u(\mathbf{x})$ exists, it is also a strong, hence weak, derivative of u. Hence, in view of this fact and the uniqueness of $v^{\boldsymbol{\alpha}}$ and $u^{\boldsymbol{\alpha}}$, if u has a weak derivative $w^{\boldsymbol{\alpha}}$ we use the notation $w^{\boldsymbol{\alpha}} = D^{\boldsymbol{\alpha}}u = v^{\boldsymbol{\alpha}}$; when distinctions are necessary, the type of derivative needed is generally made clear in the context.

It is of fundamental importance to realize that, while u may be locally integrable, neither u nor its weak derivatives $D^{\boldsymbol{\alpha}}u$ may be in $L_p(\Omega)$. We show in the next section that often, when $u(\mathbf{x})$ is in $L_p(\Omega)$, its weak derivatives are also strong derivatives.

2.10 MOLLIFIERS AND MEAN FUNCTIONS

Thus far we have seen in this chapter how notions of convergence and differentiation can be established for quantities that cannot even be defined in a pointwise sense. Now we consider a converse problem: the use of a function u that is only integrable to generate an infinitely smooth function that is arbitrarily "close" to u. To accomplish this for any locally integrable function $u \in L_p(\Omega)$ an operator J_ρ can be introduced, called a *mollifier*, which maps u into a special smooth function arbitrarily close to u in an L_p sense.

Consider an open bounded domain Ω in $\mathbf{R}^n$ and let $\mathbf{x}=(x_1,x_2,\ldots,x_n)$ be an arbitrary point in $\mathbf{R}^n$. We now introduce a special function $\omega_\rho(\mathbf{x})$ which, for any choice of a real number $\rho \geqslant 0$, has the following properties:

(i) $\omega_\rho(\mathbf{x})$ has continuous derivatives of all orders on $\mathbf{R}^n$; i.e., $\omega_\rho(\mathbf{x}) \in C^\infty(\mathbf{R}^n)$.

(ii) $\omega_\rho(\mathbf{x})=0$ for $|\mathbf{x}| \geqslant \rho$ and $\omega_\rho(\mathbf{x})>0$ for $|\mathbf{x}|<\rho$; i.e., $\omega_\rho(\mathbf{x})$ and all its partial derivatives vanish outside a ball $B_\rho(\mathbf{0})$ of radius ρ centered at $\mathbf{x}=\mathbf{0}$: $\omega_\rho(\mathbf{x}) \in C_0^\infty(\overline{B_\rho(\mathbf{0})})$ and within $B_\rho(\mathbf{0})$ $\omega_\rho(\mathbf{x})$ is positive.

(iii)

$$\rho^{-n}\int_{\mathbf{R}^n} \omega_\rho(\mathbf{x})\,dx=1 \tag{2.91}$$

Such functions do exist, e.g., the function

$$\omega_\rho(\mathbf{x})=\begin{cases} c_0 \exp \dfrac{\rho^2}{|\mathbf{x}|^2-\rho^2} & \text{for } |\mathbf{x}|<\rho \\ 0 & \text{for } |\mathbf{x}| \geqslant \rho \end{cases} \tag{2.92}$$

where $|\mathbf{x}|^2=\sum_{i=1}^n x_i^2$, satisfies (i) and (ii), and the constant c_0 can be selected so that (iii) is satisfied. Indeed, if $r=|\mathbf{x}|$ is the length of the position vector $\mathbf{x}$ emanating from the origin, we can always choose $\omega_\rho(\mathbf{x})=\psi(r)=\psi(|\mathbf{x}|)$, where ψ is a C^∞ function on $\mathbf{R}$ with corresponding properties (ii) and (iii).

Now let u be an arbitrary function locally integrable over Ω. The *mollifier operator* J_ρ is defined for each $\rho \geqslant 0$ by the integral

$$J_\rho u(\mathbf{x})=\rho^{-n}\int_\Omega u(\mathbf{y})\omega_\rho(\mathbf{x}-\mathbf{y})\,dy \tag{2.93}$$

for any function u locally integrable on Ω. The function

$$u_\rho(\mathbf{x})=J_\rho u(\mathbf{x}) \tag{2.94}$$

is called the *mean function* corresponding to u. As we demonstrate below, $u_\rho(\mathbf{x})$ behaves much like u but is very (infinitely) smooth. Notice that, if $\mathbf{x}$ is on the interior of Ω, ρ can be chosen so that the support of $u_\rho(\mathbf{x})$ is compact in Ω, i.e., $u_\rho(\mathbf{x}) \in C_0^\infty(\Omega)$. This can be done by choosing those $\mathbf{x}$ at a distance greater than ρ from the boundary $\partial\Omega$ of Ω; i.e., at all points

$\mathbf{x} \in \Omega$ such that

$$\operatorname{dist}(\mathbf{x}, \partial\Omega) > \rho$$

where $\operatorname{dist}(\mathbf{x}, A)$ denotes the euclidean distance from $\mathbf{x}$ to the closest member of the set A.

THEOREM 2.6. Let u be a function that is locally integrable on a bounded open domain $\Omega \subset \mathbb{R}^n$. Then the mean function $u_\rho(\mathbf{x})$ associated with u by (2.93) [and (2.94)] is continuous and has continuous partial derivatives of all orders in $\mathbb{R}^n$; i.e., $u_\rho(\mathbf{x}) \in C^\infty(\mathbb{R}^n)$.

PROOF. (Cf. [2.3]) Since $\omega_\rho(\mathbf{x})$ is continuous, for every $\varepsilon > 0$ there is a $\delta(\varepsilon) > 0$ such that

$$|\omega_\rho(\mathbf{x}+\Delta\mathbf{x}-\mathbf{y}) - \omega_\rho(\mathbf{x}-\mathbf{y})| < \varepsilon \qquad \text{for } |\Delta\mathbf{x}| < \delta$$

Thus, whenever $|\Delta\mathbf{x}| < \delta$,

$$\begin{aligned}
&|u_\rho(\mathbf{x}+\Delta\mathbf{x}) - u_\rho(\mathbf{x})| \\
&\qquad = \left|\rho^{-n}\int_\Omega u(\mathbf{y})\big[\omega_\rho(\mathbf{x}+\Delta\mathbf{x}-\mathbf{y}) - \omega_\rho(\mathbf{x}-\mathbf{y})\big]\,dy\right| \\
&\qquad \leqslant \rho^{-n}\int_\Omega |u(\mathbf{y})||\omega_\rho(\mathbf{x}+\Delta\mathbf{x}-\mathbf{y}) - \omega_\rho(\mathbf{x}-\mathbf{y})|\,dy \\
&\qquad \leqslant \rho^{-n}\varepsilon\int_\Omega |u(\mathbf{y})|\,dy
\end{aligned}$$

Since $u(\mathbf{y})$ is, by hypothesis, integrable, the last integral is bounded. Thus $|u_\rho(\mathbf{x}+\Delta\mathbf{x}) - u_\rho(\mathbf{x})|$ is $< \varepsilon_1$ for some ε_1 whenever $|\Delta\mathbf{x}| < \delta$. Therefore $u_\rho(\mathbf{x})$ is continuous.

Similarly, let $\Delta\mathbf{x}_j = (0, 0, \ldots, 0, \Delta x_j, 0, \ldots, 0)$. Then

$$\begin{aligned}
\frac{1}{\Delta x_j}(u_\rho(\mathbf{x}+\Delta\mathbf{x}_j) - u_\rho(\mathbf{x})) = \rho^{-n}\int_\Omega u(\mathbf{y})\frac{1}{\Delta x_j}\big[&\omega_\rho(\mathbf{x}+\Delta\mathbf{x}_j-\mathbf{y}) \\
&- \omega_\rho(\mathbf{x}-\mathbf{y})\big]\,dy
\end{aligned}$$

or, in the limit as $\Delta x_j \to 0$,

$$\frac{\partial u_\rho(\mathbf{x})}{\partial x_j} = \rho^{-n}\int_\Omega u(\mathbf{y})\frac{\partial \omega_\rho(\mathbf{x}-\mathbf{y})}{\partial x_j}\,dy$$

Since $\partial \omega_\rho / \partial x_j$ is continuous, we can repeat the arguments given previously to show that $\partial u_\rho / \partial x_j$ is also continuous. Continuing in this manner, we can obviously show that $\forall \boldsymbol{\alpha} \in Z_+^n$,

$$D_{\mathbf{x}}^{\boldsymbol{\alpha}} u_\rho(\mathbf{x}) = \rho^{-n} \int_\Omega u(\mathbf{y}) D^{\boldsymbol{\alpha}} \omega_\rho(\mathbf{x}-\mathbf{y})\, dy \tag{2.95}$$

Since $D^{\boldsymbol{\alpha}} \omega_\rho(\mathbf{x}-\mathbf{y})$ is, by definition, continuous, so also is $D^{\boldsymbol{\alpha}} u_\rho(\mathbf{x})$. ■

THEOREM 2.7. Let $u(\mathbf{x}) \in L_p(\Omega)$. Then

(i)
$$\|u_\rho\|_{L_p(\Omega)} \leqslant \|u\|_{L_p(\Omega)} \tag{2.96}$$

(ii)
$$\lim_{\rho \to 0} \|u - u_\rho\|_{L_p(\Omega)} = 0 \tag{2.97}$$

PROOF (i) Using Hölder's inequality, we have

$$|u_\rho(\mathbf{x})|^p = \rho^{-np} \left| \int_\Omega u(\mathbf{y}) \left[\omega_\rho(\mathbf{x}-\mathbf{y}) \right]^{1/p} \left[\omega_\rho(\mathbf{x}-\mathbf{y}) \right]^{1/q} dy \right|^p$$
$$\leqslant \rho^{-np} \int_\Omega |u(\mathbf{y})|^p |\omega_\rho(\mathbf{x}-\mathbf{y})|\, dy \left[\int_\Omega |\omega_\rho(\mathbf{x}-\mathbf{y})|\, dy \right]^{p/q}$$

where $1/p + 1/q = 1$. Using (2.91) we have

$$|u_\rho(\mathbf{x})|^p \leqslant \rho^{-np(1-1/q)} \int_\Omega |u(\mathbf{y})|^p |\omega_\rho(\mathbf{x}-\mathbf{y})|\, dy$$

Hence, recalling (2.91), we integrate both sides to obtain

$$\int_\Omega |u_\rho(\mathbf{x})|^p\, dx \leqslant \int_\Omega |u(\mathbf{y})|^p \left[\rho^{-n} \int_\Omega |\omega_\rho(\mathbf{x}-\mathbf{y})|\, dx \right] dy$$
$$\leqslant \|u\|_{L_p(\Omega)}^p$$

(ii) To prove (ii), we extend u by zero to all of $\mathbf{R}^n$ [i.e., $u(\mathbf{x}) = 0$ if $\mathbf{x} \notin \Omega$]. Then

$$\|u - u_\rho\|_{L_p(\mathbf{R}^n)}^p = \int_{\mathbf{R}^n} \left| \left\{ \rho^{-n} \int_{\mathbf{R}^n} \left[u(\mathbf{x}) - u(\mathbf{y}) \right] \omega_\rho(\mathbf{x}-\mathbf{y}) dy \right\} \right|^p dx$$

where we have again used the fact that $\rho^{-n}\int_{\mathbb{R}^n}\omega_\rho dy=1$. For the inner integral, we set $\xi=\mathbf{y}-\mathbf{x}$, $dy=d\xi$, for fixed $\mathbf{x}$. Then, using Hölder's inequality, we observe that the inner integral satisfies the inequality

$$\left|\int_{\mathbb{R}^n}[u(\mathbf{x})-u(\mathbf{x}+\xi)]\omega_\rho(\xi)\,d\xi\right|^p$$

$$\leqslant \int_{|\xi|\leqslant\rho}|u(\mathbf{x})-u(\mathbf{x}+\xi)|^p\,d\xi\left[\int_{|\xi|\leqslant\rho}|\omega_\rho(\xi)|^q\,d\xi\right]^{p/q}$$

$$\leqslant C\rho^{np/q}\int_{|\xi|\leqslant\rho}|u(\mathbf{x})-u(\mathbf{x}+\xi)|^p\,d\xi$$

Here we have used the fact that $|\omega_\rho(\xi)|^q<C_1$, where C_1 is a constant independent of ρ, and $\int_{|\xi|\leqslant\rho}d\xi=C_2\rho^n$. Collecting results and changing the order of integration, we have

$$\|u-u_\rho\|^p_{L_p(\mathbb{R}^n)}\leqslant C\rho^{-n}\int_{|\xi|\leqslant\rho}\int_{\mathbb{R}^n}|u(\mathbf{x})-u(\mathbf{x}+\xi)|^p\,dx\,d\xi$$

and we have used the fact that if $\mathbf{x}\notin\Omega$, the above integrals are zero. However, owing to the continuity of functions in $L_p(\Omega)$ with respect to the L_p norm, we know that for every $\varepsilon>0$ there is a $\delta(\varepsilon)>0$ such that, whenever $|\xi|<\delta$,

$$\int_\Omega|u(\mathbf{x})-u(\mathbf{x}+\xi)|^p\,dx<\varepsilon^p$$

Therefore, picking $0<\rho<\delta$,

$$\|u-u_\rho\|^p_{L_p(\Omega)}\leqslant C\rho^{-n}\varepsilon^p\int_{|\xi|\leqslant\rho}dx=\tilde{C}\varepsilon^p$$

where C and $\tilde{C}$ are positive constants. Since ε is arbitrary, we can pick it so that δ and consequently ρ are made as small as desired. Hence (2.97) holds. ■

EXAMPLE 2.26. An interesting example of a mean function is obtained by imposing the mollifier J_ρ on the unit function $u(\mathbf{x})=1$, $\mathbf{x}\in\Omega$, $u(\mathbf{x})=0$, $\mathbf{x}\notin\Omega$:

$$i_\rho(\mathbf{x})\equiv J_\rho(1)=\rho^{-n}\int_\Omega\omega_\rho(\mathbf{x}-\mathbf{y})dy \tag{2.98}$$

The function $i_\rho(\mathbf{x})$ obtained in this way can be shown to have the following properties:

(i) $i_\rho(\mathbf{x}) \in C^\infty(\mathsf{R}^n)$.
(ii) $i_\rho(\mathbf{x}) = 1,\ \mathbf{x} \in \Omega \subset \mathsf{R}^n$.
(iii) $i_\rho(\mathbf{x}) = 0,\ \mathbf{x} \in \mathsf{R}^n \dot{-} \Omega_0$, where $\Omega \subset \Omega_0$.

Figure 2.2 illustrates the smooth unit function $i_\rho(\mathbf{x})$ for R^2. ■

THEOREM 2.8. Let $u(\mathbf{x}) \in L_p(\Omega)$ have weak partial derivatives in $L_p(\Omega)$ of all orders $\leqslant m$. Then, for each $\mathbf{x} \in \Omega$ such that $\operatorname{dist}(\mathbf{x}, \partial\Omega) > \rho$, the classical partial derivatives of the mean function u_ρ of order $\boldsymbol{\alpha}$ coincide with the mean function corresponding to the weak derivatives of u of order $\boldsymbol{\alpha}$, for $|\boldsymbol{\alpha}| \leqslant m$, i.e.,

$$D^{\alpha} J_\rho u(\mathbf{x}) = J_\rho D^{\alpha} u(\mathbf{x}) \qquad |\boldsymbol{\alpha}| \leqslant m \tag{2.99}$$

PROOF. Let $D_{\mathbf{x}}^{\alpha} = \partial^{|\alpha|}/\partial x_1^{\alpha_1} \cdots \partial x_n^{\alpha_n}$. Then [see (2.95)]

$$D_{\mathbf{x}}^{\alpha} J_\rho u(\mathbf{x}) = \rho^{-n} \int_\Omega u(\mathbf{y}) D_{\mathbf{x}}^{\alpha} \omega_\rho(\mathbf{x}-\mathbf{y})\, dy$$

$$= (-1)^{|\alpha|} \rho^{-n} \int_\Omega u(\mathbf{y}) D_{\mathbf{y}}^{\alpha} \omega_\rho(\mathbf{x}-\mathbf{y})\, dy$$

or, from the definition of weak derivatives,

$$D^{\alpha} J_\rho u(\mathbf{x}) = \rho^{-n} \int_\Omega \omega_\rho(\mathbf{x}-\mathbf{y}) D_{\mathbf{y}}^{\alpha} u(\mathbf{y})\, dy = J_\rho D^{\alpha} u(\mathbf{x}) \quad ■$$

It can now be shown to follow that, when u has weak derivatives of order $|\boldsymbol{\alpha}|$ in $L_p(G)$, G being a compact subset of Ω, these are also strong derivatives; i.e., the concepts of weak and strong derivatives are then equivalent.

THEOREM 2.9. Let G be a subset of an open bounded domain $\Omega \subset \mathsf{R}^n$ such that G is strictly inside Ω (i.e., $\overline{G}$ is compact and $\overline{G} \subset \Omega$; alternatively, G is a positive distance from $\partial\Omega$). Let $u(\mathbf{x}) \in L_p(\Omega)$ have weak derivatives $D^{\alpha} u \in L_p(\Omega)$ of order $|\boldsymbol{\alpha}| \leqslant m$ and define the corresponding mean function as $u_\rho(\mathbf{x}) = \rho^{-n} \int_\Omega u(\mathbf{y}) \omega_\rho(\mathbf{x}-\mathbf{y})\, dy$. Then

$$\lim_{\rho \to 0} \|u - u_\rho\|_{L_p(G)} = 0 \tag{2.100}$$

and

$$\lim_{\rho \to 0} \|D^{\alpha} u - D^{\alpha} u_\rho\|_{L_p(G)} = 0 \tag{2.101}$$

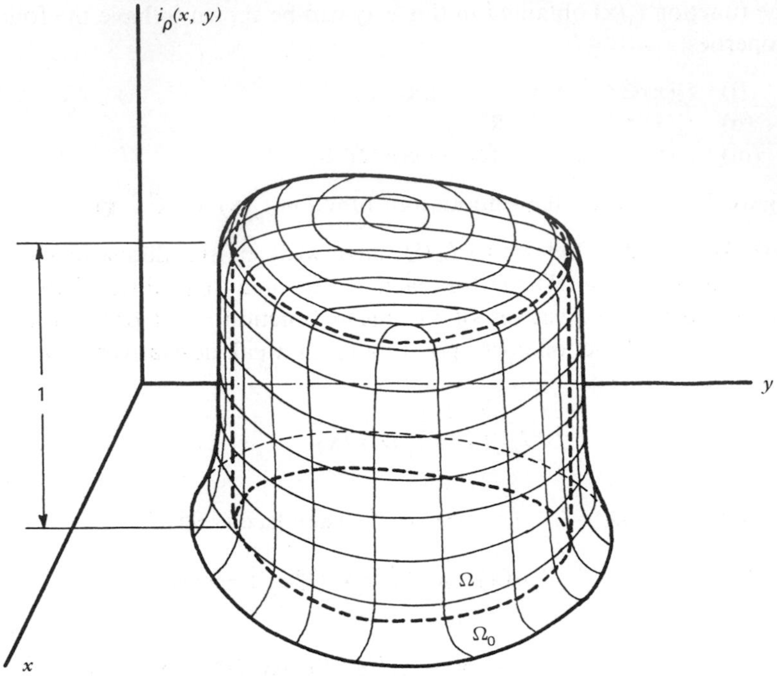

Figure 2.2. The function $i_\rho(x,y)$ corresponding to a domain Ω in $\mathbf{R}^2$.

PROOF. The proof follows immediately by a repetition of the arguments given in the proofs of Theorems 2.7 and 2.8. ■

The choice of $G \subset\subset \Omega$ (strictly inside Ω) is worth some additional comments. Certainly (2.100) also holds for $L_p(\Omega)$ by virtue of (2.97). However, (2.97) was derived by extending $u(\mathbf{x})$ by zero to all of $\mathbf{R}^n$; moreover, (2.97) holds only when $\text{dist}(\mathbf{x}, \partial\Omega) > \rho$. It is not clear that the functions $D^\alpha u_\rho(\mathbf{x})$ will be convergent to $D^\alpha u(\mathbf{x})$ in $L_p(\Omega)$, because of problems encountered in extending them to the entire space $\mathbf{R}^n$. In short, for arbitrary Ω, the extended function $u(\mathbf{x})$ may not necessarily have the corresponding extended weak derivatives in $\mathbf{R}^n$.

We now demonstrate that, for any function $u(\mathbf{x})$ with weak derivatives $D^\alpha u(\mathbf{x})$, $|\boldsymbol{\alpha}| \leqslant m$, in $L_p(\Omega)$, a sequence of functions $\{u_k\}$, m-times continuously differentiable in $\bar{\Omega}$, exists such that $\lim u_k = u$ and $\lim D^\alpha u_k = D^\alpha u$ as $k \to \infty$, provided Ω is *star-shaped*. A domain Ω is star-shaped if there exists an interior point $\mathbf{x}_0$ such that every ray drawn from $\mathbf{x}_0$ cuts the boundary of Ω only once. By a function $u_k(\mathbf{x})$ m-times continuously differentiable in the closed domain $\bar{\Omega}$, we mean that $u_k(\mathbf{x})$ is m-times continuously differentiable inside Ω, continuous on $\bar{\Omega}$, and for which all derivatives $D^\alpha u_k(\mathbf{x})$ for

$|\alpha| \leqslant m$ can be defined on the boundary $\partial\Omega$ of Ω in a way that produces continuous derivatives of the desired order in the closure $\bar{\Omega}$.

THEOREM 2.10. Let Ω be a bounded star-shaped domain in $\mathbf{R}^n$ and let $u(\mathbf{x})$ denote a function which, together with its generalized derivatives $D^{\alpha}u(\mathbf{x})$, $|\alpha| \leqslant m$, are in $L_p(\Omega)$, $p \geqslant 1$. Then there exists a sequence $\{u_k\} \in C^m(\bar{\Omega})$ of functions m-times continuously differentiable in Ω such that

$$\lim_{k\to\infty} \|u - u_k\|_{L_p(\Omega)} = 0 \qquad \text{and} \qquad \lim_{k\to\infty} \|D^{\alpha}u - D^{\alpha}u_k\|_{L_p(\Omega)} = 0 \quad (2.102)$$

for every α such that $|\alpha| \leqslant m$.

PROOF. Our proof involves two basic considerations. First, to make use of the results developed thus far we must somehow define extensions of functions on Ω to some larger domain Ω' in which Ω is "strictly inside" (i.e., $\Omega \subset \Omega'$ and the distance from any point $\mathbf{x} \in \Omega$ to the boundary of Ω' is positive). Second, we make use of the mean functions described earlier, since they are not only sufficiently smooth but also can be designed to converge to the correct elements in $L_p(\Omega)$.

Suppose that $u(\mathbf{x})$ is an element of $L_p(\Omega)$, where Ω is star-shaped with respect to the origin $\mathbf{x}_0 = \mathbf{0}$. Then the sequence of functions $u[(k-1)\mathbf{x}/k]$, $k = 2,3,4,\ldots$, are defined on domains Ω_k, each of which contains Ω strictly inside itself. [Consider, for example, a function $f(x)$ defined on $\{x: -1 \leqslant x \leqslant 1\}$. By introducing a change in variables, $y = x/2$, we obtain a function $g(x) = f(y)$ defined on the interval $[-2,2]$. Similarly, setting $y_{(k)} = (k-1)x/k$, $k = 3,4,\ldots$, we obtain a function $f(y_{(k)})$ defined on $[-k/(k-1), k/(k-1)]$. Clearly, as $k \to \infty$, we recover the original interval $[-1,1]$. In the case of functions defined on two- or higher-dimensional domains, the same type of construction can be made along all straight lines emanating from $\mathbf{x}$ (rays), provided the domain is star-shaped with respect to $\mathbf{x}$. Then Ω_k approachs Ω as k tends to infinity.]

It is important to note that the sequence of functions $u_{[k]} \equiv u[(k-1)\mathbf{x}/k] \in L_p(\Omega_k)$ and their generalized derivatives $D^{\alpha}u_{[k]}$ are convergent to $u(\mathbf{x})$ and $D^{\alpha}u(\mathbf{x})$, respectively. In fact,

$$\begin{aligned}\|D^{\alpha}u - D^{\alpha}u_{[k]}\|^p_{L_p(\Omega)} &= \int_{\Omega} \left| D^{\alpha}u(\mathbf{x}) - \left(\frac{k-1}{k}\right)^{|\alpha|} D^{\alpha}u\left(\frac{(k-1)\mathbf{x}}{k}\right)\right|^p dx \\ &\leqslant \int_{\Omega} \left| D^{\alpha}u(\mathbf{x}) - D^{\alpha}u\left(\frac{(k-1)\mathbf{x}}{k}\right)\right|^p dx \\ &\quad + \left[1 - \left(\frac{k-1}{k}\right)^{|\alpha|}\right]^p \left\| D^{\alpha}u\left(\frac{(k-1)\mathbf{x}}{k}\right)\right\|^p_{L_p(\Omega)}\end{aligned}$$

As $k \to \infty$, $(k-1)/k \to 1$, so that both terms on the right side of this inequality clearly vanish. Hence $D^\alpha u_{[k]} \to D^\alpha u$ in $L_p(\Omega)$. A similar argument shows also that $u_{[k]} \to u$ in $L_p(\Omega)$.

Now hold k fixed and observe that Ω lies strictly in Ω_k. Hence Theorem 2.9 is applicable [with G replaced by Ω and Ω by Ω_k in (2.100) and (2.101)]. Next, we construct a mean function $u_{\rho[k]}(\mathbf{x}) = J_\rho u_{[k]}(\mathbf{x})$ which is, we recall, infinitely differentiable and convergent to $u_{[k]}(\mathbf{x})$ as $\rho \to 0$. Since

$$\|u - u_{\rho[k]}\|_{L_p(\Omega)} = \|u - u_{[k]} + u_{[k]} - u_{\rho[k]}\|_{L_p(\Omega)}$$

$$\leqslant \|u - u_{[k]}\|_{L_p(\Omega)} + \|u_{[k]} - u_{\rho[k]}\|_{L_p(\Omega)}$$

it follows that

$$\lim_{\substack{k \to \infty \\ \rho \to 0}} \|u - u_{\rho[k]}\|_{L_p(\Omega)} = 0$$

Therefore the functions $u_{\rho[k]}$ can be taken to be the functions u_k mentioned in the theorem. Likewise, a similar argument shows that

$$\lim_{\substack{k \to \infty \\ \rho \to 0}} \|D^\alpha u - D^\alpha u_{\rho[k]}\|_{L_p(\Omega)} = 0$$

if $|\boldsymbol{\alpha}| \leqslant m$, and a check of the definition shows that $u_{\rho[k]} \in C^m(\Omega_k)$. Thus the theorem is proved. A glance at this proof also leads to an additional result; every conclusion still holds if we pick sequences $\{u_k\}$ in $C^\infty(\Omega)$ instead of $C^m(\Omega)$. ■

REMARK. In the next chapter we make use of a construction that generalizes the use of rays in star-shaped domains described in the above proof. This involves simply the assumption that Ω can be represented as the union of a countably infinite number of open sets Ω_k such that $\overline{\Omega}_k \subset \Omega_{k+1}$. While these sets are similar to the sets Ω_k described in the theorem here, this process is obviously more general. See Theorem 3.2. ■

REFERENCES

2.1. Schwartz, L., *Theorie des Distributions*, Vol. I, Hermann, Paris, 1950.

2.2. Yosida, K., *Functional Analysis*, 3rd ed., Springer-Verlag, New York, 1971.

2.3. Smirnov, V. I., *A Course of Higher Mathematics, V. Integration and Functional Analysis*, Translated from the 1960 Russian edition by D. E. Brown, Pergamon Press, Oxford, 1964.

3

THEORY OF SOBOLEV SPACES

3.1 INTRODUCTION

The mathematical foundations of finite-element theory, as it applies to the approximate solution of linear boundary-value problems, derives from modern Hilbert space theory and, particularly, the notions of Sobolev spaces, traces, and interpolation spaces. This chapter is devoted to a study of certain of these notions pertaining to Sobolev spaces. We prove here the Sobolev integral identity which leads us directly to the important Sobolev embedding theorems. These prove to be very useful in the study of certain aspects of convergence of Galerkin finite-element approximations.

3.2 THE SOBOLEV SPACE $W_p^m(\Omega)$

We begin by considering once again an arbitrary open bounded domain Ω in n-dimensional euclidean space R^n, the points of which are denoted $\mathbf{x}=(x_1,x_2,\ldots,x_n)$. We recall that the Lebesgue space $L_p(\Omega)$ consists of equivalence classes of functions u defined on Ω whose absolute values have pth powers which are Lebesgue-integrable on Ω, where $p \geqslant 1$; i.e., $u \in L_p(\Omega)$ if u is measurable and if its $L_p(\Omega)$ norm is finite:

$$\|u\|_{L_p(\Omega)}=\left(\int_\Omega |u|^p\,dx\right)^{1/p}<\infty. \tag{3.1}$$

Here Lebesgue integration is implied and $dx=dx_1\,dx_2\cdots dx_n$. We noted in the introduction that $L_p(\Omega)$ is a complete normed linear space, i.e., a Banach space, endowed with the norm defined in (3.1).

About 1950, S. L. Sobolev showed that the idea of $L_p(\Omega)$ spaces can be expanded in a natural way by considering functions $u(\mathbf{x})$ that are not only in $L_p(\Omega)$ but which have all weak partial derivatives up to some order m which are also in $L_p(\Omega)$ (see Sobolev [3.1]). When all weak partial derivatives of $u(\mathbf{x})$ of order $\leqslant m$, m being an integer $\geqslant 0$, are in $L_p(\Omega)$, we say that $u(\mathbf{x})$ belongs to a space denoted $W_p^m(\Omega)$, referred to as a *Sobolev space of order m,p on* Ω; i.e.,

$$W_p^m(\Omega) = \{u\colon D^{\alpha}u \in L_p(\Omega);\ \forall \boldsymbol{\alpha} \text{ such that } |\boldsymbol{\alpha}| \leqslant m\} \tag{3.2}$$

Clearly,

$$W_p^0(\Omega) = L_p(\Omega) \tag{3.3}$$

A variety of different but equivalent norms can be associated with $W_p^m(\Omega)$ (see [3.2]). Since, for each $\boldsymbol{\alpha}$ such that $0 \leqslant |\boldsymbol{\alpha}| \leqslant m$, $D^{\alpha}u$ is in $L_p(\Omega)$, the sum of the L_p norms of all the weak derivatives of u of order $\leqslant m$ obviously satisfies the norm axioms and also is suggested naturally by the definition (3.2). Thus we may introduce for each $u \in W_p^m(\Omega)$ the norm

$$\|u\|_{W_p^m(\Omega)} = \left(\int_{\Omega} \sum_{|\boldsymbol{\alpha}| \leqslant m} |D^{\alpha}u|^p \, dx \right)^{1/p}$$

$$= \left(\sum_{|\boldsymbol{\alpha}| \leqslant m} \|D^{\alpha}u\|_{L_p(\Omega)}^p \right)^{1/p} \tag{3.4}$$

Hereafter, whenever we refer to the Sobolev space $W_p^m(\Omega)$, we mean the normed space consisting of the linear space of functions given in (3.2) together with the Sobolev norm (3.4).

EXAMPLE 3.1. Let Ω denote an open interval on the real line, $\Omega = (a,b) \subset \mathbb{R}$. The space $W_3^2(a,b)$ consists of functions $u(x)$ with weak derivatives up to second order in $L_3(a,b)$. The associated Sobolev norm of u is given by

$$\|u\|_{W_3^2(a,b)} = \left[\int_a^b \left(|u|^3 + \left| \frac{du}{dx} \right|^3 + \left| \frac{d^2u}{dx^2} \right|^3 \right) dx \right]^{1/3}$$

Likewise, if $u(x) \in W_2^1(a,b)$, then

$$\|u\|_{W_2^1(a,b)} = \left\{ \int_a^b \left[u^2 + \left(\frac{du}{dx} \right)^2 \right] dx \right\}^{1/2} < \infty \quad \blacksquare$$

EXAMPLE 3.2. Suppose that $\Omega=(a,b)\times(c,d)\subset \mathbb{R}^2$ and $u(x,y)\in W_p^2(\Omega)$, $p\geqslant 1$. Then the Sobolev norm of u is

$$\|u\|_{W_p^2(\Omega)}=\left[\int_a^b\int_c^d\left(|u|^p+\left|\frac{\partial u}{\partial x}\right|^p+\left|\frac{\partial u}{\partial y}\right|^p+\left|\frac{\partial^2 u}{\partial x^2}\right|^p\right.\right.$$
$$\left.\left.+\left|\frac{\partial^2 u}{\partial x\,\partial y}\right|^p+\left|\frac{\partial^2 u}{\partial y^2}\right|^p\right)dx\,dy\right]^{1/p} \quad \blacksquare$$

We emphasize that, at this point, it is important to interpret the partial derivatives appearing in the above examples as weak derivatives in the distributional sense described in the previous chapter; i.e., the statement that "$u\in W_p^m(\Omega)\Rightarrow D^\alpha u\in L_p(\Omega), |\alpha|\leqslant m$," means that $D^\alpha u$ is an equivalence class of functions to be associated with a continuous linear functional l on the space of test functions $\mathfrak{D}(\Omega)$ such that $l(\phi)=(-1)^{|\alpha|}\int_\Omega uD^\alpha\phi\,dx$. Since $D^\alpha u$ is in $L_p(\Omega)$, this linear functional is actually generated by (i.e., is equivalent to) a class of measurable functions in $L_p(\Omega)$. Moreover, in view of Theorem 2.10, such weak derivatives $D^\alpha u$ may be, in this case, also strong derivatives in the sense described in Section 2.9.

REMARK. A comprehensive account of Sobolev spaces and their properties has been compiled by Adams [3.3]. ■

3.3 PARTITIONS OF UNITY, BOUNDARIES, AND CONE CONDITIONS

Before we can investigate in much depth the basic properties of the Sobolev spaces $W_p^m(\Omega)$, we need to introduce several auxiliary concepts. Some of these are useful in the construction of proofs of theorems to be given later; others deal with more precise definitions of the domain Ω, e.g., the degree of smoothness of the boundary $\partial\Omega$ and the identification of a class of domains for which the embedding theorems, to be described in Section 3.6, hold. We begin with the concept of a partition of unity.

Partition of Unity

Let Ω be an open bounded domain in $\mathbb{R}^n$ and let $\{G_i\}$ be a countable collection of open sets such that $\Omega\subset\bigcup_{i=1}^\infty G_i$ (i.e., $\{G_i\}$ is a countable open covering of Ω). A collection of functions $\psi_i(\mathbf{x})$ such that

(i) $\psi_i(\mathbf{x})\in C_0^\infty(G_i)$,
(ii) $\sum_{i=1}^\infty\psi_i(\mathbf{x})=1$ and $\psi_i(\mathbf{x})\geqslant 0$ in Ω, and

(iii) every compact subset of Ω intersects only a finite number of the supports of the $\psi_i(\mathbf{x})$

is called a *partition of unity subordinate to the covering* $\{G_i\}$.

That functions $\psi_i(\mathbf{x})$ do exist that have these properties is established in the following theorem.

THEOREM 3.1. Let Ω be an open bounded domain in $\mathbb{R}^n$ which is the union of a countable collection of open sets G_i: $\Omega = \cup_{i=1}^{\nu} G_i$, $1 \leqslant \nu \leqslant \infty$. Then there exist functions $\psi_i(\mathbf{x}) \in C_0^\infty(G_i)$ such that $\Sigma_{i=1}^{\nu} \psi_i(\mathbf{x}) = 1$ and $\psi_i(\mathbf{x}) \geqslant 0$ for $\mathbf{x} \in \Omega$, $1 \leqslant i \leqslant \infty$. Moreover, if $\nu = \infty$ and G^* is a compact subset of Ω, G^* intersects only a finite number of supports of the ψ_i.

PROOF. We only outline the essential features of the proof. We begin by showing the existence of a partition of unity for a compact set $A \subset \mathbb{R}^n$. For compact A, we know from the Heine–Borel theorem that there exists a finite number m of open sets $\{E_i\}_{i=1}^m$ that covers A. Moreover, it can be shown that, for any such set, there exists a set $\{F_i\}_{i=1}^m$ of compact sets such that $F_i \subset E_i$ and such that the collection of open interiors $\{\operatorname{int} F_i\}_{i=1}^m$ also forms a covering of A.

Next, we introduce for each F_i a C^∞ function $\alpha_i(\mathbf{x})$ such that $\alpha_i(\mathbf{x}) > 0$ for $\mathbf{x} \in F_i$ and $\alpha_i(\mathbf{x}) = 0$ for $\mathbf{x}$ outside of some closed set contained in E_i. Such functions exist. For example, let $\beta_i(\mathbf{x}) = 1$, $\mathbf{x} \in F_i$, and $\beta_i(\mathbf{x}) = 0$, $\mathbf{x} \notin F_i$, $1 \leqslant i \leqslant m$, and let $\rho_i < \inf d(F_i, \partial E_i) =$ distance from F_i to the boundary ∂E_i of E_i. Then we may *mollify* $\beta_i(\mathbf{x})$ and obtain the smooth unit functions $i_{\rho_i}(\mathbf{x}) = J_{\rho_i} \beta_i(\mathbf{x})$ of the type in Example 2.26, which can be then used as the functions $\alpha_i(\mathbf{x})$.

Next, we introduce a smooth unit function $i_\Omega(\mathbf{x}) \in C_0^\infty(\mathbb{R}^n)$ such that $i_\Omega(\mathbf{x}) = 1$, $\mathbf{x} \in \Omega$, and $i_\Omega(\mathbf{x}) = 0$ outside some open set $\mathcal{O}$ containing Ω. Then it is easily verified that the set of functions $\{\psi_i(\mathbf{x})\}_{i=1}^m$ is the desired partition of unity, where

$$\psi_i(\mathbf{x}) = i_\Omega(\mathbf{x}) \frac{\alpha_i(\mathbf{x})}{\sum_{j=1}^{m} \alpha_j(\mathbf{x})} \qquad 1 \leqslant i \leqslant m \tag{3.5}$$

Now, return to our open bounded domain Ω, which we regard as the union of a countably infinite collection of sets H_i, $\Omega = \cup_{i=1}^{\infty} H_i$, where for example,

$$H_i = \left\{ \mathbf{x}\colon |\mathbf{x}| \leqslant \frac{1}{i}\,;\ d(\mathbf{x}, \partial\Omega) \geqslant \frac{1}{i} \right\}$$

We shall demonstrate the proof for the choice of sets

$$G_i = \mathcal{O} \cap (\text{int}\, H_{i+2} - H_{i-2})$$

$$K_i = H_i - \text{int}\, H_{i-1} \qquad 1 \leqslant i \leqslant \nu$$

where $\mathcal{O}$ is an open set containing Ω. Then $\{G_i\}$ is an open cover of the compact set K_i. From what was said previously, we know that there exists a partition of unity $\{\beta_j^{(i)}(\mathbf{x})\}_{j=1}^{m_i}$ of K_i subordinate to G_i. Moreover, for each $\mathbf{x} \in \Omega$, the sum $\psi(\mathbf{x}) \equiv \Sigma_i \Sigma_j \beta_j^{(i)}(\mathbf{x})$ is finite, because for $\mathbf{x} \in H_i$ we have $\beta_j^{(i)}(\mathbf{x}) = 0$ for $j \geqslant i+2$. Finally, for each function $\beta_j^{(i)}(\mathbf{x})$ define $\psi_i(\mathbf{x}) = \beta_j^{(i)}(\mathbf{x})/\psi(\mathbf{x})$. The set of functions $\psi_i(\mathbf{x})$ so defined is the desired partition of unity.

The compact intersection property (property (iii)) can be shown to follow from the fact that the K_i are compact, hence there exists a finite ε net $\mathbf{x}^1, \mathbf{x}^2, \ldots, \mathbf{x}^{m_r}$ such that $K_i \subset \cup_{j=1}^{m_r} B_r(\mathbf{x}^j)$, where $B_r(\mathbf{x}^j)$ is an open ball of radius r centered at $\mathbf{x}^j$. Hence any such compact subset can intersect only a finite number of the G_i's. For additional details, see, e.g., [3.4] or [3.5]. ■

As noted earlier, it is also necessary occasionally to describe precisely the degree of smoothness of the boundary $\partial\Omega$ of a domain $\Omega \subset \mathbf{R}^n$. We adopt a rather standard means for doing this. Noting that $\bar{\Omega} = \Omega \cup \partial\Omega \subset \mathbf{R}^n$, i.e., $\bar{\Omega}$ is the closure of Ω, we pick an arbitrary point $\mathbf{x}_0 \in \partial\Omega$ which is viewed as the center of an open ball $B_\varepsilon(\mathbf{x}_0)$ of radius ε:

$$B_\varepsilon(\mathbf{x}_0) = \{\mathbf{x}\colon \mathbf{x} \in \mathbf{R}^n;\ |\mathbf{x} - \mathbf{x}_0| < \varepsilon\}$$

We next introduce a change in coordinates,

$$\xi_i = F_i(\mathbf{x}) \qquad i = 1, 2, \ldots, n$$

such that the "segment" $\partial\Omega \cap B_\varepsilon(\mathbf{x}_0)$ can be represented in the form

$$\xi_j = f(\xi_1, \xi_2, \ldots, \xi_{j-1}, \xi_{j+1}, \ldots, \xi_n) \tag{3.6}$$

for some j.

The degree of smoothness of $\partial\Omega$ at $\mathbf{x}_0$ is thus measured by the differentiability of f at $\mathbf{x}_0$. If $f \in C^m$ for *every* point $\mathbf{x}_0 \in \partial\Omega$, we say that $\partial\Omega$ is of class C^m. When f of (3.6) is Lipschitz-continuous (i.e., $\forall \mathbf{x}_0, \mathbf{y}_0 \in \partial\Omega$, there is a constant K such that $|f(\xi(\mathbf{x}_0)) - f(\xi(\mathbf{y}_0))| < K|\mathbf{x}_0 - \mathbf{y}_0|$), $\partial\Omega$ is said to be *lipschitzian*.

Many of the results to be established in this chapter hold for quite irregular domains. Indeed, the embedding theorems and most of the associated Sobolev inequalities associated with them require no more smoothness of Ω than is needed to satisfy the cone condition. ■

Cone Condition

An open bounded domain Ω in $\mathbf{R}^n$ is said to satisfy the *cone condition* (or to have the *cone property*) if there exists a cone $\mathcal{C}_\rho(\mathbf{x})$ of fixed dimensions, with radius ρ and vertex $\mathbf{x}$, such that, when any point $\mathbf{x}\in\Omega$ is used as the vertex of the cone, the cone can be oriented so that all its points lie in Ω.

When testing whether or not a given domain Ω has the cone property, it is convenient to think of $\mathcal{C}_\rho(\mathbf{x})$ as a right spherical cone of fixed radius ρ and solid angle θ. Then a fixed cone $\mathcal{C}_\rho(\mathbf{x})$ is used as template and is moved from point to point in Ω, each point being used as a vertex. For each vertex, if $\mathcal{C}_\rho(\mathbf{x})$ can be positioned so that $\mathcal{C}_\rho(\mathbf{x})\subset\Omega$, Ω has the cone property.

Consider, for example, the shaded two-dimensional domains shown in Fig. 3.1*a*. In each case we can use each point $\mathbf{x}$ in the domain as the vertex of a small cone of fixed dimensions, each point of which lies in Ω. Domains with cusps, however, such as those shown in Fig. 3.1*b*, do *not* satisfy the cone condition, as it is impossible to use the apex of the cusp as the vertex of a cone *all* of which lies inside the region.

In subsequent developments in which we make use of the cone condition, we regard the cone $\mathcal{C}_\rho(\mathbf{x})$ as a *cone with respect to a ball*; i.e., the generators of the cone intersect a spherical ball, the points of which are contained in $\mathcal{C}_\rho(\mathbf{x})$. For example, to construct such a cone in $\mathbf{R}^3$, we pick a ball $B_a(\mathbf{x}_1)=\{\mathbf{x}: |\mathbf{x}-\mathbf{x}_1|\leqslant a\}$ and any point $\mathbf{x}_0\not\in B_a(\mathbf{x}_1)$. We then construct a conical envelope of rays emanating from $\mathbf{x}_0$, those tangent to $B_a(\mathbf{x}_1)$ forming the generators of the cone. The closure of the entire collection of points inside these generators and in the ball $B_a(\mathbf{x}_1)$ constitutes the cone with vertex $\mathbf{x}_0$. Geometrically, such a cone resembles an ice cream cone.

The Space $\overline{C}^m(\Omega)$

Throughout the remainder of this chapter, we shall denote by $\overline{C}^m(\Omega)$ the space of functions u which have bounded and continuous derivatives $D^{\boldsymbol{\alpha}}u$ of orders $|\boldsymbol{\alpha}|\leqslant m$ on Ω. This space is a Banach space under the norm

$$\|u\|_{\overline{C}^m(\Omega)}=\max_{|\boldsymbol{\alpha}|\leqslant m}\ \sup_{\mathbf{x}\in\Omega}|D^{\boldsymbol{\alpha}}u(\mathbf{x})|$$

Clearly $\overline{C}^m(\Omega)\subset\hat{C}^m(\Omega)$ (see p. 61), but $C^m(\overline{\Omega})\subset\overline{C}^m(\Omega)$ since the functions in $^{-m}(\Omega)$ need not be uniformly continuous.

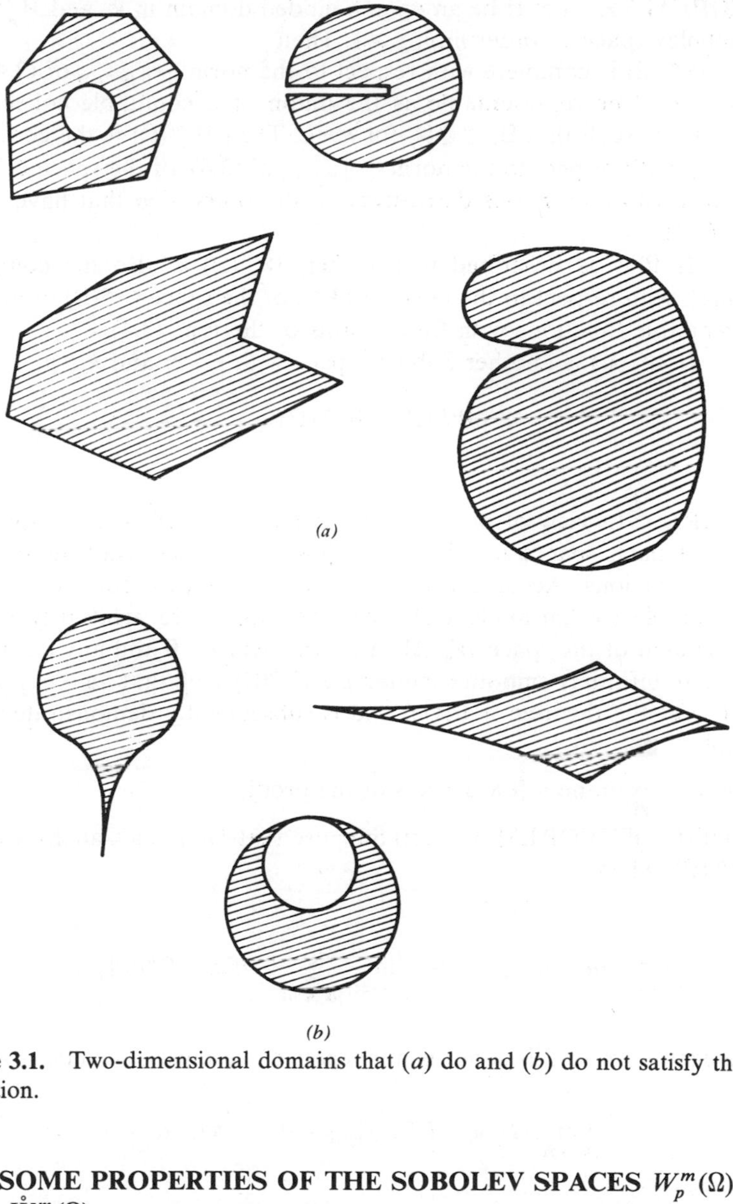

(a)

(b)

Figure 3.1. Two-dimensional domains that (*a*) do and (*b*) do not satisfy the cone condition.

3.4 SOME PROPERTIES OF THE SOBOLEV SPACES $W_p^m(\Omega)$ AND $\mathring{W}_p^m(\Omega)$

We now return to the study of the spaces $W_p^m(\Omega)$ defined by (3.2) with norms of the form (3.4). Some of the basic properties of these spaces are established in the following theorem.

THEOREM 3.2. Let Ω be an open bounded domain in $\mathbf{R}^n$ and $W_p^m(\Omega)$ be the Sobolev space of order m,p, $p \geqslant 1$. Then

(i) $W_p^m(\Omega)$ is complete with respect to the norm $\|\cdot\|_{W_p^m(\Omega)}$ of (3.4).

(ii) Let Ω be representable as the union of a countable collection of open sets Ω_k such that $\Omega_k \subset\subset \Omega_r$ for $r > k$. Then $W_p^m(\Omega)$ is the completion (closure), with respect to the norm $\|\cdot\|_{W_p^m(\Omega)}$ of (3.4), of the space $\hat{C}^m(\Omega)$ of functions with continuous derivatives of all orders $\leqslant m$ that have a finite norm (3.4).

(iii) If Ω is as described in (ii), then $W_p^m(\Omega)$ is also the completion (closure), with respect to the norm in (3.4), of the space $\hat{C}^\infty(\Omega)$ of infinitely differentiable functions with finite norms of the type (3.4).

(iv) If $W_k^r(\Omega)$ is another Sobolev space with $r \geqslant m$ and $k \geqslant p$,

$$W_k^r(\Omega) \subseteq W_p^m(\Omega) \tag{3.7}$$

■

REMARK. In view of properties (i) through (iii), all Sobolev spaces are in fact Banach spaces in which it is possible to construct dense sets of smooth functions. According to (ii) or (iii), there exist functions m-times differentiable (or infinitely differentiable) which are arbitrarily close to each element of the space $W_p^m(\Omega)$; i.e., for every $\varepsilon > 0$ and any $u \in W_p^m(\Omega)$, there is an infinitely smooth function $v \in C^\infty(\Omega)$ such that $\|u - v\|_{W_p^m(\Omega)} < \varepsilon$. Part (iv) of the theorem is trivial and is subsequently demonstrated by an example. ■

We now examine a few aspects of the proof.

PROOF OF THEOREM 3.2. (i) Suppose that $\{u_k\}$ is a Cauchy sequence in $W_p^m(\Omega)$. Then

$$\lim_{j,k\to\infty} \|u_j - u_k\|^p_{W_p^m(\Omega)} = \lim_{j,k\to\infty} \sum_{|\alpha|\leqslant m} \|D^\alpha u_j - D^\alpha u_k\|^p_{L_p(\Omega)} = 0$$

Since each term in this series is nonnegative, we must conclude that

$$\lim_{j,k\to\infty} \|D^\alpha u_j - D^\alpha u_k\|_{L_p(\Omega)} = 0 \qquad \forall \boldsymbol{\alpha}, |\boldsymbol{\alpha}| \leqslant m$$

In other words, for each index $\boldsymbol{\alpha}$ such that $|\boldsymbol{\alpha}| \leqslant m$, the sequence $\{D^\alpha u_k\}$ is a Cauchy sequence in $L_p(\Omega)$.

We recall that $L_p(\Omega)$ is complete. Hence the limit of the Cauchy sequence $\{u_k\}$ is a function $u(\mathbf{x}) \in L_p(\Omega)$, and the limit of $\{D^\alpha u_k\}$ is a function $\psi^\alpha \in L_p(\Omega)$.

Let $\phi(\mathbf{x}) \in \mathfrak{D}(\Omega)$. Then according to the definition of weak derivatives and to Theorem 2.5,

$$\int_\Omega \psi^\alpha \phi\, dx = \lim_{k\to\infty} \int_\Omega D^\alpha u_k \phi\, dx$$

$$= (-1)^{|\alpha|} \lim_{k\to\infty} \int_\Omega u_k D^\alpha \phi\, dx$$

$$= (-1)^{|\alpha|} \int_\Omega u D^\alpha \phi\, dx$$

Hence, $\psi^\alpha = D^\alpha u$ and each u has a weak $\boldsymbol{\alpha}$th derivative equal to the weak limit of $\{D^\alpha u_k\}$ in $L_p(\Omega)$.

(ii) and (iii) (cf. [3.6] or [3.5].) Let

$$K_p^m(\Omega) = \text{completion of } \hat{C}^m(\Omega) \text{ in the Sobolev norm } \|\cdot\|_{W_p^m(\Omega)} \text{ of (3.4)} \tag{3.8}$$

It is obvious that each $u \in K_p^m(\Omega)$ is in $W_p^m(\Omega)$; i.e., $K_p^m(\Omega) \subset W_p^m(\Omega)$. What we must show is that $W_p^m(\Omega) \subset K_p^m(\Omega)$ (note that this follows immediately for star-shaped Ω from Theorem 2.10). In other words, if $u \in W_p^m(\Omega)$, we must show that for every $\varepsilon > 0$ there is a function $w \in \hat{C}^m(\Omega)$ such that

$$\|D^\alpha w - D^\alpha u\|_{L_p(\Omega)} < \varepsilon \qquad \forall \boldsymbol{\alpha}, 0 \leqslant |\boldsymbol{\alpha}| \leqslant m \tag{3.9}$$

We next regard the domain Ω as the union of a countable collection of open sets Ω_k such that Ω_k is strictly inside Ω_r for $r > k$; i.e., set $\Omega_0 = \varnothing$ and

$$\Omega = \bigcup_{k=1}^{\infty} \Omega_k \qquad \bar{\Omega}_k \subset \Omega_{k+1}$$

One way of visualizing such a construction for a smooth simply connected domain is to consider a point $\mathbf{x}$ in the interior of Ω and to construct an open ball around $\mathbf{x}$ using $\mathbf{x}$ as its center. Denote this ball by Ω_1. Then pick a slightly larger open set Ω_2 so that $\bar{\Omega}_1 \subset \Omega_2$; proceeding in this way produces a representation of the type sought (see Fig. 3.2*a*).

Now define the countable class of open sets $\{G^i\}$, $0 \leqslant i \leqslant \infty$, by

$$G_i = \Omega_{i+2} - \bar{\Omega}_i$$

The class $\{G_i\}$ is an open covering of Ω. The ideas are illustrated in Fig. 3.2*b*. Let $\psi_i(\mathbf{x})$ be a partition of unity subordinate to this covering. Then,

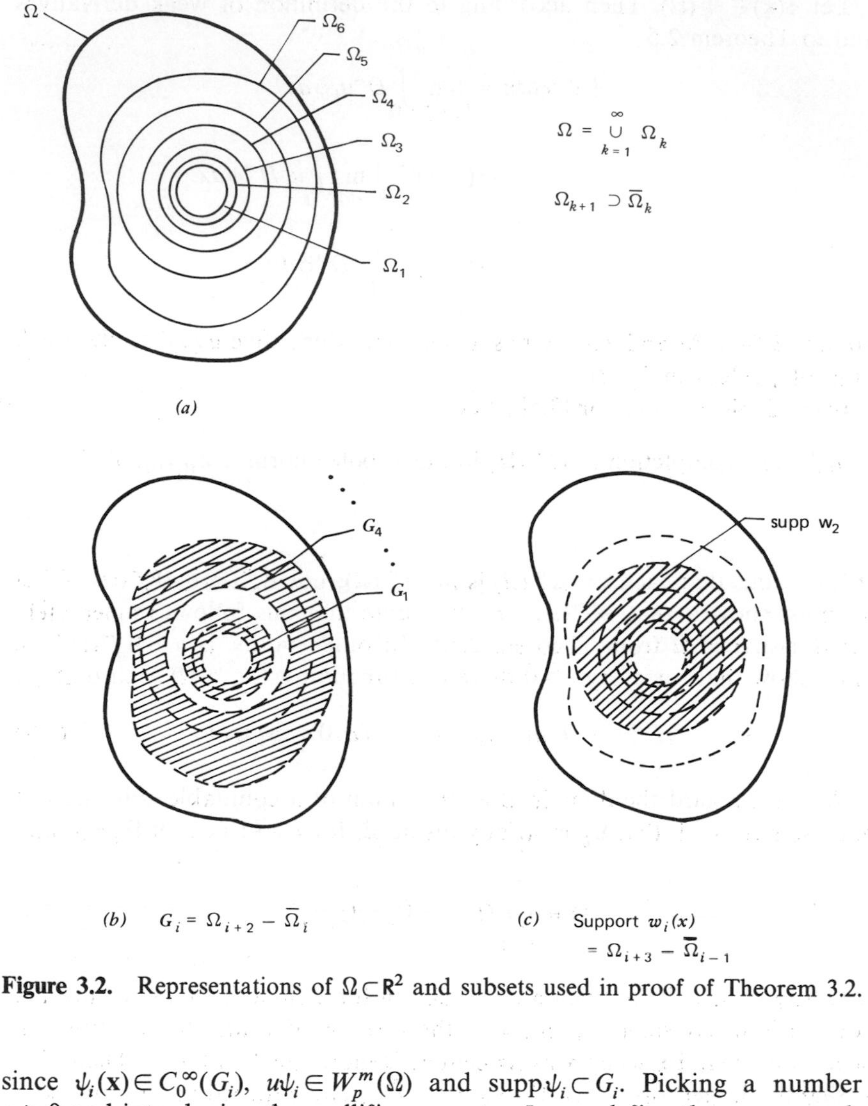

(b) $G_i = \Omega_{i+2} - \overline{\Omega}_i$ (c) Support $w_i(x) = \Omega_{i+3} - \overline{\Omega}_{i-1}$

Figure 3.2. Representations of $\Omega \subset \mathbf{R}^2$ and subsets used in proof of Theorem 3.2.

since $\psi_i(\mathbf{x}) \in C_0^\infty(G_i)$, $u\psi_i \in W_p^m(\Omega)$ and $\operatorname{supp}\psi_i \subset G_i$. Picking a number $\rho_i > 0$ and introducing the mollifier operator J_{ρ_i}, we define the corresponding mean functions

$$w_i = J_{\rho_i}(u\psi_i)$$

We choose ρ_i so that the support of each such mean function is in $\Omega_{i+3} - \overline{\Omega}_{i-1}$ as indicated in Fig. 3.2c (set $\Omega_0 = \varnothing$). We know from Theorem

2.9 [see (2.101)] that

$$\lim_{\rho_i \to 0} \| D^\alpha (u\psi_i) - D^\alpha J_{\rho_i}(u\psi_i)\|_{L_p(G_i)} = 0$$

Thus we can pick ρ_i such that

$$\|w_i - u\psi_i\|_{W_p^m(\Omega)} < \frac{\varepsilon}{2^i} \tag{3.10}$$

for each $i = 1, 2, \ldots$. Now set

$$w = \sum_{i=1}^{\infty} w_i$$

and compute

$$\begin{aligned} D^\alpha w &= \sum_{i=1}^{\infty} D^\alpha (w_i - u\psi_i) + \sum_{i=1}^{\infty} D^\alpha (u\psi_i) \\ &= \sum_{i=1}^{\infty} D^\alpha (w_i - u\psi_i) + D^\alpha u \end{aligned}$$

In this last step, we used the fact that $\Sigma_i D^\alpha (u\psi_i) = D^\alpha \Sigma_i u\psi_i = D^\alpha u \Sigma_i \psi_i = D^\alpha u$. Thus

$$\int_\Omega |D^\alpha w - D^\alpha u|^p \, dx \leqslant \int_\Omega \sum_i |D^\alpha (w_i - u\psi_i)|^p \, dx$$

and, from this result and (3.10) we obtain (3.9). Hence (ii) is proved. However, a glance at the proof also reveals that each $u \in W_p^m(\Omega)$ is in the space obtained by completing the $\hat{C}^\infty(\Omega)$ functions with finite Sobolev norm (3.4). Hence (iii) is also proved.

(iv) If all derivatives $D^\alpha u$ of u up to order r are in $L_p(\Omega)$, by definition, those of order $m \leqslant r$ are in $L_p(\Omega)$. Hence $W_p^r(\Omega) \subseteq W_p^m(\Omega)$. In addition, $L_p(\Omega) \subseteq L_q(\Omega)$ if $q \leqslant p$, $1 \leqslant q \leqslant p < \infty$. Thus the proof of the theorem is complete. ■

REMARK. When we modify part (ii) of the theorem to include functions m-times differentiable on the *closed* domain $\overline{\Omega}$, it is necessary to also make some stipulations as to the smoothness of the boundary. Indeed, it can be shown (see, e.g., [3.5, p. 18]) that, if $\tilde{K}_p^m(\Omega)$ is the completion of $C^m(\overline{\Omega})$ in $\|\cdot\|_{W_p^m(\Omega)}$ and if $\partial\Omega$ is in C^m, $\tilde{K}_p^m(\Omega) = K_p^m(\Omega)$, where $K_p^m(\Omega)$ is defined in (3.8). ■

EXAMPLE 3.3. Consider the piecewise polynomial functions $u(x)$ and $v(x)$ shown in Fig. 3.3. It is clear that u and its first and second derivatives are integrable over the interval $[a,b]$; i.e., $u \in W_1^2(a,b)$. However, $v(x)$ has derivatives of order $\leqslant 4$ in $L_1(a,b)$; i.e., $v \in W_1^4(a,b)$. Clearly $v^{\text{iv}}(x)$ is of the form of $u''(x)$. Hence $W_1^4(a,b) \subset W_1^2(a,b)$. ■

Another type of Sobolev space is encountered in the study of boundary-value problems with homogeneous boundary conditions. This space, denoted $\mathring{W}_p^m(\Omega)$, is defined as the completion, in the Sobolev norm $\|\cdot\|_{W_p^m(\Omega)}$,

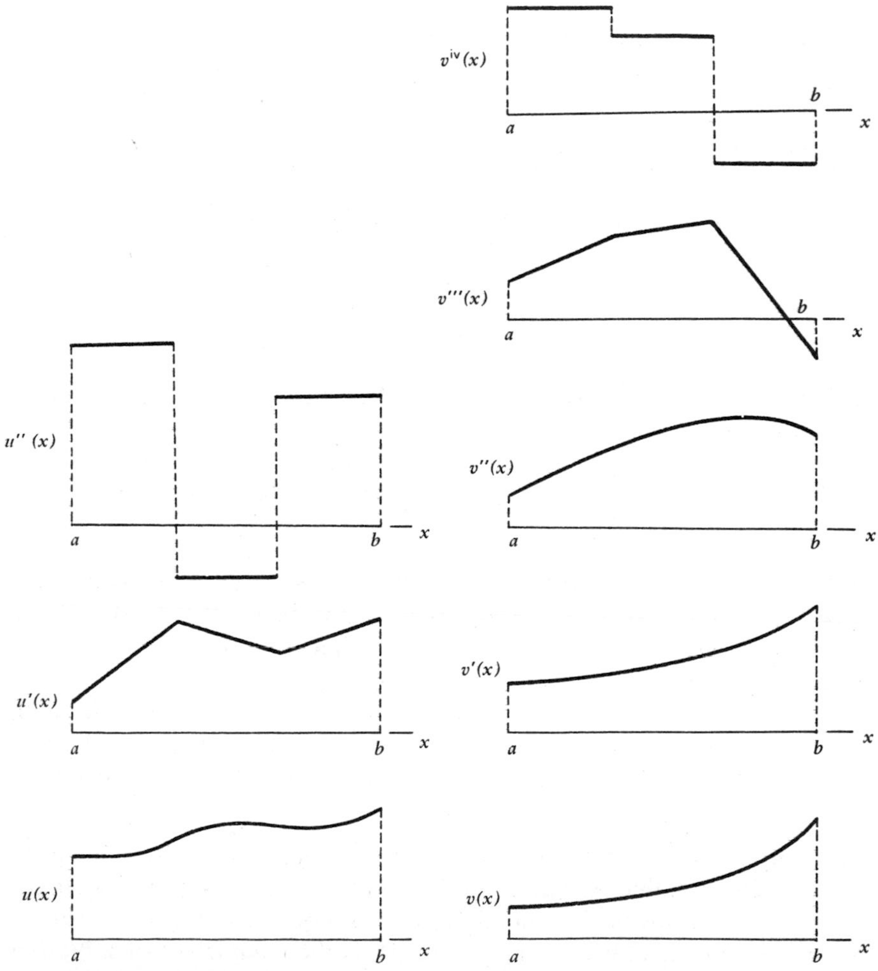

Figure 3.3. Two functions $u(x)$ and $v(x)$ and their various derivatives defined in a piecewise manner on $[a,b]$.

of the space $\hat{C}_0^m(\Omega)$ of functions with continuous derivatives of order $\leqslant m$ which have compact support in Ω and whose Sobolev norms are finite. Several properties of $\mathring{W}_p^m(\Omega)$ are given in the following theorem.

THEOREM 3.3. Let Ω be an open bounded domain in $\mathbf{R}^n$ with a sufficiently smooth boundary $\partial\Omega$ and $\mathring{W}_p^m(\Omega)$ be the normed space defined above. Then

(i) $\mathring{W}_p^m(\Omega)$ is the completion of $\hat{C}_0^\infty(\Omega)$ in the Sobolev norm $\|\cdot\|_{W_p^m(\Omega)}$ of (3.4); i.e., every $u(\mathbf{x})\in\mathring{W}_p^m(\Omega)$ can be approximated arbitrarily closely by an infinitely differentiable function which has compact support in Ω.

(ii) $\mathring{W}_p^m(\Omega)\subset W_p^m(\Omega)$.

(iii) For each $u(\mathbf{x})\in\mathring{W}_p^m(\Omega)$, $p\geqslant 1$, there is an element $v(\mathbf{x})$ in $W_q^m(\Omega)$ $(1/p+1/q=1)$ such that

$$\int_\Omega D^\alpha uv\,dx=(-1)^{|\alpha|}\int_\Omega uD^\alpha v\,dx$$

for $|\alpha|\leqslant m$.

(iv) A function $u(\mathbf{x})\in W_p^m(\Omega)$ is also in $\mathring{W}_p^m(\Omega)$ if and only if

$$D^\alpha u(\mathbf{x})=0 \qquad \mathbf{x}\in\partial\Omega;\ \forall\alpha \text{ such that } |\alpha|\leqslant m-1 \tag{3.11}$$

where $\partial\Omega$ is the boundary of Ω.

PROOF. Part (i) of the theorem is proved in essentially the same way as part (ii) of Theorem 3.2 (see also Theorem 2.9). Part (ii) is obvious from the definition of $\mathring{W}_p^m(\Omega)$ and the properties of $W_p^m(\Omega)$ described earlier. Part (iii) is merely an integration by parts formula. Let $\{\phi_n(\mathbf{x})\}_{n=1}^\infty$ be a sequence of functions in $\hat{C}_0^m(\Omega)$ whose limit is $u\in\mathring{W}_p^m(\Omega)$. Then, since each $\phi_n(\mathbf{x})$ has compact support in Ω,

$$\int_\Omega D^\alpha\phi_n v\,dx=(-1)^{|\alpha|}\int_\Omega \phi_n D^\alpha v\,dx$$

for $v\in W_p^m(\Omega)'$. Then, taking the limit as $n\to\infty$ gives the desired result. We shall only sketch the proof of (iv), which follows from the fact that the "trace" operators $\gamma_j u=(\partial^j u/\partial n^j)|_{\partial\Omega}$, $u\in C^\infty(\bar{\Omega})$, can be extended to continuous operators on $W_p^m(\Omega)$ for $0\leqslant j\leqslant m-1$ [we prove this assertion in Chapter 4 for $p=2$; see Theorems 4.9, 4.10, and 4.17]. Thus, if $\phi_n\to u$ in $W_p^m(\Omega)$, then $\gamma_j\phi_n\to\gamma_j u=0$ for $0\leqslant j\leqslant m-1$, from which (3.11) follows. ■

3.5 THE SOBOLEV INTEGRAL IDENTITY

We now direct our attention to the development of an important integral identity due to Sobolev [3.1, pp. 50–56], which forms the basis for many fundamental theorems connected with Sobolev spaces. We first develop the identity for functions in $\bar{C}^m(\Omega)$ and then show conditions under which it holds for functions in $W_p^m(\Omega)$. The integral identity can also be interpreted as a type of polynomial (Taylor) expansion of functions in $\bar{C}^m(\Omega)$ in powers of $\mathbf{x}$ of degree $\leqslant m-1$, with a remainder in the form of a weighted integral of derivatives of order m.

THEOREM 3.4. [The Sobolev Integral Identity for $u\in\bar{C}^m(\Omega)$]. Let Ω be a bounded subset of $\mathbf{R}^n$ which satisfies the cone condition, and let $u(\mathbf{x})$ be an arbitrary function in $\bar{C}^m(\Omega)$. Then $u(\mathbf{x})$ can be represented in the form

$$u(\mathbf{x})=\sum_{|\alpha|\leqslant m-1} l_\alpha(u)\mathbf{x}^\alpha+\int_\Omega \frac{1}{r^{n-m}}\sum_{|\alpha|=m} Q_\alpha(\mathbf{x},\mathbf{y})D^\alpha u(\mathbf{y})\,dy \tag{3.12}$$

where $l_\alpha(u)$ are linear functionals on $\bar{C}^m(\Omega)$ given by

$$l_\alpha(u)=\int_\Omega \zeta_\alpha(\mathbf{y})u(\mathbf{y})\,dy \tag{3.13}$$

the $\zeta_\alpha(\mathbf{y})$ being continuous, bounded functions of $\mathbf{y}$, $|\alpha|\leqslant m-1$, r is the euclidean distance

$$r=|\mathbf{x}-\mathbf{y}|=\left(\sum_{i=1}^n |x_i-y_i|^2\right)^{1/2}$$

$\mathbf{x},\mathbf{y}\in\Omega$, and $Q_\alpha(\mathbf{x},\mathbf{y})$, $|\alpha|=m$, are bounded infinitely differentiable functions of $\mathbf{x}$ and $\mathbf{y}$.

PROOF. We immediately take advantage of the fact that Ω satisfies the cone condition. Pick a spherical cone $\mathcal{C}_\rho(\mathbf{x})$ of radius ρ and vertex $\mathbf{x}$ such that, when any $\mathbf{x}$ in Ω is used as a vertex of $\mathcal{C}_\rho(\mathbf{x})$, the cone can be oriented so that all its points lie in Ω. Let $\mathbf{x}$ be a fixed point in Ω and, for the fixed cone $\mathcal{C}_\rho(\mathbf{x})$, introduce the spherical coordinate system

$$\begin{aligned}
x_1&=r\sin\theta_{n-1}\sin\theta_{n-2}\cdots\sin\theta_2\sin\theta_1\\
x_2&=r\sin\theta_{n-1}\sin\theta_{n-2}\cdots\sin\theta_2\cos\theta_1\\
x_3&=r\sin\theta_{n-1}\sin\theta_{n-2}\cdots\sin\theta_3\cos\theta_2\\
&\ \ \vdots\\
x_{n-1}&=r\sin\theta_{n-1}\cos\theta_{n-2}\\
x_n&=r\cos\theta_{n-1}
\end{aligned}$$

In this coordinate system, we write for the point $\mathbf{x}$ the symbols $(r,\boldsymbol{\theta})$. A volume element in such coordinates is given by

$$dv = r^{n-1}\,dr\,d\sigma \tag{3.14}$$

where, with $\theta_1 \in [0,2\pi]$ and $\theta_j \in [0,\pi]$, $j \geq 2$,

$$d\sigma = \sin^{n-2}\theta_{n-1}\sin^{n-3}\theta_{n-2}\cdots\sin\theta_2\,d\theta_1\,d\theta_2\cdots d\theta_{n-1} \tag{3.15}$$

If $\mathbf{y}$ is a point in $\mathcal{C}_\rho(\mathbf{x})$, a function of $f(\mathbf{y})$ is denoted

$$f(\mathbf{y}) = \bar{f}(\mathbf{x};\, r,\boldsymbol{\theta}) \tag{3.16}$$

However, $\mathbf{x}$ is only a label here, since it is a fixed point. Likewise, the chain rule of partial differentiation takes the form

$$\frac{\partial \bar{f}}{\partial r} = \sum_{i=1}^{n} \frac{\partial f}{\partial y_i}\cdot\frac{\partial y_i}{\partial r} = \sum_{i=1}^{n} \frac{\partial f}{\partial y_i}\,\mu_i(\mathbf{y}) \tag{3.17}$$

where $\mu_i(\mathbf{y})$ are the direction cosines

$$\mu_i(\mathbf{y}) = \cos(r, y_i) = \frac{y_i - x_i}{r} \tag{3.18}$$

With these preliminary notations established, we now seek a function $\chi(\mathbf{y}) = \bar{\chi}(\mathbf{x};\, r,\boldsymbol{\theta})$ which has the following properties:

$$\left.\begin{array}{ll}
\text{(i)} & \chi(\mathbf{y}) \in C^\infty(\Omega) \\[2ex]
\text{(ii)} & \dfrac{\partial^k \bar{\chi}(\mathbf{x};\, 0,\boldsymbol{\theta})}{\partial r^k} = 0 \qquad \text{for } 0 \leq k < m-1 \\[2ex]
\text{(iii)} & \dfrac{\partial^k \bar{\chi}(\mathbf{x};\, \infty,\boldsymbol{\theta})}{\partial r^k} = 0 \qquad \forall k \geq 0 \\[2ex]
\text{(iv)} & \chi(\mathbf{y}) \equiv 0 \qquad \mathbf{y} \not\in \mathcal{C}_\rho(\mathbf{x})
\end{array}\right\} \tag{3.19}$$

A function with precisely these properties is given by

$$\bar{\chi}(\mathbf{x};\, r,\boldsymbol{\theta}) = \frac{r^{m-1}}{(m-1)!}\int_r^\infty \bar{\omega}_a(\mathbf{x};\, s,\boldsymbol{\theta})\,s^{n-1}\,ds \tag{3.20}$$

where $\bar{\omega}_a(\mathbf{x};\, r,\boldsymbol{\theta})=\omega_a(\mathbf{x}-\mathbf{y})$ is the infinitely differentiable function used in the definition of the mollifier {see (2.92); e.g., $\omega_a(\mathbf{x}-\mathbf{y})=c_0\exp[r^2/(r^2-a^2)]$, $r<a$, $=0$ for $r\geqslant a$}, and a is the radius of the sphere terminating the cone $\mathcal{C}_\rho(\mathbf{x})$. It is clear that the integral in (3.20) produces a function in $C^\infty(\mathsf{R}^n)$ which vanishes on all rays drawn from $\mathbf{x}$ that do intersect the terminating sphere S_a of $\mathcal{C}_\rho(\mathbf{x})$. Moreover, for rays that do not intersect S_a, the function is zero at points outside $\mathcal{C}_\rho(\mathbf{x})$. Thus properties (i), (iii), and (iv) of (3.19) are satisfied. The coefficient $r^{m-1}/(m-1)!$ guarantees that (ii) is satisfied. Notice also that

$$\int_0^\infty \bar{\omega}_a(\mathbf{x};\, s,\boldsymbol{\theta})s^{n-1}\,ds=c_0 \tag{3.21}$$

where c_0 is some constant; i.e., this integral is independent of $\mathbf{x}$. Indeed, $\mathbf{x}$ is an arbitrary vertex in Ω for a *fixed* cone $\mathcal{C}_\rho(\mathbf{x})$. Hence this integral must assume the same value for any choice of $\mathbf{x}$.

Now we select an arbitrary function $u(\mathbf{x})\in\bar{C}^m(\Omega)$ and, using the function χ of (3.20), define

$$V(\mathbf{x};r,\boldsymbol{\theta})=\bar{u}\,\frac{\partial^{m-1}\bar{\chi}}{\partial r^{m-1}}-\frac{\partial\bar{u}}{\partial r}\,\frac{\partial^{m-2}\bar{\chi}}{\partial r^{m-2}}+\cdots+(-1)^{m-1}\,\frac{\partial^{m-1}\bar{u}}{\partial r^{m-1}}\,\bar{\chi}$$

so that

$$\frac{\partial V}{\partial r}=\bar{u}\,\frac{\partial^m\bar{\chi}}{\partial r^m}+(-1)^{m-1}\,\frac{\partial^m\bar{u}}{\partial r^m}\,\bar{\chi}$$

Then, holding $\mathbf{x}$ and $\boldsymbol{\theta}$ fixed,

$$\begin{aligned}\int_0^\infty \frac{\partial V}{\partial r}\,dr&=\int_0^\infty\left[\bar{u}\,\frac{\partial^m\bar{\chi}}{\partial r^m}+(-1)^{m-1}\frac{\partial^m\bar{u}}{\partial r^m}\bar{\chi}\right]dr\\&=V(\mathbf{x};\infty,\boldsymbol{\theta})-V(\mathbf{x};0,\boldsymbol{\theta})\\&=0-u(\mathbf{x})\int_0^\infty\bar{\omega}_a(\mathbf{x};\, s,\boldsymbol{\theta})s^{n-1}\,ds\end{aligned}$$

Using (3.21) we have

$$u(\mathbf{x})=-c_0^{-1}\int_0^\infty\frac{\partial V}{\partial r}\,dr=-c_0^{-1}\int_0^\infty\left[\bar{u}\,\frac{\partial^m\bar{\chi}}{\partial r^m}+(-1)^{m-1}\,\frac{\partial^m\bar{u}}{\partial r^m}\,\bar{\chi}\right]dr$$

Recalling that an element of volume is given by (3.14), we multiply both sides of the above equation by $d\sigma$, integrate, and obtain

$$u(\mathbf{x}) = \kappa_0 \int_\Omega \bar{u} \frac{\partial^m \bar{\chi}}{\partial r^m} \frac{dv}{r^{n-1}} + \kappa_0(-1)^{m-1} \int_\Omega \frac{\partial^m \bar{u}}{\partial r^m} \bar{\chi} \frac{dv}{r^{n-1}} \tag{3.22}$$

where $\kappa_0 = -1/c_0 \int_{\mathcal{C}_\rho(\mathbf{x})} d\sigma = -(c_0 A_\sigma)^{-1}$, where $A_\sigma \equiv \int_{\mathcal{C}_\rho(\mathbf{x})} d\sigma$ and, since $\bar{\chi}$ vanishes outside $\mathcal{C}_\rho(\mathbf{x})$, we have extended the integration over all of Ω.

We must now interpret these two integrals. Using the chain rule (3.17) and Leibniz's formula for differentiating integrals, we first note the following relations:

$$\frac{\partial}{\partial r} \int_r^\infty \bar{\omega}_a(\mathbf{x}; s, \boldsymbol{\theta}) s^{n-1} ds = -\bar{\omega}_a(\mathbf{x}; r, \boldsymbol{\theta}) r^{n-1}$$

$$\frac{\partial \bar{\omega}_a}{\partial r} = \sum_{i=1}^n \frac{\partial \omega_a}{\partial y_i} \mu_i = \frac{1}{r} \sum_{i=1}^n \frac{\partial \omega_a}{\partial y_i} (y_i - x_i)$$

$$\frac{\partial^k \bar{\omega}_a}{\partial r^k} = \sum_{|\boldsymbol{\alpha}|=k} C_{\boldsymbol{\alpha}} D_{\mathbf{y}}^{\boldsymbol{\alpha}} \omega_a \boldsymbol{\mu}^{\boldsymbol{\alpha}} = \frac{1}{r^k} \sum_{|\boldsymbol{\alpha}|=k} C_{\boldsymbol{\alpha}} (\mathbf{y}-\mathbf{x})^{\boldsymbol{\alpha}} D_{\mathbf{y}}^{\boldsymbol{\alpha}} \omega_a$$

where the $C_{\boldsymbol{\alpha}}$ are constants. Hence

$$\begin{aligned}
\frac{\partial^m \bar{\chi}}{\partial r^m} &= \sum_{i=1}^m \frac{1}{(m-1)!} \binom{m}{i} \frac{\partial^{m-i}}{\partial r^{m-i}} r^{m-1} \frac{\partial^i}{\partial r^i} \int_r^\infty \bar{\omega}_a(\mathbf{x}; s, \boldsymbol{\theta}) s^{n-1} ds \\
&= \sum_{i=1}^m \sum_{j=0}^{i-1} A_{ij} r^{n-1+j} \frac{\partial^j \bar{\omega}_a}{\partial r^j} \\
&= \sum_{|\boldsymbol{\alpha}| < m-1} r^{n-1} \kappa_0^{-1} \zeta_{\boldsymbol{\alpha}}(\mathbf{y}) \mathbf{x}^{\boldsymbol{\alpha}}
\end{aligned} \tag{3.23}$$

where the A_{ij} are merely constants depending on i, j, etc., but not r, $\mathbf{x}$, or $\mathbf{y}$, and the $\zeta_{\boldsymbol{\alpha}}(\mathbf{y})$ are bounded continuous functions of $\mathbf{y}$ of the form

$$\zeta_{\boldsymbol{\alpha}}(\mathbf{y}) = \kappa_0 \sum_{\substack{|\boldsymbol{\beta}|=m-1 \\ \alpha_i < \beta_i}} C_{\boldsymbol{\alpha}}^{\boldsymbol{\beta}} D_{\mathbf{y}}^{\boldsymbol{\beta}} \omega_a \mathbf{y}^{\boldsymbol{\beta}-\boldsymbol{\alpha}} \tag{3.24}$$

C_α^β being constants. Substituting (3.23) into the first integral in (3.22) gives

$$\kappa_0 \int_\Omega \bar{u} \frac{\partial^m \bar{\chi}}{\partial r^m} \frac{dv}{r^{n-1}} = \int_\Omega \bar{u} \sum_{|\alpha| \leqslant m-1} \zeta_\alpha(\mathbf{y}) \mathbf{x}^\alpha \, dv$$

$$= \sum_{|\alpha| \leqslant m-1} l_\alpha(u) \mathbf{x}^\alpha$$

where $l_\alpha(u)$ are precisely the linear functionals defined in (3.13). Thus the first integral in (3.22) is a polynomial of degree $\leqslant m-1$ in the coordinates of the point $\mathbf{x}$.

Turning now to the second term, we have

$$\frac{1}{r^{n-1}} \frac{\partial^m \bar{u}}{\partial r^m} \bar{\chi} = \frac{1}{r^{n+m-1}} \bar{\chi} \sum_{|\alpha|=m} C_\alpha (\mathbf{y}-\mathbf{x})^\alpha D_{\mathbf{y}}^\alpha u(\mathbf{y})$$

$$= \sum_{|\alpha|=m} \frac{1}{r^{n+m-1}} \frac{r^{m-1}}{(m-1)!} \left(\int_r^\infty \bar{\omega}_a s^{n-1} \, ds \right) r^m \boldsymbol{\mu}^\alpha D_{\mathbf{y}}^\alpha u(\mathbf{y})$$

$$= \frac{1}{r^{n-m}} \sum_{|\alpha|=m} \kappa_0^{-1} (-1)^{m-1} Q_\alpha(\mathbf{x},\mathbf{y}) D^\alpha u(\mathbf{y}) \tag{3.25}$$

where $\boldsymbol{\mu} = (\mu_1, \mu_2, \ldots, \mu_n)$ are the direction numbers in (3.18) and $Q_\alpha(\mathbf{x},\mathbf{y})$ are the bounded infinitely differentiable functions

$$Q_\alpha(\mathbf{x},\mathbf{y}) = \frac{(-1)^{m-1} \kappa_0 \boldsymbol{\mu}^\alpha}{(m-1)!} \int_r^\infty \bar{\omega}_a(\mathbf{x}; s, \boldsymbol{\theta}) s^{n-1} \, ds \qquad |\alpha| = m \tag{3.26}$$

Introducing (3.25) into (3.22) yields the second integral in (3.12) and thus completes the proof. ■

We next add to Theorem 3.4 an important related theorem.

THEOREM 3.5. Let $u(\mathbf{x}) \in \bar{C}^m(\Omega)$, where Ω staisfies the cone condition. Then

$$D^\beta u(\mathbf{x}) = \sum_{|\alpha| \leqslant m-1} \beta! \binom{\alpha}{\beta} l_\alpha(u) \mathbf{x}^{\alpha-\beta} + \int_\Omega \frac{1}{r^{n-m+|\beta|}} \sum_{|\alpha|=m} N_\alpha^\beta(\mathbf{x},\mathbf{y}) \cdot D^\alpha u(\mathbf{y}) \, dy \tag{3.27}$$

where $N_\alpha^\beta(\mathbf{x},\mathbf{y})$ are bounded functions of $\mathbf{x}$ and $\mathbf{y}$, and $|\boldsymbol{\beta}| < m$.

PROOF. (Cf. [3.1, p. 71].) Since the first term on the right side is obvious, we concentrate on the βth derivatives of the integral in (3.12). In view of (3.25)

$$D^{\beta}\int_{\Omega}\frac{1}{r^{n-m}}\sum_{|\alpha|=m} Q_{\alpha}(\mathbf{x},\mathbf{y})D^{\alpha}u(\mathbf{y})\,dy$$

$$=\int_{\Omega}\sum_{|\alpha|=m} D_{\mathbf{x}}^{\beta}\left[\frac{1}{r^{n-m}}\frac{(-1)^{m-1}\kappa_0\boldsymbol{\mu}^{\alpha}}{(m-1)!}\int_r^{\infty}\bar{\omega}_a s^{n-1}\,ds\,D_{\mathbf{y}}^{\alpha}u(\mathbf{y})\right]dy$$

$$=\int_{\Omega}\sum_{|\alpha|=m} D_{\mathbf{y}}^{\alpha}u(\mathbf{y})\left[\bar{\psi}(\mathbf{x};r,\boldsymbol{\theta})b_m D_{\mathbf{x}}^{\beta}\frac{(\mathbf{y}-\mathbf{x})^{\alpha}}{r^n}\right.$$

$$\left.+b_m\frac{(\mathbf{y}-\mathbf{x})^{\alpha}}{r^n}D_{\mathbf{x}}^{\beta}\bar{\psi}(\mathbf{x};r,\boldsymbol{\theta})\right]dy$$

where $b_m=\kappa_0(-1)^{m-1}/(m-1)!$ and $\bar{\psi}(\mathbf{x};r,\boldsymbol{\theta})=\int_r^{\infty}\bar{\omega}_a s^{n-1}\,ds$. We need only show that each differentiation increases by one the negative exponent on r. Clearly,

$$\frac{\partial}{\partial x_j}\frac{(\mathbf{y}-\mathbf{x})^{\alpha}}{r^n}=\frac{\alpha_j(\mathbf{y}-\mathbf{x})^{\alpha}}{\mu_j r^{n+1}}+\frac{n(\mathbf{y}-\mathbf{x})^{\alpha}}{\mu_j r^{n+1}}$$

etc., so for the first term in brackets this is obvious. For the second term, notice that

$$\frac{\partial}{\partial x_j}\bar{\psi}=\frac{\partial}{\partial x_j}\int_r^{\infty}\omega_a\left[\mathbf{x}+\frac{s}{r}(\mathbf{y}-\mathbf{x})\right]s^{n-1}\,ds$$

$$=-\omega_a(\mathbf{y})r^{n-1}\mu_j+\int_r^{\infty}\sum_{i=1}^{n}\frac{\partial\omega_a}{\partial y_i}\left(\delta_{ij}+\frac{s}{r}\,\mu_i\mu_j-\delta_{ij}\frac{s}{r}\right)s^{n-1}\,ds$$

$$=\frac{1}{r}\left[-\int_r^{\infty}\frac{\partial\omega_a}{\partial y_j}s^n\,ds+\sum_{i=1}^{n}\mu_i\mu_j\int_r^{\infty}\frac{\partial\omega_a}{\partial y_i}s^n\,ds\right]$$

$$+\omega_a(\mathbf{y})r^{n-2}(x_j-y_j)+\int_r^{\infty}\frac{\partial\omega_a}{\partial y_j}s^{n-1}\,ds$$

Thus, $\partial\bar{\psi}/\partial x_j$ contains a factor of $1/r$ times some bounded integrals plus some suitably differentiable functions of **x** and **y**. Each additional differentiation only increases this polarity by 1. Thus collecting these results and incorporating a more compact notation leads to a formula of the form (3.27). ■

Theorems 3.4 and 3.5 hold for functions having bounded continuous derivatives of order $\leqslant m$ on some domain $\Omega \subset \mathbf{R}^n$ satisfying the cone condition. We can also show that in many cases the integral identities (3.12) and (3.27) hold for $u(\mathbf{x}) \in W_p^m(\Omega)$.

That the first term in (3.12) [or (3.27)] holds for $u \in W_p^m(\Omega)$ is easily seen; for a given $u(\mathbf{x})$ in $W_p^m(\Omega)$ construct its mean function $u_\rho(\mathbf{x})$ with the aid of (2.93). Since $u_\rho(\mathbf{x}) \in C_0^\infty(\Omega)$, for some ρ, it is also in $\bar{C}^m(\Omega)$. Consequently, (3.12) holds for $u_\rho(\mathbf{x})$. Since $\lim_{\rho \to 0} u_\rho = u$, in the sense of strong L_p convergence, it follows that u_ρ also converges weakly to u; i.e.,

$$\lim_{\rho \to 0} l_\alpha(u_\rho) = l_\alpha(u) \tag{3.28}$$

Hence the first term in (3.12) also applies to $u \in W_p^m(\Omega)$.

It is the second term in (3.12) that requires further study. Exactly how this integral behaves as $\rho \to 0$ depends on the term $1/r^{n-m}$, and the study of such integrals is traditionally in the province of potential theory. We now record two lemmas from potential theory on integrals of the type in (3.12), which not only resolve the present question but also lead directly to the embedding theorems described in the next section.

LEMMA 3.1. Let $k(\mathbf{x},\mathbf{y})$ be any bounded function of **x** and **y**, continuous for all $\mathbf{x} \neq \mathbf{y}$, and let f be an arbitrary element in $L_p(\Omega)$, $p>1$, $\Omega \subset \mathbf{R}^n$. Then the integral operator K given by

$$K(f) = \int_\Omega |\mathbf{x}-\mathbf{y}|^{-\lambda} k(\mathbf{x},\mathbf{y}) f(\mathbf{y})\, dy \tag{3.29}$$

where λ is a real number such that

$$\frac{n}{q} > \lambda > 0 \qquad \frac{1}{p} + \frac{1}{q} = 1 \tag{3.30}$$

is a continuous operator from $L_p(\Omega)$ into the space $\bar{C}(\Omega)$ of bounded functions continuous on Ω. That is, if

$$u = K(f) \tag{3.31}$$

then $u \in \bar{C}(\Omega)$ and there exists a constant $C_1 > 0$ such that

$$|u(\mathbf{x})| \leqslant C_1 \|f\|_{L_p(\Omega)} \tag{3.32}$$

Moreover, if the domain of K is bounded [i.e., if $\|f\|_{L_p(\Omega)} < C_2$, $\forall f \in D(K)$], the functions u are uniformly bounded and equicontinuous, i.e., K is a compact operator.

PROOF. Suppose $\lambda < n/q$ and denote $|\mathbf{x}-\mathbf{y}| = r$. Then using (3.29) and Hölder's inequality,

$$|u(\mathbf{x})| = \left|\int_\Omega r^{-\lambda} k f\, dy\right| \leqslant \left(\int_\Omega r^{-\lambda q}\, dy\right)^{1/q} \left(\int_\Omega |kf|^p\, dy\right)^{1/p}$$

Since $k(\mathbf{x},\mathbf{y})$ is bounded, there is a constant $M > 0$ such that $|k(\mathbf{x},\mathbf{y})| < M$. Moreover, let $B_R(\mathbf{0})$ be any open ball of radius R that contains Ω. Then

$$|u(\mathbf{x})| \leqslant M\|f\|_{L_p(\Omega)} \left(\int_{B_R(\mathbf{0})} r^{-\lambda q} r^{n-1}\, dr\, d\sigma\right)^{1/q}$$

$$= M\|f\|_{L_p(\Omega)} A_\sigma^{1/q} \left(\int_0^R r^{n-\lambda q-1}\, dr\right)^{1/q}$$

$$= M\left(\frac{A_\sigma}{n-\lambda q}\right)^{1/q} (R)^{\frac{n}{q}-\lambda} \|f\|_{L_p(\Omega)} \tag{3.33}$$

Thus, if (3.30) holds, so does (3.32). Note that, if f is in a bounded set in $W_p^m(\Omega)$, then the right-hand side of (3.32) is less than some positive number. Then the $u(\mathbf{x}) \in \mathfrak{R}(K)$ are uniformly bounded.

We next show that $u(\mathbf{x})$ is continuous. Consider a small ball $B_\delta(\mathbf{x})$ of radius δ centered at $\mathbf{x}$; $B_\delta(\mathbf{x}) = \{\mathbf{y}\colon |\mathbf{y}-\mathbf{x}| < \delta\}$. The kernel $k(\mathbf{x},\mathbf{y})$ is bounded by M inside this ball, but it may be discontinuous at $\mathbf{x}=\mathbf{y}$. Outside $B_\delta(\mathbf{x})$, $k(\mathbf{x},\mathbf{y})$ is both continuous and bounded. Hence we compute $|u(\mathbf{x}+\Delta\mathbf{x}) - u(\mathbf{x})|$ and break up the integrals as follows:

$$|u(\mathbf{x}+\Delta\mathbf{x}) - u(\mathbf{x})|$$

$$= \left|\int_\Omega \left[|\mathbf{x}+\Delta\mathbf{x}-\mathbf{y}|^{-\lambda} k(\mathbf{x}+\Delta\mathbf{x},\mathbf{y}) - |\mathbf{x}-\mathbf{y}|^{-\lambda} k(\mathbf{x},\mathbf{y})\right] f(\mathbf{y})\, dy\right|$$

$$\leqslant \int_{\Omega - B_\delta(\mathbf{x})} \left||\mathbf{x}+\Delta\mathbf{x}-\mathbf{y}|^{-\lambda} k(\mathbf{x}+\Delta\mathbf{x},\mathbf{y}) - |\mathbf{x}-\mathbf{y}|^{-\lambda} k(\mathbf{x},\mathbf{y})\right| |f(\mathbf{y})|\, dy$$

$$+ M\int_{B_\delta(\mathbf{x})} |\mathbf{x}+\Delta\mathbf{x}-\mathbf{y}|^{-\lambda} |f(\mathbf{y})|\, dy \; + M\int_{B_\delta(\mathbf{x})} |\mathbf{x}-\mathbf{y}|^{-\lambda} |f(\mathbf{y})|\, dy$$

In the first integral, $k(\mathbf{x},\mathbf{y})$ is continuous. Hence, for each $\varepsilon>0$ there is a $\delta>0$ such that the difference appearing in this integral is $<\varepsilon$ for $|\Delta\mathbf{x}|<\delta$. Thus, using Hölder's inequality, we find that the first integral is, for $|\Delta\mathbf{x}|<\delta$, less than or equal to

$$\varepsilon\left(\int_\Omega dy\right)^{1/q}\left(\int_\Omega |f|^p\,dy\right)^{1/p}=\varepsilon|\Omega|^{1/q}\|f\|_{L_p(\Omega)}$$

where $|\Omega|$ is the measure of Ω. The remaining two integrals are of the form (3.29); hence (3.33) can be applied to them with $R=\delta$ in the third integral and $R=2\delta$ in the second, $|\Delta\mathbf{x}|<\delta$. Summing up our results, we have

$$|u(\mathbf{x}+\Delta\mathbf{x})-u(\mathbf{x})|\leqslant\left(A_0\varepsilon+A_1\delta^{n/q-\lambda}\right)\|f\|_{L_p(\Omega)} \tag{3.34}$$

where A_0 and A_1 are constants. Hence $u(\mathbf{x})$ is continuous at $\mathbf{x}$. If $\|f\|_{L_p(\Omega)}<A_2$, then $u(\mathbf{x})$ belongs to a class of equicontinuous functions in $C(\Omega)$. The proof is thus complete. ■

For future reference, we record another related lemma for integral operators of the potential type described by (3.29).

LEMMA 3.2. Let $k(\mathbf{x},\mathbf{y})$ be a bounded continuous function of $\mathbf{x}$ and $\mathbf{y}$ for $\mathbf{x}\neq\mathbf{y}$ and let $f(\mathbf{x})\in L_p(\Omega)$, $p>1$, $\Omega\subset\mathsf{R}^n$. Then the integral operator K defined in (3.29), where λ is a real number such that

$$n>\lambda\geqslant\frac{n}{q}\qquad\frac{1}{p}+\frac{1}{q}=1 \tag{3.35}$$

is a continuous operator from $L_p(\Omega)$ into $L_l(\Gamma_s)$, where Γ_s is an s-dimensional manifold in Ω, and

$$n-p(n-\lambda)<s\leqslant n\qquad l<\frac{sp}{n-p(n-\lambda)}\equiv j \tag{3.36}$$

Consequently, there exists a constant $C_3>0$ such that

$$\|u\|_{L_l(\Gamma_s)}\leqslant C_3\|f\|_{L_p(\Omega)} \tag{3.37}$$

Moreover, if the domain of K is bounded [i.e., $\|f\|_{L_p(\Omega)}<C$, $\forall f\in\mathfrak{D}(K)$], the functions $u(\mathbf{x})\in L_l(\Gamma_s)$ are uniformly bounded and equicontinuous in the $L_l(\Gamma_s)$ norm; i.e., K is a compact operator.

PROOF. If $n-p(n-\lambda)<s$,

$$\frac{[n-p(n-\lambda)]p}{n-p(n-\lambda)}=p<j$$

Thus, $l<j$ and $p<j$ so that

$$\frac{s}{l}>\frac{s}{j} \qquad \text{or} \qquad \frac{s}{l}>\frac{n-p(n-\lambda)}{p}=\frac{n}{p}-n+\lambda$$
$$=n\left(\frac{1}{p}-1\right)+\lambda$$
$$=\lambda-\frac{n}{q}$$

Thus $(\lambda-n/q)-s/l$ is a negative number, say -2μ, where $\mu>0$; then

$$\lambda=\frac{s}{l}+\frac{n}{q}-2\mu \tag{3.38}$$

Now returning to the integral operator, if M is an upper bound on $k(\mathbf{x},\mathbf{y})$, with $r=|\mathbf{x}-\mathbf{y}|$ and $u(\mathbf{x})=K(f)$, and if we temporarily assume $l>p$, then

$$|u(\mathbf{x})|\leqslant M\int_\Omega r^{-s/l+\mu}r^{-n/q+\mu}|f(\mathbf{y})|^{p/l}|f(\mathbf{y})|^{p(1/p-1/l)}dy$$
$$\leqslant M\left(\int_\Omega |f(\mathbf{y})|^p r^{-s+\mu l}dy\right)^{1/l}\left(\int_\Omega |f(\mathbf{y})|^p dy\right)^{1/p-1/l}$$
$$\cdot\left(\int_\Omega r^{-n+\mu q}dy\right)^{1/q}$$

Here we have used an extended Hölder's inequality; i.e.,

$$\int_\Omega f_1 f_2 f_3\,dx\leqslant\left(\int_\Omega |f_1|^\alpha dx\right)^{1/\alpha}\left(\int_\Omega |f_2|^\beta dx\right)^{1/\beta}\left(\int_\Omega |f_3|^\gamma dx\right)^{1/\gamma}$$
$$\frac{1}{\alpha}+\frac{1}{\beta}+\frac{1}{\gamma}=1 \qquad \alpha,\beta,\gamma>0 \tag{3.39}$$

Suppose $B_R(\mathbf{0})$ is a ball of radius R containing Ω and centered at the origin. Then $\int_\Omega |f|^\alpha dx\leqslant\int_{B_R(\mathbf{0})}|f|^\alpha dx$. Thus

$$\int_\Omega r^{-n+\mu q}dv\leqslant\int_{B_R(\mathbf{0})} r^{-n+\mu q}r^{n-1}dr\,d\sigma=\left(\frac{A_B}{\mu q}\right)(R)^{q\mu}$$

where $A_B=\int_{B_R(\mathbf{0})}d\sigma$. Collecting results, we have

$$|u(\mathbf{x})|\leqslant M\left(\frac{A_B}{\mu q}\right)^{1/q}(R)^{\mu}\|f\|_{L_p(\Omega)}^{1-p/l}\left(\int_\Omega |f|^p r^{l\mu-s}dy\right)^{1/l}$$

Now let $\mathbf{x}\in\Gamma_s$ and $dv_\Gamma = a^{s-1}\,da\,d\sigma_\Gamma$ be a volume element in Γ_s. Then

$$\int_{\Gamma_s} |u|^l\,dv_\Gamma \leqslant M^l\left(\frac{A_B}{q\mu}\right)^{l/q}(R)^{\mu l}\|f\|^{l-p}_{L_p(\Omega)}$$

$$\times\int_{\Gamma_s}\int_{\Omega}|f|^p r^{l\mu-s}\,dy\,dv_\Gamma$$

Observe that

$$\int_{\Gamma_s} r^{\mu l-s}\,dv_\Gamma \leqslant \int_{B_R(\mathbf{0})} a^{\mu l-s}a^{s-1}\,da\,d\sigma_\Gamma = \frac{A_\Gamma(R)^{l\mu}}{l\mu}$$

where $\int_{\Gamma_s} d\sigma_\Gamma = A_\Gamma$, because $\mu l - s<0$, $r\geqslant a$, and $B_R(\mathbf{0})\supset\Gamma_s$. Collecting results, we have

$$\|u\|_{L_l(\Gamma_s)} \leqslant \overline{C}R^{s/l+n/q-\lambda}\|f\|_{L_p(\Omega)} \tag{3.40}$$

where $\overline{C}=M(A_B/q\mu)^{1/q}(A_\Gamma/l\mu)^{1/l}$. The boundedness of K is thus established. If we now assume that $l<p$, and take $p<k<j$, we may use the fact that $\|u\|_{L_l(\Gamma_s)}\leqslant|\Gamma_s|^{1-l/k}\|u\|_{L_k(\Gamma_s)}$ to show that an inequality of the same form as (3.37) is obtained, except that C_3 contains as an additional factor $|\Gamma_s|^{1-l/k}$, where $|\Gamma_s|$ is the measure of Γ_s. Thus boundedness of K again follows. Notice that if the functions $f\in L_p(\Omega)$ are bounded, the functions $u=K(f)\in L_l(\Gamma_s)$ are uniformly bounded.

We now show that $u(\mathbf{x})=K(f)(\mathbf{x})$ is continuous in the $L_l(\Gamma_s)$ norm. Let $B_\delta(\mathbf{x})$ be an open ball of radius δ centered at $\mathbf{x}$, and break the integral $K(f)$ into parts as indicated in the previous lemma [see (3.33)]. The same arguments lead us again to the inequality (3.34). Then, multiplying both sides by dv_Γ and integrating the lth powers leads to the inequality

$$\|u(\mathbf{x}+\Delta\mathbf{x})-u(\mathbf{x})\|_{L_l(\Gamma_s)}\leqslant\left(B_0\varepsilon+B_1\delta^{2\mu}\right)\|f\|_{L_p(\Omega)} \tag{3.41}$$

where B_0 and B_1 are constants, μ is given by (3.38), and ε and δ are as in (3.34). Hence $u(\mathbf{x})$ is continuous. Since (3.41) holds $\forall u\in\mathfrak{R}(K)$, the class of functions $u=K(f)$ is equicontinuous. ■

We now can go directly to the Sobolev identity for $u\in W_p^m(\Omega)$.

THEOREM 3.6. (The Sobolev Integral Identity for $u\in W_p^m(\Omega)$). Let $\Omega\subset\mathbf{R}^n$ be bounded, satisfy the cone condition, and let $u(\mathbf{x})$ be an arbitrary element in $W_p^m(\Omega)$. Then $u(\mathbf{x})$ obeys the integral identity (3.12).

PROOF. When $m>n/p$, the exponent $\lambda=n-m$ of r in the integral in (3.12) is $<n/p$. Hence the integral in (3.12) is then a continuous operator

by Lemma 3.1. Recalling (3.28), we observe that, by introducing the mean function $u_\rho(\mathbf{x})$ into (3.12) and taking the limit as $\rho \to 0$, we find that this limit satisfies (3.12).

We use a similar argument for the case $m \leqslant n/p$. Suppose $\lambda = n - m \geqslant n/q$, or $m \leqslant n/p$. We can then make use of Lemma 3.2 with $s > n-(n-\lambda)p$. Then $u(\mathbf{x}) \in L_l(\Gamma_s)$, l satisfying (3.36), and u_ρ converges to u in $L_l(\Gamma_s)$. This completes the proof. ■

We also have from Theorem 3.5, the following corollary.

THEOREM 3.7. Let $\Omega \subset \mathbb{R}^n$ be bounded, satisfy the cone condition, and let $|\boldsymbol{\beta}| \leqslant m-1$. Then an arbitrary element $u(\mathbf{x}) \in W_p^m(\Omega)$ satisfies (3.27). ■

3.6 THE SOBOLEV EMBEDDING THEOREMS

With the integral identities of the previous section now established, we can turn to the investigation of several fundamental properties of the Sobolev space $W_p^m(\Omega)$. Basically, these properties are connected with the degree of smoothness that can be expected of functions in certain Sobolev spaces. Collectively, the relations to which we refer are called *embedding theorems*, because they describe the way a given space $W_p^m(\Omega)$ can be mapped continuously into (i.e., embedded in) other spaces.

The first embedding theorem states that, under certain mild conditions, every function in $W_p^m(\Omega)$ is equivalent to a continuous function; i.e., $W_p^m(\Omega)$ is embedded in $\bar{C}(\Omega)$.

THEOREM 3.8. Let $u(\mathbf{x}) \in W_p^m(\Omega)$, Ω being a bounded domain in $\mathbb{R}^n$ satisfying the cone condition, and $mp > n$. Then there exists a constant $C > 0$, independent of u, such that

$$\sup_{\mathbf{x} \in \Omega} |u(\mathbf{x})| \leqslant C \|u\|_{W_p^m(\Omega)} \qquad (3.42)$$

Moreover, the embedding E: $W_p^n(\Omega) \to \bar{C}(\Omega)$ is compact.

PROOF. This follows immediately from Theorem 3.6 and Lemma 3.1. Since u satisfies (3.12) and $\lambda = n - m < n/q$, the integral operator in (3.12) is a compact operator from $L_p(\Omega)$ into $\bar{C}(\Omega)$. The polynomial term $\Sigma_{|\alpha| \leqslant m-1} l_\alpha(u)\mathbf{x}^\alpha$ is obviously continuous. Since it is the image of a continuous mapping in a finite-dimensional space, this mapping must also be compact. ■

The following result is a generalization of Theorem 3.8, and can be established along similar lines.

COROLLARY 3.8.1. Let $u(\mathbf{x}) \in W_p^m(\Omega)$, $\Omega \subset \mathbf{R}^n$ bounded and satisfying the cone condition, and k be an integer such that $(m-k)p > n$. Then $u(\mathbf{x}) \in \bar{C}^k(\Omega)$, and there exists a constant M, independent of u, such that

$$\|u\|_{C^k(\Omega)} \leqslant M \|u\|_{W_p^m(\Omega)} \tag{3.43}$$

Moreover, the embedding of $W_p^m(\Omega)$ into $\bar{C}^k(\Omega)$ is compact. ■

When $mp \leqslant n$, we can still make use of similar arguments, provided we call on Lemma 3.2 rather than 3.1. Since (3.37) holds whenever $\lambda \geqslant n/q$, we see in (3.12) that $\lambda = n - m \geqslant n/q = m \leqslant n/p$. Then $u(\mathbf{x})$ of Theorem 3.6 is in $L_l(\Gamma_s)$, where $l \leqslant sp/[n-p(m-\lambda)] = sp/[n-p(n-n+m)] = sp/(n-mp)$, and $s > n - p(n-\lambda) = n - mp$. Summing up, we have the following.

THEOREM 3.9. Let $u(\mathbf{x}) \in W_p^m(\Omega)$, $\Omega \subset \mathbf{R}^n$ bounded and satisfying the cone condition, with $p > 1$ and $mp \leqslant n$. Let Γ_s be a smooth s-dimensional manifold of Ω such that $n - mp < s \leqslant n$. Then

(i) $u(\mathbf{x}) \in L_l(\Gamma_s)$ for any l such that $l < ps/(n-mp)$.
(ii) the embedding E: $W_p^m(\Omega) \to L_l(\Gamma_s)$ is compact. ■

We obtain another collection of embedding theorems by using Theorem 3.7 as a starting point rather than 3.6. The exponent λ is now $n - m + |\boldsymbol{\beta}|$, so if functions $D^{\boldsymbol{\beta}}u \in W_p^m(\Omega)$ are mapped into continuous functions $D^{\boldsymbol{\beta}}u \in \bar{C}(\Omega)$, we must have $\lambda - n - m + |\beta| < n(1 - 1/p)$. Since we wish to include the possibility that $|\boldsymbol{\beta}| = 0$, we must have $mp > n$. Also, $|\boldsymbol{\beta}|$ must be less than $n(1-1/p) + m - n = m - n/p$. This means that derivatives of order $\sup|\boldsymbol{\beta}| - 1$, or $m - [n/p] - 1$, where $[n/p]$ denotes the integral part of n/p, are continuous. We can also use Lemma 3.1 to extend the arguments used in proving Theorem 3.9. Collecting all these results, we arrive at the following embedding theorems (again $\Omega \subset \mathbf{R}^n$ satisfies the cone condition).

THEOREM 3.10. Let $u(\mathbf{x})$ be an arbitrary function in $W_p^m(\Omega)$, Ω bounded in $\mathbf{R}^n$ and satisfying the cone condition. Then

(i) If $mp > n$ and $0 < |\boldsymbol{\beta}| < m - (n/p)$, then $D^{\boldsymbol{\beta}}u(\mathbf{x})$ is continuous and a constant C_0 can be found such that

$$\sup_{\mathbf{x} \in \Omega} |D^{\boldsymbol{\beta}}u(\mathbf{x})| \leqslant C_0 \|u\|_{W_p^m(\Omega)} \tag{3.44}$$

(ii) The imbedding of $W_p^m(\Omega)$ into $\bar{C}^{|\beta|}(\Omega)$ is compact
(iii)

$$W_p^m(\Omega) \subset \bar{C}^{m-[n/p]-1}(\Omega) \tag{3.45}$$

■

THEOREM 3.11. Let $u(\mathbf{x})$ be an arbitrary function in $W_p^m(\Omega)$ and suppose $|\boldsymbol{\beta}| \geqslant m-(n/p)$ and

$$s>n-p(m-|\boldsymbol{\beta}|)$$

Then, on every s-dimensional manifold Γ_s of Ω, there exists a constant C_1 such that

(i)

$$\|D^{\boldsymbol{\beta}}u\|_{L_l(\Gamma_s)} \leqslant C_1 \|u\|_{W_p^m(\Omega)} \tag{3.46}$$

where $l<sp/[n-p(m-|\boldsymbol{\beta}|)]$.

(ii) The embedding of $W_p^m(\Omega)$ into $L_l(\Gamma_s)$ is compact.

(iii)

$$W_p^m(\Omega) \subset W^{|\boldsymbol{\beta}|}_{np/[n-p(m-|\boldsymbol{\beta}|)]}(\Omega) \quad \blacksquare \tag{3.47}$$

REMARKS. 1. Note that part (iii) of the last theorem follows from part (ii) on setting $s=n$, $\Gamma_s=\Omega$.

2. It can be shown that the Sobolev norm (3.4) is topologically equivalent to a variety of other norms. For example, for any choice of continuous linear functionals $l_{\alpha}(u)$ such as those in the Sobolev integral identity (3.12), which do not simultaneously vanish on polynomials of degree $\leqslant m-1$, the norm

$$\|u\|^p_{\tilde{W}_p^m(\Omega)} = \sum_{|\alpha| \leqslant m-1} |l_{\alpha}(u)|^p + \int_{\Omega} \sum_{|\alpha|=m} |D^{\alpha}u|^p\,dx \tag{3.48}$$

is equivalent to the norm in (3.4).

3. The interpretation of continuous functions in $W_p^m(\Omega)$ such as those in (3.42) and (3.44) deserves a comment. Recall that the elements of $W_p^m(\Omega)$ are not actually functions but rather are equivalence classes of functions equal almost everywhere on Ω. Thus, (3.42), for example, indicates that the equivalence class $u \in W_p^m(\Omega)$ contains an element in $\bar{C}(\Omega)$ that is bounded by $C\|u\|_{W_p^m(\Omega)}$. $\blacksquare$

EXAMPLE 3.4. Use of (3.48) and the embedding theorems leads to several useful inequalities. For example, let $p=2$, $m=1$, and observe that

$$\|u\|_{L_2(\Omega)} \leqslant \|u\|_{W_2^1(\Omega)}$$

Since the norm (3.4) on $W_2^1(\Omega)$ is equivalent to the norm $W_2^1(\Omega)$ of (3.48), we can set

$$l_{(0)}(u)=\int_{\Omega} u\,dx$$

in (3.4), all other $l_{\alpha}=0$, and conclude that the norms

$$\|u\|^2_{W_2^1(\Omega)}=\int_{\Omega}|\text{grad}\,u|^2\,dx+\int_{\Omega}u^2\,dx$$

and

$$\|u\|^2_{\tilde{W}_2^1(\Omega)}=\int_{\Omega}|\text{grad}\,u|^2\,dx+\left(\int_{\Omega}u\,dx\right)^2$$

are equivalent, wherein $\text{grad}\,u\equiv\Sigma_{|\alpha|=1}D^{\alpha}u$. Thus

$$\int_{\Omega}u^2\,dx\leqslant C_0\int_{\Omega}|\text{grad}\,u|^2\,dx+C_0\left(\int_{\Omega}u\,dx\right)^2 \tag{3.49}$$

where C_0 is a constant. This result is known as *Poincare's inequality*.

When $u\in W_2^1(\Omega)$ vanishes on $\partial\Omega$, then

$$\int_{\Omega}u^2\,dx\leqslant C\int_{\Omega}(\text{grad}\,u)^2\,dx \tag{3.50}$$

a result known as *Friedrichs' inequality*. ■

3.7 THE DECOMPOSITION OF $W_p^m(\Omega)$

Consider again the Sobolev space $W_p^m(\Omega)$ and let $\mathcal{P}_{m-1}(\Omega)$ denote the subspace of $W_p^m(\Omega)$ consisting of polynomials of degree $\leqslant m-1$. We denote by Π a projection operator that casts each u in $W_p^m(\Omega)$ into some polynomial in $\mathcal{P}_{m-1}(\Omega)$, i.e.,

$$\Pi\colon\ W_p^m(\Omega)\rightarrow\mathcal{P}_{m-1}(\Omega);\ \Pi^2=\Pi \tag{3.51}$$

Once such a projection is defined, we effectively decompose $W_p^m(\Omega)$ into the direct sum of two subspaces, $\mathcal{P}_{m-1}(\Omega)$ and a space $M_p^m(\Omega)=W_p^m(\Omega)-\mathcal{P}_{m-1}(\Omega)$. In other words, if $u\in W_p^m(\Omega)$, set

$$\hat{\Pi}u=u-\Pi u=(I-\Pi)u \tag{3.52}$$

Clearly, $\hat{\Pi}^2=(I-\Pi)(I-\Pi)=I-2\Pi+\Pi^2=I-\Pi=\hat{\Pi}$. Thus $\hat{\Pi}\colon\ W_p^m(\Omega)\rightarrow W_p^m(\Omega)-\mathcal{P}_{m-1}(\Omega)=M_p^m(\Omega)$ is a projection, and each u in $W_p^m(\Omega)$ is of the form

$$u = \Pi u + \hat{\Pi} u \qquad \Pi u \in \mathcal{P}_{m-1}(\Omega) \qquad \hat{\Pi} u \in M_p^m(\Omega) \tag{3.53}$$

i.e.,

$$W_p^m(\Omega) = \mathcal{P}_{m-1}(\Omega) \oplus M_p^m(\Omega) \tag{3.54}$$

The Seminorm Space $L_p^m(\Omega)$

In several applications of Sobolev space theory it is convenient to introduce a special space of functions which is essentially the same as (i.e., equivalent to) the space $M_p^m(\Omega)$ described above. We denote it by $L_p^m(\Omega)$ and define it by

$$L_p^m(\Omega) = \{ u(\mathbf{x}) \colon D^\alpha u(\mathbf{x}) \in L_p(\Omega);\ |\alpha| = m \} \tag{3.55}$$

In other words, $L_p^m(\Omega)$ consists of those functions whose generalized derivatives of order m are in $L_p(\Omega)$. We associate with each element u in $L_p^m(\Omega)$ a real number $\|u\|_{L_p^m(\Omega)}$ given by

$$\|u\|_{L_p^m(\Omega)}^p = \sum_{|\alpha| = m} \|D^\alpha u\|_{L_p(\Omega)}^p \tag{3.56}$$

EXAMPLE 3.5. Suppose $u(x) \in C^2[a,b]$. Then

$$\|u\|_{W_2^2(a,b)}^2 = \int_a^b \left[u^2 + \left(\frac{du}{dx} \right)^2 + \left(\frac{d^2u}{dx^2} \right)^2 \right] dx$$

whereas

$$\|u\|_{L_2^2(a,b)}^2 = \int_a^b \left(\frac{d^2u}{dx^2} \right)^2 dx \quad \blacksquare$$

The space $L_p^m(\Omega)$ is *not* a normed linear space. It is true that two of the norm properties are satisfied by (3.56); e.g.,

$$\|\alpha u\|_{L_p^m(\Omega)} = |\alpha|\, \|u\|_{L_p^m(\Omega)}$$

and

$$\|u_1 + u_2\|_{L_p^m(\Omega)} \leqslant \|u_1\|_{L_p^m(\Omega)} + \|u_2\|_{L_p^m(\Omega)}$$

However, if

$$\|u\|_{L_p^m(\Omega)} = 0 \tag{3.57}$$

it does not necessarily follow that $u(\mathbf{x}) = 0$; i.e., $\|\cdot\|_{L_p^m(\Omega)}$ is a seminorm. Indeed, (3.57) is satisfied by any nonzero polynomial of degree $m-1$.

The latter observation deserves a closer examination. Suppose $u(\mathbf{x})$ is a function in $L_p^m(\Omega)$ whose derivatives of order m are zero. Then, for a given $\rho > 0$, the generalized derivatives of the mean function of u_ρ of order m are also zero:

$$D^\alpha u_\rho(\mathbf{x}) = \rho^{-n} \int_\Omega \omega_\rho(\mathbf{y};\, \mathbf{x}) D^\alpha u(\mathbf{y})\, dy$$

$$= \rho^{-n} \int_\Omega \omega_\rho(\mathbf{y};\, \mathbf{x})(0)\, dy = 0 \qquad |\alpha| = m \tag{3.58}$$

But $D^\alpha u_\rho(\mathbf{x}) \in C^\infty(\mathbf{R}^n)$. Hence on successive integrations of (3.58) we discover that the mean function $u_\rho(\mathbf{x})$ is a complete polynomial of degree $m-1$; i.e., $u_\rho(\mathbf{x})$ is of the form

$$u_\rho(\mathbf{x}) = \sum_{|\alpha| \leqslant m-1} a_\alpha \mathbf{x}^\alpha \tag{3.59}$$

Sequences of functions of the form (3.59) can be constructed only by altering the choices of the coefficients a_α; hence the limit of any convergent sequence of polynomials of degree $\leqslant m-1$ can only be a polynomial of degree $\leqslant m-1$. Since the mean functions $u_\rho(\mathbf{x})$ converge to $u(\mathbf{x})$ as $\rho \to 0$, $u(\mathbf{x})$ must itself be of the form (3.59). We have thus established the following result.

THEOREM 3.12. Let $u(\mathbf{x}) \in L_p^m(\Omega)$ be such that its generalized derivatives of order m are zero. Then $u(\mathbf{x})$ is almost everywhere a polynomial in $\mathbf{x}$ of degree $\leqslant m-1$. ■

All the discussion thus far leads us to a rather important property of $L_p^m(\Omega)$. Suppose that $\mathcal{P}_{m-1}(\Omega)$ is the space of all polynomials of degree $\leqslant m-1$ defined on the domain $\Omega \subseteq \mathbf{R}^n$. Now introduce an equivalence relation R on $W_p^m(\Omega)$ such that two functions $u(\mathbf{x})$ and $v(\mathbf{x})$ are related under $R, (uRv)$, if and only if $u(\mathbf{x}) - v(\mathbf{x}) \in \mathcal{P}_{m-1}(\Omega)$. It is easily verified that R is reflexive, transitive, and symmetric; hence R effects a partition of $W_p^m(\Omega)$ into equivalence classes of elements. For example, if $[u]$ denotes a equivalence class of functions associated with a function u in $W_p^m(\Omega)$, $[u]$ consists of functions v such that $v - u \in \mathcal{P}_{m-1}(\Omega)$. The space of all such equivalence classes of functions is of course the quotient space, which is a linear space, denoted $Q_p^m(\Omega) = W_p^m(\Omega)/\mathcal{P}_{m-1}(\Omega)$. We introduce a norm on

$Q_p^m(\Omega)$ according to

$$\begin{aligned}\|[u]\|_{Q_p^m(\Omega)} &= \inf_{v\in[u]} \|v\|_{W_p^m(\Omega)} \\ &= \inf_{v-u\in\mathcal{P}_{m-1}(\Omega)} \|v\|_{W_p^m(\Omega)} \\ &= \inf_{w\in\mathcal{P}_{m-1}(\Omega)} \|u+w\|_{W_p^m(\Omega)} \end{aligned} \tag{3.60}$$

We now show that the function (3.56) on $L_p^m(\Omega)$ [i.e., the seminorm in $W_p^m(\Omega)$] is a norm on $Q_p^m(\Omega)$ equivalent to (3.60). To this end we need the following theorem.

THEOREM 3.13. Let Ω be a bounded domain in $\mathbf{R}^n$ satisfying the cone condition. Then there exists a polynomial q_{m-1} of degree $\leqslant m-1$ for each $u(\mathbf{x})\in W_p^m(\Omega)$ such that

$$\|u-q_{m-1}\|_{L_p^k(\Omega)} \leqslant c\rho^{m-k}\|u\|_{L_p^m(\Omega)} \qquad 0\leqslant k\leqslant m-1 \tag{3.61}$$

where c is a constant independent of the radius ρ of a ball containing Ω.

PROOF. Existence of a polynomial of degree $\leqslant m-1$ follows directly from the Sobolev integral identity (3.12), with

$$q_{m-1} = \sum_{|\alpha|\leqslant m-1} l_\alpha(u(\mathbf{x}))\mathbf{x}^\alpha \tag{3.62}$$

Then

$$u(\mathbf{x})-q_{m-1}(\mathbf{x}) = \int_\Omega \frac{1}{r^{n-m}} \sum_{|\alpha|=m} Q_\alpha(\mathbf{x},\mathbf{y})D^\alpha u(\mathbf{y})\,dy$$

and

$$D^\beta(u-q_{m-1}) = \int_\Omega \frac{1}{r^{n-m+k}} \sum_{|\alpha|=m} N_\alpha^\beta(\mathbf{x},\mathbf{y})D^\alpha u(\mathbf{y})\,dy \qquad |\beta|=k$$

$$\begin{aligned}|D^\beta(u-q_{m-1})| &\leqslant \int_\Omega \frac{1}{r^{n-m+k}}\left| \sum_{|\alpha|=m} N_\alpha^\beta(\mathbf{x},\mathbf{y})D^\alpha u(\mathbf{y})\right| dy \\ &\leqslant \left(\int_\Omega r^{(m-n-k)q}\,dv\right)^{1/q}\left[\int_\Omega \left| \sum_{|\alpha|=m} N_\alpha^\beta(\mathbf{x},\mathbf{y})D^\alpha u(\mathbf{y})\right|^p dy\right]^{1/p}\end{aligned}$$

where we used Hölder's inequality ($1/p+1/q=1$). Using Hölder's inequality for sums, the second term on the right can be further reduced:

$$|D^{\beta}(u-q_{m-1})| \leqslant \left(\int_{\Omega} r^{(m-n-k)q}\,dv\right)^{1/q}$$

$$\cdot\left\{\int_{\Omega}\left[\sum_{|\alpha|=m}|N_{\alpha}^{\beta}(\mathbf{x},\mathbf{y})|^{q}\right]^{p/q}\left[\sum_{|\alpha|=m}|D^{\alpha}u(\mathbf{y})|^{p}\right]dy\right\}^{1/p}$$

Since $N_{\alpha}^{\beta}(\mathbf{x},\mathbf{y})$ are bounded functions, there exists a constant K, depending on m, n, p, and Ω, such that

$$|D^{\beta}(u-q_{m-1})| \leqslant K\|u\|_{L_p^m(\Omega)}\left(\int_{\Omega} r^{(m-n-k)q}\,dv\right)^{1/q}$$

$$= K\|u\|_{L_p^m(\Omega)}\left(\int_{\Theta}\int_0^{\rho} r^{(m-n-k)q+n-1}\,dr\,d\sigma\right)^{1/q}$$

$$= K\|u\|_{L_p^m(\Omega)}\left(\int_{\Theta}\int_0^{\rho} r^{(m-k-n/p)q-1}\,dr\,d\sigma\right)^{1/q}$$

$$= \overline{K}\|u\|_{L_p^m(\Omega)}\rho^{m-k-n/p}$$

where $\int_{\Omega}dv=\int_{\Theta}\int_0^{\rho}r^{n-1}dr\,d\sigma$, and

$$\overline{K}=\frac{K\left(\int_{\Theta}d\sigma\right)^{1/q}}{m-k-n/p} \qquad m-k-n/p>0$$

Thus

$$\left[\int_{\Omega}\sum_{|\beta|=k}|D^{\beta}(u-q_{m-1})|^{p}\,dv\right]^{1/p}$$

$$= \overline{K}\|u\|_{L_p^m(\Omega)}\rho^{m-k-n/p}\left(\sum_{|\beta|=k}\int_{\Theta}\int_0^{\rho} r^{n-1}\,dr\,d\sigma\right)^{1/p}$$

$$\leqslant C(m,n,p,k,\Omega)\rho^{m-k}\|u\|_{L_p^m(\Omega)}$$

or

$$\|D^\beta (u-q_{m-1})\|_{L_p(\Omega)} \leqslant C\rho^{m-k}\|u\|_{L_p^m(\Omega)}$$

This completes the proof of the theorem. ■

REMARK. For economy and consistency in notation, we henceforth follow a fairly standard practice of denoting seminorms by only single strokes $|\cdot|$ to distinguish them from norms $\|\cdot\|$. Thus we write

$$|u|_{W_p^m(\Omega)} \equiv \|u\|_{L_p^m(\Omega)} = \left(\sum_{|\alpha|=m} \|D^\alpha u\|^p_{L_p(\Omega)} \right)^{1/p} \tag{3.63}$$

■

THEOREM 3.14. Let Ω satisfy the cone condition, let $\mathcal{P}_k(\Omega)$ be the space of polynomials of degree not greater than k, for some $k \geqslant 1$, and let $Q_p^{k+1}(\Omega) = W_p^{k+1}(\Omega)/\mathcal{P}_k(\Omega)$ be the quotient space with respect to $\mathcal{P}_k(\Omega)$ and the norm $\|\cdot\|_{Q_p^{k+1}(\Omega)}$ defined by (3.60). Then the seminorm $|u|_{W_p^{k+1}(\Omega)} (\equiv \|u\|_{L_p^{k+1}(\Omega)})$ is a norm on $Q_p^{k+1}(\Omega)$ equivalent to $\|[u]\|_{Q_p^{k+1}(\Omega)}$. That is, there exists a constant $\hat{C} > 0$ independent of u but depending on Ω, such that

$$|u|_{W_p^{k+1}(\Omega)} \leqslant \|[u]\|_{Q_p^{k+1}(\Omega)} \leqslant \hat{C}|u|_{W_p^{k+1}(\Omega)} \qquad \forall u \in W_p^{k+1}(\Omega) \tag{3.64}$$

PROOF. Note that, for $u \in W_p^{k+1}(\Omega)$ and $q \in \mathcal{P}_k(\Omega)$, we have

$$\|u+q\|_{W_p^{k+1}(\Omega)} = \left[\sum_{|\alpha| \leqslant k+1} |D^\alpha (u+q)|^p_{L_p(\Omega)} \right]^{1/p}$$

$$= \left(\sum_{|\alpha| \leqslant k+1} |u+q|^p_{W_p^{|\alpha|}(\Omega)} \right)^{1/p}$$

Choose q such that $q = -q_k$, where $q_k \in \mathcal{P}_k(\Omega)$ is given by (3.62) with $m-1=k$. Using the inequality (3.61) on each $|u+q|_{W_p^{|\alpha|}(\Omega)}$ for $0 \leqslant |\alpha| \leqslant k$, we obtain

$$\|u+q\|_{W_p^{k+1}(\Omega)} \leqslant \hat{C}(k,n,p,\Omega)|u|_{W_p^{k+1}(\Omega)} \tag{3.65}$$

where

$$\hat{C} = \max(C^p, 1) \sum_{|\alpha| \leqslant k+1} \rho^{(k+1-|\alpha|)p}$$

C is the constant in (3.61), and ρ is the radius of the ball containing Ω. Then

$$\|[u]\|_{Q_p^{k+1}(\Omega)} = \inf_{\hat{q} \in \mathcal{P}_k(\Omega)} \|u+\hat{q}\|_{W_p^{k+1}(\Omega)} \leqslant \|u+q\|_{W_p^{k+1}(\Omega)}$$
$$\leqslant \hat{C}|u|_{W_p^{k+1}(\Omega)}$$

The left-hand inequality in (3.64) is obtained as follows:

$$|u+q|_{W_p^{k+1}(\Omega)} = |u|_{W_p^{k+1}(\Omega)} \leqslant \inf_{\hat{q} \in \mathcal{P}_k(\Omega)} \|u+\hat{q}\|_{W_p^{k+1}(\Omega)} = \|[u]\|_{Q_p^{k+1}(\Omega)}$$

This completes the proof. ■

REFERENCES

3.1. Sobolev, S. L., *Applications of Functional Analysis in Mathematical Physics*, Translated from the 1950 Russian edition by F. Browder, Translations of Mathematical Monographs, Vol. 7, American Mathematical Society, Providence, R. I., 1963.

3.2. Smirnov, V. I., *A Course of Higher Mathematics, V. Integration and Functional Analysis*, Translated from the 1960 Russian edition by D. E. Brown, Pergamon Press, Oxford, 1964.

3.3. Adams, R. A., *Sobolev Spaces*, Academic Press, New York, 1975.

3.4. Spivak, M., *Calculus on Manifolds*, W. A. Benjamin, Menlo Park, Calif., 1965.

3.5. Friedman, A., *Partial Differential Equations*, Holt, New York, 1969.

3.6. Meyers, N. G. and Serrin, J., H=W, *Proc. Nat. Acad. Sci. U. S.*, Vol. 51, pp. 1055–1056, 1967.

4

HILBERT SPACE THEORY OF TRACES AND INTERMEDIATE SPACES

4.1 INTRODUCTION

If we are to construct a very deep theory of linear boundary-value problems, we will need a much finer resolution of the theory of Sobolev spaces presented in the previous chapter. For such purposes we shall concentrate on the $W_2^m(\Omega)$ spaces, since these are precisely those encountered in most linear boundary-value problems. These, we shall soon see, are Hilbert spaces endowed with special kinds of inner products. An intriguing aspect of such spaces, however, is that some of the most elementary boundary-value problems give rise to instances in which there is good reason to define spaces $W_2^m(\Omega)$ (or $H^m(\Omega)$, as we shall call them) in which m is an arbitrary real number—indeed, even a negative number.

Perhaps the first hint that Sobolev spaces of arbitrary real order have meaning in boundary-value problems is encountered in the solution of special problems (with infinite or semi-infinite domains) by Fourier transforms. There we can define on the transformed variables norms in which it is possible to treat exponents as nonintegral, but which, when interpreted in terms of the natural variables, represent norms involving derivatives of nonintegral order. It so happens that the very same situations are encountered locally near smooth boundaries in virtually any boundary-value problem, and therefore the theory of spaces of fractional order forms a fundamental part of the general theory of partial differential equations. Moreover, this theory is particularly relevant when we wish to describe properties of extensions of functions and their derivatives (traces) defined on open sets Ω onto functions defined on their boundaries $\partial\Omega$.

In this chapter we embark on an introductory study of this circle of ideas, which collectively might be called modern Hilbert space theory. However, many of the basic notions of intermediate spaces and interpolation spaces can be developed for Banach spaces as well, and we choose to present the subject in this more general setting in some of the early portions of this chapter. Our aim is to present the essential features of the theory of fractional Sobolev spaces, intermediate and interpolation spaces, Hilbert scales and several related ideas, and finally the fundamentally important trace theorem for functions in Hilbert spaces $W_2^m(\Omega)$. All these concepts prove to be indispensible tools in studying properties of solutions of linear elliptic boundary-value problems—especially their approximation. For additional details on these subjects, the references listed at the end of the chapter can be consulted; in particular, see the works of Lions and Magenes [4.1], Krein and Petunin [4.2], Berezanskii [4.3], and Nečas [4.4], and relevant references therein.

4.2 HILBERT SPACES $H^m(\Omega)$ OF INTEGER ORDER

Let m be an integer $\geqslant 1$ and consider the space $H^m(\Omega)$ of functions $u(\mathbf{x})$ on Ω defined by

$$H^m(\Omega)=\{u: D^{\boldsymbol{\alpha}}u\in L_2(\Omega);\ \forall\boldsymbol{\alpha}\quad \text{such that}\quad |\boldsymbol{\alpha}|\leqslant m\} \tag{4.1}$$

where Ω is a bounded open set in $\mathbf{R}^n$, and $D^{\boldsymbol{\alpha}}u$ denotes the weak $\boldsymbol{\alpha}$th derivative of u. We provide $H^m(\Omega)$ with the inner product

$$(u,v)_{H^m(\Omega)}=\sum_{|\boldsymbol{\alpha}|\leqslant m}(D^{\boldsymbol{\alpha}}u,D^{\boldsymbol{\alpha}}v)_{L_2(\Omega)}=\sum_{|\boldsymbol{\alpha}|\leqslant m}\int_\Omega D^{\boldsymbol{\alpha}}u\,\overline{D^{\boldsymbol{\alpha}}v}\,dx \tag{4.2}$$

and the associated norm

$$\|u\|_{H^m(\Omega)}=\left(\sum_{|\boldsymbol{\alpha}|\leqslant m}\|D^{\boldsymbol{\alpha}}u\|^2_{L_2(\Omega)}\right)^{1/2}=\left[(u,u)_{H^m(\Omega)}\right]^{1/2}<\infty \tag{4.3}$$

In view of definitions (3.2) and (3.4), we see that

$$H^m(\Omega)=W_2^m(\Omega) \tag{4.4}$$

The space $H^m(\Omega)$ is thus a Sobolev space of order $m, 2$, and we refer to such spaces henceforth as simply Sobolev spaces of order m. In view of identity (4.4) and Theorem 3.2, we can immediately record the following properties of $H^m(\Omega)$.

THEOREM 4.1. Let $H^m(\Omega)$ be a Sobolev space of order m defined by (4.1), where Ω is an open bounded subset of $\mathbb{R}^n$. Then

(i) $H^m(\Omega)$ is complete with respect to the norm in (4.3).
(ii) $H^m(\Omega)$ is the completion of the space $\hat{C}^\infty(\Omega)$ in the norm (4.3).
(iii) $H^m(\Omega)$ is also the completion of $\hat{C}^m(\Omega)$ in the norm (4.3).
(iv) If $H^{m_1}(\Omega)$ and $H^{m_2}(\Omega)$ are two Sobolev spaces, $m_1 \geqslant m_2 > 0$, then

$$H^{m_1}(\Omega) \subseteq H^{m_2}(\Omega) \subset H^0(\Omega) = L_2(\Omega) \tag{4.5}$$

■

Note that it also follows from the definitions that, whenever $m_1 \geqslant m_2$, and $u \in H^{m_1}(\Omega)$,

$$\|u\|_{H^{m_1}(\Omega)} \geqslant \|u\|_{H^{m_2}(\Omega)}$$

We can also use the spaces $\mathring{W}_p^m(\Omega)$ described previously to define Hilbert spaces when $p = 2$:

$$H_0^m(\Omega) \equiv \mathring{W}_2^m(\Omega) \tag{4.6}$$

That is, $H_0^m(\Omega)$ is defined as the completion in the norm (4.3) of the space $\hat{C}_0^m(\Omega)$ of m-times continuously differentiable functions with compact support in Ω. In view of Theorem 3.3, this space has the following properties.

THEOREM 4.2. Let $H_0^m(\Omega)$ be the completion of $\hat{C}_0^m(\Omega)$ in the norm $\|u\|_{H^m(\Omega)}$ of (4.3). Then

(i) $H_0^m(\Omega)$ is also the completion of $\hat{C}_0^\infty(\Omega)$ in the norm (4.3). In other words, every $u(\mathbf{x}) \in H^m(\Omega)$ can be approximated arbitrarily closely by an infinitely differentiable function which has compact support in Ω.
(ii) $H_0^m(\Omega) \subset H^m(\Omega)$.
(iii) A function $u(\mathbf{x}) \in H^m(\Omega)$ is in $H_0^m(\Omega)$ if and only if

$$D^\alpha u(\mathbf{x}) = 0 \qquad \mathbf{x} \in \partial\Omega; \qquad \forall \alpha \quad \text{such that} \quad |\alpha| \leqslant m-1$$

where $\partial\Omega$ is the boundary of Ω. ■

In subsequent work we also need the corresponding ideas of semi-normed spaces associated with the spaces $H^m(\Omega)$. The seminorm $|u|_{H^m(\Omega)}$

of a function $u \in H^m(\Omega)$ is defined by

$$|u|_{H^m(\Omega)} = \left(\sum_{|\alpha|=m} \|D^\alpha u\|^2_{L_2(\Omega)} \right)^{1/2} \tag{4.7}$$

Clearly $|u|_{H^m(\Omega)} = \|u\|_{L_2^m(\Omega)}$ [see (3.63)], so that all the properties of the spaces $L_2^m(\Omega)$ described earlier are applicable. Note also that

$$\|u\|^2_{H^m(\Omega)} = \sum_{k=0}^{m} |u|^2_{H^k(\Omega)} \tag{4.8}$$

We also remark that all results on representations and embeddings described in the previous chapter now hold for spaces $H^m(\Omega) = W_2^m(\Omega)$.

4.3 HILBERT SPACES $H^s(\mathbf{R}^n)$ FOR REAL $s \geqslant 0$

In (4.1) and in Theorems 4.1 and 4.2 it was understood that the Hilbert spaces $H^m(\Omega)$ are defined when m is a positive integer. We show in this section that, when $\Omega = \mathbf{R}^n$, spaces $H^s(\mathbf{R}^n)$ can be defined for any real number $s \geqslant 0$, which coincide with $H^m(\mathbf{R}^n)$ of (4.1) whenever $s = m$ is a positive integer.

Our principal tool in this analysis is the Fourier transform, discussed in Section 2.8. We first recall a few properties. According to (2.61) and (2.62), if $u(\mathbf{x}) \in L_2(\mathbf{R}^n)$, the Fourier transform of u is the function $\hat{u}(\mathbf{y})$ in $L_2(\mathbf{R}^n)$ given by

$$\hat{u}(\mathbf{y}) = \mathcal{F}(u) = \frac{1}{(2\pi)^{n/2}} \int_{\mathbf{R}^n} u(\mathbf{x}) \exp(-i\mathbf{x}\mathbf{y})\, dx$$

and the inverse is

$$u(\mathbf{x}) = \mathcal{F}^{-1}(\hat{u}) = \frac{1}{(2\pi)^{n/2}} \int_{\mathbf{R}^n} \hat{u}(\mathbf{y}) \exp(i\mathbf{x}\mathbf{y})\, dy$$

We can also describe Fourier transforms of distributions by introducing the space $\mathcal{S}(\mathbf{R}^n)$ of (2.63) of functions of rapid decay,

$$\mathcal{S}(\mathbf{R}^n) = \{\phi : \mathbf{x}^\beta D^\alpha \phi \in L_2(\mathbf{R}^n); \quad \forall \alpha \text{ and } \beta \in Z_+^n\}$$

We observed in Section 2.8 that, with the convergence properties in $\mathcal{S}$ implied by (2.64), every $\phi \in \mathcal{S}$ has a Fourier transform $\hat{\phi}$ in $\mathcal{S}$, and ϕ and

$\hat{\phi}$ satisfy the identities (2.78) through (2.83) of Theorem 2.3. A continuous linear functional q on $\mathcal{S}(\mathsf{R}^n)$ is called a tempered distribution, and we can define Fourier transforms of such distributions according to the rule, (2.70):

$$\langle \hat{q}, \phi \rangle = \langle q, \hat{\phi} \rangle$$

If $u \in L_2(\mathsf{R}^n)$, it generates a tempered distribution q and the Fourier transform $\hat{u}$ of u coincides with the distribution q. Finally, in Theorem 2.4, we established that

$$\mathcal{F}(D_{\mathbf{x}}^{\alpha} q) = (i\mathbf{y})^{\alpha} \mathcal{F}(q)$$

$$D_{\mathbf{y}}^{\alpha} \mathcal{F}(q) = \mathcal{F}((-i\mathbf{x})^{\alpha} q)$$

for any tempered distribution $q \in \mathcal{S}(\mathsf{R}^n)'$, where $\mathcal{F}$ and D^{α} are interpreted in a distributional sense. Moreover, for $u \in L_2(\mathsf{R}^n)$, we have Plancherel's equality [see (2.83)],

$$\|u\|_{L_2(\mathsf{R}^n)} = \|\mathcal{F}(u)\|_{L_2(\mathsf{R}^n)}$$

We can now turn to the central question of this section—the spaces $H^s(\mathsf{R}^n)$. Their key properties are exposed in the following theorem.

THEOREM 4.3. Let $\hat{H}^m(\mathsf{R}^n)$ denote the space defined by

$$\hat{H}^m(\mathsf{R}^n) = \left\{ u : u \in \mathcal{S}(\mathsf{R}^n)'; \quad (1 + |\mathbf{y}|^2)^{m/2} \hat{u} \in L_2(\mathsf{R}^n) \right\} \tag{4.9}$$

which is supplied with the norm

$$|||u|||_{\hat{H}^m(\mathsf{R}^n)} = \|(1 + |\mathbf{y}|^2)^{m/2} \hat{u}\|_{L_2(\mathsf{R}^n)} \tag{4.10}$$

where $|\mathbf{y}|^2 = y_1{}^2 + y_2{}^2 + \cdots + y_n{}^2$. Then

(i) $\hat{H}^m(\mathsf{R}^n)$ is a Hilbert space.

(ii) The norm (4.10) is equivalent to the norm (4.3).

PROOF. First, an inner product can be defined in $\hat{H}^m(\mathsf{R}^n)$ by means of the formula

$$(u, v)_{\hat{H}^m(\mathsf{R}^n)} = \left((1 + |\mathbf{y}|^2)^{m/2} \hat{u}, \quad (1 + |\mathbf{y}|^2)^{m/2} \hat{v} \right)_{L_2(\mathsf{R}^n)} \tag{4.11}$$

Next, we must show that $\hat{H}^m(\mathsf{R}^n)$ is complete. Suppose that $\{u_k\}$ is a

Cauchy sequence in $\hat{H}^m(\mathsf{R}^n)$. Then

$$|||u_i - u_j|||_{\hat{H}(\mathsf{R}^n)} = \|(1+|\mathbf{y}|^2)^{m/2}(\hat{u}_i - \hat{u}_j)\|_{L_2(\mathsf{R}^n)}$$
$$\geqslant \|\hat{u}_i - \hat{u}_j\|_{L_2(\mathsf{R}^n)}$$

because $(1+|\mathbf{y}|^2)^{m/2} \geqslant 1$. Therefore $\{\hat{u}_k\}$ is a Cauchy sequence in $L_2(\mathsf{R}^n)$. Since $L_2(\mathsf{R}^n)$ is complete, there exists a $\hat{u} \in L_2(\mathsf{R}^n)$ such that $\lim_{k\to\infty} \hat{u}_k = \hat{u}$ in $L_2(\mathsf{R}^n)$, and therefore $\lim_{k\to\infty}(1+|\mathbf{y}|^2)^{m/2}\hat{u}_k = (1+|\mathbf{y}|^2)^{m/2}\hat{u}$ in $L_2(\mathsf{R}^n)$. But this means that $\lim_{k\to\infty} |||u_k - u|||_{\hat{H}^m(\mathsf{R}^n)} = 0$, where $u \in \hat{H}^m(\mathsf{R}^n)$. Therefore $\hat{H}^m(\mathsf{R}^n)$ is complete.

To prove (ii) we must establish the inequality

$$C_1|||u|||_{\hat{H}^m(\mathsf{R}^n)} \leqslant \|u\|_{H^m(\mathsf{R}^n)} \leqslant C_2|||u|||_{\hat{H}^m(\mathsf{R}^n)} \tag{4.12}$$

for some positive constants C_1 and C_2. Since, by hypothesis, $(i\mathbf{y})^\alpha \hat{u} \in L_2(\mathsf{R}^n)$ for $|\alpha| \leqslant m$, $\|(i\mathbf{y})^\alpha \hat{u}\|_{L_2(\mathsf{R}^n)} < \infty$. Thus, using (2.84) and the Plancherel equality (2.83), we have

$$\|D^\alpha u\|_{L_2(\mathsf{R}^n)} = \|\mathbf{y}^\alpha \hat{u}\|_{L_2(\mathsf{R}^n)}, \qquad |\alpha| \leqslant m \tag{4.13}$$

Thus, $\hat{H}^m(\mathsf{R}^n) \subset H^m(\mathsf{R}^n)$. Next, if $u \in H^m(\mathsf{R}^n)$, then $D^\alpha u \in L_2(\mathsf{R}^n)$ for $|\alpha| \leqslant m$. Thus $D^\alpha u$ generates a tempered distribution on test functions in $\mathcal{S}(\mathsf{R}^n)$. We may conclude that $\hat{H}^m(\mathsf{R}^n)$ and $H^m(\mathsf{R}^n)$ contain the same functions.

Next, we call upon (4.13) and note that

$$\|u\|^2_{H^m(\mathsf{R}^n)} = \sum_{|\alpha| \leqslant m} \|D^\alpha u\|^2_{L_2(\mathsf{R}^n)}$$
$$= \sum_{|\alpha| \leqslant m} \int_{\mathsf{R}^n} |\mathbf{y}^\alpha \hat{u}(\mathbf{y})|^2 \, d\mathbf{y}$$
$$\leqslant \int_{\mathsf{R}^n} \Bigg(\sum_{|\alpha| \leqslant m} |\mathbf{y}|^{2|\alpha|} \Bigg) |\hat{u}(\mathbf{y})|^2 \, d\mathbf{y} \tag{4.14}$$

Clearly, $C_1(1+|\mathbf{y}|^2)^m \leqslant \sum_{|\alpha| \leqslant m} |\mathbf{y}|^{2|\alpha|}$ for some $C_1 = C_1(m)$. However, even for $|\mathbf{y}| \to \infty$, we can always find a constant C_2 such that $\sum_{|\alpha| \leqslant m} |\mathbf{y}|^{2|\alpha|} \leqslant C_2(1+|\mathbf{y}|^2)^m$. Therefore

$$C_1(1+|\mathbf{y}|^2)^m \leqslant \sum_{|\alpha| \leqslant m} |\mathbf{y}|^{2|\alpha|} \leqslant C_2(1+|\mathbf{y}|^2)^m \tag{4.15}$$

Substituting these inequalities into (4.14) and using definition (4.10) gives (4.12). This completes the proof of the theorem. ■

We see from this theorem that the Fourier transform defines an isomorphism of $H^m(\mathbf{R}^n)$ onto $\hat{H}^m(\mathbf{R}^n)$. However, the norm (4.10) makes sense for any choice of m, even nonintegral. Thus *we define the Hilbert space* $H^s(\mathbf{R}^n)$, *for any real number* $s \geqslant 0$, as follows:

$$H^s(\mathbf{R}^n) = \left\{ u : u \in \mathcal{S}(\mathbf{R}^n)'; \qquad (1+|\mathbf{y}|^2)^{s/2}\hat{u} \in L_2(\mathbf{R}^n) \right\} \tag{4.16}$$

The norm in $H^s(\mathbf{R}^n)$ is given by

$$\|u\|_{H^s(\mathbf{R}^n)} = \|(1+|\mathbf{y}|^2)^{s/2}\hat{u}\|_{L_2(\mathbf{R}^n)} \tag{4.17}$$

THEOREM 4.4. The space $H^m(\mathbf{R}^n)$ is dense in $H^k(\mathbf{R}^n)$ for $m \geqslant k$.

PROOF. It suffices to use the spaces $\hat{H}^r(\mathbf{R}^n)$. We first show that $\hat{H}^s(\mathbf{R}^n)$ is dense in $L_2(\mathbf{R}^n)$; recall that $\mathcal{D}(\mathbf{R}^n) \subset \mathcal{S}(\mathbf{R}^n)$ (almost everywhere), and $L_2(\mathbf{R}^n)$ is the closure of $C_0^\infty(\mathbf{R}^n)$ in the L_2-norm. Hence $\mathcal{S}(\mathbf{R}^n)$ is dense in $L_2(\mathbf{R}^n)$. Thus there exists a L_2-Cauchy sequence $\{\phi_k\}_{k=1}^\infty \subset \mathcal{S}(\mathbf{R}^n)$ which converges to a function $v \in L_2(\mathbf{R}^n)$. In fact, $\hat{\phi}_k \in \mathcal{S}(\mathbf{R}^n) \cap L_2(\mathbf{R}^n)$, $\forall k$.

Now since $\hat{\phi}_k(\mathbf{y}) \in \mathcal{S}(\mathbf{R}^n)$, we know that $|\mathbf{y}|^\alpha \hat{\phi}_k(\mathbf{y}) \in L_2(\mathbf{R}^n)$ $\forall \boldsymbol{\alpha}$. Thus $A_s\hat{\phi}_k \equiv [(1+|\mathbf{y}|^2)^{s/2}\hat{\phi}_k] \in L_2(\mathbf{R}^n)$. But this means that $\phi_k \in \hat{H}^s(\mathbf{R}^n)$. Thus, for each $v \in L_2(\mathbf{R}^n)$ we can find a sequence $\{\phi_k\}$ in $\hat{H}^s(\mathbf{R}^n)$ that converges to v in $L_2(\mathbf{R}^n)$.

To prove that $\hat{H}^{s+r}(\mathbf{R}^n)$ is dense in $\hat{H}^r(\mathbf{R}^n)$, let $u \in \hat{H}^r(\mathbf{R}^n)$. Then $A_r\hat{u} \equiv (1+|\mathbf{y}|^2)^{r/2}\hat{u} \equiv \tilde{u} \in \hat{L}_2(\mathbf{R}^n)$. Since $\hat{H}^s(\mathbf{R}^n)$ is dense in $L_2(\mathbf{R}^n)$ then, $\forall \varepsilon > 0$, there exists a $\psi \in \hat{H}^s(\mathbf{R}^n)$ such that $\|\hat{\psi} - \tilde{u}\|_{\hat{L}_2(\mathbf{R}^n)} < \varepsilon$. Clearly $A_s\hat{\psi} \in L_2(\mathbf{R}^n)$ and $\hat{\psi} \in L_2(\mathbf{R}^n)$. Let ϕ be a function such that $\hat{\phi} \equiv A_{-r}\hat{\psi}$. Then $A_{r+s}\hat{\phi} = A_s\hat{\psi} \in L_2(\mathbf{R}^n)$. Therefore $\phi \in H^{r+s}(\mathbf{R}^n)$.

Finally we note that

$$\begin{aligned}\|\phi - u\|_{\hat{H}^r(\mathbf{R}^n)} &= \|A_r\hat{\phi} - A_r\hat{u}\|_{L_2(\mathbf{R}^n)} \\ &= \|\hat{\psi} - \tilde{u}\|_{L_2(\mathbf{R}^n)} < \varepsilon\end{aligned}$$

Therefore $\hat{H}^{r+s}(\mathbf{R}^n)$ is dense in $\hat{H}^r(\mathbf{R}^n)$. ■

REMARKS. 1. The spaces $H^s(\mathbf{R}^n)$, $s \geqslant 0$, so defined are often called *fractional order Sobolev spaces* (see Lions and Magenes [4.1] and Aronszajn [4.5], and Sloboditskii [4.6]). We see later that the fractional order Sobolev spaces $H^s(\Omega)$ can be defined on bounded domains; our methods are those of Lions and Magenes [4.1] who describe fractional spaces in terms of intermediate spaces. Alternate methods, however, do exist. The norm in $H^s(\Omega)$ can also be defined without employing the Fourier transform, for nonintegral $s=[s]+\sigma$, $s>0$, $0<\sigma<1$, where $[s]=$integral part of s, as follows:

$$\|u\|^2_{H^s(\Omega)} = \|u\|^2_{H^{[s]}(\Omega)} + \sum_{|\alpha|=[s]} \|D^\alpha u\|^2_{H^\sigma(\Omega)} \tag{4.18}$$

where

$$\|u\|^2_{H^\sigma(\Omega)} = \int_\Omega \int_\Omega \frac{[u(\mathbf{x})-u(\boldsymbol{\xi})]^2}{|\mathbf{x}-\boldsymbol{\xi}|^{n+2\sigma}}\, dx\, d\xi$$

and $|\mathbf{x}|$ denotes the usual euclidean norm in $\mathbf{R}^n$. Such spaces are sometimes called Sloboditskii spaces. For additional information, see Adams [4.7].

2. It can also be shown that the space $H^m(\Omega)$ is dense in $H^k(\Omega)$, $m \geqslant k$, for smooth bounded domains, and we use this fact repeatedly throughout the remainder of this book. The proof of density for bounded domains follows easily from Theorem 4.4 and the use of local map concepts described later in Section 4.9. ■

4.4 DUALS OF HILBERT SPACES

In subsequent studies of Hilbert spaces we make frequent use of the concept of duality of spaces already discussed briefly in Chapter 2. Because of the important role of duality in Hilbert space theory, we consider in this and in the next section several additional properties of dual spaces.

Recall that the dual of a Hilbert space $\mathcal{H}$ is the space of continuous linear functionals on $\mathcal{H}$, and is denoted $\mathcal{H}'$. In other words, if $\mathbf{R}$ is the real number field, and $\mathcal{L}(\mathcal{H},\mathcal{K})$ is the space of continuous linear operators from a real Hilbert space $\mathcal{H}$ into a space $\mathcal{K}$, then

$$\mathcal{H}' = \mathcal{L}(\mathcal{H},\mathbf{R}) \tag{4.19}$$

(Obviously, if $\mathcal{H}$ is complex, $\mathcal{H}'=\mathcal{L}(\mathcal{H},\mathbf{C})$ where $\mathbf{C}$ is the complex number field). If l is a continuous linear functional on $\mathcal{H}$, we use the following notations for its value at a point $u \in \mathcal{H}$:

$$l(u) = \langle l,u\rangle \qquad l\in\mathcal{H}'; \qquad u\in\mathcal{H} \tag{4.20}$$

The bilinear form $\langle l,u\rangle$ is said to describe the *duality pairing on* $\mathcal{H}'\times\mathcal{H}$; i.e., $\langle l,u\rangle$ is a bilinear functional on the product space $\mathcal{H}'\times\mathcal{H}$.

Since l is a bounded operator on $\mathcal{H}$, we can use the definition of operator norms to define the norm of elements in $\mathcal{H}'$. Indeed, $\mathcal{H}'$ is supplied with the norm

$$\|l\|_{\mathcal{H}'}=\sup_{\substack{u\in\mathcal{H}\\u\neq 0}}\frac{|\langle l,u\rangle|}{\|u\|_{\mathcal{H}}} \tag{4.21}$$

Therefore

$$|\langle l,u\rangle|\leqslant\|l\|_{\mathcal{H}'}\|u\|_{\mathcal{H}} \tag{4.22}$$

The Riesz representation theorem establishes a close relationship between Hilbert spaces and their duals; in fact, according to this theorem, for any $l\in\mathcal{H}'$, there exists a unique $v_l\in\mathcal{H}$ such that

$$\langle l,u\rangle=(v_l,u)_{\mathcal{H}}\qquad\forall u\in\mathcal{H} \tag{4.23}$$

where $(\cdot,\cdot)_{\mathcal{H}}$ is the inner product in $\mathcal{H}$. Stated in another way, there exists a one-to-one and onto correspondence of elements of $\mathcal{H}$ with elements of $\mathcal{H}'$. If we introduce an operator K to describe this correspondence, we may write

$$l=Kv_l \tag{4.24}$$

Moreover, we also have by the Riesz theorem that $\|v_l\|_{\mathcal{H}}=\|l\|_{\mathcal{H}'}$, so that K is an *isometry* from $\mathcal{H}$ onto $\mathcal{H}'$; i.e., K is distance preserving: $\|v_l-v_q\|_{\mathcal{H}}=\|l-q\|_{\mathcal{H}'}$. We refer to K as the *canonical isometry* from $\mathcal{H}$ onto $\mathcal{H}'$. From (4.23) and (4.24) we see that

$$(v_l,u)_{\mathcal{H}}=\langle Kv_l,u\rangle\qquad\forall u\in\mathcal{H} \tag{4.25}$$

Since we can also find a linear functional Ku in $\mathcal{H}'$ corresponding to u, the dual space $\mathcal{H}'$ is also an inner-product space relative to the inner product

$$(l,q)_{\mathcal{H}'}=(K^{-1}l,K^{-1}q)_{\mathcal{H}}\qquad l,q\in\mathcal{H}' \tag{4.26}$$

Likewise, the norm

$$\|l\|_{\mathcal{H}'}=\sqrt{(l,l)_{\mathcal{H}'}} \tag{4.27}$$

is precisely the operator norm defined in (4.21). To see this, set $l = Ku$ in (4.21). Then, for a fixed u,

$$\frac{\langle l,u\rangle}{\|u\|_{\mathcal{H}}} = \frac{\langle Ku,u\rangle}{\|u\|_{\mathcal{H}}}$$
$$= \frac{(u,u)_{\mathcal{H}}}{\|u\|_{\mathcal{H}}}$$
$$= \sqrt{(u,u)_{\mathcal{H}}}$$
$$= \sqrt{(K^{-1}l, K^{-1}l)_{\mathcal{H}}}$$
$$= \|l\|_{\mathcal{H}'}$$

It is well known that a Hilbert space $\mathcal{H}$ is reflexive; that is to say,

$$(\mathcal{H}')' = \mathcal{H} \tag{4.28}$$

because each $u \in \mathcal{H}$ generates a continuous linear functional on $\mathcal{H}'$. Thus, in correspondence with (4.21), we also have

$$\|u\|_{\mathcal{H}} = \sup_{\substack{l \in \mathcal{H}' \\ l \neq 0}} \frac{|\langle l,u\rangle|}{\|l\|_{\mathcal{H}'}} \tag{4.29}$$

It is natural to identify $\mathcal{H}$ with its dual $\mathcal{H}'$; then we have to identify K with the identity mapping. That is, the inner product $(u,v)_{\mathcal{H}}$ on $\mathcal{H}$ must be identified with the bilinear form $\langle f,v\rangle$ on $\mathcal{H}' \times \mathcal{H}$. In this case the space $\mathcal{H}$ is called the *pivot space*. The reason for the use of the term "pivot" becomes apparent in the next section. It is often convenient, however, to retain the notion of a duality pairing. For example, let $\mathcal{H}_1$ and $\mathcal{H}_2$ denote two Hilbert spaces, and $\mathcal{H}_1'$ and $\mathcal{H}_2'$ their duals. The duality pairings are denoted $\langle f,u\rangle_1$ and $\langle g,v\rangle_2$, respectively. Then, if A is a continuous linear operator from $\mathcal{H}_1$ into $\mathcal{H}_2$, its *transpose* A': $\mathcal{H}_2' \to \mathcal{H}_1'$ is given by

$$\langle g, Au\rangle_2 = \langle A'g, u\rangle_1 \qquad \text{for any } g \in \mathcal{H}_2' \quad \text{and } u \in \mathcal{H}_1$$

Now if $\mathcal{H}_2 = \mathcal{H}_2'$ (i.e., if we identify $\mathcal{H}_2$ with its dual), the adjoint A^* of A is given by

$$\langle g, Au\rangle_2 = (g, Au)_2 = (A^*g, u)_1 \qquad u \in \mathcal{H}_1,\ g \in \mathcal{H}_2$$

where $(\cdot,\cdot)_1$ and $(\cdot,\cdot)_2$ now denote the inner products in $\mathcal{H}_1$ and $\mathcal{H}_2$, respectively. If Λ and K are the canonical isometries from $\mathcal{H}_1$ onto $\mathcal{H}_1'$ and from $\mathcal{H}_2$ onto $\mathcal{H}_2'$, respectively, the adjoint A^* is related to A' by

$$\begin{aligned}\langle A'g,u\rangle_1 &= \langle g,Au\rangle_2 = (K^{-1}g,Au)_2 = (A^*K^{-1}g,u)_1\\ &= \langle \Lambda A^*K^{-1}g,u\rangle_1\end{aligned}$$

or

$$A' = \Lambda A^* K^{-1} \tag{4.30}$$

The following example, given by Aubin [4.8], clarifies this point.

EXAMPLE 4.1. Suppose $\mathcal{H} = \mathbf{R}^n$, and let the inner product in $\mathcal{H}$ be defined by

$$(u,v)_{\mathcal{H}} = \sum_{i,j=1}^{n} g_{ij}u^i v^j$$

Here g_{ij} are elements in an nth-order symmetric, positive definite, matrix $\mathbf{G}$. Let us also consider the inner product associated with the identity matrix

$$(u,v) = \sum_{i=1}^{n} u_i v_i$$

which we identify with its duality pairing [i.e., $(u,v) = \langle u,v\rangle$]. The inner products are thus related by

$$(u,v)_{\mathcal{H}} = (\mathbf{G}u,v)$$

The dual space of $\mathcal{H}$ is the space $\mathbf{R}^n$ supplied with the inner product

$$(u,v)_{\mathcal{H}'} = \sum_{i,j=1}^{n} g^{ij}u_i v_j$$

where g^{ij} are the entries of the matrix $\mathbf{G}^{-1}$.

If $\mathbf{A} = [a_{ij}]$ is a matrix, its transpose $\mathbf{A}'$ with respect to the duality pairing $\langle u,v\rangle$ is the matrix $\mathbf{A}' = [a_{ji}]$, and the adjoint $\mathbf{A}^*$ is the matrix $\mathbf{A}^* = \mathbf{G}\mathbf{A}'\mathbf{G}^{-1}$. ■

The most important examples of duals of Hilbert spaces involve cases in which two Hilbert spaces are encountered, one being densely embedded in the other. For example, if $\mathcal{U}$ and $\mathcal{H}$ are Hilbert spaces such that

(i) $\mathcal{U} \subset \mathcal{H}$

(ii) a constant $C > 0$ exists such that

$$\|u\|_{\mathcal{H}} \leq C\|u\|_{\mathcal{U}} \qquad \forall u \in \mathcal{U}$$

then the association $\Pi : \mathcal{U} \to \mathcal{H}$ is continuous, and we say that $\mathcal{U}$ is (continuously) *embedded* in $\mathcal{H}$. If $\mathcal{U}$ is also dense in $\mathcal{H}$, $\mathcal{U}$ is densely embedded in $\mathcal{H}$. The correspondence $u = \Pi u$ of elements in $\mathcal{H}$ and $\mathcal{U}$ is then referred to as a *continuous and dense injection* from $\mathcal{U}$ into $\mathcal{H}$.

The important fact we are concerned with here is that, whenever a Hilbert space $\mathcal{U}$ is densely embedded in a Hilbert space $\mathcal{H}$, the dual of $\mathcal{H}$ is densely embedded in the dual of $\mathcal{U}$. We set forth this property as a theorem.

THEOREM 4.5. Let $\mathcal{U}$ and $\mathcal{H}$ be Hilbert spaces such that $\mathcal{U}$ is densely embedded in $\mathcal{H}$ and let $\mathcal{U}'$ and $\mathcal{H}'$ denote the duals of $\mathcal{U}$ and $\mathcal{H}$, respectively. Then there exists a continuous linear mapping Π' of $\mathcal{H}'$ into $\mathcal{U}'$ such that $\Pi'\mathcal{H}'$ is dense in $\mathcal{U}'$; i.e., $\mathcal{H}'$ is densely embedded in $\mathcal{U}'$.

PROOF. Let $\langle\cdot,\cdot\rangle_{\mathcal{U}}$ and $\langle\cdot,\cdot\rangle_{\mathcal{H}}$ denote the duality pairings on $\mathcal{U}'\times\mathcal{U}$ and $\mathcal{H}'\times\mathcal{H}$, respectively. Then, if $\Pi : \mathcal{U} \to \mathcal{H}$, we have

$$l(u) = \langle l, u\rangle_{\mathcal{U}} \qquad l \in \mathcal{U}'$$

$$q(u) = \langle q, u\rangle_{\mathcal{H}} = \langle q, \Pi u\rangle_{\mathcal{H}} \qquad q \in \mathcal{H}'$$

This establishes a correspondence Π' of q and l which we describe by

$$\langle \Pi' q, u\rangle_{\mathcal{U}} = \langle q, \Pi u\rangle_{\mathcal{H}} \qquad \forall u \in \mathcal{U}$$

It is clear that Π' is the transpose of Π and that it is linear. Π' is also continuous; indeed,

$$\begin{aligned}\|\Pi' q\| &= \sup \frac{|\langle \Pi' q, u\rangle_{\mathcal{U}}|}{\|u\|_{\mathcal{U}}} = \sup \frac{|\langle q, \Pi u\rangle_{\mathcal{H}}|}{\|u\|_{\mathcal{U}}} \leq \|q\|_{\mathcal{H}'} \sup \frac{\|\Pi u\|_{\mathcal{H}}}{\|u\|_{\mathcal{U}}}\\ &\leq \|\Pi\|\,\|q\|_{\mathcal{H}'}\end{aligned}$$

Π' is also injective, because $\Pi' q = 0 \Rightarrow \langle q, u\rangle_{\mathcal{U}} = 0 \, \forall u \in \mathcal{U}$, which, in turn implies $q = 0$.

It remains to be shown that $\Pi'\mathcal{H}'$ is dense in $\mathcal{U}'$. This requires two steps: (i) we first show that any subspace $\mathcal{M}$ of a Hilbert space $\mathcal{H}$ is dense in $\mathcal{H}$ if and only if any continuous linear functional vanishing identically on $\mathcal{M}$ also vanishes identically on $\mathcal{H}$. (ii) we show that, if an element u in $\mathcal{U}$ is used to produce a vanishing functional on $\Pi'\mathcal{H}'$, u must itself be zero.

We first prove (i). The necessity is obvious. Indeed, if $\langle q,u\rangle_{\mathcal{H}}=0$ for every $u\in\mathcal{M}$ and if $\mathcal{M}$ is dense in $\mathcal{H}$, then we have $\langle q,u\rangle_{\mathcal{H}}=0$ for every $u\in\mathcal{H}$. To prove the sufficiency, let q be a continuous linear functional on $\mathcal{H}(q\in\mathcal{H}')$ vanishing identically on $\mathcal{M}$. Then it must also vanish on its closure $\overline{\mathcal{M}}$. Let P be the orthogonal projector on $\overline{\mathcal{M}}$, $P:\mathcal{H}\to\overline{\mathcal{M}}$. Then $P'q=0$ implies that $q=0$, since $\langle P'q,u\rangle_{\overline{\mathcal{M}}}=\langle q,Pu\rangle_{\mathcal{H}}=\langle q,u\rangle_{\mathcal{H}}=0$. Thus, $P'=I$ and $P=I$. This implies that $\overline{\mathcal{M}}=\mathcal{H}$; i.e., $\mathcal{M}$ is dense in $\mathcal{H}$.

We now prove (ii). First note that, since $\mathcal{U}$ is a Hilbert space, a linear form on $\mathcal{U}'$ is an element of $(\mathcal{U}')'=\mathcal{U}$. Let $u\in(\mathcal{U}')'=\mathcal{U}$ be a linear form on $\mathcal{U}'$ vanishing identically on the subspace $P'\mathcal{H}'$. Then $\langle P'q,u\rangle_{\mathcal{U}}=\langle q,Pu\rangle_{\mathcal{H}}=\langle q,u\rangle_{\mathcal{H}}=0$ for any $q\in\mathcal{H}'$. This implies that $u=0$ and, in view of (i), $P'\mathcal{H}'$ is dense in $\mathcal{U}'$. ■

REMARKS. 1. The role of the Hahn–Banach theorem in the proof is worth mentioning. The mapping Π' corresponds to an injection of $\mathcal{H}'$ into $\mathcal{U}'$. If $\langle q,u\rangle_{\mathcal{H}}$ describes the duality pairing on $\mathcal{H}'\times\mathcal{H}$, the duality pairing $\langle l,u\rangle_{\mathcal{U}}$ on $\mathcal{U}'\times\mathcal{U}$ is identified with the unique extension of the functional $q(u)$. In other words, whenever $\mathcal{U}$ is densely embedded in $\mathcal{H}$, $\mathcal{H}'$ is (identified with) a dense subspace of $\mathcal{U}'$, and the duality pairing on $\mathcal{U}'\times\mathcal{U}$ is identified with the extension of $\langle q,u\rangle_{\mathcal{H}}$.

2. A fundamentally important case arises when we choose to identify $\mathcal{H}$ with its dual; i.e., choose $\mathcal{H}$ to be a pivot space,

$$\mathcal{H}=\mathcal{H}'$$

Then suppose $\mathcal{U}$ is a densely embedded subspace of $\mathcal{H}$. Then both $\mathcal{U}$ and $\mathcal{H}$ are (or can be identified with) dense subspaces of $\mathcal{U}'$, and we have

$$\mathcal{U}\subset\mathcal{H}=\mathcal{H}'\subset\mathcal{U}' \tag{4.31}$$

Moreover, in light of Remark 1, the duality pairing on $\mathcal{U}'\times\mathcal{U}$ can be identified with the unique extension of the inner product in $\mathcal{H}$, $(\cdot,\cdot)_{\mathcal{H}}$.

3. Our discussion here follows closely that of Aubin [4.8, pp. 47–60]. For a more elaborate account, see Trevĕs [4.9]. ■

EXAMPLE 4.2. (Cf. Aubin [4.8, p. 52].) Consider the following spaces of functions defined on an open bounded interval $(a,b)\subset \mathbf{R}$:

$\mathcal{D}(a,b)=$ the usual space of test functions supplied with the topology described in Chapter 2.

$$L_2(a,b)=H^0(a,b), \qquad \text{with} \quad (u,v)_{L_2(a,b)}=\int_a^b u(x)v(x)\,dx.$$

The space $\mathcal{D}(a,b)$ is dense in $L_2(a,b)$ and, in view of (4.31),

$$\mathcal{D}(a,b)\subset L_2(a,b)\subset(\mathcal{D}(a,b))' \tag{4.32}$$

where $(\mathcal{D}(a,b))'$ is the space of distributions. However, a continuous embedding of the type described earlier does not exist in this case. Note that

$$f(\phi)=\langle f,\phi\rangle \qquad f\in(\mathcal{D}(a,b))'; \quad \phi\in\mathcal{D}(a,b)$$

can be identified with the extension of the inner product $(u,v)_{L_2(a,b)}$. ■

EXAMPLE 4.3. As a continuation of Example 4.2, consider any Hilbert space $\mathcal{V}$ that contains the space of test functions $\mathcal{D}(a,b)$ as a densely embedded subspace. Furthermore, suppose $\mathcal{V}$ is dense in $L_2(a,b)$. Then, by identifying $L_2(a,b)$ as the pivot space, we have, from (4.31),

$$\mathcal{D}(a,b)\subset\mathcal{V}\subset L_2(a,b)\subset\mathcal{V}'\subset(\mathcal{D}(a,b))' \tag{4.33}$$

We return to this important relation in the next section. ■

It is convenient at this point to make some observations on duals of products of Hilbert spaces. Indeed, if $\mathcal{H}_1, \mathcal{H}_2, \dots, \mathcal{H}_N$ is a set of N Hilbert spaces, we can define a product Hilbert space $\mathcal{H}$ given by

$$\mathcal{H}=\prod_{i=1}^{N}\mathcal{H}_i=\mathcal{H}_1\times\mathcal{H}_2\times\cdots\times\mathcal{H}_N \tag{4.34}$$

If the component space $\mathcal{H}_j$ is supplied with the inner product $(\cdot,\cdot)_j$, $\mathcal{H}$ is a Hilbert space with respect to the inner product

$$(u,v)=\sum_{j=1}^{N}(u_j,v_j)_j \tag{4.35}$$

where $u=(u_1,u_2,\dots,u_N)$ and u_j denotes an element of $\mathcal{H}_j$. If $\mathcal{H}_j'$ denotes the dual of $\mathcal{H}_j$ for the duality pairing $\langle f_j, v_j\rangle_j$, it can be easily verified that the space

$$\mathcal{H}'=\prod_{i=1}^{N}\mathcal{H}_i'=\mathcal{H}_1'\times\mathcal{H}_2'\times\cdots\times\mathcal{H}_N' \tag{4.36}$$

is a dual of $\mathcal{H}$, and

$$\langle f,v\rangle=\sum_{j=1}^{N}\langle f_j,v_j\rangle_j \tag{4.37}$$

is the associated duality pairing. Moreover, $\mathcal{H}'$ is a Hilbert space with respect to the inner product

$$(f,g)_{\mathcal{H}'}=\sum_{j=1}^{N}\left(K_j^{-1}f_j,K_j^{-1}g_j\right)_j \tag{4.38}$$

where K_j is the canonical isometry from $\mathcal{H}_j$ onto $\mathcal{H}_j'$. The canonical isometry K from $\mathcal{H}$ onto $\mathcal{H}'$ is defined by

$$Ku=(K_1u_1,K_2u_2,\dots,K_Nu_N) \tag{4.39}$$

We conclude this section with a comment on duals of domains of operators. Let $\mathcal{U}$ be a vector space and consider a finite number N of operators A_j mapping $\mathcal{U}$ into Hilbert spaces $\mathcal{H}_j$ with inner products $(\cdot,\cdot)_j$. The space $\mathcal{U}$ is said to be the *domain of the operators* A_j when $\mathcal{U}$ is supplied with the "graph inner product"

$$(u,v)=\sum_{j=1}^{N}(A_ju,A_jv)_j$$

We now define the graph $G(\mathcal{U})$ (note that this is different from the graph of an operator) of the linear space $\mathcal{U}$ by

$$G(\mathcal{U})=A(\mathcal{U})\subset\prod_j\mathcal{H}_j$$

where $A(\mathcal{U})=\{Au\colon Au=(A_1u,A_2u,\dots)\in\prod_j\mathcal{H}_j\}$. We often identify $\mathcal{U}$ with its graph $G(\mathcal{U})$. The *dual* of $G(\mathcal{U})$ is defined to be the space (see Aubin [4.8, p. 56])

$$\mathcal{U}'\approx G(\mathcal{U})'=\prod_j\mathcal{H}_j'/G(\mathcal{U}')^{\perp}$$

where $G(\mathcal{U}')^{\perp}$ is the orthogonal complement of $G(\mathcal{U}')$. Thus is $f \in G(\mathcal{U})'$, we can write

$$\langle f,u\rangle_{\mathcal{U}} = \sum_j \langle f_j, A_j u\rangle_j \qquad \forall u \in \mathcal{U} \tag{4.40}$$

where $f_j \in \mathcal{H}_j'$, $A_j u \in \mathcal{H}_j$; but since

$$\langle f,u\rangle_{\mathcal{U}} = \sum_j \langle f_j, A_j u\rangle_j = \left\langle \sum_j A_j' f_j, u \right\rangle_{\mathcal{U}}$$

we have

$$f = \sum_j A_j' f_j \tag{4.41}$$

This equation provides a means for representing functionals on spaces considered to be domains of certain operators A_j.

4.5 DUALS OF SPACES $H^s(\mathbf{R}^n)$ AND $H^m(\Omega)$

The groundwork is now laid for the introduction of *negative* Sobolev spaces. For the spaces $H^s(\mathbf{R}^n)$ of functions defined on $\mathbf{R}^n$, with s any nonnegative real number, this job is simple; we denote by $H^{-s}(\mathbf{R}^n)$ the dual of $H^s(\mathbf{R}^n)$. That is, the Sobelev space $H^{-s}(\mathbf{R}^n)$, $s \geqslant 0$, is defined by

$$H^{-s}(\mathbf{R}^n) = (H^s(\mathbf{R}^n))' \tag{4.42}$$

where $H^s(\mathbf{R}^n)$ is the space defined in (4.16).

We can now use (4.40) and (4.41) to describe the form of elements in $H^{-s}(\mathbf{R}^n)$. Take, for example, the space $H^m(\mathbf{R})$, where m is an integer >0, and observe that $H^m(\mathbf{R})$ can be considered the domain of operators $A_j = D^j = d^j/dx^j$, $0 \leqslant j \leqslant m$, mapping $\mathcal{U} = H^m(\mathbf{R})$ into $\mathcal{H}_j = L_2(\mathbf{R})$. According to (4.41), each functional $q(=f)$ in $(H^m(\mathbf{R}))'$ can be represented by the sum $q = \sum_{k=0}^{m} D^k q_k$, where $q_k \in (L_2(\mathbf{R}))' = L_2(\mathbf{R})$.

Moreover, let $\{\phi_n\} \in \mathcal{D}(\mathbf{R})$ and suppose that in $H^m(\mathbf{R})$ $\lim_{n\to\infty}\phi_n = u$. Let q be a linear functional on $H^m(\mathbf{R})$. Then, by the Riesz theorem, there is a $v \in H^m(\mathbf{R})$ such that

$$\begin{aligned} q(\phi_n) &= \langle q, \phi_n\rangle = (v, \phi_n)_{H^m(\mathbf{R})} \\ &= \int_{\mathbf{R}} \sum_{k=0}^{m} D^k v\, \overline{D^k \phi_n}\, dx \\ &= \left\langle \sum_{k=0}^{m} (-1)^k D^{2k} v, \phi_n \right\rangle \end{aligned}$$

Thus,

$$\lim_{n\to\infty} q(\phi_n) = q(u) = \langle q, u\rangle = \left\langle \sum_{k=0}^{m} (-1)^k D^{2k} v, u \right\rangle$$

Therefore

$$q = \sum_{k=0}^{m} (-1)^k D^{2k} v$$

In other words, the operator $\Sigma_k(-1)^k D^{2k}$ defines the canonical isometry of $H^m(\mathbb{R})$ onto its dual $H^{-m}(\mathbb{R})$.

We sum up these observations in the following theorem.

THEOREM 4.6. The canonical isometry from $H^m(\mathbb{R})$ onto $H^{-m}(\mathbb{R})$ is a differential operator K given by

$$Ku = \sum_{j=0}^{m} (-1)^j D^{2j} u \tag{4.43}$$

Moreover, a distribution q is in $H^{-m}(\mathbb{R})$ if and only if there exists $m+1$ functions of $q_j \in L_2(\mathbb{R})$ such that

$$q = \sum_{j=0}^{m} D^j q_j \tag{4.44}$$

■

EXAMPLE 4.4. This theorem suggests that a useful interpretation of the negative Sobolev spaces $H^{-m}(\mathbb{R})$ can be obtained by considering those functions $u(x)$ whose derivatives of various orders are in $L_2(\mathbb{R}) = H^0(\mathbb{R})$. For example, consider the function $u(x)$ which has the property that its second derivative is piecewise constant; e.g., if $\tilde{h}(x)$ is the piecewise constant function shown in Fig. 4.1*a*, we may define $u(x)$ by

$$D^2 u(x) = \tilde{h}(x)$$

where $D = d/dx$. If we denote the *antiderivative* by D^{-1},

$$D^{-1} D^2 u = Du = \int \tilde{h}(x)\, dx$$

which is (within a constant) the piecewise linear function shown in Fig.

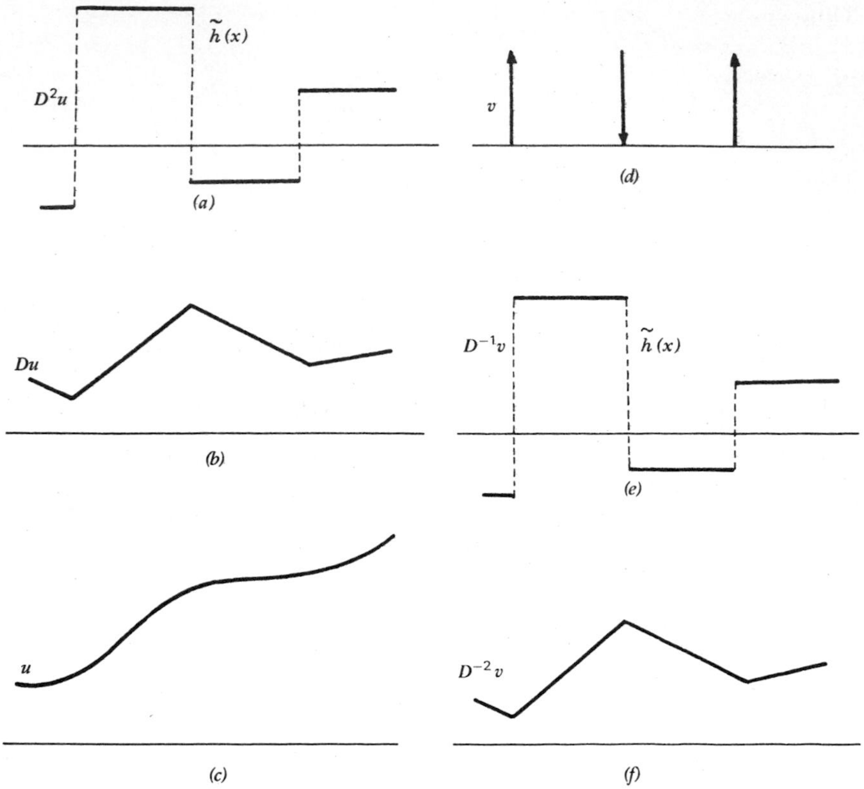

Figure 4.1 Functions $u(x)$ and $v(x)$ in $H^2(\mathsf{R})$, and $H^{-1}(\mathsf{R})$, respectively.

4.1b. Likewise,

$$D^{-1}Du = u = \int \int \tilde{h}(x)\,dx\,dx$$

is the piecewise quadratic in Fig. 4.1c.

Now it is clear that D^2u is smooth enough to be in $L_2(\mathsf{R})$; the square of the piecewise constant function shown is obviously integrable—the integral is in fact the area under the curve $[\tilde{h}(x)]^2$. The derivative of u is piecewise continuous, and the second derivative is bounded on R. Since D^2u is in $L_2(\mathsf{R})$ and Du and u are even smoother than D^2u, they are also in $L_2(\mathsf{R})$. Hence u is a function with all derivatives of order $\leqslant 2$ in $L_2(\mathsf{R})$, which means that

$$u \in H^2(\mathsf{R})$$

Now consider a function $v(x)$ which is much less smooth than $u(x)$; $v(x)$ in fact is defined to be so discontinuous that we must integrate it one time before we obtain anything in $L_2(\mathsf{R})$. For example, take

$$D^{-1}v = \tilde{h}(x)$$

To represent v graphically, we must show delta functions at the points at which infinite jumps in $D\tilde{h}(x)$ occur. Technically, this is impossible, but we depict the deltas pictorially in Fig. 4.1*d* as concentrated forces (arrows) at the appropriate points. $D^{-1}v$ is shown in Fig. 4.1*e*. Clearly, $D^{-2}v$ is piecewise linear (Fig. 4.1*f*), $D^{-3}v$ is piecewise quadratic, etc. This function v, whose -1 derivative (i.e., its first integral) is in $L_2(\mathsf{R})$ is in $H^{-1}(\mathsf{R})$:

$$v \in H^{-1}(\mathsf{R})$$

The pattern is now clear. If, for example,

$$D^{-m}w = \tilde{h}(x)$$

then

$$w \in H^{-m}(\mathsf{R})$$

and

$$\tilde{h} = \int\int \cdots \int w(x)\,dx\,dx \cdots dx \qquad (m \text{ times})$$

If

$$D^{m}w = \tilde{h}(x)$$

then w is in $H^m(\mathsf{R})$. ■

We note that, in view of (4.21),

$$\|v\|_{H^{-s}(\mathsf{R}^n)} = \sup_{u \in H^s(\mathsf{R}^n)} \frac{|\langle v, u \rangle|}{\|u\|_{H^s(\mathsf{R}^n)}}, \qquad u \neq 0 \qquad (4.45)$$

and, in view of (4.31), we have

$$\begin{aligned} \cdots \subset H^{m+1}(\mathsf{R}^n) \subset H^m(\mathsf{R}^n) \subset \cdots \\ \cdots H^1(\mathsf{R}^n) \subset H^0(\mathsf{R}^n) \subset H^{-1}(\mathsf{R}^n) \subset \cdots \\ \cdots \subset H^{-m}(\mathsf{R}^n) \subset H^{-(m+1)}(\mathsf{R}^n) \subset \cdots \end{aligned} \qquad (4.46)$$

wherein all inclusions describe a dense embedding (or continuous injection) of one space into another. The pivot space, in this case, is $H^0(\mathbf{R}^n) = L_2(\mathbf{R}^n)$, and we see that, for the space $\mathfrak{D}(\mathbf{R}^n)$ of test functions,

$$\mathfrak{D}(\mathbf{R}^n) \subset H^s(\mathbf{R}^n) \subset H^0(\mathbf{R}^n) \subset H^{-s}(\mathbf{R}^n) \subset (\mathfrak{D}(\mathbf{R}^n))' \tag{4.47}$$

for every $s \geqslant 0$, the embeddings being, once again, dense.

The Spaces $H^{-m}(\Omega)$

The situation is a bit more complicated for bounded domains, because the space of test functions $C_0^\infty(\Omega)$ is dense in the space $H_0^m(\Omega)$ of (4.6) but not in the larger space $H^m(\Omega)$ (Ω is, as usual, an open bounded domain in $\mathbf{R}^n$). In fact, $H_0^m(\Omega)$ is a closed subspace of $H^m(\Omega)$ consisting of those elements $u(\mathbf{x})$ of $H^m(\Omega)$ such that $D^\alpha u(\mathbf{x}) = 0$ for $\mathbf{x} \in \partial\Omega$ and $|\boldsymbol{\alpha}| \leqslant m-1$. We are thus led to the following definition. Let m be an integer $\geqslant 0$; then the negative Sobolev space $H^{-m}(\Omega)$ is defined as the dual of $H_0^m(\Omega)$:

$$H^{-m}(\Omega) = (H_0^m(\Omega))' \tag{4.48}$$

Since

$$H_0^m(\Omega) \subset H^m(\Omega) \subset H^0(\Omega) \tag{4.49}$$

we again have the inclusion properties

$$H^m(\Omega) \subset H^0(\Omega) \subset H^{-m}(\Omega) \tag{4.50}$$

for every $m \geqslant 0$. In addition, we can still use the ideas of Theorem 4.6 to characterize elements of $H^{-m}(\Omega)$. For example, if $\Omega = (a,b) \subset \mathbf{R}$, a distribution q belongs to $H^{-m}(a,b)$ if and only if there exist $m+1$ functions q_j, $j = 0, 1, \ldots, m$ in $H^0(a,b) = L_2(a,b)$ such that (4.44) holds. Moreover, the canonical isometry from $H^m(a,b)$ into $H^{-m}(a,b)$ is again a differential operator of the type in (4.43).

EXAMPLE 4.5. Consider the Sobolev space $H_0^1(-1,1)$. We first ask whether or not it makes sense to define a Dirac measure δ, i.e., a δ distribution on $H_0^1(-1,1)$. Recall that δ denotes a continuous linear functional on the space of test functions $C_0^\infty(\mathbf{R})$ which has the property $\delta(\phi) = \phi(0)$, $\forall \phi \in C_0^\infty(\mathbf{R})$. We use the same symbol for the linear functional

$$\delta(v) = v(0) \qquad \forall v(x) \in H_0^1(-1,1) \tag{4.51}$$

It so happens that this is a well-defined functional, because the value of

$v(x)$ at $x=0$ is a uniquely defined quantity. This fact follows from the Sobolev embedding theorem, Theorem 3.8, since when $m>n/p$ (and in this example $m=1$, $n=1$, $p=2$, so $1>\frac{1}{2}$), each $v(x)\in H^1(-1,1)$ is equivalent to a continuous function and, according to (3.42),

$$\sup_{x\in(-1,1)} |v(x)| \equiv \|v\|_\infty \leqslant C\|v\|_{H^1(-1,1)}$$

Thus

$$\|\delta\|_{H^{-1}(-1,1)} = \sup_{v\in H_0^1(-1,1)} \frac{|v(0)|}{\|v\|_{H^1(-1,1)}} < C \qquad v\neq 0$$

which means that the delta distribution (4.51) exists and is in $H^{-1}(-1,1)$.

Now in some applications it is important to define the *smallest* space in which δ belongs. It is certainly in $H^{-1}(-1,1)$ and not in $H^0(-1,1)$; is it in a smaller space between $H^0(-1,1)$ and $H^{-1}(-1,1)$? Again, the answer follows from the Sobolev embedding theorem: $v(x)\in H_0^\mu(-1,1)$ is continuous whenever $\mu>n/2=\frac{1}{2}$. Thus take $\mu=\frac{1}{2}+\varepsilon$, where ε is any positive number, and note that $\delta\in H^{-\mu}(-1,1)$; i.e.,

$$\delta\in H^{-1/2-\varepsilon}(-1,1) \qquad \varepsilon>0 \quad \blacksquare$$

EXAMPLE 4.6. This is a continuation of the previous example. We wish to demonstrate here the nature of the canonical isometry of $H_0^1(-1,1)$ and $H^{-1}(-1,1)$. In particular, for every continuous linear functional q on $H_0^1(-1,1)$ $[q\in H^{-1}(-1,1)]$ is there a unique $v_q\in H_0^1(-1,1)$ such that $q(u)=(v_q,u)_{H^1(-1,1)}$, $\forall u\in H_0^1(-1,1)$, as required? Indeed, what is the function $v_\delta(x)$ in $H_0^1(-1,1)$ corresponding to δ of (4.51)? Theorem 4.6 (or a form of it appropriate for bounded domains) provides the answer; according to (4.43), $v_\delta(x)$ is the solution of the distributional differential equation

$$Kv_\delta = v_\delta(x) - v_\delta''(x) = \delta(x)$$

where $v_\delta''=D^2v_\delta$. Since $v_\delta\in H_0^1(-1,1)$, we must also have

$$v_\delta(-1)=0 \qquad v_\delta(1)=0$$

This equation is easily solved to give

$$v_\delta(x)=\begin{cases}\alpha_0\sinh(1+x) & x\leqslant 0\\ \alpha_0\sinh(1-x) & x\geqslant 0\end{cases} \tag{4.52}$$

where $\alpha_0 = 1/[2\cosh(1)]$. A check reveals that

$$\int_{-1}^{1}(v_\delta u + v_\delta' u')\,dx = (u,v)_{H^1(-1,1)} = 2u(0)\alpha_0\cosh(1) = u(0)$$

Thus $v_\delta(x)$ of (4.52) has first derivatives in $L_2(-1,1)$, but it uniquely corresponds to $\delta(x)$ in $H^{-1}(-1,1)$. Moreover,

$$\begin{aligned}\|\delta\|_{H^{-1}(-1,1)} &= \|v_\delta\|_{H^1(-1,1)}\\ &= \left[\int_{-1}^{1}(v_\delta^2 + v_\delta'^2)\,dx\right]^{1/2}\\ &= \frac{1}{2\cosh 1}\sqrt{\sinh 2}\end{aligned}$$

■

REMARK. The observation made in the Example 4.5 concerning the existence of a Dirac delta distribution in $H^m(\Omega)$ deserves further comment. The question is, when can there exist a functional $\delta(v) = v(0)$ in the negative Sobolev space $H^{-m}(\Omega)$? As in the above examples, the answer is furnished by the Sobolev embedding theorem, Theorem 3.8. When $m > n/2$, each $v(\mathbf{x}) \in H^m(\Omega)$, Ω satisfying the cone condition, is such that $\sup|v(\mathbf{x})| \leqslant C\|v\|_{H^m(\Omega)}$, $\mathbf{x}\in\Omega$. Then for some point $\mathbf{x}_0 \in \Omega$,

$$\|\delta_{\mathbf{x}_0}\|_{H^{-m}(\Omega)} = \sup_{v\in H_0^m(\Omega)} \frac{|v(\mathbf{x}_0)|}{\|v\|_{H^m(\Omega)}} \qquad v \neq 0 \tag{4.53}$$

This means that δ_{x_0} does *not* belong to $H^{-1}(\Omega)$ when $\Omega\subset \mathsf{R}^2$; then $m=1 = n/2$. However, δ_{x_0} does lie in $H^{-2}(\Omega)$ for $n=2$. In fact, $\delta_{x_0}\in H^{-1-\varepsilon}(\Omega)$, $\varepsilon>0$. ■

We remarked earlier that $C_0^\infty(\Omega)$ is dense in $H_0^m(\Omega)$ but not in $H^m(\Omega)$. The description of the dual of $H^m(\Omega)$, which is contained in $H^{-m}(\Omega)$, generally involves complications at the boundary. For example, we find that surface or boundary distributions must be introduced to characterize $(H^m(\Omega))'$ completely. To illustrate this, we reproduce an example of Aubin [4.8, p. 60] for the one-dimensional case in which $\Omega = (a,b)\subset \mathsf{R}$. The principal result is given in the following theorem.

THEOREM 4.7. The dual $(H^m(a,b))'$ of $H^m(a,b)$ is the space of functionals q of the form

$$q = \sum_{k=0}^{m} D^k q_k + \sum_{k=0}^{m-1}\left(a_k\delta_a^{(k)} + b_k\delta_b^{(k)}\right) \tag{4.54}$$

where q_k are functions in $L_2(a,b)$, a_k and b_k are constants, and $\delta_\xi^{(k)}$ is the kth distributional derivative of the Dirac delta at point ξ:

$$\delta_\xi^{(k)}(\phi)=(-1)^k D^k\phi(\xi) \tag{4.55}$$

PROOF. We sketch only the essential features of the proof. Aubin's proof of (4.54) makes use of the following fact. If $A\in\mathcal{L}(\mathcal{U},\mathcal{V})$, $\mathcal{U}$ and $\mathcal{V}$ being Hilbert spaces, its transpose $A'\in\mathcal{L}(\mathcal{V}',\mathcal{U}')$ is one-to-one if and only if $A(\mathcal{U})$ is dense in $\mathcal{V}$, and the range $A(\mathcal{U})$ is closed if and only if $A'(\mathcal{V}')$ is closed. We can then construct a mapping B from $H^m(a,b)$ onto its closed range in $H^m(\mathsf{R})$ and be sure that the range of B' is the dual space $(H^m(a,b))'$. We then need only to characterize B' to obtain (4.54).

Indeed, let u be an arbitrary function in $H^m(a,b)$ and construct a function v in $H^m(\mathsf{R})$ that (i) coincides with u in (a,b), (ii) has derivatives of order $\leqslant m-1$ which coincide with those of u at $x=a$ and $x=b$, and (iii) has compact support in $(a-\varepsilon,b+\varepsilon)$ for some $\varepsilon>0$. We then write, for example,

$$v(x)=Bu(x)=\begin{cases}\sum_{k=0}^{m-1} D^k u(a)\psi_k(x) & x\leqslant a\\ u(x) & x\in(a,b)\\ \sum_{k=0}^{m-1} D^k u(b)\phi_k(x) & x\geqslant b\end{cases}$$

where $\psi_k(x)$ and $\phi_k(x)$ are any functions, vanishing for $x\leqslant a-\varepsilon$ and $x\geqslant b+\varepsilon$, respectively, such that $D^k\psi_j(a)=\delta_j^k, D^k\phi_j(b)=\delta_j^k$, where δ_j^k is the Kronecker delta. Thus B is a linear continuous injection from $H^m(a,b)$ onto its range in $H^m(\mathsf{R})$; it is the right inverse of the restriction r of $v(x)\in H^m(\mathsf{R})$ to $H^m(a,b)$. From what was said earlier, B' takes $H^{-m}(\mathsf{R})$ onto $(H^m(a,b))'$. Therefore, if $q\in H^{-m}(\mathsf{R})$,

$$\begin{aligned}\langle q,v\rangle&=\langle q,Bu\rangle\\ &=\langle q,u\rangle|_{(a,b)}+\sum_{k=0}^{m-1}\left(a_k D^k u(a)+b_k D^k u(b)\right)\end{aligned}$$

where

$$a_k=q(\psi_k)\qquad\text{and}\qquad b_k=q(\phi_k)$$

From (4.44),

$$\langle q,u\rangle|_{(a,b)}=\left\langle \sum_{k=0}^{m} D^k q_k,u\right\rangle, \qquad q_k\in L_2(a,b),\quad 0\leqslant k\leqslant m.$$

Hence

$$\langle q,v\rangle=\left\langle \sum_{k=0}^{m} D^k q_k+\sum_{k=0}^{m-1}\left(a_k\delta_a^{(k)}+b_k\delta_b^{(k)}\right),u\right\rangle$$

which establishes (4.54). ■

4.6 THE TRACE THEOREM FOR $H^m(\mathsf{R}^n_+)$

In this section we take a fundamentally important step—the transition from $H^m(\mathsf{R}^n)$ to $H^m(\mathsf{R}^n_+)$, where R^n_+ is the half space

$$\mathsf{R}^n_+=\{\mathbf{x}:\mathbf{x}=(x_1,x_2,\ldots,x_{n-1},x_n)\in\mathsf{R}^n;\ x_n>0\}$$

What makes this step crucial is that R^n_+ is the simplest domain with a boundary, and when we consider introducing a smooth boundary into the picture, many interesting properties of functions in $H^m(\mathsf{R}^n_+)$ manifest themselves. Chief among these properties are those appearing in the trace theorem, which provides a cornerstone for much of the material covered in the next chapter. Since all the basic ideas can be demonstrated for the two-dimensional case ($n=2$), our approach is to develop them for R^2_+ and then write down the results for R^n_+.

The Space $H^m(\mathsf{R}^n_+)$

We begin by describing a few properties of $H^m(\mathsf{R}^2_+)$. We denote R^2_+ by $\mathsf{R}_x\times(0,\infty)=\{(x,t):\ x\in\mathsf{R}_x\equiv\mathsf{R}^1;\ t>0\}$; then $\partial\Omega=\mathsf{R}_x$ is the boundary of $\Omega=\mathsf{R}^2_+$. As usual, $u(x,t)\in H^m(\mathsf{R}^2_+)$ for integer m means that u has generalized derivatives of order $\leqslant m$ in $L_2(\mathsf{R}^2_+)$. But we can easily describe $H^s(\mathsf{R}^2_+)$ for any real s by regarding the functions $H^s(\mathsf{R}^2_+)$ as restrictions to R^2_+ of functions in $H^s(\mathsf{R}^2)$. This requires a little care at the boundary and prompts us to impose certain conditions on the normal derivatives of $u(x,t)$ there. For the moment, let m be an integer. For $0\leqslant j<m$, we define the trace operators γ_j by

$$(\gamma_j u)(x)=\left.\frac{\partial^j u(x,t)}{\partial n^j}\right|_{t=0} \tag{4.56}$$

where n is a normal to the boundary. In this case, of course,

$$D_n^j u = \frac{\partial^j u}{\partial n^j} \equiv -\frac{\partial^j u}{\partial t^j} = -D_t^j u$$

Consider a function $u(x,t)$ in $H^m(\mathbf{R}^2_+)$, whose *trace* on $\mathbf{R}_x = \partial\Omega$ is defined by (4.56). We can consider another function $v(x,t)$ in $H^m(\mathbf{R}^2)$, which coincides with $u(x,t)$ for $t>0$. Then $u(x,t)$ is a restriction of $v(x,t)$ to $\mathbf{R}^2_+$, written $u=r_+(v)$, and $v(x,t)$ is an extension of $u(x,t)$ to $\mathbf{R}^2$. This extension, however, is not necessarily continuous; as $t \to 0$ there is no guarantee that the normal derivatives of u and v of order $j<m$ will match. We can resolve this by a more judicious choice of v. For example, if we pick

$$v(x,t) = \begin{cases} u(x,t) & t>0 \\ \sum_{k=1}^{m} a_k u(x,-kt) & t<0 \end{cases}$$

where the coefficients a_k are chosen so that

$$\sum_{k=1}^{m} (-k)^s a_k = 1 \qquad 0 \leqslant s \leqslant m-1$$

then a simple calculation reveals that

$$D^{(\alpha_1,\alpha_2)} v(x,t) = \begin{cases} D^{(\alpha_1,\alpha_2)} u(x,t) & t>0 \\ \sum_{k=1}^{m} (-k)^{\alpha_2} a_k D_x^{\alpha_1} D_t^{\alpha_2} u(x,-kt) & t<0 \end{cases}$$

so that, in the limit as $t \to 0$,

$$\gamma_j v = \gamma_j u \tag{4.57}$$

In this case, $v(x,t)$ is a *continuous* extension of $u(x,t)$, and we write

$$v(x,t) = \mathcal{E} u(x,t) \tag{4.58}$$

The operator $\mathcal{E}$ is a continuous right inverse of the restriction r_+.

All these notions carry directly to $\mathbf{R}^n$, and we can conclude that it is always possible to construct a continuous extension operator $\mathcal{E}$ such that (4.57) holds. Moreover, there exists a constant $C>0$ such that

$$\|v\|_{H^m(\mathbf{R}^n)} = \|\mathcal{E} u\|_{H^m(\mathbf{R}^n)} \leqslant C \|u\|_{H^m(\mathbf{R}^n_+)} \tag{4.59}$$

There is still another way of describing $H^m(\mathbf{R}^n_+)$, which often proves to be convenient. We observe that, if $\mathcal{H}$ is any Hilbert space and (a,b) is any open interval in R (even an infinite interval), the space $L_2(a,b;\mathcal{H})$ of measurable functions from (a,b) into $\mathcal{H}$ such that

$$\|u\|_{L_2(a,b;\,\mathcal{H})}=\left(\int_a^b \|u(t)\|_{\mathcal{H}}^2\,dt\right)^{1/2}<\infty \tag{4.60}$$

is also a Hilbert space. Indeed, (4.60) is the norm in $L_2(a,b;\mathcal{H})$ associated with the inner product

$$(u,v)_{L_2(a,b;\,\mathcal{H})}=\int_a^b (u(t),v(t))_{\mathcal{H}}\,dt \tag{4.61}$$

For example, a function $u(x,t)$ in $L_2(0,\infty;H^m(\mathsf{R}_x))$ is, by definition, square-integrable in t while $D_x^k u$ is in $L_2(\mathsf{R}_x)$ for $k\leqslant m$. But a simple change in the order of integration (Fubini's theorem) reveals that $L_2(0,\infty;\mathsf{R}_x)$ $=L_2(\mathsf{R}_+^2)$. Therefore $D_x^k u\in L_2(\mathsf{R}_+^2)$ for $k\leqslant m$. In other words, $u(x,t)$ is in $H^m(\mathsf{R}_x)$ for almost every t, and $u\in H^r(\mathsf{R}_+^2)$, where $r\leqslant m$ depends on how smooth u is with respect to t. Suppose $D_t^j\, u(x,t)\in L_2(0,\infty;H^{m-j}(\mathsf{R}_x))$ for $0\leqslant j\leqslant m-1$. Then this is equivalent to saying that $D_x^{m-j}(D_t^j u)\in L_2(\mathsf{R}_+^2)$. But this means that u is in $H^m(\mathsf{R}_+^2)$. Therefore each u in $L_2(0,\infty;H^m(\mathsf{R}_x))$ for which $D_t^j u\in L_2(0,\infty;H^{m-j}(\mathsf{R}_x))$ is also in $H^m(\mathsf{R}_+^2)$. Conversely, each $u\in H^m(\mathsf{R}_+^2)$ is such that $D_t^j u\in L_2(0,\infty;H^{m-j}(\mathsf{R}_x))$.

The extension of these observations to R_+^n is obvious, and we summarize them in the following theorem.

THEOREM 4.8. Let $\mathcal{H}$ denote the space of functions in $L_2(\mathsf{R}_+,$ $H^m(\mathsf{R}^{n-1}))$ such that $D_n^j u\in L_2(\mathsf{R}_+,\ H^{m-j}(\mathsf{R}^{n-1}))$, where $\mathsf{R}_+=(0,\infty)$, $D_n^j u=\partial^j u/\partial x_n^j$, $\mathbf{x}=(x_1,\dots,x_{n-1},x_n)$, and $0\leqslant j\leqslant m$. Then $\mathcal{H}$ coincides with $H^m(\mathsf{R}_+^n)$. ■

EXAMPLE 4.7. The function u shown in Fig. 4.2a is piecewise linear in x and piecewise constant in t. Clearly, $u\in L_2(0,\infty;H^1(\mathsf{R}_x))$. Since $D_t u\in$ $H^{-1}(\mathsf{R}_+^2)$, $u\in H^0(\mathsf{R}_+^2)$. However, the function w in Fig. 4.2b is also in $L_2(0,\infty;H^1(\mathsf{R}_x))$, but $D_t w$ is in $H^0(\mathsf{R}_+^2)$; i.e., $w(x,t)$ is in $H^1(\mathsf{R}_+^2)$. ■

The Trace Theorem

We now focus our attention on the trace operator γ_j of (4.56). Our principal aim at this point is to resolve the following question. If $u\in$ $H^m(\mathsf{R}_+^2)$, can γ_j be extended to a continuous operator from $H^m(\mathsf{R}_+^2)$ onto some "boundary space" $H^\mu(\partial\Omega)\equiv H^\mu(\mathsf{R}_x)$? Indeed, if the answer to this question is affirmative (and it so happens that it is), what is the largest μ that admits a continuous mapping from $H^m(\mathsf{R}_+^2)$ into $H^\mu(\mathsf{R}_x)$? The answers to these questions require a bit of preliminary work.

First, suppose that $v(x,t)$ is, for each x, a function of t which rapidly

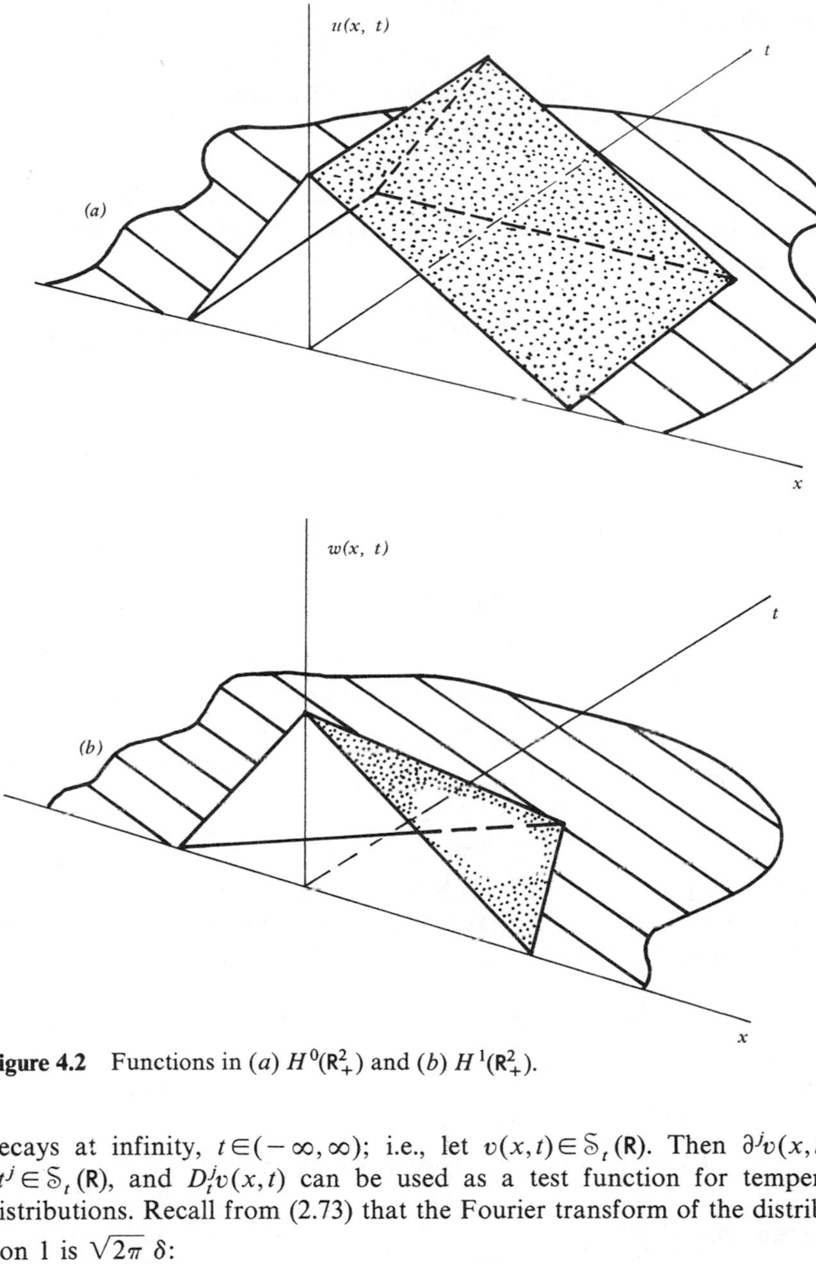

Figure 4.2 Functions in (a) $H^0(\mathsf{R}^2_+)$ and (b) $H^1(\mathsf{R}^2_+)$.

decays at infinity, $t\in(-\infty,\infty)$; i.e., let $v(x,t)\in\mathcal{S}_t(\mathsf{R})$. Then $\partial^j v(x,t)/\partial t^j\in\mathcal{S}_t(\mathsf{R})$, and $D_t^j v(x,t)$ can be used as a test function for tempered distributions. Recall from (2.73) that the Fourier transform of the distribution 1 is $\sqrt{2\pi}\,\delta$:

$$\langle\hat{1},\phi\rangle=\langle 1,\hat{\phi}\rangle=\int_{-\infty}^{\infty}\hat{\phi}(s)\,ds=\sqrt{2\pi}\,\phi(0)$$

Therefore

$$\sqrt{2\pi}\, D_t^j v(x,0) = \int_{-\infty}^{\infty} s^j \bar{v}(x,s)\, ds$$

where $\bar{v}(x,s) = \mathcal{F}_t v(x,t)$ is the Fourier transform of $v(x,t)$ with respect to t. Using the notation of (4.56) and definition (4.10) of $H^\mu(\mathsf{R}_x)$, we see that

$$\|\gamma_j v\|^2_{H^\mu(\mathsf{R}_x)} = \int_{\mathsf{R}_x} (1+|y|^2)^\mu \frac{1}{2\pi} \left(\int_{-\infty}^{\infty} s^j \hat{v}(y,s)\, ds \right)^2 dy \tag{4.62}$$

Next we observe that

$$\begin{aligned} \int_{-\infty}^{\infty} s^j \hat{v}(y,s)\, ds &\leqslant \int_{-\infty}^{\infty} |s^j \hat{v}(y,s)|\, ds \\ &= \int_{-\infty}^{\infty} |s^j \hat{v}|(1+|y|^2+|s|^2)^{m/2} \cdot (1+|y|^2+|s|^2)^{-m/2}\, ds \\ &\leqslant \left[\int_{-\infty}^{\infty} |\hat{v}(y,s)|^2 (1+|y|^2+|s|^2)^m\, ds \right]^{1/2} \\ &\quad \cdot \left[\int_{-\infty}^{\infty} (1+|y|^2+|s|^2)^{-m} s^{2j}\, ds \right]^{1/2} \end{aligned} \tag{4.63}$$

where, in this step, we have used Schwarz's inequality.

To evaluate the last integral in (4.63), we introduce for fixed y the change of variables

$$p = s(1+|y|^2)^{-1/2}, \qquad dp = ds(1+|y|^2)^{-1/2}$$

Then

$$\int_{-\infty}^{\infty} (1+|y|^2+|s|^2)^{-m} s^{2j}\, ds = c(m,j)(1+|y|^2)^{j-m+1/2}$$

wherein

$$c(m,j) = \int_{-\infty}^{\infty} (1+|p|^2)^{-m} p^{2j}\, dp > 0, \qquad j \leqslant m-1$$

Combining this result and (4.63), we see that

$$\|\gamma_j v\|^2_{H^\mu(\mathbf{R}_x)} \leqslant \frac{1}{2\pi} c(m,j) \int_{-\infty}^{\infty} (1+|y|^2)^{\mu-m+j+1/2} \int_{-\infty}^{\infty} (1+|y|^2$$

$$+|s|^2)^m |\hat{v}(y,s)|^2 \, ds\, dy$$

Now if $\mu - m + j + 1/2 \leqslant 0$, then $(1+|y|^2)^{\mu-m+j+1/2} \leqslant 1$. Thus, for any $\mu \leqslant m-j-1/2$, we have

$$\|\gamma_j v\|^2_{H^\mu(\mathbf{R}_x)} \leqslant c^2 \int_{-\infty}^{\infty}\int_{-\infty}^{\infty} (1+|y|^2+|s|^2)^m |\hat{v}(y,s)|^2 \, ds\, dy$$

$$= c^2 \|v\|^2_{H^m(\mathbf{R}^2)} \tag{4.64}$$

where $c = \sqrt{c(m,j)/2\pi}$, and the *largest* μ for which this inequality holds is

$$\mu = m - j - \frac{1}{2} \tag{4.65}$$

Next, we remove the smoothness requirements on $\gamma_j v(x,t)$ by noting that $\mathcal{S}_t$ is dense in $H^m(\mathbf{R}_t)$. Hence, we can use $\gamma_j v$ to construct sequences of traces of smooth functions converging in $H^m(\mathbf{R}_t)$. We also use the notation $\gamma_j v$ to describe the limit of this sequence so that by γ_j we now mean an *extension* of $D_t^j|_{t=0}$ to a continuous operator from $H^m(\mathbf{R}^2)$ into $H^\mu(\mathbf{R}_x)$. Since v is then an arbitrary element in $H^m(\mathbf{R}^2)$, we can choose it as the continuous extension $\mathcal{E}u$ of a function u in $H^m(\mathbf{R}^2_+)$. Collecting all of these observations, we have for $u \in H^m(\mathbf{R}^2_+)$,

$$\|\gamma_j u\|_{H^{m-j-1/2}(\mathbf{R}_x)} \leqslant C \|u\|_{H^m(\mathbf{R}^2_+)} \tag{4.66}$$

The extension of these results to $H^m(\mathbf{R}^n_+)$ is obvious, and we have therefore proved the trace theorem.

THEOREM 4.9 (The Trace Theorem). The trace operators γ_j of (4.56) can be extended to continuous operators from $H^m(\mathbf{R}^n_+)$ into $H^{m-j-1/2}(\mathbf{R}^{n-1})$, with $0 \leqslant j \leqslant m-1$. ■

There are some additional comments that should be made about the extended trace operators γ_j. They are, for example, surjective mappings i.e., they map $H^m(\mathbf{R}^n_+)$ *onto* $H^{m-j-1/2}(\mathbf{R}^{n-1})$. To see this, we need only to examine an arbitrary element $w_j(x)$ of $H^{m-j-1/2}(\mathbf{R}_x)$ and see if it is the image under γ_j of some element in $H^m(\mathbf{R}^2_+)$. Let $\tilde{w}_j(y) \in \hat{H}^{m-j-1/2}(\mathbf{R}_y)$

denote the Fourier transform of $w_j(x)$ and define

$$\tilde{u}_j(y,t)=|y|^{-j}\tilde{w}_j(y)\phi_j(|y|t)$$

where $\phi_j(t)$ is a test function with the property $D_t^j\phi_j(0)=1$. Then $D_t^k\tilde{u}_j(y,t)\in L_2(\mathsf{R}_+,H^{m-k}(\mathsf{R}_x))$ and

$$\gamma_j\tilde{u}_j(y)=|y|^{j-j}\tilde{w}_j(y)\gamma_j\phi_j(0)=\tilde{w}_j(y)$$

The fact that $D_t^k\tilde{u}_j(y,t)\in L_2(\mathsf{R}_+,H^{m-k}(\mathsf{R}_x))$ means that $\tilde{u}_j$ is an element of $H^m(\mathsf{R}_+^2)$.

Next, observe that

$$D_t\phi_j=D_\lambda\phi_j|y| \qquad \lambda=|y|t$$

and

$$D_t^k\phi_j=D_\lambda^k\phi_j|y|^k$$

Thus

$$\begin{aligned}
&\int_0^\infty\int_{\mathsf{R}_x}|y|^{2(m-k)}|D_t^k\tilde{u}_j(y,t)|^2\,dy\,dt\\
&\qquad=\int_0^\infty\int_{\mathsf{R}_x}|y|^{2(m-j)}|\tilde{w}_j(y)|^2|D_\lambda^k\phi_j(\lambda)|^2(|y|^{-1}d\lambda)\,dy\\
&\qquad=\int_0^\infty|D_\lambda^k\phi_j(\lambda)|^2\,d\lambda\int_{\mathsf{R}_x}|y|^{2m-2j-1}|\tilde{w}_j(y)|^2\,dy\\
&\qquad\leqslant C\int_{\mathsf{R}_x}(1+|y|^2)^{2(m-j-1/2)}|\tilde{w}_j(y)|^2\,dy\\
&\qquad=C\|(1+|y|^2)^{m-j-1/2}\tilde{w}_j\|^2_{L_2(\mathsf{R}_x)}\\
&\qquad\leqslant\overline{C}\|w_j\|^2_{H^{m-j-1/2}(\mathsf{R}_x)}
\end{aligned}$$

Recalling Theorem 4.8, we see that $u_j(x,t)$ is precisely the function in $H^m(\mathsf{R}_+^2)$ for which $w_j=\gamma_ju_j$. Thus by passing to R_+^n we have proved the following.

THEOREM 4.10. The extended trace operators γ_j map $H^m(\mathsf{R}_+^n)$ onto $H^{m-j-1/2}(\mathsf{R}^{n-1})$. ■

Next, we expand Theorems 4.9 and 4.10 slightly and give an interpretation of the kernel of the family of trace operators γ_j, $j=0,1,\ldots,m-1$.

THEOREM 4.11 (The Trace Theorem—Extended). Let γ denote the ordered collection of trace operators

$$\gamma=(\gamma_0,\gamma_1,\ldots,\gamma_{m-1}) \qquad \gamma_k=\left.\frac{\partial^k}{\partial x_n^k}\right|_{x_n=0} \qquad k=0,1,2,\ldots,m-1 \quad (4.67)$$

Then

(i) γ maps $H^m(\mathbf{R}^n_+)$ onto the product space

$$\prod_{j=0}^{m-1} H^{m-j-1/2}(\mathbf{R}^{n-1}) \equiv H^{m-1/2}(\mathbf{R}^{n-1})\times H^{m-1-1/2}(\mathbf{R}^{n-1}) \times\cdots\times H^{1-1/2}(\mathbf{R}^{n-1}) \quad (4.68)$$

(ii) The kernel of γ coincides with the space $H_0^m(\mathbf{R}^n_+)$.

PROOF. Our proof, which follows that of Aubin [4.8], makes use of the methodology already used in Theorems 4.9 and 4.10.

(i) Let $(w_0(\mathbf{x}),w_1(\mathbf{x}),\ldots,w_{m-1}(\mathbf{x}))\in\prod_{j=0}^{m-1}H^{m-j-1/2}(\mathbf{R}^{n-1})$, where $\mathbf{x}=(x_1,x_2,\ldots,x_{n-1})$, and let u_j be the inverse Fourier transform of functions $\tilde{u}_j(\mathbf{y},x_n)$ of the type used in Theorem 4.10. This done, consider the functions $U_j(\mathbf{x},x_n)$ defined by

$$U_j(\mathbf{x},x_n)=\sum_{r=1}^{m} a_r^j u_j(\mathbf{x},rx_n)$$

Then $D_n^k U_j(\mathbf{x},0)=\delta_{kj}w_j(\mathbf{x})$, provided $\sum_{r=1}^m r^k a_r^j=\delta_{kj}$ (δ_{kj} is the Kronecker delta), $0\leqslant k\leqslant m-1$. Hence $\sum_{0\leqslant j\leqslant m-1}U_j(\mathbf{x},x_n)\in H^m(\mathbf{R}^n_+)$ and satisfies $\gamma_j u_j=w_j$ for $0\leqslant j\leqslant m-1$.

(ii) The proof of (ii) is given in Aubin [4.8, p. 199] and it suffices to only sketch a few details here. Suppose $\mathfrak{N}(\gamma)$ denotes the kernel of γ. Recall that $H_0^m(\mathbf{R}^n_+)$ is the closure of $C_0^\infty(\mathbf{R}^n_+)$ in $H^m(\mathbf{R}^n_+)$. Let $\{\phi_n\}$ be a sequence of functions in $C_0^\infty(\mathbf{R}^n_+)$ and let u be the limit of this sequence in $H_0^m(\mathbf{R}^n_+)$. Then in $H^{m-j-1/2}(\mathbf{R}^{n-1})$, $\gamma u=\lim_{n\to\infty}\gamma\phi_n=0$, because $\phi_n|_{\partial\Omega}=0$ and γ is continuous, as we proved earlier. Thus $H_0^m(\mathbf{R}^n_+)\subset\mathfrak{N}(\gamma)$. Thus, we must prove that $\mathfrak{N}(\gamma)$ is contained in $H_0^m(\mathbf{R}^n_+)$. Let $u\in\mathfrak{N}(\gamma)$. Then, by an extended Taylor formula, u is representable as the integral

$$u(\bar{\mathbf{x}},x_n)=\int_0^{x_n}\frac{(x_n-s)^{m-1}}{(m-1)!}D_n^m u(\bar{\mathbf{x}},s)\,ds$$

where $\bar{\mathbf{x}}=(x_1,\ldots,x_{n-1})$. Next introduce an infinitely differentiable function $\zeta(x_n)$ such that $\zeta(x_n)=1$ for $|x_n|\leqslant 1$, $0\leqslant\zeta(x_n)\leqslant 1$ for $1\leqslant|x_n|\leqslant 2$, and $\zeta(x_n)=0$ for $|x_n|>2$. Then for $\rho>0$, the function

$$u_\rho(x_n)=\zeta\left(\frac{x_n}{\rho}\right)u(\mathbf{x})$$

converges to zero as $\rho\to 0$ in the $H^m(\mathsf{R}^n_+)$ norm whenever $u\in\mathfrak{N}(\gamma)$. Therefore, $u-u_\rho$ converges to zero on the boundary because $\zeta=1$ for $|x_n|<\rho$. Hence, $\mathfrak{N}(\gamma)\subset H^m_0(\mathsf{R}^n_+)$. ■

EXAMPLE 4.8. Consider the space $H^1(\mathsf{R}^2_+)$. According to Theorem 4.9, the trace operator ($m=1$, $n=2$, $0\leqslant j\leqslant m-1=0$)

$$\gamma_0 u=u(x,0)$$

can be extended to a continuous operator mapping $H^1(\mathsf{R}^2_+)$ into $H^{1/2}(\mathsf{R}_x)$. However, since j in Theorems 4.9 and 4.10 must be $\leqslant m-1$, the operator

$$\gamma_1 u=\frac{\partial u(x,0)}{\partial n}$$

cannot be extended to a continuous operator from $H^1(\mathsf{R}^2_+)$ into $L_2(\mathsf{R}_x)$. According to Theorem 4.11, the kernel of γ_0 is the space $H^1_0(\mathsf{R}^2_+)$. ■

REMARKS. 1. It is also possible to show that D_t^j can be extended to a continuous operator mapping $H^m(\mathsf{R}^n_+)$ into the space $B(\overline{\mathsf{R}}_+, H^{m-j-1/2}(\mathsf{R}^{n-1}))$ of bounded continuous functions, vanishing at infinity, from $\overline{\mathsf{R}}_+$ into $H^{m-j-1/2}(\mathsf{R}^{n-1})$, for $0\leqslant j\leqslant m-1$ (cf. Aubin [4.8]). This in fact is essentially what we did in obtaining (4.66). Theorem 4.9 then follows by setting $\gamma_j u(x)=D_t^j u(x,0)$.

2. The space $H^{m-j-1/2}(\mathsf{R}_x)$ is an example of a Hilbert space of functions defined on smooth boundaries $\partial\Omega$. We study such spaces in more detail in Section 4.9. It is worth mentioning here, however, that $H^{m-1/2}(\mathsf{R}_x)$ is essentially the space of functions defined on $\mathsf{R}_x=\partial\Omega$ that have the values in R_x of a function in $H^m(\mathsf{R}^2_+)$. Indeed, if $w\in H^{m-1/2}(\mathsf{R}_x)$, then (see Berezanskii [4.3])

$$\|w\|_{H^{m-1/2}(\mathsf{R}_x)}=\inf_{u\in W}\|u\|_{H^m(\mathsf{R}^2_+)}$$

where the infimum is taken over the set $W=\{u\in H^m(\mathsf{R}^2_+),\ \gamma_0 u=w\}$.

3. If M is the space of functions in $H^m(\mathsf{R}^2_+)$ that vanish on R_x, we can argue (e.g., [4.3]) that $H^{m-1/2}(\mathsf{R}_x)$ can be associated with the quotient space $H^m(\mathsf{R}_x)/M$. ■

4.7 INTERMEDIATE AND INTERPOLATION SPACES

Much of what we have established thus far in this chapter points to a natural ordering of Sobolev spaces $H^m(\Omega)$ or $H^s(\mathbf{R}^n)$ suggested, for example, by the inclusion properties (4.33) and (4.46). For example, $C_0^\infty(\mathbf{R}^n)$ is a relatively small space, contained in a slightly bigger space $H^s(\mathbf{R}^n)$. Likewise, $H^s(\mathbf{R}^n)$ is contained in $H^{s-1}(\mathbf{R}^n)$ which itself is contained in $H^{-s}(\mathbf{R}^n)$, $s \geqslant 1$, etc. This suggests that for any two numbers s_1, $s_2 > 0$, it may make sense to speak of spaces $H^s(\mathbf{R}^n)$ *between* $H^{s_1}(\mathbf{R}^n)$ and $H^{s_2}(\mathbf{R}^n)$, $s_1 < s < s_2$.

It so happens that it is not only possible to define such intermediate spaces, but that they form a fundamental building block in the construction of a theory of linear boundary-value problems. In this section, we develop several basic properties of intermediate and interpolation spaces within the more general framework of Banach spaces. In subsequent sections, we apply these concepts to the Sobolev spaces $H^m(\Omega)$ described earlier.

Let $\mathcal{U}$ and $\mathcal{V}$ be two Banach spaces and let the intersection $\mathcal{U} \cap \mathcal{V}$ be dense in $\mathcal{U}$ and $\mathcal{V}$.

$$\mathcal{U} + \mathcal{V} = \{w: w = u + v;\ u \in \mathcal{U};\ v \in \mathcal{V}\} \tag{4.69}$$

where the decomposition of $w \in \mathcal{U} + \mathcal{V}$ into $u \in \mathcal{U}$ and $v \in \mathcal{V}$ is not unique. We define the norm in $\mathcal{U} \cap \mathcal{V}$ by

$$\|u\|_{\mathcal{U} \cap \mathcal{V}} = \max(\|u\|_{\mathcal{U}}, \|u\|_{\mathcal{V}}) \tag{4.70}$$

and the norm in $\mathcal{U} + \mathcal{V}$ by

$$\|w\|_{\mathcal{U} + \mathcal{V}} = \inf_{u+v=w} (\|u\|_{\mathcal{U}} + \|v\|_{\mathcal{V}}) \tag{4.71}$$

Here $\|\cdot\|_{\mathcal{U}}$ and $\|\cdot\|_{\mathcal{V}}$ denote the norms in $\mathcal{U}$ and $\mathcal{V}$, respectively. The spaces $\mathcal{U} \cap \mathcal{V}$ and $\mathcal{U} + \mathcal{V}$ are also Banach spaces with respect to the norms (4.70) and (4.71), respectively.

Other equivalent norms can be defined. For instance, the following norms in $\mathcal{U} \cap \mathcal{V}$ and $\mathcal{U} + \mathcal{V}$ can also be used:

$$\begin{aligned} \|u\|_{\mathcal{U} \cap \mathcal{V}} &= \left(\|u\|_{\mathcal{U}}^2 + \|u\|_{\mathcal{V}}^2\right)^{1/2} \\ \|w\|_{\mathcal{U} + \mathcal{V}} &= \inf_{u+v=w} \left(\|u\|_{\mathcal{U}}^2 + \|u\|_{\mathcal{V}}^2\right)^{1/2} \end{aligned} \tag{4.72}$$

An intermediate space between $\mathcal{U}$ and $\mathcal{V}$, denoted $[\mathcal{U}, \mathcal{V}]$, is a Banach space such that

$$\mathcal{U} \cap \mathcal{V} \subset [\mathcal{U}, \mathcal{V}] \subset \mathcal{U} + \mathcal{V} \tag{4.73}$$

The intermediate space $[\mathcal{U}, \mathcal{V}]$ can be also provided with an appropriate norm. We describe a technique for defining such norms subsequently. At this point, some simple examples are in order.

EXAMPLE 4.9. Let us ignore, for the moment, the requirement that $\mathcal{U} \cap \mathcal{V}$ be dense in $\mathcal{U}$ and $\mathcal{V}$. Then consider as Banach spaces the planes $\mathcal{U}$ and $\mathcal{V}$ shown in Fig. 4.3. If $(\mathbf{i}, \mathbf{j}, \mathbf{k})$ denote the usual unit triad,

$$\mathcal{U} = \left\{ \mathbf{u}\text{: } \mathbf{u} = \alpha\mathbf{m} + \beta\mathbf{i};\ \mathbf{m} = \frac{1}{\sqrt{2}}(-\mathbf{j} + \mathbf{k}) \right\}$$

$$\mathcal{V} = \{ \mathbf{v}\text{: } \mathbf{v} = \lambda\mathbf{i} + \mu\mathbf{j} \}$$

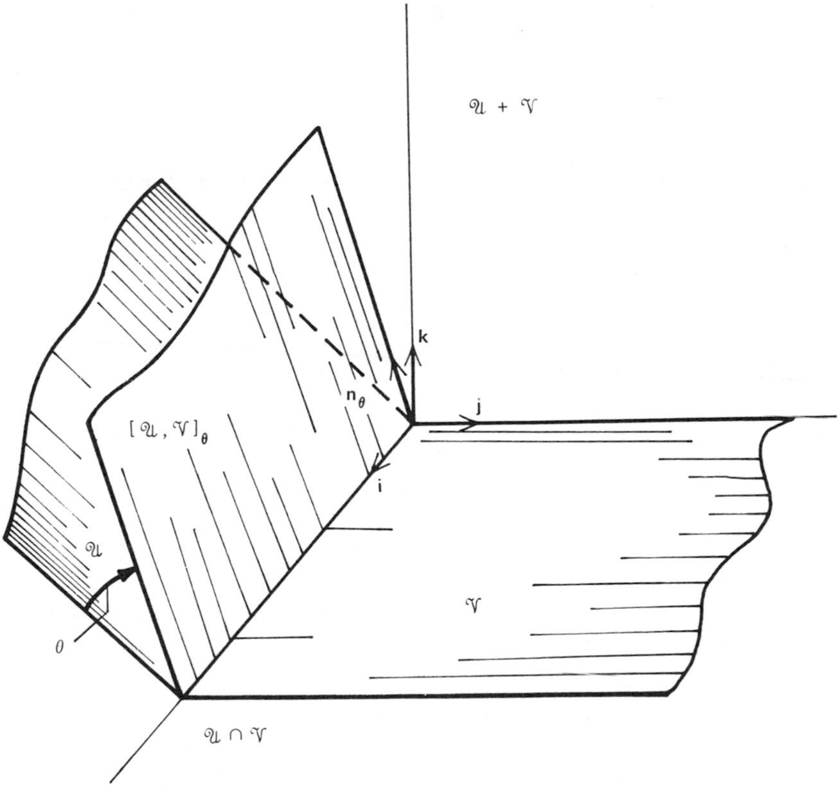

Figure 4.3 An intermediate space $[\mathcal{U}, \mathcal{V}]_\theta$.

where α, β, λ, and μ are scalars. Clearly,

$$\mathcal{U} + \mathcal{V} = \{\mathbf{w}: \mathbf{w} = \hat{\alpha}\mathbf{i} + \hat{\beta}\mathbf{j} + \hat{\gamma}\mathbf{k}\} = \mathbf{R}^3$$

$$\mathcal{U} \cap \mathcal{V} = \{\mathbf{w}: \mathbf{w} = \tilde{\alpha}\mathbf{i}\}$$

The space

$$[\mathcal{U}, \mathcal{V}]_\theta = \left\{\mathbf{w}: \mathbf{w} = \alpha\mathbf{n}_\theta + \beta\mathbf{i};\ \mathbf{n}_\theta = -\cos\left(\frac{\pi}{4} + \theta\right)\mathbf{j} + \sin\left(\frac{\pi}{4} + \theta\right)\mathbf{k}\right\}$$

clearly satisfies (4.73) for any choice of θ, $0 \leqslant \theta \leqslant 2\pi$. Geometrically, $[\mathcal{U}, \mathcal{V}]_\theta$ is the plane oriented at an angle θ relative to $\mathcal{U}$ shown in the figure. Observe that $[\mathcal{U}, \mathcal{V}]_0 = \mathcal{U}$ and $[\mathcal{U}, \mathcal{V}]_{3\pi/4} = \mathcal{V}$. ■

EXAMPLE 4.10. Let $\mathcal{U} = H^2(\Omega)$ and $\mathcal{V} = H^0(\Omega) = L_2(\Omega)$. Then $H^1(\Omega)$ is an intermediate space between $\mathcal{U} = H^2(\Omega)$ and $\mathcal{V} = H^0(\Omega)$, since $H^2(\Omega) \cap H^0(\Omega) = H^2(\Omega)$ is dense in itself and dense in $H^0(\Omega)$, and

$$H^2(\Omega) \subset H^1(\Omega) \subset H^0(\Omega)$$

In general, $H^s(\Omega)$ is an intermediate space between $H^{m_1}(\Omega)$ and $H^{m_2}(\Omega)$ for $m_2 > s > m_1 > 0$, since

$$H^{m_2}(\Omega) \subset H^s(\Omega) \subset H^{m_1}(\Omega)$$

where s is a real number > 0. ■

Interpolation Spaces

We now pass on to the closely related concept of interpolation spaces. This requires that we first describe some special classes of operators. Suppose that $\mathcal{U}$ and $\mathcal{V}$ are Banach spaces and that $\mathcal{L}(\mathcal{U} + \mathcal{V}, \mathcal{U} + FV)$ denotes, as usual, the space of bounded linear operators from $\mathcal{U} + \mathcal{V}$ into itself. Consider next a subspace $\hat{\mathcal{L}}(\mathcal{U}, \mathcal{V})$ of $\mathcal{L}(\mathcal{U} + \mathcal{V}, \mathcal{U} + \mathcal{V})$ which has the property that each element $A \in \hat{\mathcal{L}}(\mathcal{U}, \mathcal{V})$ maps $\mathcal{U}$ continuously into $\mathcal{U}$ and $\mathcal{V}$ continuously into $\mathcal{V}$. When such an operator is regarded as a mapping from $\mathcal{U}$ into $\mathcal{U}$, we can compute its norms, denoted $|A|_{\mathcal{U}}$. Likewise, $|A|_{\mathcal{V}}$ denotes its norm when considered as a mapping from $\mathcal{V}$ into $\mathcal{V}$. For each $A \in \hat{\mathcal{L}}(\mathcal{U}, \mathcal{V})$, we define

$$\|Aw\|_{\mathcal{U}+\mathcal{V}} = \inf_{w=u+v} (\|Au\|_{\mathcal{U}} + \|Av\|_{\mathcal{V}})$$

$$\leqslant \inf_{w=u+v} (|A|_{\mathcal{U}} \|u\|_{\mathcal{U}} + |A|_{\mathcal{V}} \|v\|_{\mathcal{V}})$$

This implies that

$$\|A\|_{\mathcal{L}(\mathcal{U}+\mathcal{V},\,\mathcal{U}+\mathcal{V})} \leqslant \max(|A|_{\mathcal{U}}, |A|_{\mathcal{V}})$$

and, for $u \in \mathcal{U} \cap \mathcal{V}$,

$$\|Au\|_{\mathcal{U}\cap\mathcal{V}} = \max(\|Au\|_{\mathcal{U}}, \|Au\|_{\mathcal{V}})$$

Hence

$$\|A\|_{\mathcal{L}(\mathcal{U}\cap\mathcal{V},\,\mathcal{U}\cap\mathcal{V})} \leqslant \max(|A|_{\mathcal{U}}, |A|_{\mathcal{V}})$$

It can be verified that $\hat{\mathcal{L}}(\mathcal{U}, \mathcal{V})$ is a Banach space with respect to the norm

$$\|A\|_{\hat{\mathcal{L}}(\mathcal{U},\mathcal{V})} = \max(|A|_{\mathcal{U}}, |A|_{\mathcal{V}})$$

A Banach space $\mathcal{W}$ is called an *interpolation space* between $\mathcal{U}$ and $\mathcal{V}$ if $A(\mathcal{W}) \subset \mathcal{W}$ for every $A \in \hat{\mathcal{L}}(\mathcal{U}, \mathcal{V})$.

EXAMPLE 4.11. Consider again the two planes $\mathcal{U}$ and $\mathcal{V}$ described in Example 4.9. An arbitrary element $\mathbf{u} \in \mathcal{U}$ can be represented by

$$\mathbf{u} = \alpha_1\mathbf{i} + \alpha_2\mathbf{j} + \alpha_3\mathbf{k}$$

where $\alpha_2 = -\alpha_3$. Similarly, an element $\mathbf{v} \in \mathcal{V}$ is of the form

$$\mathbf{v} = \beta_1\mathbf{i} + \beta_2\mathbf{j} + \beta_3\mathbf{k} \qquad \beta_3 = 0$$

Then $\mathcal{U} + \mathcal{V}$ is the space of elements $\mathbf{w}$ of the form

$$\mathbf{w} = \gamma_1\mathbf{i} + \gamma_2\mathbf{j} + \gamma_3\mathbf{k}$$

Clearly, $\mathbf{w} \in \mathcal{U}$ if $\gamma_2 = -\gamma_3$, and $\mathbf{w} \in \mathcal{V}$ if $\gamma_3 = 0$.

Now consider the linear transformation $A \in \mathcal{L}(\mathcal{U} + \mathcal{V}, \mathcal{U} + \mathcal{V})$ represented, relative to the triad $(\mathbf{i}, \mathbf{j}, \mathbf{k})$, by the matrix

$$\mathbf{A} = \begin{bmatrix} a & b & b \\ 0 & c & 0 \\ 0 & 0 & c \end{bmatrix}$$

where a, b, and c are positive constants. Note that, for $\mathbf{u} = \alpha_1\mathbf{i} + \alpha_2\mathbf{j} + \alpha_3\mathbf{k} \in$

$\mathcal{U}$, we have

$$\mathbf{Au} = \begin{Bmatrix} a\alpha_1 \\ c\alpha_2 \\ c\alpha_3 \end{Bmatrix} \in \mathcal{U} \qquad \text{implying } A : \mathcal{U} \to \mathcal{U}$$

Also, for $\mathbf{v} = \beta_1\mathbf{i} + \beta_2\mathbf{j} + \beta_3\mathbf{k} \in \mathcal{V}$ with $\beta_3 = 0$, we have

$$\mathbf{Av} = \begin{Bmatrix} a\beta_1 + b\beta_2 \\ c\beta_2 \\ c\beta_3 \end{Bmatrix} \in \mathcal{V} \qquad \text{implying } A : \mathcal{V} \to \mathcal{V}$$

Further, for $\mathbf{w} = \gamma_1\mathbf{i} + \gamma_2\mathbf{j} + \gamma_3\mathbf{k} \in \mathcal{U} + \mathcal{V}$, we have

$$\mathbf{Aw} = \begin{Bmatrix} a\gamma_1 + b(\gamma_2 + \gamma_3) \\ c\gamma_2 \\ c\gamma_3 \end{Bmatrix} \in \mathcal{U} + \mathcal{V}$$

which implies that $A : \mathcal{U} + \mathcal{V} \to \mathcal{U} + \mathcal{V}$. Since A is finite-dimensional, it is bounded. Thus A is an element of the space $\hat{\mathcal{L}}(\mathcal{U}, \mathcal{V})$ defined earlier.

Now consider the plane

$$\mathcal{W}_\theta = \{\mathbf{w}: \mathbf{w} = \alpha\mathbf{i} + \beta\mathbf{j} + \gamma\mathbf{k}; \alpha, \gamma \geqslant 0; \beta = \gamma\cot(3\pi/4 - \theta); 0 \leqslant \theta < 3\pi/4\}$$

It can be shown that A maps $\mathcal{W}_\theta$ into $\mathcal{W}_\theta$ for θ between 0 and $3\pi/4$, and $A(\mathcal{W}_\theta) \subset \mathcal{W}_\theta$. In this case

$$\mathcal{W}_\theta = [\mathcal{U}, \mathcal{V}]_\theta, \qquad 0 \leqslant \theta < \frac{3\pi}{4}$$

where $[\mathcal{U}, \mathcal{V}]_\theta$ is the space described in Example 4.9. ■

Banach Scales and Interpolation Properties

We next introduce the concept of Banach scales and study some properties of interpolation spaces. First, we recall once again that a Banach space B_1 is said to be embedded in a Banach space B_0 if $B_1 \subset B_0$ and if there is a constant C such that $\|u\|_{B_0} \leqslant C\|u\|_{B_1}$ for all $u \in B_1$. The space B_1 is densely embedded in B_0 if B_1 is dense in B_0, and it is compactly embedded in B_0 if any bounded set in the norm of B_1 is compact in B_0. A family of Banach spaces B_θ, θ being a real parameter such that $\theta_0 \leqslant \theta \leqslant \alpha$, with norms $\|u\|_\theta$, is called a *scale of spaces* if

(i) B_α is densely embedded in B_θ when $\alpha > \theta$, and

$$\|u\|_\theta \leqslant C(\theta,\alpha)\|u\|_\alpha \tag{4.74}$$

where C is a positive number depending on θ and α but not on u.

(ii) There exists a function $C(\theta,\alpha,\beta)$, finite at all points of the domain $\theta_0 \leqslant \theta < \alpha < \beta \leqslant \theta_1$, such that

$$\|u\|_\alpha \leqslant C(\theta,\alpha,\beta)\|u\|_\theta^{(\beta-\alpha)/(\beta-\theta)}\|u\|_\beta^{(\alpha-\theta)/(\beta-\theta)} \tag{4.75}$$

for all $u \in B_\beta$, where $C(\theta,\alpha,\beta)$ is a positive number depending only on θ, α, and β but not on u.

Consider a family of Banach spaces B_θ, $\theta_0 \leqslant \theta \leqslant \theta_1$, with the property that B_α is densely embedded in B_θ for $\alpha > \theta$. The family B_θ is said to have the *interpolation property relative to the family of Banach spaces* $\hat{B}_\mu$, $\mu_0 \leqslant \mu \leqslant \mu_1$, if every linear operator T defined on B_{θ_1} and bounded from B_{θ_0} to $\hat{B}_{\mu_0}$ and from B_{θ_1} to $\hat{B}_{\mu_1}$ is also a bounded operator from B_θ to $\hat{B}_\mu$, where

$$\frac{\theta-\theta_0}{\theta_1-\theta_0} = \frac{\mu-\mu_0}{\mu_1-\mu_0} \tag{4.76}$$

The family B_θ has the *normal interpolation property relative to* $\hat{B}_\mu$ if the norm of the operator T satisfies the inequality

$$\|T\|_{\mathcal{L}(B_\theta,\hat{B}_\mu)} \leqslant \|T\|^\sigma_{\mathcal{L}(B_{\theta_0},\hat{B}_{\mu_0})}\|T\|^\nu_{\mathcal{L}(B_{\theta_1},\hat{B}_{\mu_1})} \tag{4.77}$$

where $\sigma = (\theta_1-\theta)/(\theta_1-\theta_0)$, and $\nu = (\theta-\theta_0)/(\theta_1-\theta_0)$. The family B_θ has the *strict interpolation property relative to the family* $\hat{B}_\mu$ if every subfamily of it whose indices run over an arbitrary interval $[\hat{\theta}_0,\hat{\theta}_1] \subset [\theta_0,\theta_1]$ has the normal interpolation property relative to the corresponding subfamily of $\hat{B}_\mu$ with the correspondence given by (4.76).

EXAMPLE 4.12. Let $B_p = \hat{B}_p = L_p$, $1 \leqslant p \leqslant \infty$, and consider the integral operator

$$Au(t) = \int_0^1 K(t,s)u(s)\,ds$$

from $C(0,1)$ into $C(0,1)$, which is linear and bounded.

$$\|A\|_{\mathcal{L}(C,C)} = \max_{0 \leqslant s \leqslant 1} \int_0^1 |K(t,s)|\,dt$$

This operator is a bounded operator from $L_1(0,1)$ into $L_1(0,1)$ and has the norm

$$\|A\|_{\mathcal{L}(L_1,L_1)} = \max_{0 \leqslant t \leqslant 1} \int_0^1 |K(t,s)|\,ds$$

Now if the operator A is considered an operator from $L_p(0,1)$ into $L_q(0,1)$, the inequality

$$\|A\|_{\mathcal{L}(L_p,L_q)} \leqslant \left[\max_{0 \leqslant t \leqslant 1} \int_0^1 |K(t,s)|^q\,ds \right]^{1/q}$$

$$\times \left[\max_{0 \leqslant s < 1} \int_0^1 |K(t,s)|^p\,dt \right]^{1/p}$$

where $1/p + 1/q = 1$ holds for its norm. ■

REMARK. We remark that, in general, if a linear operator A is bounded as an operator from $L_{p_1}(0,1)$ into $L_{q_1}(0,1)$ and as an operator from $L_{p_2}(0,1)$ into $L_{q_2}(0,1)$, it is bounded as an operator from $L_p(0,1)$ into $L_q(0,1)$, where

$$\frac{1}{p} = \frac{1-\theta}{p_1} + \frac{\theta}{p_2}, \qquad \frac{1}{q} = \frac{1-\theta}{q_1} + \frac{\theta}{q_2}, \qquad 0 \leqslant \theta \leqslant 1$$

Furthermore,

$$\|A\|_{\mathcal{L}(L_p,L_q)} \leqslant \|A\|^{1-\theta}_{\mathcal{L}(L_{p_1},L_{q_1})} \|A\|^{\theta}_{\mathcal{L}(L_{p_2},L_{q_2})} \tag{4.78}$$

That is, the family of Banach spaces L_p, $1 \leqslant p \leqslant \infty$, has the normal interpolation property. ■

Suppose B_{θ_0} and B_{θ_1} are two intermediate spaces, with $\theta_0 < \theta_1$, between $\mathfrak{U}$ and $\mathfrak{V}$. Then the intermediate space B_μ between B_{θ_0} and B_{θ_1} is given by the correspondence (4.76).

$$B_\mu = B_{(1-\theta)\theta_0 + \theta\theta_1}, \qquad 0 < \theta < 1$$

or, in terms of intermediate spaces,

$$\left[[U,V]_{\theta_0}, [U,V]_{\theta_1} \right]_\theta = [U,V]_{(1-\theta)\theta_0 + \theta\theta_1} \tag{4.79}$$

Property (4.79) is called *reiteration property*. For various intermediate spaces of a certain type and for various values of θ, we use (4.79) to recover a space of the same type.

If a Banach space B is embedded in one of the spaces B_ν of a family B_θ that has the interpolation property relative to the family $\hat{B}_\nu$, the operator T induces a bounded linear operator on B with range in the corresponding space $\hat{B}_\nu$. If B is normally embedded in B_ν,

$$\|T\|_{\mathcal{L}(B,\hat{B}_\nu)} \leqslant \|T\|_{\mathcal{L}(B_\nu,\hat{B}_\nu)} \tag{4.80}$$

THEOREM 4.12. Let $\{B_\alpha\}$ be a family with the normal interpolation property relative to the family $\{\hat{B}_\alpha\}$, $0 \leqslant \alpha \leqslant 1$, described above. If T is a bounded operator from B_0 to $\hat{B}_0$ and from B_1 to $\hat{B}_1$ and is compact with respect to at least one of these pairs of spaces, T is a compact operator from B_α to $\hat{B}_\alpha$ for $0 < \alpha < 1$.

PROOF. Suppose T is compact from B_1 to $\hat{B}_1$. This implies that the set $T(S_1)$, where S_1 is the unit ball in B_1, is compact in $\hat{B}_1$. Let $\{P_n\}$ be a sequence of injections such that $P_n u = u \forall u \in \hat{B}_1$ and $\|P_n\|_{\mathcal{L}(\hat{B}_0,\hat{B}_1)} < K_0 \forall n$. That is, the operators $\{P_n T\}$ converge to T in the norm $\|\cdot\|_{\mathcal{L}(B_1,\hat{B}_1)}$. Then from (4.78), as $n \to \infty$,

$$\|T - P_n T\|_{\mathcal{L}(B_\alpha,\hat{B}_\alpha)} \leqslant \|T - P_n T\|^{1-\alpha}_{\mathcal{L}(B_0,\hat{B}_0)} \|T - P_n T\|^{\alpha}_{\mathcal{L}(B_1,\hat{B}_1)}$$

$$\leqslant (1 + K_0)^{1-\alpha} \|T\|^{1-\alpha}_{\mathcal{L}(B_0,\hat{B}_0)} \|T - P_n T\|^{\alpha}_{\mathcal{L}(B_1,\hat{B}_1)} \to 0$$

Since T is the limit in the operator norm of compact operators, it is a compact operator from B_α to $\hat{B}_\alpha$ (see, e.g., [4.10]). ■

4.8 INTERPOLATION THEORY IN HILBERT SPACES

We now consider some special results of the interpolation theory developed for Banach spaces in the previous section applied to Hilbert spaces. Let $\mathcal{U}$ and $\mathcal{V}$ denote two Hilbert spaces such that $\mathcal{U} \subset \mathcal{V}$, the injection being dense and continuous, and let $(\cdot,\cdot)_{\mathcal{U}}$ and $(\cdot,\cdot)_{\mathcal{V}}$ be the inner products in $\mathcal{U}$ and $\mathcal{V}$, and $\langle\cdot,\cdot\rangle_{\mathcal{U}}$ and $\langle\cdot,\cdot\rangle_{\mathcal{V}}$ the duality pairings on $\mathcal{U}' \times \mathcal{U}$ and $\mathcal{V}' \times \mathcal{V}$, respectively. We denote by i the injection of $\mathcal{U}$ into $\mathcal{V}$ ($iu = u$), by i' its transpose ($i' : \mathcal{V}' \to \mathcal{U}'$), and by $Z_{\mathcal{U}}$ and $Z_{\mathcal{V}}$ the Riesz maps of $\mathcal{U}$ onto $\mathcal{U}'$ and $\mathcal{V}$ onto $\mathcal{V}'$, respectively. In summary, we have

$$\left.\begin{array}{c} \mathcal{U} \subset \mathcal{V} \subset \mathcal{V}' \subset \mathcal{U}' \\ \langle v', iu\rangle_{\mathcal{V}} = \langle i'v', u\rangle_{\mathcal{U}}, \qquad v' \in \mathcal{V}', \qquad i'v' \in \mathcal{U}' \\ \langle Z_{\mathcal{V}} w, v\rangle_{\mathcal{V}} = (w,v)_{\mathcal{V}} \qquad \forall v \in \mathcal{V}; \qquad \langle Z_{\mathcal{U}} z, u\rangle_{\mathcal{U}} = (z,u)_{\mathcal{U}} \qquad \forall u \in \mathcal{U} \end{array}\right\} \tag{4.81}$$

When relations such as this hold for a given $v \in \mathcal{V}$, it is possible to select a linear functional $u' \in \mathcal{U}'$ such that for some $v \in \mathcal{V}$,

$$u'(u) = (u, v)_{\mathcal{V}} \qquad \forall u \in \mathcal{U} \tag{4.82}$$

Indeed

$$(u, v)_{\mathcal{V}} = (iu, v)_{\mathcal{V}} = \langle Z_{\mathcal{V}} v, iu \rangle_{\mathcal{V}} = \langle i' Z_{\mathcal{V}} v, u \rangle_{\mathcal{U}}$$

so we can take

$$u' = i' Z_{\mathcal{V}} v$$

Now u' is a linear functional on $\mathcal{U}$. Hence, there is a $z \in \mathcal{U}$ such that $u' = Z_{\mathcal{U}} z$ and

$$u'(u) = \langle Z_{\mathcal{U}} z, u \rangle_{\mathcal{U}} = (z, u)_{\mathcal{U}} \qquad \forall u \in \mathcal{U} \tag{4.83}$$

Comparing (4.82) and (4.83), we find that there exists a map $A : \mathcal{V} \to \mathcal{U}$ such that

$$(u, Av)_{\mathcal{U}} = (u, v)_{\mathcal{V}} \qquad \forall u \in \mathcal{U} \tag{4.84}$$

In fact, it is clear that

$$A = Z_{\mathcal{U}}^{-1} i' Z_{\mathcal{V}} \tag{4.85}$$

A number of properties of this operator are summarized in the following theorem.

THEOREM 4.13. The operator A appearing in (4.84) is a linear, bounded, positive-definite operator from $\mathcal{V}$ into $\mathcal{U}$.

PROOF. A is linear and bounded because it is the composition of linear bounded operators by virtue of (4.85). It is also positive definite in the following sense: take $v = u$ in (4.84); then, for $u \in \mathcal{U}$

$$(u, Au)_{\mathcal{U}} = (u, u)_{\mathcal{V}} > 0 \qquad \forall u \neq 0$$

Finally, if u and v are in $\mathcal{U}$, we have $(u, v)_{\mathcal{V}} = (v, Au)_{\mathcal{U}} = (u, Av)_{\mathcal{U}}$. Hence, in this sense, A is self-adjoint. ■

We also note that it follows from (4.84) that there exists an operator B from $\mathcal{U}$ into $\mathcal{V}$ such that $v = Bw$, and $A^{-1} = B$. Since B is not defined on all of $\mathcal{U}$, it is an unbounded operator. Since A is positive definite, its positive square root exists. Further, using the spectral decomposition of self-adjoint operators, the powers A^{θ}, $0 \leqslant \theta \leqslant 1$, can be defined. In particular, if we set

$$T = A^{-1/2} = B^{1/2} \tag{4.86}$$

T maps $\mathcal{U}$ into $\mathcal{V}$,

$$(u,v)_{\mathcal{U}} = (u,Bv)_{\mathcal{V}} = (B^{1/2}u, B^{1/2}v)_{\mathcal{V}} = (Tu,Tv)_{\mathcal{V}}$$

for $u, v \in \mathcal{U}$, and $A^{1/2}$ maps $\mathcal{V}$ into $\mathcal{U}$ with

$$(u,v)_{\mathcal{V}} = (A^{1/2}u, A^{1/2}v)_{\mathcal{U}} = (T^{-1}u, T^{-1}v)_{\mathcal{U}}$$

From the spectral theory of linear operators, we know that A can be decomposed according to

$$A = T^{-2} = \sum_n \lambda_n P_n \tag{4.87}$$

where λ_n are the eigenvalues of A, all real, and P_n are the orthogonal projection operators onto the subspaces of eigenfunctions $\{\phi_n\}$ of A. We have

$$T = \sum_n \mu_n P_n \qquad \mu_n = \lambda_n^{-1/2} \tag{4.88}$$

We also know that $0 < \lambda_{n+1} \leqslant \lambda_n$ (or $\mu_{n+1} \geqslant \mu_n$) and $\lim_{n\to\infty}\lambda_n = 0$ (or $\lim_{n\to\infty}\mu_n = \infty$).

EXAMPLE 4.13. Let $\mathcal{U} = H_0^1(0,1)$ and $\mathcal{V} = H^0(0,1) = L_2(0,1)$. Obviously, from what we have established thus far, $H^1(0,1)$ is dense in $H^0(0,1)$. Define

$$(u,v)_{\mathcal{U}} = (u,v)_1 = \int_0^1 (uv + u'v')\,dx \qquad u,v \in H_0^1(0,1)$$

$$(u,v)_{\mathcal{V}} = (u,v)_0 = \int_0^1 uv\,dx \qquad u,v \in H^0(0,1)$$

Let ϕ_n be a sequence in $C_0^\infty(0,1)$ with a limit v in $L_2(0,1)$. Then

$$(u,A\phi_n)_{\mathcal{U}} \equiv \int_0^1 (uA\phi_n + u'(A\phi_n)')\,dx \qquad (u,\phi_n)_{\mathcal{V}} = \int_0^1 u\phi_n\,dx$$

Since A is self-adjoint, we have (distributionally) $\forall\phi_n \in C_0^\infty(0,1)$,

$$\langle A(I-D^2)u, \phi_n\rangle = \langle u, \phi_n\rangle = \int_0^1 u\phi_n\, dx$$

This relation also holds for $v = \lim_{n\to\infty}\phi_n$. Thus, $A(I-D^2)=I$ which means that, in the sense of distributions,

$$B = I - D^2$$

The inverse is given by

$$(B^{-1}v)(x) = Av(x) = \int_0^1 k(x,\xi)v(\xi)\,d\xi \qquad v \in H^0(0,1) \tag{4.89}$$

where

$$k(x,\xi) = \begin{cases} \dfrac{\sinh(\xi)\sinh(1-x)}{\sinh 1} & 0 \leqslant \xi \leqslant x \\[2ex] \dfrac{\sinh(x)\sinh(1-\xi)}{\sinh 1} & x \leqslant \xi \leqslant 1 \end{cases}$$

is the *Green's function* associated with B for the boundary conditions $u(0)=u(1)=0$. Indeed, we have, for $v \in H^0(0,1) = L_2(0,1)$,

$$u(x) = \int_0^1 k(x,\xi)v(\xi)\,d\xi = \frac{\sinh(1-x)}{\sinh 1}\int_0^x \sinh(\xi)v(\xi)\,d\xi$$
$$+ \frac{\sinh x}{\sinh 1}\int_x^1 \sinh(1-\xi)v(\xi)\,d\xi$$

It is clear that $u(0)=u(1)=0$, and

$$u'(x) = -\frac{\cosh(1-x)}{\sinh 1}\int_0^x \sinh(\xi)v(\xi)\,d\xi$$
$$+ \frac{\cosh(x)}{\sinh 1}\int_x^1 \sinh(1-\xi)v(\xi)\,d\xi$$

etc. and

$$u''(x) = -v(x) + u(x)$$

Thus $Bu(x) = u(x) - u''(x) = v(x)$, and we have proved that $B^{-1} = A$ is given by (4.89). Also, A is self-adjoint, since $k(x,\xi)=k(\xi,x)$. Since $k(x,\xi)$ is continuous and bounded,

$$\int_0^1\int_0^1 |k(x,\xi)|^2\,dx\,d\xi<\infty$$

it can be shown that A is compact. Since B is self-adjoint, we can decompose it into $B=(B^{1/2})^*(B^{1/2})$, where $(B^{1/2})^*$ is the adjoint of $B^{1/2}$. It can be easily verified that

$$B^{1/2}u=u-\frac{du}{dx}\qquad \text{and}\qquad B^{-1/2}(v)(x)=A^{1/2}u(x)=\int_0^1 g(x,\xi)v(\xi)\,d\xi$$

where

$$g(x,\xi)=\begin{cases}-e^{(x-\xi)} & 0\leqslant\xi\leqslant x\\ 0 & 0\leqslant x\leqslant\xi\leqslant 1\end{cases}$$

The eigenvalues and eigenfunctions associated with the operators B are

$$\mu_n=1+(n\pi)^2,\qquad \phi_n=2^{-1/2}\sin n\pi x$$

so that B can be represented by

$$Bu=\sum_{n=1}^{\infty}\mu_n(u,\phi_n)_0\phi_n$$

and A by

$$Au=\sum_{n=1}^{\infty}\lambda_n(u,\phi_n)_0\phi_n\qquad \lambda_n=\mu_n^{-1}$$

■

Intermediate Spaces

The intermediate space $H_\theta=[\mathcal{U},\mathcal{V}]_{1-\theta}$ between the Hilbert spaces $\mathcal{U}$ and $\mathcal{V}$ which satisfies the condition (4.73) is defined to be the domain of operator T^θ, T being defined in (4.86):

$$H_\theta=\{u:u\in\mathcal{D}(T^\theta);\ 0\leqslant\theta\leqslant 1\}\tag{4.90}$$

Clearly, we have $H_1=\mathcal{U}$ and $H_0=\mathcal{V}$. It can be shown that the inclusions

$H_1 \subset H_\theta \subset H_0$ are dense. We provide H_θ with an inner product

$$(u,v)_\theta = (T^\theta u, T^\theta v)_0 \qquad \|u\|_\theta = \|T^\theta u\|_0 < \infty \tag{4.91}$$

where $(\cdot,\cdot)_\theta$ and $(\cdot,\cdot)_0$ denote the inner products in H_θ and H_0. Then we have the following.

THEOREM 4.14. The space H_θ of (4.90) and (4.91) is a Hilbert space.

PROOF. It is sufficient to show that H_θ is complete in the norm (4.91). Let $\{u_k\}$ be a Cauchy sequence for this norm. Then for fixed θ, it follows from

$$\|u_k - u_j\|_\theta = \|T^\theta u_k - T^\theta u_j\|_0$$

that $\{T^\theta u_k\}$ is a Cauchy sequence in H_0. Since H_0 is a Hilbert space, it is complete, and therefore there exists an element $v \in H_0$ such that $T^\theta u_k \to v$ in H_0. Then $\lim A^{\theta/2} T^\theta u_k = A^{\theta/2} v = u \in H_\theta$; i.e., $\lim u_k$ is in H_θ. ■

Recall that the fractional order Sobolev spaces H^θ (with superscript) are Hilbert spaces. We proved that the spaces H_θ (with subscript) are also Hilbert spaces with respect to (4.91). We show later that these two spaces are equivalent.

Recall from the spectral analysis of self-adjoint operators that T can be decomposed into

$$Tu = \sum_{n=1}^{\infty} \mu_n P_n u = \sum_{n=1}^{\infty} \mu_n (\phi_n, u)_0 \phi_n$$

and

$$T^\theta u = \sum_{n=1}^{\infty} \mu_n^\theta (\phi_n, u)_0 \phi_n$$

whence we note that $\{T^\theta \phi_n\}$ is an orthogonal basis of H_θ. The norm (4.91) can be expressed in terms of ϕ_n, according to

$$\|u\|_\theta = \left(\sum_{n=1}^{\infty} \mu_n^{2\theta} |a_n|^2 \right)^{1/2} \qquad a_n = (\phi_n, u)_0 \tag{4.92}$$

which can be written in the alternate form

$$\|u\|_\theta = \left(\sum_{n=1}^{\infty} \mu_n^{2\theta} |a_n|^{2\theta} |a_n|^{2(1-\theta)} \right)^{1/2}$$

Using Hölder's inequality for infinite sums, we obtain

$$\|u\|_\theta^2 \leqslant \left(\sum_n^\infty |\mu_n a_n|^{2\theta p} \right)^{1/p} \left(\sum_n^\infty |a_n|^{2(1-\theta)q} \right)^{1/q}$$

where $1/p + 1/q = 1$. Let $1/p = \theta$ and $1/q = 1-\theta$, so that

$$\|u\|_\theta \leqslant \left(\sum_{n=1}^\infty |\mu_n a_n|^2 \right)^{\theta/2} \left(\sum_{n=1}^\infty |a_n|^2 \right)^{(1-\theta)/2}$$

But we have from (4.90) and (4.91) that

$$\|u\|_{\mathcal{U}} = \|Tu\|_{\mathcal{V}} = \|u\|_1 = \left(\sum_n^\infty \mu_n^2 |a_n|^2 \right)^{1/2}$$

$$\|u\|_{\mathcal{V}} = \|T^0 u\|_{\mathcal{V}} = \|u\|_0 = \left(\sum_n^\infty |a_n|^2 \right)^{1/2}$$

Thus $\|u\|_\theta$ satisfies the interpolation inequality

$$\|u\|_\theta \leqslant \|u\|_{\mathcal{U}}^\theta \|u\|_{\mathcal{V}}^{1-\theta} \tag{4.93}$$

In view of (4.93), H_θ is often denoted by

$$H_\theta = \mathcal{U}^\theta \mathcal{V}^{1-\theta} \tag{4.94}$$

The spaces $\mathcal{U}^\theta$ and $\mathcal{V}^{1-\theta}$ are called the *interpolation spaces* between $\mathcal{U}$ and $\mathcal{V}$, and the space H_θ whose norm satisfies the interpolation inequality (4.93) is called a *space of order* $1-\theta$.

EXAMPLE 4.14. This is a continuation of Example 4.13. Using the results obtained in that example, we have

$$T^\theta u = B^{\theta/2} u = \sum_{n=1}^\infty \frac{1}{\sqrt{2}} (1 + n^2\pi^2)^{\theta/2} a_n \sin n\pi x$$

where

$$a_n = (u, \phi_n)_0 = \frac{1}{\sqrt{2}} \int_0^1 u(x) \sin nx \, dx$$

Thus

$$\|u\|_{H^{\theta}(0,1)} = \|T^{\theta}u\|_{H^{0}(0,1)}$$

$$= \left[\int_0^1 \sum_{n=1}^{\infty} \tfrac{1}{2}(1+n^2\pi^2)^{\theta} a_n^2 \sin^2 n\pi x\, dx\right]^{1/2}$$

$$= \left[\sum_{n=1}^{\infty} (1+n^2\pi^2)^{\theta} a_n^2\right]^{1/2}$$

Observe that, when $\theta=0$, $\|u\|^2_{H^{\theta}(0,1)} = \Sigma_n a_n^2 = \|u\|^2_{H^0(0,1)}$, as expected. ■

EXAMPLE 4.15. (Lions and Magenes [4.1]). Let $\mathcal{U}$ and $\mathcal{V}$ be two Hilbert spaces with $\mathcal{U}$ dense in $\mathcal{V}$. Then the correspondence

$$(u,v)_{\mathcal{U}} = (Bu,v)_{\mathcal{V}}$$

defines an unbounded, self-adjoint, and strictly positive operator S on $\mathcal{V}$. Now let $\Lambda = B^{1/2} = T^{-1}$. The intermediate space $H_{1-\theta}$ is defined to be the domain of $\Lambda^{1-\theta}$ and is denoted, alternatively by $[\mathcal{U},\mathcal{V}]_{\theta}$, $0 \leqslant \theta \leqslant 1$. The norm in $[\mathcal{U},\mathcal{V}]_{\theta}$ is defined by the norm of the graph $G_r(\Lambda^{1-\theta})$ of $\Lambda^{1-\theta}$:

$$\|u\|_{[\mathcal{U},\mathcal{V}]_{\theta}} = \left(\|u\|^2_{\mathcal{V}} + \|\Lambda^{1-\theta}u\|^2_{\mathcal{V}}\right)^{1/2}$$

Clearly, $[\mathcal{U},\mathcal{V}]_0 = \mathcal{U}$ and $[\mathcal{U},\mathcal{V}]_1 = \mathcal{V}$. Moreover, $[\mathcal{U},\mathcal{V}]_{\theta}$ is a space of order θ between $\mathcal{U}$ and $\mathcal{V}$. To see this, first note that $\|u\|_{\mathcal{U}} = \|\Lambda u\|_{\mathcal{V}}$ for $u \in \mathcal{U}$, and $\|u\|_{\mathcal{V}} = \|\Lambda^{-1}u\|_{\mathcal{U}}$ for $u \in \mathcal{V}$. We have, from (4.93),

$$\|\Lambda^{1-\theta}u\|_{\mathcal{V}} \leqslant \|\Lambda u\|^{1-\theta}_{\mathcal{V}} \|u\|^{\theta}_{\mathcal{U}}$$

Consequently,

$$\|u\|_{[\mathcal{U},\mathcal{V}]_{\theta}} = \left(\|u\|^2_{\mathcal{V}} + \|\Lambda^{1-\theta}u\|^2_{\mathcal{V}}\right)^{1/2}$$

$$\leqslant \left(\|u\|^{2(1-\theta)}_{\mathcal{V}} + \|u\|^{2(1-\theta)}_{\mathcal{U}}\right)^{1/2} \|u\|^{\theta}_{\mathcal{V}}$$

$$\leqslant \|u\|^{\theta}_{\mathcal{V}} \left(\|\Lambda^{-1/2}\|^{2(1-\theta)} + 1\right)^{1/2} \|u\|^{1-\theta}_{\mathcal{U}}$$

$$\leqslant C\|u\|^{\theta}_{\mathcal{V}} \|u\|^{1-\theta}_{\mathcal{U}}$$

where $C = \sqrt{\|\Lambda^{-1/2}\|^{2(1-\theta)} + 1}$, and $\|\Lambda^{-1/2}\|$ is the operator norm induced by the topology of Λ. ■

Hilbert Scale of Spaces

Let H_0 and H_1 be two Hilbert spaces, with $H_1 \subset H_0$, H_1 be dense in H_0, and T be the unbounded, self-adjoint, positive definite operator from H_1 into H_0 defined by (4.86). We showed in Theorem 4.14 that the space $H_\theta (\theta \geqslant 0)$ defined in (4.90) is a Hilbert space with respect to the inner product (4.91). An inner product can be introduced in H_0 for $\theta < 0$ using the formula (4.89). Then the completion of H_0 with respect to the corresponding norm is denoted $H_\theta (\theta < 0)$.

The continuum of Hilbert spaces $\{H_\theta\}$, $-\infty < \theta < \infty$, is called a *Hilbert scale of spaces*. Moreover, the Hilbert scale generated by an operator, connecting two Hilbert spaces H_0 and H_1, $H_1 \subset H_0$ and H_1 being dense in H_0, is uniquely defined.

We list in the following theorem several properties of Hilbert scales (see e.g. [4.1]).

THEOREM 4.15. Let $\{H_\alpha\}$, $-\infty < \alpha < \infty$, be a Hilbert scale of spaces. Then it has the following properties.

(i) If $\alpha < \beta$, $H_\beta \subset H_\alpha$ with H_β dense in H_α, and

$$\|u\|_\alpha \leqslant \|u\|_\beta$$

(ii) If $\alpha < \beta < \gamma$, the norms in H_α, H_β, and H_γ satisfy the inequality

$$\|u\|_\beta \leqslant \|u\|_\alpha^{(\gamma-\beta)/(\gamma-\alpha)} \|u\|_\gamma^{(\beta-\alpha)/(\gamma-\alpha)} \qquad \forall u \in H_\gamma$$

(iii) The spaces H_α and $H_{-\alpha}$ are mutually conjugate (or dual) with respect to the inner product in H_0:

$$|(u,v)_0| \leqslant \|u\|_\alpha \|v\|_{-\alpha}, \qquad u \in H_\alpha,\ v \in H_{-\alpha}$$

(iv) The following *reiteration property* holds (see Example 4.14 and Lions and Magenes [4.1]):

$$\left([H_1, H_0]_{\theta_0}, [H_1, H_0]_{\theta_1} \right)_\theta = [H_1, H_0]_{\theta\theta_1 + (1-\theta)\theta_0} \tag{4.95}$$

where $[H_1, H_0]_{1-\theta} \equiv H_\theta$, and $\theta \in [0,1]$. ■

EXAMPLE 4.16. Suppose $H_1 = L_2(\mathsf{R}^n)$, and let $\hat{u}$ denote the Fourier transform of a function $u \in L_2(\mathsf{R}^n)$. Define the operator T by

$$Tu(\xi) = (1 + |\xi|^2)^{1/2} \hat{u}(\xi)$$

Then the Hilbert scale H_α for $\alpha \geqslant 0$ generated by T consists of functions $u(\mathbf{x})$ for which

$$\|u\|_\alpha^2 = \int_{\mathbf{R}^n} (1+|\xi|^2)^\alpha |\hat{u}(\xi)|^2 d\xi < \infty$$

For $\alpha < 0$ the Hilbert spaces H_α are obtained by the completion of $L_2(\mathbf{R}^n)$ with respect to the above norm. ■

It is clear from (4.9) and (4.10) that H_α in the above example coincides with $\hat{H}^\alpha(\mathbf{R}^n)$ which, for $\alpha > 0$, is equivalent to the Sobolev space $H^\alpha(\mathbf{R}^n)$. These observations lead to the following theorem (see Berezanskii [4.3, p. 243]).

THEOREM 4.16. For each $m > 0$, there exists a Hilbert scale of spaces $\{H_\alpha\}$, $-\infty < \alpha < \infty$, such that for $-m \leqslant s \leqslant m$ the norms in $H_{s/m}$ and $H^s(\mathbf{R}^n)$ are equivalent. ■

Hilbert Spaces $H^s(\Omega)$, Real s

Now we can define the Hilbert spaces $H^s(\Omega)$ for real values of s on a smooth bounded domain Ω using the reiteration property (iv) of Theorem 4.15. Since $H^m(\Omega)$ and $H^0(\Omega) = L_2(\Omega)$ are defined for $\Omega \subset \mathbf{R}^n$, we set $H_0 = H^m(\Omega)$, $H_1 = H^0(\Omega)$,

$$H^s(\Omega) \equiv \left[H^m(\Omega), H^0(\Omega)\right]_{1-\theta} \qquad s = \theta m \tag{4.96}$$

where m is an integer. It can be shown that Theorem 4.16 also holds for $H^s(\Omega)$ (see [4.3]).

As usual, we associate $H^{-s}(\Omega)$, $s \geqslant 0$, with the dual of $H_0^s(\Omega)$:

$$H^{-s}(\Omega) = (H_0^s(\Omega))' \tag{4.97}$$

4.9 HILBERT SPACES $H^s(\partial\Omega)$

It is also possible to define Hilbert spaces containing functions defined on the boundary $\partial\Omega$ of an open bounded domain Ω, provided Ω is sufficiently smooth. Recall that $\partial\Omega$ is said to be *of class* C^m if for each point $\mathbf{x}_0 \in \partial\Omega$ there exists a ball $B_\epsilon(\mathbf{x}_0)$ with center at $\mathbf{x}_0$ such that $\partial\Omega \cap B_\epsilon(\mathbf{x}_0)$ can be represented in the form

$$x_n = f_n(x_1, x_2, \ldots, x_{n-1})$$

with $f_n \in C^m$. We say Ω is *smooth* if it is an open bounded subset of $\mathbf{R}^n$, if its boundary $\partial\Omega$ is a $(n-1)$-dimensional infinitely differentiable manifold, and if it lies locally on one side of $\partial\Omega$.

Local Mappings

Consider a smooth domain $\Omega \subset \mathbf{R}^n$ with boundary $\partial\Omega$. There exists an open covering of $\partial\Omega$ by a finite number of bounded open subsets Θ_j, $1 \leqslant j \leqslant \nu$, of $\mathbf{R}^n$ and an infinitely differentiable mapping

$$\phi_j: \Theta_j \to Q = \{\mathbf{y} = \phi_j(\mathbf{x}): \mathbf{y} = (\mathbf{y}_0, y_n) \in \mathbf{R}^n \quad \text{such that} \quad |\mathbf{y}_0| < 1; -1 < y_n < 1\}$$

and the inverse ϕ_j^{-1} of ϕ_j is an infinitely differentiable mapping $\phi_j^{-1}: Q \to \Theta_j$. The mapping ϕ_j has the properties

$$\begin{aligned} &\text{(i)} \qquad \phi_j: \Theta_j \cap \Omega \to Q_+ = \{\mathbf{y} \in Q; y_n > 0\} \\ &\text{(ii)} \qquad \phi_j: \Theta_j \cap \partial\Omega \to Q_0 = \{\mathbf{y} \in Q; y_n = 0\} \end{aligned} \tag{4.98}$$

Further, if $\Theta_j \cap \Theta_i$ is not empty, there exists an infinitely differentiable homeomorphism J_{ij} of $\phi_i(\Theta_i \cap \Theta_j)$ onto $\phi_j(\Theta_i \cap \Theta_j)$, with positive Jacobian J_{ij}, such that

$$\phi_j(\mathbf{x}) = J_{ij}(\phi_i(\mathbf{x})) \qquad \forall \mathbf{x} \in \Theta_i \cap \Theta_j$$

There exists also an open subset Θ_0 of Ω, contained in Ω $(\overline{\Theta}_0 \subset \Omega)$, such that $\{\Theta_i\}_{i=0}^{\nu}$ is a covering of $\overline{\Omega}$ (see Fig. 4.4).

Now let $\{\psi_j\}_{j=0}^{\nu}$ denote a partition of unity subordinate to the covering $\{\Theta_j\}_{j=0}^{\nu}$; i.e.,

$$\begin{aligned} &\text{(i)} \qquad \psi_j(\mathbf{x}) \in C_0^\infty(\Theta_j) \qquad 0 \leqslant j \leqslant \nu \\ &\text{(ii)} \qquad \sum_{j=0}^{\nu} \psi_j(\mathbf{x}) = 1 \qquad \text{and} \qquad \psi_j(\mathbf{x}) \geqslant 0; \mathbf{x} \in \Theta_j \end{aligned}$$

Then, if u is an element of $H^m(\Omega)$, we have

$$u = \sum_{j=0}^{\nu} \psi_j u \qquad \text{and} \qquad u^2 = \sum_{j=0}^{\nu} \left(\sqrt{\psi_j}\, u\right)^2$$

We next introduce some special notation to reflect the fact that certain

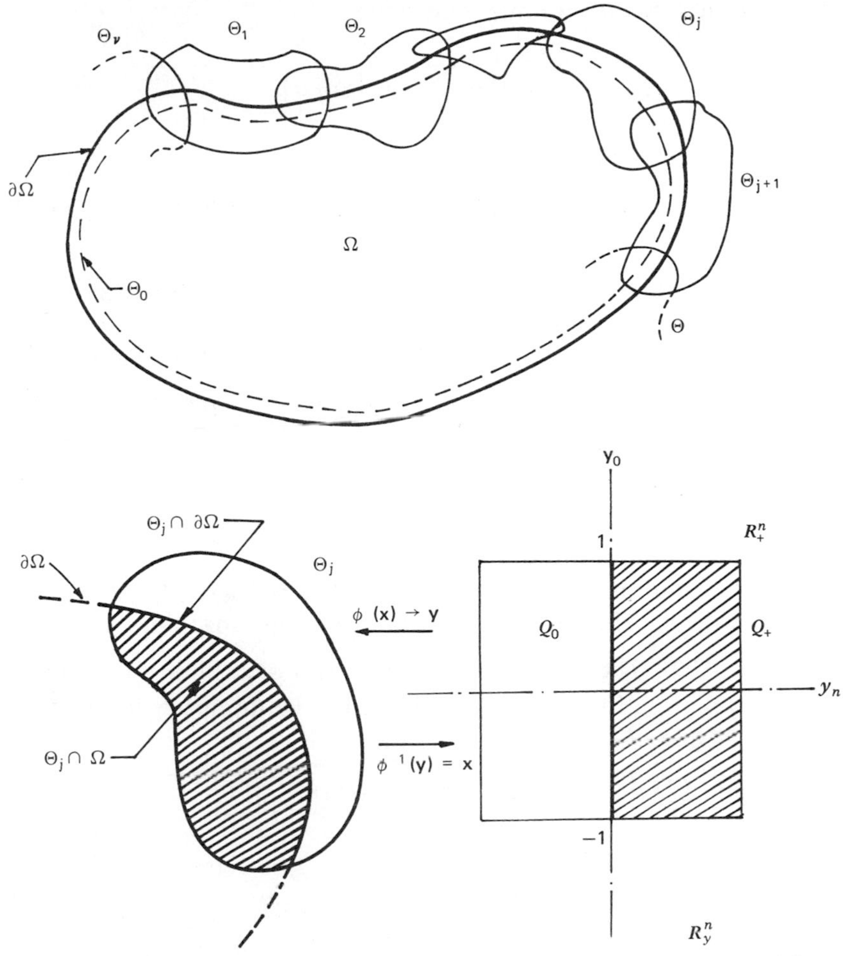

Figure 4.4 Domains of local maps of segments of a smooth boundary $\partial\Omega$ of a smooth region Ω.

terms in (4.98) involve functions of points restricted to the sets $\bar{\Omega}\cap\Theta_j$. Indeed, if $f(\mathbf{x})$ is a function of such points, we describe this fact by means of a function

$$w_j(f)(\mathbf{y})\equiv f\left(\phi_j^{-1}(\mathbf{y})\right) \tag{4.99}$$

$\mathbf{y}\in Q$. Then, if $g(\mathbf{y})$ is a function of points in Q, which are images of points $\mathbf{x}$ in $\bar{\Omega}\cap\Theta_j$, we write

$$w_j^*(g)(\mathbf{x})\equiv g(\phi_j(\mathbf{x})) \tag{4.100}$$

Collecting these results, we may now write

$$u=\sqrt{\psi_0}\left(\sqrt{\psi_0}\,u\right)+\sum_{j=1}^{\nu}\sqrt{\psi_j}\left(\sqrt{\psi_j}\,u\right)$$

$$=\sqrt{\psi_0}\left(\sqrt{\psi_0}\,u\right)+\sum_{j=1}^{\nu}\sqrt{\psi_j}\,w_j^*\left[w_j\left(\sqrt{\psi_j}\,u\right)(\mathbf{y})\right] \tag{4.101}$$

Since ψ_j has compact support in Θ_j, $w_j\left(\sqrt{\psi_j}\,u\right)$ has compact support in Q. We can therefore define $w_j\left(\sqrt{\psi_j}\,u\right)$ on the whole of $\mathbf{R}^n_+$ by extending it by zero outside Q; i.e., $w_j\left(\sqrt{\psi_j}\,u\right)$ belongs to $H^m(\mathbf{R}^n_+)$ with compact support in $Q_+=\{\mathbf{y}:\mathbf{y}\in Q;\, y_n\geqslant 0\}$. Then the operator

$$\psi u\equiv\left(\sqrt{\psi_0}\,u, w_1\sqrt{\psi_1}\,u,\ldots,w_\nu\sqrt{\psi_\nu}\,u\right)$$

is seen to be an isomorphism from $H^m(\mathbf{R}^n)$ into $H^m(\mathbf{R})\times\prod_{j=1}^{\nu}H^m(\mathbf{R}^n_+)$.

We now define Hilbert space $H^s(\partial\Omega)$ using the mappings defined in (4.97) and (4.99). The collection of functions u such that $w_j(\sqrt{\psi_j}\,u)(\mathbf{y})\in H^s(\mathbf{R}^{n-1}_{\mathbf{y}_0})$, $1\leqslant j\leqslant\nu$, for all $\mathbf{y}\in Q_0$, is defined to be the space $H^s(\partial\Omega)$:

$$H^s(\partial\Omega)=\left\{u:\ w_j\left(\sqrt{\psi_j}\,u\right)(\mathbf{y}_0,0)\in H^s\left(\mathbf{R}^{n-1}_{\mathbf{y}_0}\right);\, j=1,\ldots,\nu\right\} \tag{4.102}$$

Here, s is any real number. The norm in $H^s(\partial\Omega)$ is defined by

$$\|u\|_{H^s(\partial\Omega)}=\left[\sum_{j=1}^{\nu}\left\|w_j\left(\sqrt{\psi_j}\,u\right)(\mathbf{y}_0,0)\right\|^2_{H^s\left(\mathbf{R}^{n-1}_{\mathbf{y}_0}\right)}\right]^{1/2} \tag{4.103}$$

We remark here that definition (4.102) does not depend on the choice of Θ_j, ϕ_j, and ψ_j, a fact that is not obvious from the definition. However, the norm (4.103) does depend on the system $\{\Theta_j,\phi_j,\psi_j\}$, and different norms (4.103) are equivalent. It can be easily verified that $H^s(\partial\Omega)$ is a Hilbert space with respect to the norm (4.103). We also note that $C_0^\infty(\partial\Omega)$ is dense in $H^s(\partial\Omega)$ for $s\geqslant 0$.

Analogous to the definition (4.97), we have

$$(H^s(\partial\Omega))'=H^{-s}(\partial\Omega)\qquad s\geqslant 0 \tag{4.104}$$

and a counterpart of (4.95) of the form

$$[H^{s_1}(\partial\Omega), H^{s_2}(\partial\Omega)]_{1-\theta} = H^{\theta s_1 + (1-\theta)s_2}(\partial\Omega), \qquad s_1 > s_2 \quad (4.105)$$

Proofs of above assertions can be found in Lions and Magenes [4.1, p. 36]; see also Aubin [4.8].

EXAMPLE 4.17. (see [4.1]) Suppose that $\Delta_{\partial\Omega}$ is the Laplace operator defined on the boundary $\partial\Omega$. Then $H^2(\partial\Omega)$ is defined by

$$H^2(\partial\Omega) = \{u : u \in L_2(\partial\Omega);\ \Delta_{\partial\Omega} u \in L_2(\partial\Omega)\}$$

Norms in $H^2(\partial\Omega)$ can be defined by

$$\|u\|^2_{H^2(\partial\Omega)} = \|u\|^2_{L_2(\partial\Omega)} + \|\Delta_{\partial\Omega} u\|^2_{L_2(\partial\Omega)}$$

which is equivalent to (4.103), with $s = 2$.

Now we can define $H^s(\partial\Omega)$ for any real s using the reiteration formula (4.105) and (4.95): $H^s(\partial\Omega) = \text{domain of } (-\Delta_{\partial\Omega})^s$. Let e_j be the eigenfunctions of the problem $-\Delta_{\partial\Omega} e_j = \lambda_j e_j$, where e_j are orthonormalized in $H^0(\partial\Omega)$. If u is a distribution on $\partial\Omega$ ($u \in (C_0^\infty(\partial\Omega))'$) and a_j is its Fourier coefficient relative to $\{e_j\}$, then

$$H^s(\partial\Omega) = \left\{ u : u \in (C_0^\infty(\partial\Omega))'; \quad \left(\sum_{j=1}^{\infty} \lambda_j^{2s} |a_j|^2 \right)^{1/2} < \infty \right\}$$

■

4.10 THE TRACE THEOREM FOR $H^s(\Omega)$

Recall that we proved the trace theorem, Theorem 4.11, in Section 4.6 for $\Omega = \mathbf{R}^n_+$. We now use the local map concept to extend it to the case in which Ω is a smooth bounded subset of $\mathbf{R}^n$.

First, we give the trace theorem for integer-order Hilbert spaces $H^m(\Omega)$. Let γ_k denote the trace operator $\gamma_k u \equiv (\partial^k u / \partial n^k)(\mathbf{x}_0, x_n)|_{x_n = 0}$ defined on $\partial\Omega \subset \mathbf{R}^{n-1}$, n being the vector normal to the boundary $\partial\Omega$.

THEOREM 4.17 (The Trace Theorem for $H^m(\Omega)$). Let Ω be a smooth bounded open subset of $\mathbf{R}^n$ and $\partial\Omega$ be its boundary. Then the trace operators γ_k, $0 \leqslant k \leqslant m-1$, can be extended to continuous linear operators mapping $H^m(\Omega)$ *onto* $H^{m-k-1/2}(\partial\Omega)$. Further, the operator $\gamma = (\gamma_0, \gamma_1, \ldots, \gamma_{m-1})$ is a continuous linear mapping of $H^m(\Omega)$ *onto* $\Pi_{k=0}^{m-1} H^{m-k-1/2}(\partial\Omega)$.

Moreover, there exists a continuous linear (right) inverse γ^{-1} mapping $\Pi_{k=0}^{m-1} H^{m-k-1/2}(\partial\Omega)$ into $H^m(\Omega)$.

PROOF. Suppose $u \in \hat{C}^\infty(\Omega)$. Using the local mappings defined in (4.98) and (4.99), we choose the local coordinates such that

$$\frac{\partial}{\partial n}\left(\sqrt{\psi_j}\, u\right) = \frac{\partial}{\partial y_n}\left(w_j \sqrt{\psi_j}\, u\right)\bigg|_{y_n=0}$$

Now using (4.98), the definition of $H^s(\partial\Omega)$, and Theorem 4.11, we can prove that γ_k is a continuous linear mapping. Since $\hat{C}^\infty(\Omega)$ is dense in $H^m(\Omega)$, γ_k can be extended to $H^m(\Omega)$. The surjectivity follows from Theorem 4.11.

To prove that γ has a right inverse, note that $H^m(\Omega)$ can be decomposed into

$$H^m(\Omega) = H^m_\perp \oplus \mathcal{N}(\gamma)$$

where $\mathcal{N}(\gamma)$ is the null space of γ, and $H^m_\perp(\Omega)$ is the orthogonal complement of $\mathcal{N}(\gamma)$ with respect to the inner product in $H^0(\Omega)$:

$$H^m_\perp(\Omega) = \{u\colon u \in H^m(\Omega);\ (u,v)_0 = 0;\ \forall v \in \mathcal{N}(\gamma)\}$$

The mapping

$$\gamma : H^m_\perp(\Omega) \to \prod_{j=0}^{m-1} H^{m-j-1/2}(\partial\Omega)$$

is one-to-one. Hence the right inverse of γ exists. The continuity of γ^{-1} follows from the Banach theorem (see Theorem 5.2). ■

We have the following corollary to Theorem 4.17.

COROLLARY 4.17.1. Let the conditions of Theorem 4.17 hold. Then there exist constants C_1 and C_2 such that

$$C_1 \|u\|_{H^m(\Omega)} \leqslant \|\gamma_j u\|_{H^{m-j-1/2}(\partial\Omega)} \leqslant C_2 \|u\|_{H^m(\Omega)} \tag{4.106}$$

for all j, $0 \leqslant j \leqslant m-1$. ■

For real s, we defined Hilbert spaces $H^s(\Omega)$ and $H^s(\partial\Omega)$ in the previous sections [see (4.96) and (4.102)]. For functions in $H^s(\Omega)$, the trace theorem takes the alternate form.

THEOREM 4.18 (The Trace Theorem for $H^s(\Omega)$). Let Ω be a smooth open bounded domain in $\mathbb{R}^n$ with boundary $\partial\Omega$. Then the mapping $u \to \{\gamma_j u\colon j=0,1,\ldots,\mu\}$ extends by continuity to a continuous linear mapping γ of

$$H^s(\Omega) \quad \text{onto} \quad \prod_{j=0}^{\mu} H^{s-j-1/2}(\partial\Omega) \tag{4.107}$$

where μ is the greatest integer such that

$$\mu < s - \tfrac{1}{2} \tag{4.108}$$

Moreover, the mapping γ is surjective, and there exists a right inverse of

$$\prod_{j=0}^{\mu} H^{s-j-1/2}(\partial\Omega) \quad \text{onto} \quad H^s(\Omega) \tag{4.109}$$

PROOF. The proof, involving the use of partition of unity, local mappings, and Fourier transforms, is given in Lions and Magenes [4.1, p. 42]. ■

COROLLARY 4.18.1. There exist constants $C_1, C_2 > 0$ such that

$$C_1\|u\|_{H^s(\Omega)} \leqslant \|\gamma_j u\|_{H^{s-j-1/2}(\partial\Omega)} \leqslant C_2\|u\|_{H^s(\Omega)} \tag{4.110}$$

for all j, $0 \leqslant j \leqslant \mu < s - \frac{1}{2}$. ■

EXAMPLE 4.18. Let $u \in H^s(\Omega)$, $s > \frac{1}{2}$. Then there exists a trace of the function u on $\partial\Omega$ and $C > 0$ such that

$$\|\gamma_0 u\|_{H^{s-1/2}(\partial\Omega)} \leqslant C\|u\|_{H^s(\Omega)} \tag{4.111}$$

Further, if $s > \frac{3}{2}$, there exists a trace $\partial u/\partial n$ on $\partial\Omega$ and

$$\left\|\frac{\partial u}{\partial n}\right\|_{H^{s-3/2}(\partial\Omega)} \leqslant C\|u\|_{H^s(\Omega)} \tag{4.112}$$

Clearly, (4.111) and (4.112) are special cases of (4.110), with $j=0$ and $j=1$, respectively. ■

REFERENCES

4.1. Lions, J. L. and Magenes, E., *Non-Homogeneous Boundary Value Problems and Applications I*, Translated from a revision of the 1968 French edition by P. Kenneth, Springer-Verlag, New York, 1972.

4.2. Krein, S. G. and Petunin, Y. I., "Scales of Banach Spaces," *Russ. Math. Surv.*, Vol. 21, No. 2, pp. 85–160, 1966.

4.3. Berezanskii, Ju. M., *Expansions in Eigenfunctions of Self-adjoint Operators*, Translations of Mathematical Monographs, Vol. 17, American Mathematical Society, Providence, R. I., 1968.

4.4. Nečas, J., *Les Méthodes Directes en Théorie des Équations Elliptiques*, Masson, Paris, 1967.

4.5. Aronszajn, N., "Boundary Value of Functions with Finite Dirichlet Integral," Conference on Partial Differential Equations, No. 14, University of Kansas, Lawrence, 1955.

4.6. Sloboditskii, M. I., "Generalized Sobolev Spaces and their Application to Boundary Problems for Partial Differential Equations," *Amer. Math. Soc. Trans.*, Vol. 57, No. 2, pp. 207–275, 1966.

4.7. Adams, R. A., *Sobolev Spaces*, Academic Press, New York, 1975.

4.8. Aubin, J. P., *Approximation of Elliptic Boundary-Value Problems*, Wiley-Interscience, New York, 1972.

4.9. Treves, F., *Topological Vector Spaces, Distributions, and Kernels*, Academic Press, New York, 1967.

4.10. Yosida, K., *Functional Analysis*, 3rd ed., Springer-Verlag, New York, 1971.

5

SOME ELEMENTS OF ELLIPTIC THEORY

5.1 INTRODUCTION

This chapter is devoted to an introduction to the modern theory of elliptic boundary-value problems. We concentrate on those aspects of the theory that are useful in approximations, including certain results on the regularity of solutions. The summary account given here is based on the treatises of Lions and Magenes [5.1] and Berezanskii [5.2], wherein a more detailed development of the theory can be found.

In particular, we wish to consider a class of boundary-value problems characterized by linear partial differential equations of the form

$$\begin{aligned} Au(\mathbf{x}) &= f(\mathbf{x}) \quad \mathbf{x}\in\Omega \\ Bu(\mathbf{x}) &= g(\mathbf{x}) \quad \mathbf{x}\in\partial\Omega \end{aligned} \tag{5.1}$$

where A is a linear partial differential operator of order k in n variables:

$$\begin{aligned} A \equiv A(\mathbf{x}, D) &= \sum_{i=0}^{k} \sum_{|\alpha|=i} a_{(\alpha_1,\alpha_2,\ldots,\alpha_n)}(\mathbf{x}) D_1^{\alpha_1} D_2^{\alpha_2} \cdots D_n^{\alpha_n} \\ &= \sum_{|\alpha|\leqslant k} a_{\alpha}(\mathbf{x}) D^{\alpha} \end{aligned} \tag{5.2}$$

In (5.1), $\mathbf{x}=(x_1,x_2,\ldots,x_n)$ is a point in an open bounded set $\Omega\subset\mathbf{R}^n$ which is the domain of the real- (or sometimes complex-) valued function $u(\mathbf{x})$. The "nonhomogeneous" part of (5.1) is represented by the prescribed functions $f(\mathbf{x})\in\mathfrak{R}(A)$, $g(\mathbf{x})\in\mathfrak{R}(B)$.

The coefficients $a_{\alpha}(\mathbf{x})$ are complex-valued functions of $\mathbf{x}$, and $D_i^{\alpha_i}(D^{\alpha})$

are partial differential operators defined (as in previous chapters) by

$$D_i^{\alpha_i} = \frac{\partial^{\alpha_i}}{\partial x_i^{\alpha_i}} \qquad D^{\alpha} = \prod_{i=1}^{n} D_i^{\alpha_i} = \frac{\partial^{\alpha_1 + \alpha_2 + \cdots + \alpha_n}}{\partial x_1^{\alpha_1} \partial x_2^{\alpha_2} \cdots \partial x_n^{\alpha_n}} \tag{5.3}$$

Also, B is a linear operator, depending on A and Ω, which enters the definition of boundary conditions on $u(\mathbf{x})$ so that $Bu(\mathbf{x})$ is required to coincide with a function $g(\mathbf{x})$ prescribed on the boundary $\partial\Omega$ of Ω.

In subsequent discussions, we limit ourselves to special forms of (5.1) that pertain to *elliptic boundary-value problems*, terms which we define below. There it is also convenient to decompose the boundary operator B into various components which may involve derivatives of differing orders.

5.2 LINEAR ELLIPTIC OPERATORS

The classification of linear partial differential operators of the type in (5.2) depends only on the properties of the coefficients of derivatives of highest order [i.e., of order k for A in (5.2)]. These terms are said to constitute the *principal part* of A, denoted A_0:

$$A_0(\mathbf{x}, D) = \sum_{|\boldsymbol{\alpha}|=k} a_{\boldsymbol{\alpha}}(\mathbf{x}) D^{\boldsymbol{\alpha}} \tag{5.4}$$

Let $\boldsymbol{\xi} = (\xi_1, \xi_2, \ldots, \xi_n)$ be a vector with real components. For a fixed point $\mathbf{x}_0 \in \Omega$, we can always associate with the principal part A_0 a homogeneous polynomial in ξ_i of degree k given by

$$A_0(\mathbf{x}_0, \boldsymbol{\xi}) = \sum_{|\boldsymbol{\alpha}|=k} a_{\boldsymbol{\alpha}}(\mathbf{x}_0) \boldsymbol{\xi}^{\boldsymbol{\alpha}} \tag{5.5}$$

where it is understood that

$$\boldsymbol{\xi}^{\boldsymbol{\alpha}} = \xi_1^{\alpha_1} \xi_2^{\alpha_2} \cdots \xi_n^{\alpha_n}$$

The operator A can be characterized by determining properties of the polynomial $A_0(\mathbf{x}_0, \boldsymbol{\xi})$:

(i) A is *elliptic* at a point $\mathbf{x}_0 \in \Omega$ if

$$A_0(\mathbf{x}_0, \boldsymbol{\xi}) \neq 0 \tag{5.6}$$

for all real vectors $\boldsymbol{\xi} \neq \mathbf{0}$.

(ii) A is *properly elliptic* at point $\mathbf{x}_0 \in \Omega$ if

(ii.1) A is of even order, $k = 2m$,

(ii.2) A is elliptic at $\mathbf{x}_0$, and

(ii.3) for any pair of linearly independent vectors $\boldsymbol{\xi}$ and $\boldsymbol{\eta}$, the polynomial equation

$$A_0(\mathbf{x}_0, \lambda\boldsymbol{\xi}+\boldsymbol{\eta})=0 \tag{5.7}$$

has exactly $m=k/2$ roots $\lambda_1, \lambda_2, \ldots, \lambda_m$ with positive imaginary parts.

(iii) A is *strongly elliptic* at the point $\mathbf{x}_0 \in \Omega$ if

(iii.1) A is properly elliptic at $\mathbf{x}_0$, and

(iii.2) there exists a constant $\mu_0 > 0$ such that

$$\left|\mathrm{Re}\left[A_0(\mathbf{x}_0, \boldsymbol{\xi})\right]\right| \geqslant \mu_0 |\boldsymbol{\xi}|^k \qquad \forall \boldsymbol{\xi} \tag{5.8}$$

where $|\boldsymbol{\xi}|$ is the euclidean norm,

$$|\boldsymbol{\xi}| = \left(\xi_1^2 + \xi_2^2 + \cdots + \xi_n^2\right)^{1/2}$$

In all the above definitions we set $A(\mathbf{x}, D) \equiv A$.

We say that the operator A is elliptic (or properly elliptic or strongly elliptic) in Ω if it is elliptic (or properly elliptic or strongly elliptic) for each $\mathbf{x} \in \Omega$. We say that A is *uniformly strongly elliptic* if (5.8) is satisfied for every $\mathbf{x} \in \Omega$, with $\boldsymbol{\xi} \in \mathsf{R}^n$ and μ_0 independent of $\mathbf{x}$.

EXAMPLE 5.1. Consider the second-order partial differential operator in two variables

$$A(\mathbf{x}, D) = \sum_{|\alpha| \leqslant 2} a_{\alpha}(\mathbf{x}) D^{\alpha}$$

$$\boldsymbol{\alpha} = (\alpha_1, \alpha_2) \qquad \alpha_i = 0, 1, 2 \qquad |\boldsymbol{\alpha}| = \alpha_1 + \alpha_2$$

By expanding, we see that

$$A(\mathbf{x}, D) = a_{(0,0)}(\mathbf{x}) D^{(0,0)} + a_{(1,0)}(\mathbf{x}) D^{(1,0)} + a_{(0,1)}(\mathbf{x}) D^{(0,1)}$$
$$+ a_{(2,0)}(\mathbf{x}) D^{(2,0)} + a_{(1,1)}(\mathbf{x}) D^{(1,1)} + a_{(0,2)}(\mathbf{x}) D^{(0,2)}$$

Set $x_1 = x$, $x_2 = y$, $\mathbf{x} = (x, y)$. Then

$$D^{(0,0)} = I \qquad D^{(1,0)} = \frac{\partial}{\partial x} \qquad D^{(0,1)} = \frac{\partial}{\partial y} \qquad \cdots \qquad D^{(0,2)} = \frac{\partial^2}{\partial y^2}$$

and A becomes

$$A(x,y,D)=a_{20}(x,y)\frac{\partial^2}{\partial x^2}+a_{11}(x,y)\frac{\partial^2}{\partial x\,\partial y}+a_{02}(x,y)\frac{\partial^2}{\partial y^2}$$

$$+a_{10}(x,y)\frac{\partial}{\partial x}+a_{01}(x,y)\frac{\partial}{\partial y}+a_{00}(x,y) \tag{5.9}$$

where $a_{ij}(x,y)\equiv a_{(i,j)}(\mathbf{x})$, $i,j=0,1,2$.

Let α and β be arbitrary nonzero real numbers. Then, according to (5.6), $A(x,y,D)$ is elliptic at a point $(x,y)\in\Omega\subset\mathbf{R}^2$ if

$$A_0(x,y,(\alpha,\beta))=a_{20}(x,y)\alpha^2+a_{11}(x,y)\alpha\beta+a_{02}(x,y)\beta^2\neq 0$$

We interpret this requirement more thoroughly in the next example. ■

EXAMPLE 5.2. It is informative to carry the previous example a step further so as to review the origins of the term "elliptic equation." Recall that the equation of a conic section in analytic geometry is of the form

$$ax^2+bxy+cy^2+dx+ey+f=0$$

A simple geometric argument shows that this equation describes the following types of curves:

(i) An ellipse if $b^2-4ac<0$.
(ii) A parabola if $b^2-4ac=0$.
(iii) A hyperbola if $b^2-4ac>0$.

It so happens that a quadratic form similar to the equation for a conic section is encountered in the study of linear boundary-value problems. Using the operator A in (5.9), we consider the problem

$$A(x,y,D)u(x,y)=f(x,y), \qquad (x,y)\in\Omega\subset\mathbf{R}^2$$

$$\frac{\partial u(x,y)}{\partial n}+u(x,y)=g(x,y), \qquad (x,y)\in\Gamma\subset\mathbf{R}^2$$

Here the problem is to determine, if possible, a function $u(x,y)$ such that $Au=f$ at each point $(x,y)\in\Omega$ and which satisfies prescribed conditions of the form $\partial u/\partial n+u=g$ on a smooth curve Γ in $\mathbf{R}^2$, $\partial u/\partial n$ being the directional derivative of u normal to Γ. Of course, if Ω is bounded, Γ is not necessarily the boundary $\partial\Omega$ of Ω. The prescribed values of u and of normal derivatives of order $\leqslant k-1$, k being the highest order of derivatives in A, is called the *Cauchy data* for the operator A. In this case, $k=2$.

The difficulty with the problem as now stated is that the curve Γ cannot be picked arbitrarily if a solution $u(x,y)$ is to exist. By prescribing Cauchy data, we effectively prescribe all derivatives of u of order $\leqslant 1$ at $(x_0,y_0)\in\Gamma$. With all this information, plus the differential equation $Au=f$, can the second derivatives of u be determined unambiguously at a point $(x,y)\in\Omega$ in a neighborhood of (x_0,y_0)? To resolve this question, we pick two points (x_0,y_0) and (x_1,y_1) on Γ. Then, if $x_1-x_0=\Delta x$ and $y_1-y_0=\Delta y$, we have

$$\frac{\partial^2 u(x_0,y_0)}{\partial x^2}\Delta x+\frac{\partial^2 u(x_0,y_0)}{\partial x\,\partial y}\Delta y=\frac{\partial u(x_1,y_1)}{\partial x}-\frac{\partial u(x_0,y_0)}{\partial x}+O(\varepsilon^2)$$

$$\frac{\partial^2 u(x_0,y_0)}{\partial x\,\partial y}\Delta x+\frac{\partial^2 u(x_0,y_0)}{\partial y^2}\Delta y=\frac{\partial u(x_1,y_1)}{\partial y}-\frac{\partial u(x_0,y_0)}{\partial y}+O(\varepsilon^2)$$

$$\begin{aligned}a_{20}(x_0,y_0)\frac{\partial^2 u(x_0,y_0)}{\partial x^2}&+a_{11}(x_0,y_0)\frac{\partial^2 u(x_0,y_0)}{\partial x\,\partial y}+a_{02}(x_0,y_0)\frac{\partial^2 u(x_0,y_0)}{\partial y^2}\\&=f(x_0,y_0)-a_{10}(x_0,y_0)\frac{\partial u(x_0,y_0)}{\partial x}-a_{01}(x_0,y_0)\frac{\partial u(x_0,y_0)}{\partial y}\\&\quad-a_{00}(x_0,y_0)u(x_0,y_0)\end{aligned}$$

where $O(\varepsilon^2)$ indicates terms of quadratic order in Δx and Δy. The right sides of these equations are assumed known. Hence, on taking the limit as Δx, $\Delta y\to 0$, we see that the boundary-value problem has a unique solution whenever

$$\begin{vmatrix} dx & dy & 0\\ 0 & dx & dy\\ a_{20} & a_{11} & a_{02}\end{vmatrix}=a_{02}\,dx^2-a_{11}\,dx\,dy+a_{20}\,dy^2\neq 0$$

Thus, if $a_{20}(x_0,y_0)\neq 0$, unique solutions may not exist when Cauchy data are prescribed on curves for which

$$\frac{dy}{dx}=\frac{a_{11}\pm\sqrt{a_{11}{}^2-4a_{02}a_{20}}}{2a_{20}}$$

Such curves are called *characteristic curves* of A. As in the case of conic sections, we classify A as follows:

(i) A is *elliptic* at (x_0,y_0) if $a_{11}^2-4a_{02}a_{20}<0$ (no real characteristic curves exist).

(ii) A is *parabolic* at (x_0,y_0) if $a_{11}^2-4a_{02}a_{20}=0$ (one-parameter families of characteristic curves exist).

(iii) A is *hyperbolic* at (x_0,y_0) if $a_{11}^2-4a_{02}a_{20}>0$ (two-parameter families of characteristic curves exist).

Returning to (5.6), we see that, in the present case, $\forall$ real $\boldsymbol{\xi}\neq\mathbf{0}$,

$$A_0(\mathbf{x},\boldsymbol{\xi})=a_{20}\xi_1{}^2+a_{11}\xi_1\xi_2+a_{02}\xi_2{}^2\neq 0$$

if $a_{11}{}^2<4a_{02}a_{20}$, which is precisely the classical ellipticity condition described in (i) above. Equivalently, A is elliptic when the following matrix is positive definite:

$$\mathbf{A}=\begin{bmatrix} a_{20} & \frac{1}{2}a_{11} \\ \frac{1}{2}a_{11} & a_{02} \end{bmatrix}$$

■

EXAMPLE 5.3. According to the results in the previous example, the laplacian operator

$$\nabla^2 u=\frac{\partial^2 u}{\partial x^2}+\frac{\partial^2 u}{\partial y^2}$$

and the Helmholtz operator

$$Hu=\nabla^2 u+ku=\frac{\partial^2 u}{\partial x^2}+\frac{\partial^2 u}{\partial y^2}+ku$$

$(k>0)$ are elliptic operators, whereas the diffusion operator

$$Au=\frac{\partial u}{\partial t}-\frac{\partial^2 u}{\partial x^2}$$

is parabolic and the wave operator

$$\Box^2 u=\frac{\partial^2 u}{\partial t^2}-\frac{\partial^2 u}{\partial x^2}$$

is hyperbolic. The operator

$$Mu=\frac{\partial^2 u}{\partial x^2}+(1-y)\frac{\partial^2 u}{\partial y^2}$$

is elliptic for $y<1$, parabolic at $y=1$, and hyperbolic for $y>1$. ■

EXAMPLE 5.4. The Cauchy-Riemann operator

$$A=\frac{\partial}{\partial x}+i\frac{\partial}{\partial y}, \qquad i=\sqrt{-1}$$

is an example of a first-order operator that is elliptic in R^2. ■

EXAMPLE 5.5. The laplacian operator described in Example 5.3 is also a properly elliptic operator. Here, and in the sequel, we write $A_0(\xi)$ for $A_0(\mathbf{x},\xi)$ when A_0 is independent of $\mathbf{x}$. Note that

$$A_0(\lambda\boldsymbol{\xi}+\boldsymbol{\eta})=\lambda^2(\xi_1^2+\xi_2^2)+2\lambda(\xi_1\eta_1+\xi_2\eta_2)+(\eta_1^2+\eta_2^2)$$
$$=\lambda^2|\boldsymbol{\xi}|^2+2\lambda\boldsymbol{\xi}\cdot\boldsymbol{\eta}+|\boldsymbol{\eta}|^2=0$$

whenever

$$\lambda=\frac{-\boldsymbol{\xi}\cdot\boldsymbol{\eta}\pm\sqrt{(\boldsymbol{\xi}\cdot\boldsymbol{\eta})^2-|\boldsymbol{\xi}|^2|\boldsymbol{\eta}|^2}}{|\boldsymbol{\xi}|^2}$$

According to the Schwarz inequality, $|\boldsymbol{\xi}\cdot\boldsymbol{\eta}|\leqslant|\boldsymbol{\xi}|\,|\boldsymbol{\eta}|$; hence the polynomial has exactly $1=2/2$ root with a positive imaginary part.

The laplacian is also uniformly strongly elliptic, since

$$A_0(\xi)=\xi_1^2+\xi_2^2=1\cdot|\xi|^2$$

■

EXAMPLE 5.6. The biharmonic operator

$$\nabla^4u=\frac{\partial^4u}{\partial x^4}+2\frac{\partial^4u}{\partial x^2\partial y^2}+\frac{\partial^4u}{\partial y^4}$$

is a fourth-order elliptic operator. Observe that

$$A_0(\xi)=\xi_1^4+2\xi_1^2\xi_2^2+\xi_2^4$$
$$=(\xi_1^2+\xi_2^2)^2\neq 0 \qquad \forall\xi\neq\mathbf{0}$$

This operator is also properly elliptic. This can be seen by observing that the polynomial

$$A_0(\lambda\boldsymbol{\xi}+\boldsymbol{\eta})=\left[(\lambda\xi_1+\eta_1)^2+(\lambda\xi_2+\eta_2)^2\right]^2$$

has the same roots as those corresponding to the laplacian operator in the previous example, except that the multiplicity of each root is now 2; hence there are two roots with positive imaginary parts.

Moreover, ∇^4 is obviously uniformly strongly elliptic:

$$\operatorname{Re} A_0(\xi) = (\xi_1^2 + \xi_2^2)^2 = 1 \cdot |\xi|^4$$

■

EXAMPLE 5.7. It can be shown that the operator

$$\frac{\partial^4}{\partial x^4} + \frac{\partial^4}{\partial y^4} - \frac{\partial^4}{\partial z^4} + i\left(\frac{\partial^4}{\partial x^2 \partial z^2} + \frac{\partial^4}{\partial y^2 \partial z^2}\right)$$

is properly elliptic but *not* strongly elliptic (cf. [5.1, p. 111]). ■

REMARKS. 1. For operators on $\mathbf{R}^n$, it can be shown that every elliptic operator is of even order for $n > 2$, even if the coefficients are complex-valued. The proof of this assertion (e.g., [5.1, p. 109]) involves showing that, for operators of order $k = 2m$, the number of roots of $A_0(\mathbf{x}_0, -\lambda\boldsymbol{\xi} + \boldsymbol{\eta})$ with imaginary part < 0 is the same as the number of roots of $A_0(\mathbf{x}_0, \lambda\boldsymbol{\xi} + \boldsymbol{\eta})$ with imaginary part > 0, when $A_0(\mathbf{x}_0, \lambda\boldsymbol{\xi} - \boldsymbol{\eta}) = (-1)^{2m} A_0(\mathbf{x}_0, -\lambda\boldsymbol{\xi} + \boldsymbol{\eta})$, and the common number of such roots is precisely m. This observation in fact suggests the identification of the subclass of elliptic operators described as properly elliptic.

2. From the above remark it is clear that, if $n > 2$, every elliptic operator is also properly elliptic.

3. If we restrict ourselves to elliptic operators on $\mathbf{R}^n$ with real coefficients, then, from remark 1, these operators must be of even order when $n \geqslant 2$; i.e., all elliptic operators with real coefficients are of even order when $n \geqslant 2$. However, if complex coefficients are admitted, there exist odd-order elliptic operators on $\mathbf{R}^2$, as indicated in Example 5.4. ■

5.3 BOUNDARY CONDITIONS

The boundary conditions $Bu = g$ described in (5.1) cannot of course be specified arbitrarily. They must in some meaningful way be consistent with the operator A.

Henceforth we assume, for simplicity, that the boundary $\partial\Omega$ of Ω is an infinitely differentiable, $n-1$-dimensional linear manifold with Ω locally on one side of $\partial\Omega$. The closure $\bar{\Omega}$ is then compact, with $\partial\Omega \in C^\infty$. We are primarily concerned with elliptic boundary-value problems of order $2m$ involving operators A of the form

$$Au = \sum_{|\alpha|,|\beta| \leqslant m} (-1)^{|\alpha|} D^{\alpha}\left(a_{\alpha\beta}(\mathbf{x}) D^{\beta} u\right) \tag{5.10}$$

where $a_{\alpha\beta}(\mathbf{x})$ are infinitely differentiable. For problems involving operators of this type, we are largely concerned with a general class of differential boundary operators of the form

$$B_k(\mathbf{x}, D) = \sum_{|\alpha| \leqslant q_k} b_{k\alpha}(\mathbf{x}) D^{\alpha} \qquad \mathbf{x} \in \partial\Omega; 0 \leqslant k \leqslant m-1 \tag{5.11}$$

We assume that the coefficients $b_{k\alpha}(\mathbf{x})$ are real-valued, C^{∞} functions on $\partial\Omega$. Again, the highest derivatives in each boundary operator B_k constitute its principal part:

$$B_{k0}(\mathbf{x}, D) = \sum_{|\alpha| = q_k} b_{k\alpha}(\mathbf{x}) D^{\alpha} \tag{5.12}$$

Now the system of operators (5.11) does not necessarily describe unambiguously boundary conditions on $\partial\Omega$: no limit on the value that can be assumed by the q_k is given, nor have we eliminated the possibility that two operators B_i and B_j may be equivalent or in fact identical. These ambiguities are overcome if we impose the following additional requirements:

(i) $B_{k0}(\mathbf{x}, \mathbf{n}) \neq 0$, $\forall \mathbf{x} \in \partial\Omega$, and $\forall$ vector $\mathbf{n}$ normal to the boundary at $\mathbf{x}$
(ii) $q_i \neq q_j$, $i \neq j$, $0 \leqslant i,j \leqslant m-1$
(iii) $0 \leqslant q_k \leqslant 2m-1$ (5.13)

A system of boundary operators satisfying these conditions is said to be *normal.*

To obtain an idea of what the normal conditions (5.13) mean, consider first condition (i) for the case in which $B_{k0}(\mathbf{x}, D)$ is a first-order operator on $\partial\Omega \subset \mathbf{R}^2$, which prescribes the directional derivative of $u(\mathbf{x}) \equiv u(x,y)$ at $(x,y) \in \partial\Omega$ in the direction of a vector $\boldsymbol{\nu} = (\nu_x, \nu_y)$:

$$B_k u = B_{k0} u = \nabla u \cdot \boldsymbol{\nu} = \nu_x \frac{\partial u}{\partial x} + \nu_y \frac{\partial u}{\partial y} = g_k$$

Suppose that, in addition, $u(x,y)$ is given at a point $(x_0,y_0)=\mathbf{x}_0$ on $\partial\Omega$. If $\boldsymbol{\tau}=(\tau_x,\tau_y)$ is a unit tangent to $\partial\Omega$, and if (x_1,y_1) is a point on $\partial\Omega$ arbitrarily close to (x_0,y_0), it is clear that

$$u(x_1,y_1)-u(x_0,y_0)=\Delta x\,\frac{\partial u}{\partial x}+\Delta y\,\frac{\partial u}{\partial y}+O(\varepsilon^2)$$

or, equivalently, $du/ds=\tau_x\,\partial u/\partial x+\tau_y\,\partial u/\partial y$. Thus to take $\nu_x=\tau_x$ and $\nu_y=\tau_y$ would mean that the prescribed data g_k would conflict with the tangential rate of change of u; i.e., the stated boundary conditions cannot in general be unambiguously defined when $\boldsymbol{\nu}=\boldsymbol{\tau}$. If $\mathbf{n}$ is a unit normal to $\partial\Omega$, this means that we must have

$$B_k(\mathbf{x},\mathbf{n})=\nu_x n_x+\nu_y n_y=\boldsymbol{\nu}\cdot\mathbf{n}\neq 0$$

hence condition (i). Condition (ii) in (5.13) helps to ensure that two operators that are equivalent (or identical) are not contained in the system $\{B_0,B_1,\ldots,B_{m-1}\}$, and condition (iii) merely restricts the order of the derivatives appearing in B_k to be less than the derivatives of highest order appearing in the operator A.

For normal systems of boundary operators $\{B_k\}_{k=0}^{m-1}$ it may be possible to prescribe directional derivatives normal to $\partial\Omega$ of all orders $\leqslant m-1$. Indeed, the special normal systems in which the q_i run through exactly the set of integers 0, 1, 2, ..., $m-1$ as i goes from 0 to $m-1$ is called a *Dirichlet system* (or *Dirichlet normal system*) of order m.

EXAMPLE 5.8. Suppose $\partial\Omega$ is a smooth curve in the plane $\mathbb{R}^2$. Take $m=3$ and consider the system of boundary operators

$$B_k u=\sum_{|\alpha|\leqslant q_k} b_{k\alpha}(\mathbf{x})D^{\alpha}u,\qquad 0\leqslant k\leqslant 2$$

For each k, this system may involve derivatives of all orders $\leqslant q_k$, and it may happen that $q_k\geqslant 2m$. Thus it is unlikely that such a system will have meaning when associated with an operator A of order $2m$.

Now let us suppose that $\{B_k\}_{k=0}^{m-1}$ is a normal system. Then take, for example,

$$q_0=1,\qquad q_1=2,\qquad q_2=5\qquad 2m-1=5$$

so that we have the set of operators

$$B_0 u = b_{0(0,0)} u + b_{0(1,0)} \frac{\partial u}{\partial x} + b_{0(0,1)} \frac{\partial u}{\partial y}$$

$$B_1 u = b_{1(0,0)} u + b_{1(1,0)} \frac{\partial u}{\partial x} + b_{1(0,1)} \frac{\partial u}{\partial y} + b_{1(2,0)} \frac{\partial^2 u}{\partial x^2}$$

$$+ b_{1(1,1)} \frac{\partial^2 u}{\partial x\, \partial y} + b_{1(0,2)} \frac{\partial^2 u}{\partial y^2}$$

$$B_2 u = b_{2(0,0)} u + b_{2(1,0)} \frac{\partial u}{\partial x} + b_{2(0,1)} \frac{\partial u}{\partial y} + b_{2(2,0)} \frac{\partial^2 u}{\partial x^2}$$

$$+ b_{2(1,1)} \frac{\partial^2 u}{\partial x\, \partial y} + b_{2(0,2)} \frac{\partial^2 u}{\partial y^2}$$

$$+ b_{2(3,0)} \frac{\partial^3 u}{\partial x^3} + \cdots + b_{2(0,5)} \frac{\partial^5 u}{\partial y^5}$$

■

EXAMPLE 5.9. Consider a Dirichlet system of boundary operators $\{B_k\}_{k=0}^{m-1}$, where $m=3$ and $\partial\Omega \subset \mathsf{R}^2$, as in the previous example. In this case we take

$$q_0 = 0, \qquad q_1 = 1, \qquad q_2 = 2 \qquad m-1=2$$

and obtain the operators

$$B_0 u = b_{0(0,0)} u$$

$$B_1 u = b_{1(1,0)} \frac{\partial u}{\partial x} + b_{1(0,1)} \frac{\partial u}{\partial y} + b_{1(0,0)} u$$

$$B_2 u = b_{2(2,0)} \frac{\partial^2 u}{\partial x^2} + b_{2(1,1)} \frac{\partial^2 u}{\partial x\, \partial y} + b_{2(0,2)} \frac{\partial^2 u}{\partial y^2}$$

$$+ b_{2(1,0)} \frac{\partial u}{\partial x} + b_{2(0,1)} \frac{\partial u}{\partial y} + b_{2(0,0)} u$$

Now suppose that at any point $(x,y) \in \partial\Omega$ we construct tangential and normal coordinates $(\tau(x,y),\ n(x,y))$. Take

$$b_{0(0,0)}=b_0 \qquad b_{1(1,0)}=n_x=\frac{\partial x}{\partial n} \qquad b_{1(0,1)}=n_y=\frac{\partial y}{\partial n}$$

$$b_{1(0,0)}=0 \qquad b_{2(2,0)}=n_x^2 \qquad b_{2(1,1)}=2n_xn_y$$

$$b_{2(0,2)}=n_y^2 \qquad b_{2(1,0)}=n_x\frac{\partial n_x}{\partial x}+n_y\frac{\partial n_x}{\partial y}$$

$$b_{2(0,1)}=n_x\frac{\partial n_y}{\partial x}+n_y\frac{\partial n_y}{\partial y} \qquad b_{2(0,0)}=0$$

Then

$$B_0u=b_0u$$

$$B_1u=\frac{\partial u}{\partial x}n_x+\frac{\partial u}{\partial y}n_y=\frac{\partial u}{\partial n}$$

$$B_2u=n_x\frac{\partial}{\partial x}\frac{\partial u}{\partial n}+n_y\frac{\partial}{\partial y}\frac{\partial u}{\partial n}=\frac{\partial^2 u}{\partial n^2}$$

i.e., this Dirichlet system is the system of normal operators

$$\{B_k\}_{k=0}^2=\left\{b_0u,\ \frac{\partial u}{\partial n},\ \frac{\partial^2 u}{\partial n^2}\right\}$$

■

Some additional comments are in order on the relationship of the boundary operators B_k to the normal derivatives D_n^k. Suppose that $\partial\Omega\in C^m$, $m\geqslant 1$. That is, for any point $\mathbf{x}^0\in\partial\Omega$, let there exist a ball $B_\delta(\mathbf{x}^0)$ with center at $\mathbf{x}^0$ such that $\partial\Omega\cap B_\delta(\mathbf{x}^0)$ can be represented in the form $x_n=f(x_1,x_2,\ldots,x_{n-1})$, where $f(0,\ldots,0)=0$ and $f\in C^m$. Assume that $x_n\geqslant f(x_1,x_2,\ldots,x_{n-1})$ in $B_\delta(\mathbf{x}^0)\cap\Omega$. The transformation

$$y_j=x_j \qquad j=1,\ldots,n-1$$

$$y_n=x_n-f(x_1,x_2,\ldots,x_{n-1})$$

is a C^m one-to-one map of $B_\delta(\mathbf{x}^0)\cap\overline{\Omega}$ onto a hemisphere $\overline{S}_\delta(|\mathbf{y}|\leqslant\delta, y_n\geqslant 0)$. After such a transformation, any boundary operator $B(\mathbf{x},D)$ of order

$m-1$ of the type (5.11) takes the form

$$\hat{B}(\mathbf{y},D)=\sum_{k=0}^{m-1}\tau_k(\mathbf{y},D)D_n^k \tag{5.14}$$

where $D_n^k=\partial^k/\partial y_n^k$, and $\tau_k(\mathbf{y},D)$ is an expression in partial derivatives of order $\leqslant m-k-1$ with respect to $(y_1,\cdots,y_{n-1})$. That is, the derivative $\partial/\partial y_n$ is directed along the normal to the boundary $y_n=0$ of the region S_δ, and the remaining derivatives occurring in τ_k are tangential. Recall that, if $B(\mathbf{x},D)$ belongs to a normal system, then $\tau_k(\mathbf{y},0)\neq 0$.

Now when the original system of boundary operators is a Dirichlet system of order m with sufficiently smooth coefficients, the mapping $x_j\to y_j(1\leqslant j\leqslant n-1)$, $y_n=x_n-f(x_1,\cdots,x_{n-1})$ is invertible, and it is possible to express the operators $D_n^j=\partial^j/\partial y_n^j$ in terms of the B_j (see Lions and Magenes [5.1, p. 116]) by expansions of the form

$$D_n^k=\sum_{j=0}^{k}\hat{\tau}_{kj}(\mathbf{y},D)B_j(\mathbf{y},D);\ k=0,1,\ldots,m-1 \tag{5.15}$$

where the $\hat{\tau}_{kj}$ are functions similar in form to the coefficients τ_k of (5.14); i.e. they depend on $\mathbf{y}$ and tangential derivatives only.

The following lemma is useful in the sequel.

LEMMA 5.1. Let the system $\{B_k\}_{k=0}^{m-1}$ of boundary operators in (5.11) be a Dirichlet system, with

$$b_{k\alpha}(\mathbf{x})\in C^{m-k}(\Omega)\qquad k=0,1,\ldots,m-1$$

and $\partial\Omega$ is of class C^m. If $\{g_k\}_{k=0}^{m-1}$ is any system of functions on $\partial\Omega$, $g_k\in H^{m-k-1/2}(\partial\Omega)$, there exists a function $u(\mathbf{x})\in H^m(\Omega)$ such that

$$(B_ku)(\mathbf{x})=g_k(\mathbf{x})\qquad \mathbf{x}\in\partial\Omega\qquad 0\leqslant k\leqslant m-1$$

PROOF. We use the constructions described previously to write D_n^k in the form (5.15). Then the problem of finding $u\in H^m(\Omega)$ such that $(B_ku)(\mathbf{x})=g_k(\mathbf{x})$, $\mathbf{x}\in\partial\Omega$, is equivalent to finding a u which has traces on $\partial\Omega$ satisfying

$$D_n^ku=\sum_{j=0}^{k}\hat{\tau}_{kj}(\mathbf{y},D)\,g_j\in H^{m-k-1/2}(\partial\Omega)$$

Thus, by the trace theorem, Theorem 4.17, the assertions of the lemma follow. ■

The specification of a system of boundary operators as a normal system does not impose sufficient restrictions on the system to make it completely consistent with a given elliptic operator A of order $2m$. To obtain some idea of what sort of conditions must be imposed in order to ensure perfectly compatible boundary conditions, we first consider a simple example.

EXAMPLE 5.10. Consider the elliptic boundary-value problem

$$a\frac{\partial^2 u}{\partial x^2}+2b\frac{\partial^2 u}{\partial x\,\partial y}+c\frac{\partial^2 u}{\partial y^2}=f \qquad x\in(-\infty,\infty);\, y>0$$

$$\alpha\frac{\partial u(x,0)}{\partial x}+\beta\frac{\partial u(x,0)}{\partial y}+\gamma u(x,0)=g(x)$$

with $b^2-ac<0$ for ellipticity. Thus Ω is the upper half plane in R^2, and $\partial\Omega$ is simply the x axis. Since Ω is infinite in the x direction, the use of Fourier transforms suggests itself. Let

$$\hat{u}(\xi,y)=\mathcal{F}_x u(x,y)=\frac{1}{\sqrt{2\pi}}\int_{-\infty}^{\infty}u(x,y)\exp(-i\xi x)\,dx$$

and suppose that u, $\partial u/\partial x$, and $\partial u/\partial y$ decay to zero as $|x|\to\infty$. Then, taking the Fourier transform of the given partial differential equation and boundary conditions, we reduce the problem to the solution of the system

$$\begin{aligned}&-\frac{a}{c}\xi^2\hat{u}+2\frac{b}{c}i\xi\hat{u}'+\hat{u}''=\frac{\hat{f}}{c} \qquad y>0\\ &i\xi\alpha\hat{u}(\xi,0)+\beta\hat{u}'(\xi,0)+\gamma\hat{u}(\xi,0)=\hat{g}(\xi)\end{aligned} \tag{5.16}$$

for each ξ, where $\hat{u}'=d\hat{u}/dy$. The general solution of this second-order ordinary differential equation is

$$\hat{u}(\xi,y)=\left(C_0e^{(\mu\xi/c)y}+C_1e^{-(\mu\xi/c)y}\right)\left(\cos\frac{b\xi y}{c}-i\sin\frac{b\xi y}{c}\right)+\hat{u}_p(\xi,y)$$

where

$$\mu^2=ac-b^2>0$$

and $\hat{u}_p(\xi,y)$ is any particular solution of (5.16). Since $\hat{u}(\xi,y)$ must remain finite as $y\to\infty$, we must have $C_0=0$. The coefficient $C_1(\xi)$ must be determined from the boundary condition

$$-\frac{\xi}{c}\left[\left(\mu\beta-\frac{c\gamma}{\xi}\right)+i(b\beta-\alpha c)\right]C_1(\xi)=g^*(\xi) \tag{5.17}$$

where

$$g^*(\xi)=\hat{g}(\xi)-i\xi\alpha\hat{u}_p(\xi,0)-\beta\hat{u}_p'(\xi,0)-\gamma\hat{u}_p(\xi,0)$$

Finally, the solution of the boundary-value problem is

$$u(x,y)=\mathcal{F}_x^{-1}\hat{u}(\xi,y)=\frac{1}{\sqrt{2\pi}}\int_{-\infty}^{\infty}\hat{u}(\xi,y)\exp(i\xi x)\,d\xi$$

We are not so much interested here in the solution as we are in the way the boundary conditions interplay with the given partial differential equation. To see this, we need only consider the homogeneous problem, $f=0$, $g=0$ ($\hat{u}_p=0$). Then the differential equation

$$\hat{u}''+2i\frac{b\xi}{c}\hat{u}'-\frac{a\xi^2}{c}\hat{u}=0$$

and the boundary condition

$$\beta\hat{u}'+(\gamma+i\alpha\xi)\hat{u}=0 \qquad \text{at } y=0$$

should lead to the solution

$$\hat{u}(\xi,y)=0$$

if they are compatible.

When $g^*(\xi)=0$, however, it is clear that one might obtain a coefficient $C_1(\xi)\neq 0$ when

$$\left(\mu\beta-\frac{c\gamma}{\xi}\right)^2+(b\beta-\alpha c)^2=0$$

which can be written

$$(\alpha c-b\beta)^2+\mu^2\beta^2+\frac{c\gamma}{\xi}\left(\frac{c\gamma}{\xi}-2\mu\beta\right)=0$$

A little algebra will reveal that no real ξ exists that will make this quantity vanish identically. But we may take $\xi = c\gamma/2\mu\beta$. Thus a sufficient condition of compatibility of the boundary conditions with the given partial differential equation, which involves only the coefficients in the principal parts of the operators, is that

$$(\alpha c - b\beta)^2 + \mu^2\beta^2 \neq 0 \tag{5.18}$$

This condition is always satisfied for real a, b, c, α, and $\beta \neq 0$. ■

We now expand the methodology described in Example 5.10 so as to obtain a general test for compatible boundary conditions. This can be done by the "local map-partition of unity" method used in the previous chapter to prove the trace theorem [see (4.98) or Fig. 4.4]; i.e., we identify a finite open covering $\{\Theta_j\}_{j=0}^N$ of Ω, a corresponding set of infinitely differentiable local maps $\{\phi_j\}_{j=1}^N$ satisfying (4.98), and a partition of unity $\psi_j(\mathbf{x})$ subordinate to $\{\Theta_j\}_{j=0}^N$, so that the images of functions on $\bar{\Omega} \cap \Theta_j$ in the transformed domain $\phi_j(\Theta_j) = Q$ can essentially be treated as if they are defined on the half space $\mathbb{R}_+^n$. Within each set $\phi_j(\bar{\Omega} \cap \Theta_j)$ therefore we can perform an analysis using Fourier transforms in precisely the same way as in Example 5.10. The fact that we are only interested in the local behavior of the solution in each half plane $\phi_j(\bar{\Omega} \cap \Theta_j)$ is accounted for by the properties of the functions $\psi_j(\mathbf{x})$ in the partition of unity. Recall that $\operatorname{supp}\psi_j \subset \Theta_j$; hence $u = u\sum_{j=0}^N \psi_j$, $u\psi_j(\mathbf{x}) = 0$, if $\mathbf{x} \not\in \Theta_j$.

Thus, to establish what type of conditions on A and $\{B_k\}$ this process leads to, we need only reconsider Example 5.10. In the transformed domain, we have tangential and normal coordinates $\mathbf{y}_0$ and y_n, respectively. Given operators A and $\{B_k\}$ are transformed so that they pertain to a boundary-value problem in $\mathbb{R}_+^n$ of the type in Example 5.10.

Next, we require that the solution of the ordinary differential equation and boundary condition necessarily vanishes at $y_n = \infty$. Following the reasoning surrounding (5.17), we conclude that the system is compatible if the only decaying solution is the zero function $u(y_n) = 0$. If this were not the case, it would be impossible to determine uniquely the coefficients in the general solution of the ordinary differential equation. As indicated by (5.18), sufficient conditions for the existence of a unique solution can always be limited to conditions on only the principal parts of the operators and boundary operators involved.

All these observations can be shown to lead to the following test. Let A be an elliptic operator and $\{B_k\}$ a system of boundary operators. The system $\{B_k\}$ is *compatible* with A at a point $\mathbf{x} \in \partial\Omega$, if the only function

$u(s)$ satisfying

$$\left.\begin{aligned} &A_0\left(\mathbf{x},\boldsymbol{\tau}-i\mathbf{n}\frac{d}{ds}\right)u(s)=0 \qquad s>0 \\ &B_{k0}\left(\mathbf{x},\boldsymbol{\tau}-i\mathbf{n}\frac{d}{ds}\right)u(s)\Big|_{s=0}=0 \qquad 0\leqslant k\leqslant m-1 \end{aligned}\right\} \tag{5.19}$$

where $\boldsymbol{\tau}$ is any tangent to $\partial\Omega$ at $\mathbf{x}$, and $\mathbf{n}$ is any outward normal to $\partial\Omega$ at $\mathbf{x}$, is the function

$$u(s)\equiv 0$$

Any such compatible system of boundary operators $\{B_k\}$ is called a *covering family* of A and is said to *cover A* at the point $\mathbf{x}\in\partial\Omega$.

EXAMPLE 5.11. Consider again the differential operator

$$Au=a\frac{\partial^2 u}{\partial x^2}+2b\frac{\partial^2 u}{\partial x\,\partial y}+c\frac{\partial^2 u}{\partial y^2}$$

and the associated boundary operator

$$Bu=\alpha\frac{\partial u}{\partial x}+\beta\frac{\partial u}{\partial y}+\gamma u$$

where a, b, c, α, β, and γ are real, and $b^2-ac<0$.

Consider a point $\mathbf{x}_0\in\partial\Omega$ with normal $\mathbf{n}=(0,1)$ and tangent $\boldsymbol{\tau}=(\xi,0)$. Then we wish to solve the equations

$$A_0\left(\boldsymbol{\tau}-i\mathbf{n}\frac{d}{ds}\right)\psi(s)=-(-a\xi^2\psi+2bi\xi\psi'+c\psi'')=0 \qquad s>0$$

$$B_0\left(\boldsymbol{\tau}-i\mathbf{n}\frac{d}{ds}\right)\psi(s)=-i[\xi i\alpha\psi(0)+\beta\psi'(0)]=0$$

which, aside from constant factors, are of the same form as those in Example 5.10 [see (5.16)]. Hence, since $|\psi(s)|<\infty$, the solution is of the form

$$\psi(s)=C_1e^{-\mu\xi s/c}\left(\cos\frac{b\xi s}{c}-i\sin\frac{b\xi s}{c}\right)$$

The boundary condition leads to the requirement

$$-\frac{\xi}{c}[\mu\beta+i(b\beta-\alpha c)]C_1=0$$

where $\mu^2 = ac - b^2$. Hence $C_1 = 0$ so long as $\beta \neq 0$ or $\beta = 0$ and $\alpha c \neq 0$, or equivalently, so long as the modulus of the complex coefficient does not vanish:

$$P(\alpha, \beta, \mu) \equiv (\alpha c - b\beta)^2 + \beta^2 \mu^2 \neq 0$$

Notice that $P(\alpha, \beta, \mu) \neq 0$ is precisely the condition (5.18) obtained in the previous example by different arguments. ■

EXAMPLE 5.12. For a concrete application of the results of the previous example, consider the Laplacian

$$Au = \frac{\partial^2 u}{\partial x^2} + \frac{\partial^2 u}{\partial y^2}$$

In this case, $a = c = 1$, $b = 0$, $\mu = 1$, so that $P(\alpha, \beta, \mu)$ becomes

$$P(\alpha, \beta, \mu) = \alpha^2 + \beta^2 \neq 0$$

No real nonzero values of α and β will make $P(\alpha, \beta, \mu)$ vanish. Hence any boundary operator of the form

$$Bu = \alpha \frac{\partial u}{\partial x} + \beta \frac{\partial u}{\partial y} + \gamma u \qquad \alpha, \beta \quad \text{real} \neq 0$$

covers the laplacian operator. ■

5.4 GREEN'S FORMULAS

We now examine several generalizations of Green's formula, which prove to be particularly important in the study of existence and regularity of solutions to various boundary-value problems.

Let $\mathfrak{U}$ denote a Hilbert space of functions of some type defined on a bounded domain $\Omega \subset \mathbf{R}^n$ with smooth boundary $\partial \Omega$, and let $\mathfrak{U}'$ denote its dual. We consider an operator $A = T'T$ which maps $\mathfrak{U}$ continuously onto $\mathfrak{U}'$ and which is the composition of two operators, T and its transpose T'. The operator T maps $\mathfrak{U}$ continuously into another Hilbert space $\mathfrak{F}$. We identify $\mathfrak{F}$ as the pivot space, $\mathfrak{F}' = \mathfrak{F}$. The transpose operator maps $\mathfrak{F}'$ continuously into $\mathfrak{U}'$. If $\langle \cdot, \cdot \rangle_{\mathfrak{U}}$ and $\langle \cdot, \cdot \rangle_{\mathfrak{F}}$ denote duality pairing on on $\mathfrak{U}' \times \mathfrak{U}$ and $\mathfrak{F}' \times \mathfrak{F}$, the operators T and T' satisfy a relation of the form

$$\langle T'\mathbf{g}, u \rangle_{\mathfrak{U}} = \langle \mathbf{g}, Tu \rangle_{\mathfrak{F}} \forall u \in \mathfrak{U}, \mathbf{g} \in \mathfrak{F}' \tag{5.20}$$

Suppose $\mathcal{U}$ is densely embedded in a pivot space $\mathcal{H}=\mathcal{H}'$ and define the *formal adjoint* $\tilde{T}^*$ of T as the restriction of T' into the space $\mathcal{W}\equiv\{\mathbf{g}\in\mathcal{F}'$ such that $T'\mathbf{g}\in\mathcal{H}'\}$. Then

$$(\tilde{T}^*\mathbf{g},u)_{\mathcal{H}}=(\mathbf{g},Tu)_{\mathcal{F}}+\Gamma_{\partial\Omega}\langle u,\mathbf{g}\rangle \qquad \forall u\in\mathcal{U},\quad \mathbf{g}\in\mathcal{W} \tag{5.21}$$

where $\Gamma_{\partial\Omega}\langle u,\mathbf{g}\rangle$ is a bilinear (or sesquilinear) form on extensions of u and $\mathbf{g}$ to $\partial\Omega$. The form $\Gamma_{\partial\Omega}\langle u,\mathbf{g}\rangle$ is called the (global) *bilinear concomitant* of T, and its precise form depends on properties of T. Equation (5.20) is said to describe a *generalized Green's formula* for the operator T.

If $Z_{\mathcal{U}}:\mathcal{U}\to\mathcal{U}'$ and $Z_{\mathcal{F}}:\mathcal{F}\to\mathcal{F}'$ are the Riesz maps on $\mathcal{U}$ and $\mathcal{F}$ (now $Z_{\mathcal{F}}=I$), we observe [recall (4.30)] that the adjoint T^* of T is related to T' by

$$T^*=Z_{\mathcal{U}}^{-1}T'Z_{\mathcal{F}}=Z_{\mathcal{U}}^{-1}T' \tag{5.22}$$

Hereafter, we do not distinguish between T^* and $\tilde{T}^*$. We now record a few examples of adjoint operators to fix ideas.

EXAMPLE 5.13. Let $\mathcal{U}=H^1(a,b)$, $\mathcal{F}=L_2(a,b)$, and $T=d/dx$. Then the integration-by-parts formula verifies that

$$\int_a^b v\frac{du}{dx}\,dx=\int_a^b u\left(-\frac{dv}{dx}\right)dx+u(x)v(x)\big|_a^b$$

Thus

$$T^*=-\frac{d}{dx} \qquad \text{and} \qquad \Gamma_{\partial\Omega}(u,v)=-uv\big|_a^b=u(a)v(a)-u(b)v(b)$$

■

EXAMPLE 5.14. Let $\mathcal{U}=H^1(\Omega)$, where $\Omega\subset\mathbb{R}^2$, and let $\mathcal{F}$ denote $H^0(\Omega)\times H^0(\Omega)$, i.e., the class of vector-valued functions endowed with the inner product $\int_\Omega \mathbf{v}_1\cdot\mathbf{v}_2\,dx\,dy$. Let $Tu=\operatorname{grad} u$; then, from the classical form of Green's formula,

$$\int_\Omega \mathbf{v}\cdot\operatorname{grad}u\,dx\,dy=\int_\Omega u(-\operatorname{div}\mathbf{v})\,dx\,dy+\oint_{\partial\Omega}u\mathbf{v}\cdot\mathbf{n}\,ds$$

where $\mathbf{n}$ is a unit exterior normat to $\partial\Omega$. In this case

$$T^*=-\operatorname{div} \qquad \text{and} \qquad \Gamma_{\partial\Omega}(u,\mathbf{v})=-\oint_{\partial\Omega}u\mathbf{v}\cdot\mathbf{n}\,ds \quad ■$$

EXAMPLE 5.15. According to the divergence (Green-Gauss) theorem

$$\int_{\Omega} \nabla \cdot (w \nabla u)\, dx\, dy = \oint_{\partial\Omega} w \nabla u \cdot \mathbf{n}\, ds$$

for sufficiently smooth w and u. Thus

$$\int_{\Omega} \nabla w \cdot \nabla u\, dx\, dy = -\int_{\Omega} w \nabla^2 u\, dx\, dy + \oint_{\partial\Omega} w \nabla u \cdot \mathbf{n}\, ds$$

Taking $\mathbf{v} = \operatorname{grad} w \equiv \nabla w$ in Example 5.14 and using the above identity, we find that

$$\int_{\Omega} w \nabla^2 u\, dx\, dy = \int_{\Omega} u \nabla^2 w\, dx\, dy + \oint_{\partial\Omega} (w \nabla u \cdot \mathbf{n} - u \nabla w \cdot \mathbf{n})\, ds$$

Thus, in this case,

$$T = T^* = \nabla^2 \qquad \text{and} \qquad \Gamma_{\partial\Omega}(u, w) = -\oint_{\partial\Omega} (w \nabla u - u \nabla w) \cdot \mathbf{n}\, ds \quad \blacksquare$$

We now concentrate on the very special class of elliptic partial differential operators $T = A$ of the form (5.10). If A is given by (5.10), its adjoint is of the form

$$A^* v = \sum_{|\alpha|, |\beta| \leq m} (-1)^{|\alpha|} D^{\alpha} \overline{(a_{\beta\alpha}(\mathbf{x})}\, D^{\beta} v) \tag{5.23}$$

and a Green's formula can be constructed involving A and A^*.

To remove some of the arbitrariness from the possible forms the bilinear functional $\Gamma_{\partial\Omega}(u, v)$ associated with A can assume, we make the following assumptions concerning A and an associated system of boundary operators:

(i) A is properly elliptic on $\bar{\Omega}$ and its coefficients are in $C^{\infty}(\bar{\Omega})$.

(ii) The system of boundary operators $\{B_k\}_{k=0}^{m-1}$ is normal and covers A on $\partial\Omega$, and $b_{k\alpha}(\mathbf{x}) \in C^{\infty}(\partial\Omega)$ [recall (5.11) and (5.13)].

We next select any complementary system of boundary operators $\{S_k\}_{k=0}^{m-1}$ such that

(iii) The coefficients $s_{k\alpha}(\mathbf{x}) \in C^{\infty}(\partial\Omega)$ and the order p_k of S_k is $\leq 2m-1$.

(iv) The system

$$\{B_0, B_1, \ldots, B_{m-1}, S_0, S_1, \ldots, S_{m-1}\}$$

constitutes a Dirichlet system of order $2m$ on $\partial\Omega$. When these conditions prevail, it can be shown (see [5.1, p. 115]) that there exist two uniquely defined normal systems of m boundary operators $\{C_k\}_{k=0}^{m-1}$ and $\{T_k\}_{k=0}^{m-1}$ such that

(v) C_k is of order $2m-1-p_k$ (recall that p_k is the order of S_k) and T_k is of order $2m-1-q_k$ (q_k being the order of B_k).

(vi) The system

$$\{C_0, C_1, \dots, C_{m-1}, T_0, T_1, \dots, T_{m-1}\}$$

is a Dirichlet system of order $2m$ on $\partial\Omega$.

(vii) The systems $\{B_k\}$, $\{S_k\}$, $\{C_k\}$, and $\{T_k\}$ are such that the following generalized Green's formula holds:

$$\int_\Omega \bar{v} Au\,dx = \int_\Omega u\,\overline{A^*v}\,dx + \sum_{k=0}^{m-1}\int_{\partial\Omega} S_k u\,\overline{C_k v}\,ds - \sum_{k=0}^{m-1}\int_{\partial\Omega} B_k u\,\overline{T_k v}\,ds \tag{5.24}$$

Clearly, in this case

$$\Gamma_{\partial\Omega}(u,v) = -\sum_{k=0}^{m-1}\int_{\partial\Omega}\left(S_k u\,\overline{C_k v} - B_k u\,\overline{T_k v}\right)ds \tag{5.25}$$

EXAMPLE 5.16. As an example of the above formula, consider the general second-order partial differential operator in two variables:

$$Au = \frac{\partial}{\partial x}\left(a\frac{\partial u}{\partial x}\right) + \frac{\partial}{\partial x}\left(b_1\frac{\partial u}{\partial y}\right) + \frac{\partial}{\partial y}\left(b_2\frac{\partial u}{\partial x}\right) + \frac{\partial}{\partial y}\left(c\frac{\partial u}{\partial y}\right) + d\frac{\partial u}{\partial x} + e\frac{\partial u}{\partial y} + fu \tag{5.26}$$

where the coefficients $a, b_1, \dots, f$ are functions of points $(x,y)\in\Omega\subset\mathbf{R}^2$. By successive partial integrations, we easily find that

$$\iint_\Omega vAu\,dx\,dy = \iint_\Omega uA^*v\,dx\,dy + \oint_{\partial\Omega}(avu_x n_x - auv_x n_x + b_1 vu_y n_x - b_1 uv_x n_y + b_2 vu_x n_y - b_2 uv_y n_x + cvu_y n_y - cuv_y n_y + duvn_x + euvn_y)\,ds \tag{5.27}$$

where $u_x = \partial u/\partial x$, $u_y = \partial u/\partial y$, etc., n_x and n_y are the components of a unit outward normal $\mathbf{n}$ to $\partial\Omega$, and

$$A^*v = \frac{\partial}{\partial x}\left(a\frac{\partial v}{\partial x}\right) + \frac{\partial}{\partial y}\left(b_1\frac{\partial v}{\partial x}\right) + \frac{\partial}{\partial x}\left(b_2\frac{\partial v}{\partial y}\right)$$
$$+ \frac{\partial}{\partial y}\left(c\frac{\partial v}{\partial y}\right) - \frac{\partial}{\partial x}(dv) - \frac{\partial}{\partial y}(ev) + fv \tag{5.28}$$

Now suppose that we are given on $\partial\Omega$ boundary conditions involving a normal boundary operator of the form

$$B_0 u = \alpha u_x + \beta u_y + \gamma u$$

Since α and β are given and the boundary curve $\partial\Omega$ is given, the fact that B_0 is normal [see (5.13)] allows us to write:

$$\alpha n_x + \beta n_y = \mu_0(s) \neq 0$$

where $\mu_0(s) = \mu_0(x(s), y(s))$ is a known function defined on the boundary ($ds^2 = dx^2 + dy^2$).

Since B_0 is of first order, the operator S_0 must be of zero order if $\{B_0, S_0\}$ is to form a Dirichlet system of order 2. This means that C_0 must be a normal, first-order boundary operator and T_0 must be of zero order; i.e., the operators S_0, C_0, and T_0 must be of the form

$$S_0 u = \sigma_0 u \qquad C_0 v = \theta_1 v_x + \theta_2 v_y + \theta_0 v \qquad T_0 v = \tau_0 v$$

where $\sigma_0, \theta_1, \ldots, \tau_0$ are functions of $s = (x(s), y(s)) \in \partial\Omega$. The function σ_0 is arbitrary, but once it is specified, C_0 and T_0 are uniquely determined. We must of course keep in mind that θ_1 and θ_2 should be such that

$$\theta_1 n_x + \theta_2 n_y = \mu_1(s) \neq 0 \tag{5.29}$$

$\mu_1(s)$ being a known function of s, if C_0 is normal. We show that this property does in fact hold.

In order that these systems of operators fit into the framework of (5.24), it is necessary that the integral

$$\oint_{\partial\Omega} \left(S_0 u\,\overline{C_0 v} - B_0 u\,\overline{T_0 v}\right) ds$$

coincide with the boundary integral given in (5.27). This leads to the equations

$$\left.\begin{array}{ll} \sigma_0\theta_0 - \tau_0\gamma = dn_x + en_y & \tau_0\alpha = -an_x - b_2n_y \\ \sigma_0\theta_1 = -an_x - b_1n_y & \tau_0\beta = -cn_y - b_1n_x \\ \sigma_0\theta_2 = -cn_y - b_2n_x & \tau_0 \neq 0 \end{array}\right\} \tag{5.30}$$

We arbitrarily assign $\sigma_0 = \sigma_0(s) \neq 0$, $s \in \partial\Omega$. Then

$$\theta_1 = \frac{1}{\sigma_0}(-an_x - b_1n_y)$$

$$\theta_2 = \frac{1}{\sigma_0}(-cn_y - b_2n_x)$$

It may appear that τ_0 must satisfy two independent conditions, but in fact it is uniquely determined by (5.30) because of the requirements

$$n_x^2 + n_y^2 = 1 \quad \text{and} \quad (n_x, n_y) = \text{outward normal}$$

Solving (5.30) for n_x and n_y and using the above conditions, we find

$$\tau_0(s) = \left.\frac{ac - b_1b_2}{\left[(\alpha c - b_2\beta)^2 + (\alpha b_1 - \beta a)^2\right]^{1/2}}\right|_{(x,y)\in\partial\Omega} \tag{5.31}$$

It follows that

$$\theta_0 = \frac{1}{\sigma_0}(dn_x + en_y + \gamma\tau_0)$$

Notice also that C_0 is normal; indeed

$$C_{00}(\mathbf{n}) = \theta_1 n_x + \theta_2 n_y = -\frac{1}{\sigma_0}\left[an_x^2 + (b_1 + b_2)n_xn_y + cn_y^2\right] \neq 0$$

and the quantity in brackets cannot vanish because of the assumed ellipicity of A.

Summing up, we have

$$\left.\begin{array}{l} S_0u = \sigma_0 u \qquad T_0v = \tau_0 v \\ C_0v = -\dfrac{1}{\sigma_0}\left[(an_x + b_1n_y)v_x + (cn_y + b_2n_x)v_y - (dn_x + en_y + \gamma\tau_0)v\right] \end{array}\right\} \tag{5.32}$$

■

EXAMPLE 5.17. Suppose that A is the laplacian

$$Au = \frac{\partial^2 u}{\partial x^2} + \frac{\partial^2 u}{\partial y^2}$$

and

$$B_0 u = \frac{\partial u}{\partial n} = \nabla u \cdot \mathbf{n}$$

Then, we see that b, e, d, and γ are zero and $a = c = 1$. Choosing $\sigma_0 \neq 0$, we find that in this case $\tau_0 = -1$,

$$C_0 v = \frac{-1}{\sigma_0} \nabla v \cdot \mathbf{n} = -\frac{1}{\sigma_0}\frac{\partial v}{\partial n} \qquad T_0 v = -v$$

Hence (5.24) assumes the well-known form given in Example 5.15. ■

EXAMPLE 5.18. Consider the biharmonic operator

$$Au = \nabla^4 u = \frac{\partial^4 u}{\partial x^4} + 2\frac{\partial^4 u}{\partial x^2 \partial y^2} + \frac{\partial^4 u}{\partial y^4}$$

and consider the associated system of boundary operators

$$B_0 u = b_0 u$$

$$B_1 u = b_1 \frac{\partial u}{\partial n} \qquad \mathbf{n} \perp \partial\Omega;\ b_0, b_1 = \text{constants} > 0$$

By successive applications of Green's formula of Example 5.15, it is easily shown that

$$\int_\Omega v \nabla^4 u\, dx = \int_\Omega u \nabla^4 v\, dx + \oint_{\partial\Omega}\left(-u\frac{\partial \nabla^2 v}{\partial n} + \frac{\partial u}{\partial n}\nabla^2 v + v\frac{\partial \nabla^2 u}{\partial n} - \frac{\partial v}{\partial n}\nabla^2 u\right) ds \tag{5.33}$$

Thus ∇^4 is self-adjoint, $A^* = A = \nabla^4$, and

$$\Gamma_{\partial\Omega}(u, v) = -\oint_{\partial\Omega}\left[\frac{\partial \nabla^2 u}{\partial n} v + (-\nabla^2 u)\frac{\partial v}{\partial n}\right] ds$$
$$+ \oint_{\partial\Omega}\left[b_0 u\left(\frac{1}{b_0}\frac{\partial \nabla^2 v}{\partial n}\right) + b_1 \frac{\partial u}{\partial n}\left(-\frac{1}{b_1}\nabla^2 v\right)\right] ds \tag{5.34}$$

Comparing $\Gamma_{\partial\Omega}(u,v)$ with (5.24), we see that, in this case,

$$\left.\begin{aligned} T_0 &= \frac{1}{b_0}\frac{\partial \nabla^2}{\partial n} & T_1 &= -\frac{1}{b_1}\nabla^2 \\ S_0 &= \frac{\partial \nabla^2}{\partial n} & S_1 &= -\nabla^2 \\ C_0 &= 1 & C_1 &= \frac{\partial}{\partial n} \end{aligned}\right\} \tag{5.35}$$

Clearly $\{C_0, C_1\}$ is a normal system covering A^* (and A), and the systems

$$\{B_0, B_1, S_1, S_0\} = \left\{b_0, b_1\frac{\partial}{\partial n}, -\nabla^2, \frac{\partial \nabla^2}{\partial n}\right\}$$

$$\{C_0, C_1, T_1, T_0\} = \left\{1, \frac{\partial}{\partial n}, -\frac{1}{b_1}\nabla^2, \frac{1}{b_0}\frac{\partial \nabla^2}{\partial n}\right\}$$

are Dirichlet systems of order $2m=4$. ■

5.5 REGULARITY THEORY IN $H^s(\Omega)$, $s \geq 2m$

We now turn to questions of existence, uniqueness, and regularity of solutions to a general class of elliptic boundary-value problems of the form

$$\left.\begin{aligned} Au &= f && \text{in } \Omega \\ B_k u &= g_k && \text{on } \partial\Omega \end{aligned}\right\} \tag{5.36}$$

where $k=0, 1,\ldots,m-1$. We continue to abide by the assumptions made earlier; i.e., Ω is an open bounded domain in $\mathbb{R}^n$ with a boundary $\partial\Omega$ which is an $(n-1)$-dimensional manifold, Ω being locally on one side of $\partial\Omega$. The operator A is a properly elliptic partial differential operator of order $2m$, with infinitely differentiable coefficients, of the form given in (5.10); i.e.,

$$Au = \sum_{|\alpha|,|\beta| \leq m} (-1)^{|\alpha|} D^\alpha \left(a_{\alpha\beta}(\mathbf{x}) D^\beta u\right) \tag{5.37}$$

As before, the family of boundary operators B_k is defined by

$$B_k u = \sum_{|\alpha| \leq q_k} b_{k\alpha}(\mathbf{x}) D^\alpha u \qquad 0 \leq q_k \leq 2m-1 \tag{5.38}$$

with coefficients $b_{k\alpha}(\mathbf{x})$ infinitely differentiable on $\partial\Omega$, the system $\{B_k\}_{k=0}^{m-1}$ being a normal covering of A on $\partial\Omega$. Following standard terminology, we refer to a boundary-value problem of the type (5.36), with the definitions and assumptions concerning A, B_k, Ω, and $\partial\Omega$ stated above, as a *regularly elliptic boundary-value problem of order 2m.*

For such classes of problems, we wish to determine conditions on the data f and g_k under which solutions to regularly elliptic problems exist and are unique. Equally important, we wish to determine the regularity of the solution; i.e., if f and g_k have a certain degree of smoothness (alternatively, if f and g_k belong to specific Sobolev spaces), what degree of smoothness can we expect of the solution u (in what space does u lie)?

It is worth mentioning at this point that the techniques we employ to study existence and uniqueness parallel those encountered in studying the solvability of systems of linear equations defined on finite-dimensional linear spaces. The linear transformation $\mathbf{A}$, mapping $\mathcal{V}^m$ into $\mathcal{V}^n$, may appear in the linear system

$$\mathbf{A}\mathbf{x}=\mathbf{b}$$

where $\mathbf{A}$ is an $n\times m$ matrix and $\mathbf{x}$ and $\mathbf{b}$ are $m-$ and $n-$vectors in $\mathcal{V}^m$ and $\mathcal{V}^n$, respectively. The space $\mathcal{V}^n$ is self-dual, and the adjoint (transpose) of $\mathbf{A}$, denoted $\mathbf{A}^*$, maps $\mathcal{V}^n$ into $\mathcal{V}^m$ and has the property

$$\mathbf{y}^T\mathbf{A}\mathbf{x}=\mathbf{x}^T\mathbf{A}^*\mathbf{y}$$

where $\mathbf{y}^T$ is the transpose of $\mathbf{y}\in\mathcal{V}^n$, etc. The *adjoint problem* is

$$\mathbf{A}^*\mathbf{y}=\mathbf{c}$$

with $\mathbf{y}\in\mathcal{V}^n$, $\mathbf{c}\in\mathcal{V}^m$, and the *null space* of $\mathbf{A}^*$ is the subspace $\mathfrak{N}^*$ of $\mathcal{V}^n$ such that $\mathbf{A}^*\mathbf{z}=\mathbf{0}$, $\forall\mathbf{z}\in\mathfrak{N}^*$. Indeed, $\mathcal{V}^n=\mathfrak{N}^*\oplus\mathfrak{N}^{*\perp}$. Thus if $\mathbf{z}$ is any element of $\mathfrak{N}^*$, the *data* $\mathbf{b}$ must be orthogonal to $\mathbf{z}$; i.e.,

$$\mathbf{z}^T\mathbf{b}=\mathbf{x}^T\mathbf{A}^*\mathbf{z}=0$$

In other words, *the equation* $\mathbf{A}\mathbf{x}=\mathbf{b}$ *admits a solution only for vectors* $\mathbf{b}$ *orthogonal to* $\mathfrak{N}^*$. This condition is a natural consequence of Green's formula, $\mathbf{y}^{\mathrm{T}}\mathbf{A}\mathbf{x}=\mathbf{x}^{\mathrm{T}}\mathbf{A}^*\mathbf{y}$. Moreover, such solutions $\mathbf{x}$ are unique whenever $\mathfrak{N}=\{\mathbf{0}\}$, where $\mathfrak{N}\subset\mathcal{V}^m$ is the null space of $\mathbf{A}$.

Essentially the same concepts as these are used in the study of existence and uniqueness of general linear boundary-value problems. Indeed, conditions on the data f and g_k required for the existence of solutions to (5.36) can be deduced with the aid of Green's formula (5.24), presented in the previous section. First note that there is associated with each regularly elliptic boundary-value problem of the type in (5.36) an *adjoint problem*

$$\left.\begin{aligned} A^*u &= f^* && \text{in } \Omega \\ C_k u &= g_k^* && \text{on } \partial\Omega \end{aligned}\right\} \tag{5.39}$$

$k = 0, 1, 2, \ldots, m-1$, where A^* is the adjoint of A, defined in (5.23), and C_k is the system of normal boundary operators of order $q'_k = 2m-1-p_k$ appearing in the Green's formula (5.24) for the operator A. The form of the particular operators $\{C_k\}_{k=0}^{m-1}$ depends of course on the choice of the system $\{S_k\}_{k=0}^{m-1}$ in Green's formula, but each normal set of boundary operators $\{C_k\}_{k=0}^{m-1}$ so obtained also covers A^*. Such a system is referred to as the *adjoint system of* $\{B_k\}_{k=0}^{m-1}$. Moreover, the adjoint problem (5.39) associated with (5.36) is also a regularly elliptic boundary-value problem of order $2m$.

Since the boundary operators $\{B_k\}_{k=0}^{m-1}$ in (5.38) are intrinsically related to the operator A, it is convenient to express the entire boundary-value problem in the concise form

$$\mathcal{P}u = F \tag{5.40}$$

where $\mathcal{P}$ denotes the ordered set of operators

$$\mathcal{P} = (A;\, B_0, B_1, \ldots, B_{m-1}) \equiv (A;\, B_k) \tag{5.41}$$

and

$$F = (f;\, g_0, g_1, \ldots, g_{m-1}) \equiv (f;\, g_k) \tag{5.42}$$

We denote $\mathcal{N}$ the subspace of solutions for the problem (5.40) with $F = (0;\, 0)$:

$$\mathcal{N} = \{v\colon Av = 0 \text{ in } \Omega;\ B_k v = 0 \text{ on } \partial\Omega\} \tag{5.43}$$

Likewise, the adjoint problem (5.39) can be written:

$$\mathcal{P}^* u = F^* \tag{5.44}$$

where

$$\mathcal{P}^* = (A^*;\, C_0, C_1, \ldots, C_{m-1}) \equiv (A^*;\, C_k) \tag{5.45}$$

and the null space of $\mathcal{P}^*$ is defined by

$$\mathcal{N}^* = \{v\colon A^*v = 0 \text{ in } \Omega;\ C_k v = 0 \text{ on } \partial\Omega\} \tag{5.46}$$

if one of the two spaces $\mathfrak{N}$ or $\mathfrak{N}^*$ is finite-dimensional, the difference

$$\chi = \dim \mathfrak{N} - \dim \mathfrak{N}^*$$

is called the *index* of the operator $\mathcal{P}$.

In view of the properties of product Hilbert spaces presented in Section 4.4 [see (4.34)], the operator $\mathcal{P}$ in (5.40) can be considered a linear operator mapping $H^s(\Omega)$, $s \geqslant 2m$, into the product space

$$J^{s-2m}_{s-q_k-1/2}(\bar{\Omega}) \equiv H^{s-2m}(\Omega) \times \prod_{k=0}^{m-1} H^{s-q_k-1/2}(\partial\Omega) \tag{5.47}$$

provided with the norm

$$\|\mathcal{P}u\|^2_{J^{s-2m}_{s-q_k-1/2}(\bar{\Omega})} = \|Au\|^2_{H^{s-2m}(\Omega)} + \sum_{k=0}^{m-1} \|B_k u\|^2_{H^{s-q_k-1/2}(\partial\Omega)} \tag{5.48}$$

Now under the stated conditions on A and B_k it is not difficult to show that the operator $\mathcal{P}$ of (5.40) is continuous, and we may convince ourselves of this fact by considering two properties of such operators described in the following lemmas:

LEMMA 5.2. Let A be any partial differential operator of order $r \leqslant 2m$ on $H^s(\Omega)$, Ω being a smooth domain in $\mathbf{R}^n$, with coefficients which have bounded continuous derivatives of a sufficiently high order. Then there exist constants C_s, $C > 0$, such that for real $s > 0$,

$$\|Au\|_{H^{s-r}(\Omega)} \leqslant C_s \|u\|_{H^s(\Omega)} \tag{5.49}$$

for all $u \in H^\alpha(\Omega)$, $\alpha = \max(s, 2m)$, and, for $u \in H^r(\Omega)$,

$$\|Au\|_{H^{-k}(\Omega)} \leqslant C\|u\|_{H^{r-k}(\Omega)}, \quad 0 \leqslant k \leqslant r \tag{5.50}$$

PROOF. Inequality (5.49) is obvious for $s \geqslant r$. Let $s < r$ and note that (5.49) then follows from (5.50). Thus, it is sufficient to prove (5.50).

Let $\phi \in C_0^\infty(\Omega)$ and $u \in H^k(\Omega)$ and consider the inner product

$$(Au, \phi)_{H^0(\Omega)} = \int_\Omega \left(\sum_{j=0}^{r} \sum_{|\alpha|=j} a_\alpha(\mathbf{x}) D^\alpha u \right) \phi \, dx$$

By integrating by parts and using the fact that ϕ and its derivatives vanish on $\partial\Omega$, we can shift k derivatives from u to ϕ to obtain

$$
\begin{aligned}
|(Au,\phi)_{H^0(\Omega)}| &= \left| \sum_{j=0}^{r} \sum_{\substack{|\alpha|=j\\|\beta|=k}} (-1)^{|\beta|}\left(D^{\alpha-\beta}u, D^{\beta}(a_\alpha\phi)\right)_{H^0(\Omega)} \right| \\
&\leqslant \sum_{j=0}^{r} \sum_{\substack{|\alpha|=j\\|\beta|=k}} \|D^{\alpha-\beta}u\|_{H^0(\Omega)} \|D^{\beta}a_\alpha\phi\|_{H^0(\Omega)} \\
&\leqslant C\|u\|_{H^{r-k}(\Omega)}\|\phi\|_{H^k(\Omega)}
\end{aligned}
$$

If we consider a sequence $\{\phi_n\}$ of elements in $C_0^\infty(\Omega)$ converging in the $H^k(\Omega)$-norm to a function $v \in H_0^k(\Omega)$, then it is clear that the above inequality holds for $\phi = v \in H_0^k(\Omega)$. Then

$$
\|Au\|_{H^{-k}(\Omega)} = \sup_{\substack{v\in H_0^k(\Omega)\\ v\neq 0}} \frac{|(Au,v)_{H^0(\Omega)}|}{\|v\|_{H^k(\Omega)}} \leqslant C\|u\|_{H^{r-k}(\Omega)}
$$

which was to be proved. ■

LEMMA 5.3. Let $\{B_k\}_{k=0}^{m-1}$ be a normal family of boundary operators which cover an elliptic operator A of order $2m$ in a regularly elliptic boundary-value problem of order $2m$. Then there exist constants $M_k > 0$ such that for $s \geqslant 2m$ and any $u \in H^s(\Omega)$,

$$
\|B_k u\|_{H^{s-q_k-1/2}(\partial\Omega)} \leqslant M_k\|u\|_{H^s(\Omega)} \tag{5.51}
$$

PROOF. This follows immediately from the trace inequalities (4.106). See also Lemma 5.1. ■

Collecting all of these results, we see that the operator $\mathcal{P}$ in (5.41) is bounded. Indeed, in view of (5.48), (5.49), and (5.51),

$$
\begin{aligned}
\|\mathcal{P}u\|^2_{J^{s-2m}_{s-q_k-1/2}(\bar\Omega)} &\leqslant C_s^2\|u\|^2_{H^s(\Omega)} + \sum_{k=0}^{m-1} M_k^2\|u\|^2_{H^s(\Omega)} \\
&= \left(C_s^2 + \sum_{k=0}^{m-1} M_k^2\right)\|u\|^2_{H^s(\Omega)}
\end{aligned}
$$

or, with $C = [C_0^2 + \Sigma_{k=0}^{m-1} M_k^2]^{1/2} > 0$,

$$
\|\mathcal{P}u\|_{J^{s-2m}_{s-q_k-1/2}(\bar\Omega)} \leqslant C\|u\|_{H^s(\Omega)} \tag{5.52}
$$

Similarly, we can argue that the adjoint operator $\mathcal{P}^*$ maps the dual space $\left(J^{s-2m}_{s-q_k-1/2}(\overline{\Omega})\right)'$ continuously into $(H^s(\Omega))'$:

$$\mathcal{P}^*: \left(H^{s-2m}(\Omega)\right)' \times \prod_{k=0}^{m-1} \left(H^{s-q_k-1/2}(\partial\Omega)\right)' \to \left(H^s(\Omega)\right)' \tag{5.53}$$

Well-Posed Problems

The boundary-value problem (5.36) is said to be *well-posed* if there exists a unique solution $u(\mathbf{x}) \in H^s(\Omega)$, $s \geqslant 2m$, that depends continuously on the data

$$F = (f; g_k) \in J^{s-2m}_{s-q_k-1/2}(\overline{\Omega})$$

The existence and uniqueness of a solution hinges on whether or not $\mathcal{P}$ is one-to-one (uniqueness) and onto (existence). We discuss conditions for such solutions in the next section.

Since $u = \mathcal{P}^{-1}F$, if u is to depend continuously on F, the inverse of $\mathcal{P}$ must be continuous. Conditions for continuous inverses are given in the following two well-known theorems from linear operator theory.

THEOREM 5.1 (The Bounded Inverse Theorem). Let A be a linear operator from $\mathcal{U}$ into $\mathcal{V}$, where $\mathcal{U}$ and $\mathcal{V}$ are normed linear spaces. Moreover, let A be bounded below; i.e., let there exist a constant $M > 0$ such that

$$\|Au\|_{\mathcal{V}} \geqslant M\|u\|_{\mathcal{U}} \qquad \forall u \in \mathcal{U}$$

Then A has a continuous inverse defined on its range $\mathcal{R}(A) \subset \mathcal{V}$. Conversely, if A^{-1} exists and is continuous, then A is bounded below.

PROOF. A is clearly one-to-one. Thus A^{-1} exists on $\mathcal{R}(A)$. If $v = Au$, $\|u\|_{\mathcal{U}} = \|A^{-1}v\|_{\mathcal{V}} \leqslant 1/M\|v\|_{\mathcal{V}}$. Hence A^{-1} is continuous. The converse part of the theorem is obvious. ■

THEOREM 5.2 (The Banach Theorem). Let $\mathcal{U}$ and $\mathcal{V}$ be Banach spaces, and let A be a continuous, one-to-one, linear operator mapping $\mathcal{U}$ onto $\mathcal{V}$. Then A^{-1} exists and is linear and continuous.

PROOF. This result is a direct consequence of the open mapping theorem for linear operators, and its proof can be found in works on linear operator theory. In particular, see Dunford and Schwartz [5.3, p. 57]. ■

In view of Theorem 5.1, $\mathcal{P}$ has a continuous inverse if it is bounded below; i.e. if a solution u to (5.40) exists for any $F \in J^{s-2m}_{s-q_k-1/2}(\overline{\Omega})$ then this solution depends continuously on F only if the inequality

$$\|u\|_{H^s(\Omega)} \leqslant C \|\mathcal{P} u\|_{J^{s-2m}_{s-q_k-1/2}(\bar{\Omega})} \tag{5.54}$$

holds for some $C>0$.

Next we note that, when a solution to (5.40) exists, the null spaces $\mathfrak{N}$ and $\mathfrak{N}^*$ are necessarily finite-dimensional. Indeed, they often consist of only the zero element, $\mathfrak{N} = \mathfrak{N}^* = \{0\}$; otherwise, they contain a finite number of smooth functions. We can easily convince ourselves of this fact for special choices of A by considering the two-dimensional case ($x_1 = x, x_2 = y$) in which the operator A of (5.37) has constant coefficients and is reducible; i.e., A can be expressed as the product (or the sum of products) of $r<\infty$ linear factors of the type $\alpha D_x + \beta D_y + \gamma$. Then the homogeneous solution is necessarily a combination of a finite number $q, q \leqslant r$, of linearly independent solutions. The number of nonzero solutions to $\mathcal{P}u=0$ is further reduced on applying boundary conditions, but the important thing is that this number is finite. Arguments for the finite-dimensionality of $\mathfrak{N}$ and $\mathfrak{N}^*$ for more general cases can be found in [5.2].

We now have the groundwork laid for the following theorem.

THEOREM 5.3. Consider the operator $\mathcal{P} = (A; B_k)$ associated with a regularly elliptic boundary-value problem (5.40) of order $2m$. Then, for any $u \in H^s(\Omega), s \geqslant 2m$, there exists a constant $C>0$, independent of u, such that

$$\|u\|_{H^s(\Omega)} \leqslant C\left(\|\mathcal{P} u\|_{J^{s-2m}_{s-q_k-1/2}(\bar{\Omega})} + \|u\|_{H^0(\Omega)}\right) \tag{5.55}$$

PROOF. Let $\mathfrak{N}^\perp$ denote the L_2-orthogonal complement of $\mathfrak{N}$ in $H^s(\Omega)$; i.e.,

$$\mathfrak{N}^\perp = \left\{u : u \in H^s(\Omega); (u, v)_{L_2(\Omega)} = 0; \forall v \in \mathfrak{N}\right\}$$

Since $\mathcal{P}$ is linear, its restriction to $\mathfrak{N}^\perp$ is a one-to-one operator from $\mathfrak{N}^\perp$ onto its range $\mathfrak{R}(\mathcal{P})$ (indeed, if $u_1, u_2 \in \mathfrak{N}^\perp$ and we assume $u_1 \neq u_2$ and $\mathcal{P}(u_1) = \mathcal{P}(u_2)$, then $\mathcal{P}(u_1 - u_2) = 0$ implies $u_1 - u_2 \in \mathfrak{N}$, a contradiction; hence $u_1 = u_2$ and $\mathcal{P}$ is one-to-one). Thus by the Banach theorem (Theorem 5.2), $\mathcal{P}^{-1}$ is a bounded map from $\mathfrak{R}(\mathcal{P})$ into $\mathfrak{N}^\perp$:

$$\|\mathcal{P}^{-1} v\|_{H^s(\Omega)} \leqslant \frac{1}{C_0} \|v\|_{J^{s-2m}_{s-q_k-1/2}(\bar{\Omega})} \qquad v \in \mathfrak{R}(\mathcal{P})$$

By the bounded inverse theorem (Theorem 5.1), we have

$$\|u\|_{H^s(\Omega)} \leqslant C_0 \|\mathcal{P} u\|_{J^{s-2m}_{s-q_k-1/2}(\bar{\Omega})} \qquad u \in \mathfrak{N}^\perp$$

Since every $u \in H^s(\Omega)$ can be represented uniquely in the form $u = v + w, v \in \mathfrak{N}, w \in \mathfrak{N}^\perp$ (recall that $\mathfrak{N}$ is finite-dimensional), we have

$$\begin{aligned}\|u\|_{H^s(\Omega)} &= \|v + w\|_{H^s(\Omega)} = \|v\|_{H^s(\Omega)} + \|w\|_{H^s(\Omega)} \\ &\leqslant C_0 \|\mathcal{P} w\|_{J^{s-2m}_{s-q_k-1/2}(\bar{\Omega})} + \|v\|_{H^s(\Omega)} \\ &= C_0 \|\mathcal{P} u\|_{J^{s-2m}_{s-q_k-1/2}(\bar{\Omega})} + \|v\|_{H^s(\Omega)}\end{aligned}$$

Since $\mathfrak{N}$ is finite-dimensional, all norms on $\mathfrak{N}$ are equivalent. Therefore we can write

$$\|v\|_{H^s(\Omega)} \leqslant C_1 \|v\|_{H^0(\Omega)} \leqslant C_1 \|u\|_{H^0(\Omega)}$$

where we used the fact that $\|v\|^2_{H^0(\Omega)} = \|u\|^2_{H^0(\Omega)} - \|w\|^2_{H^0(\Omega)}$. Substituting this result into the previous inequality, we obtain (5.55). ■

By taking note of the details of the proof and of definition (5.48), we have the following.

COROLLARY 5.3.1. Let the conditions of Theorem 5.3 hold. Then there is a constant $C > 0$ such that

$$\|u\|_{H^s(\Omega)} \leqslant C \left(\|Au\|_{H^{s-2m}(\Omega)} + \sum_{k=0}^{m-1} \|B_k u\|_{H^{s-q_k-1/2}(\partial\Omega)} + \|u\|_{H^0(\Omega)} \right) \tag{5.56}$$

Moreover, if the null space $\mathfrak{N}$ of $\mathcal{P}$ is $\{0\}$, the term $\|u\|_{H^0(\Omega)}$ in (5.56) can be deleted. ■

Clearly, when $\mathfrak{N} = \{0\}$, we see that inequality (5.54) holds.

5.6 COMPATIBILITY CONDITIONS—EXISTENCE AND UNIQUENESS IN $H^s(\Omega)$, $s \geqslant 2m$

We now turn to the question of existence and uniqueness of solutions to (5.40). Specifically, we wish to determine conditions on the data (f, g_k), which will provide for the existence of solutions to regularly elliptic boundary-value problems. We refer to such conditions as *compatibility conditions*. In addition, we wish to establish conditions for a solution to be unique.

THEOREM 5.4 (Compatibility Conditions for Existence). There exists at least one solution $u \in H^s(\Omega), s \geqslant 2m$, to the problem

$$Au = f \text{ in } \Omega$$
$$B_k u = 0 \text{ on } \partial\Omega \qquad k = 0, 1, 2, \ldots, m-1$$

where A and B_k are the operators defined in (5.36), if and only if $f \in \mathfrak{N}^{*\perp}$,

$$\mathfrak{N}^{*\perp} = \left\{ v : v \in H^{s-2m}(\Omega); (v, w)_{H^0(\Omega)} = 0; \forall w \in \mathfrak{N}^* \right\}$$

PROOF. To prove necessity, let there exist a solution u_0 to the stated problem; i.e. $Au_0 = f, B_k u_0 = 0$. Let $v \in \mathfrak{N}^*$. Then

$$(v, Au_0)_{H^0(\Omega)} = (u_0, A^* v)_{H^0(\Omega)} + \sum_{k=0}^{m-1} \oint_{\partial\Omega} \left(S_k u_0 \overline{C_k v} - B_k u_0 \overline{T_k v} \right) ds$$

$$= (u_0, 0)_{H^0(\Omega)} + \sum_{k=0}^{m-1} \oint_{\partial\Omega} \left(S_k u_0 \cdot 0 - 0 \cdot \overline{T_k v} \right) ds = 0$$

Thus $(v, f)_{H^0(\Omega)} = 0 \ \forall v \in \mathfrak{N}^*$, which means that $f \in \mathfrak{N}^{*\perp}$.

To prove sufficiency, we first note that the range $\mathfrak{R}(A)$ of A is closed. This follows from the fact that A is a continuous linear operator from $\mathfrak{N}_A^\perp$ onto $\mathfrak{R}(A)$ with a continuous linear inverse $A^{-1}: \mathfrak{R}(A) \to \mathfrak{N}_A^\perp$. Hence, $\mathfrak{R}(A)^{\perp\perp} = \overline{\mathfrak{R}(A)} = \mathfrak{R}(A)$.

Now let $v \in \mathfrak{R}(A)^\perp$ and $u \in H^s(\Omega)$ with $B_k u = 0$. Then

$$(v, Au)_{H^0(\Omega)} = 0 = (u, A^* v)_{H^0(\Omega)} + \sum_{k=0}^{m-1} \oint_{\partial\Omega} S_k u \cdot \overline{C_k v} \, ds$$

from which we gather that $v \in \mathfrak{N}^*$. Thus $\mathfrak{R}(A)^\perp \subset \mathfrak{N}^*$. But this implies that $\mathfrak{N}^{*\perp} \subset \mathfrak{R}(A)^{\perp\perp} = \mathfrak{R}(A)$, which was to be proved. ■

Analogously, the adjoint problem $A^* u = f$ in Ω and $C_j u = 0$ on $\partial\Omega$ has at least one solution $u \in H^s(\Omega)$ if and only if $f \in \mathfrak{N}^\perp = \{u : u \in H^s(\Omega); (u, v)_{H^0(\Omega)} = 0; \forall v \in \mathfrak{N}\}$.

We now state the compatibility conditions for existence of solutions to the nonhomogeneous boundary-value problem (5.36).

THEOREM 5.5. Let the boundary-value problem $\mathfrak{P}u = F$ be regularly elliptic. For the equation $\mathfrak{P}u = F$ to have a solution it is necessary and sufficient that $F = (f; g_k)$ satisfy

$$(f, v)_{H^0(\Omega)} + \sum_{k=0}^{m-1} (g_k, T_k v)_{H^0(\partial\Omega)} = 0 \tag{5.57}$$

for every $v \in \mathfrak{N}^*$. Further, if (5.57) holds, there is a unique $u \in H^s(\Omega)$ such that $\mathcal{P}u = (f; g_k)$ and

$$(u, v)_{H^0(\Omega)} = 0 \tag{5.58}$$

for all $v \in \mathfrak{N}$.

PROOF. We first prove (5.58) by contradiction. Suppose that the solution u of $\mathcal{P}u = F$ is unique, and that $(u,v)_{H^0(\Omega)} \neq 0$ for all nonzero $v \in \mathfrak{N}$. This implies that $u \in H^s(\Omega) = \mathfrak{N}^\perp \oplus \mathfrak{N}$. Hence u can be represented uniquely as $u = u_0 + w$, $u_0 \in \mathfrak{N}$, $w \in \mathfrak{N}^\perp$, and $u_0 \neq 0$ [otherwise, $(u,v)_{H^0(\Omega)} = 0$]. Since u is unique, we have $Au = Aw = f \Rightarrow u = w$, which contradicts our assumption.

Necessity of (5.57) follows directly from Green's formula (5.24). To show that it is sufficient, recall from Theorem 5.3 that the operator $\mathcal{P}$ is bounded below, hence has a continuous inverse on $\mathfrak{R}(\mathcal{P})$. We must show that $(f; g_k)$ is in the range of $\mathcal{P}$. We know that we can construct a function $u_0(\mathbf{x}) \in H^s(\Omega)$ such that

$$B_k u_0 = g_k \qquad \mathbf{x} \in \partial\Omega; \; g_k \in H^{s-q_k-1/2}(\partial\Omega); \qquad k = 0, 1, 2, \ldots, m-1$$

Letting $w = u - u_0$ in (5.40), we obtain, from (5.41),

$$\mathcal{P}w = \mathcal{P}(u - u_0) = (f - Au_0, g_k - B_k u_0) = (f - Au_0, 0)$$

Thus we have reduced the nonhomogeneous problem (5.36) to an equivalent problem with homogeneous boundary conditions. But from Theorem 5.4 we know that the range of $\mathcal{P}$ is $\mathfrak{N}^{*\perp}$. Now using (5.57) and Green's formula (5.24), we obtain, for $v \in \mathfrak{N}^*$,

$$\begin{aligned}
(f - Au_0, v)_{H^0(\Omega)} &= (f, v)_{H^0(\Omega)} - (Au_0, v)_{H^0(\Omega)} \\
&= (f, v)_{H^0(\Omega)} + (u_0, A^* v)_{H^0(\Omega)} + \sum_{k=0}^{m-1} (B_k u_0, T_k v)_{H^0(\partial\Omega)} \\
&\quad - \sum_{k=0}^{m-1} (S_k u_0, C_k v)_{H^0(\partial\Omega)} \\
&= (f, v)_{H^0(\Omega)} + \sum_{k=0}^{m-1} (g_k, T_k v)_{H^0(\partial\Omega)} = 0
\end{aligned}$$

Thus $f - Au_0 \in \mathfrak{N}^{*\perp}$, which means that $(f; g_k)$ lies in the range of $\mathcal{P}$. ■

Condition (5.58) amounts to saying that the unique solution belongs to the orthogonal complement of the null space of $\mathcal{P}$.

We demonstrate the use of the above theorem with some simple examples.

EXAMPLE 5.19. Consider the Dirichlet boundary-value problem

$$-\frac{d^2u}{dx^2}=f \qquad 0<x<1$$
$$u(0)=g_0 \qquad u(1)=\hat{g}_0 \tag{5.59}$$

Solving the homogeneous problem with homogeneous boundary conditions, we obtain $u=0$. That is, $\mathfrak{N}=\{0\}$. Similarly, we find that $\mathfrak{N}^*=\{0\}$, and conditions (5.57) and (5.58) are identically satisfied. Hence, there exists a unique solution to (5.59). Moreover, if $f\in H^{s-2}(0,1)$ and $g,\hat{g}\in H^{s-1/2}(\partial\Omega)$, then $u\in H^s(0,1)$. ■

EXAMPLE 5.20. Now consider the Neumann problem

$$-\frac{d^2u}{dx^2}=f \qquad 0<x<1$$
$$u'(0)=g_1 \qquad u'(1)=\hat{g}_1 \tag{5.60}$$

In this case, the null spaces contain elements of the type $u=C$, where C is a constant. Then (5.60) has a solution $u\in H^s(0,1)$ for any $f\in H^{s-2}(0,1)$ and $(g_1,\hat{g}_1)\in H^{s-3/2}(\partial\Omega)$ that satisfy (from Example 5.17, we have $T_0=-I$)

$$\int_0^1 f\,dx-(\hat{g}_1-g_1)=0$$

For instance, suppose $f(x)=x$. Then from above we have $g_1=\hat{g}_1-\frac{1}{2}$, and the solution of (5.60) is given by

$$-u(x)=\frac{x^3}{6}+\left(\hat{g}_1-\frac{1}{2}\right)x+C_1$$

where C_1 is an arbitrary constant. Clearly, $u(x)$ satisfies (5.60), and in fact any function $v(x)=u(x)+C_2$, where C_2 is a constant, is a solution of (5.60).

The solution of u of (5.60) is unique if we choose C_1 so that (5.58) holds; i.e. $\int_0^1 uC\,dx=0\ \forall C$. Then $C_1=\hat{g}_1-1/24$. ■

EXAMPLE 5.21. Consider a generalization of Example 5.20:

$$-\nabla^2 u=f \text{ in } \Omega$$
$$\frac{\partial u}{\partial n}=g_1 \text{ on } \partial\Omega$$

The null spaces $\mathfrak{N}$ and $\mathfrak{N}^*$ contain only constant elements. The above equation has a solution $u \in H^s(\Omega)$, $s \geqslant 2$, for any $f \in H^{s-2}(\Omega)$ and $g_1 \in H^{s-3/2}(\partial\Omega)$ which satisfy

$$\int_\Omega f\,dx - \oint_{\partial\Omega} g_1\,dx = 0$$

If we constrain u so as to satisfy

$$\int_\Omega u\,dx = 0$$

then the solution u is unique. ■

EXAMPLE 5.22. Consider the fourth-order boundary-value problem

$$D^4u = f(x) \qquad 0 < x < 1$$

$$D^2u(0) = m_0 \qquad D^2u(1) = m_1$$

$$D^3u(0) = p_0 \qquad D^3u(1) = p_1$$

The student of mechanics will recognize these as the equations governing the transverse deflection u of an elastic beam subjected to an applied force f, end moments m_0 and m_1, and end shears p_0 and p_1.

Simple integration by parts produces Green's formula

$$\int_0^1 vD^4u\,dx = \int_0^1 uD^4v\,dx + vD^3u|_0^1 - DvD^2u|_0^1 + D^2vDu|_0^1 - D^3vu|_0^1$$

Thus

$$A^* = D^4 \qquad B_0u = D^2u|_0^1 \qquad B_1u = -D^3u|_0^1$$

$$T_0v = Dv|_0^1 \qquad T_1v = v|_0^1 \qquad S_0u = -u|_0^1$$

$$S_1u = Du|_0^1 \qquad C_0v = D^3v|_0^1 \qquad C_1v = D^2v|_0^1$$

The homogeneous adjoint problem is then

$$D^4v = 0$$

$$D^3v(0) = 0 \qquad D^3v(1) = 0$$

$$D^2v(0) = 0 \qquad D^2v(1) = 0$$

and its solution is $v(x)=C_0+C_1x$, where C_0 and C_1 are arbitrary constants. Therefore

$$\mathfrak{N}^*=\{v:v(x)=C_0+C_1x;0\leqslant x\leqslant 1;C_0,C_1\in\mathbf{R}\}$$

The compatibility conditions on the data of the original problem are obtained by substituting $v=C_0+C_1x$, and the data, into our Green's formula. Two independent conditions are obtained by first setting $C_1=0$ and then $C_0=0$. We find that a solution exists if and only if

$$\int_0^1 f\,dx=p_1-p_0$$

$$\int_0^1 fx\,dx=p_1-m_1+m_0$$

Again, if we think of the beam problem, these are nothing more than the equilibrium equations. The first condition states that the sum of the transverse forces must be zero, and the second condition states that the sum of the moments about the origin must vanish.

To study uniqueness, we observe that

$$\mathfrak{N}=\{u:u(x)=C_2+C_3x;0\leqslant x\leqslant 1;C_2,C_3\in\mathbf{R}\}$$

The solution is therefore not unique but is determined only to within an arbitrary linear term, C_2+C_3x. This means that the transverse deflection of our beam produced by applied forces f, p_0, p_1 and moments m_0, m_1 which satisfy the equilibrium equations (i.e., the compatibility equations) can be determined only to within an arbitrary rigid motion: a translation $u=C_2$ and an infinitesimal rotation $u=C_3x$. ■

We summarize the results of this section and the previous one in the following fundamental theorem.

THEOREM 5.6. Consider a regularly elliptic boundary-value problem (5.36) with operators $\mathcal{P}=(A;B_k)$. Let $\tilde{J}^{s-2m}_{s-q_k-1/2}(\bar{\Omega})$ be the subspace of $J^{s-2m}_{s-q_k-1/2}(\Omega)$ consisting of functions that satisfy the compatibility relations (5.57) and (5.58), and suppose that u is the solution in $H^s(\Omega), s\geqslant 2m$, whose existence is guaranteed by Theorem 5.3. Then $\mathcal{P}$ is a homeomorphism of $H^s(\Omega)$ onto $\tilde{J}^{s-2m}_{s-q_k-1/2}(\bar{\Omega})$, and the following a priori estimate holds for $f\in H^{s-2m}(\Omega), g_k\in H^{s-q_k-1/2}(\partial\Omega)$, $\mathcal{P}u=(f;g_k)$:

$$\|u\|_{H^s(\Omega)}\leqslant C\left(\|f\|_{H^{s-2m}(\Omega)}+\sum_{k=0}^{m-1}\|g_k\|_{H^{s-q_k-1/2}(\partial\Omega)}\right) \tag{5.61}$$

PROOF. In Theorem 5.3, it was proved that the mapping $\mathcal{P}$ is one-to-one and onto the range $\tilde{J}^{s-2m}_{s-q_k-1/2}(\bar{\Omega})$, and in Theorem 5.5 it was proved that $(f;g_k)\in\tilde{J}^{s-2m}_{s-q_k-1/2}(\bar{\Omega})$. That $\mathcal{P}$ is a homeomorphism follows from (5.52). The a priori estimate (5.61) is obtained by setting $Au=f$ and $B_k u=g_k$ in (5.56) and noting the fact that $u\in\mathfrak{N}^{\perp}$ allows us to eliminate the last term in (5.56). ■

EXAMPLE 5.23. Let $f\in H^k(\Omega), k\geqslant 0, g_0\in H^l(\partial\Omega), l\geqslant\frac{1}{2}$. Then there exists a unique solution $u\in H^s(\Omega)$ of the boundary-value problem

$$-\nabla^2 u+u=f \text{ in } \Omega$$

$$u=g_0 \text{ on } \partial\Omega$$

where $s=\min(k+2, l+\frac{1}{2})$. The a priori estimate (5.61), in this case, is given by

$$\|u\|_{H^s(\Omega)}\leqslant C(\|f\|_{H^k(\Omega)}+\|g_0\|_{H^l(\partial\Omega)}) \tag{5.62}$$

■

EXAMPLE 5.24. Consider the boundary-value problem

$$-\nabla^2 u+u=f \text{ in } \Omega \qquad f\in H^k(\Omega); k\geqslant 0$$

$$\frac{\partial u}{\partial n}=g_1 \text{ on } \partial\Omega \qquad g_1\in H^l(\partial\Omega); l\geqslant\tfrac{1}{2}$$

Then the solution $u\in H^s(\Omega), s=\min(k+2, l+\frac{3}{2})$, satisfies the estimate

$$\|u\|_{H^s(\Omega)}\leqslant C(\|f\|_{H^k(\Omega)}+\|g_1\|_{H^l(\partial\Omega)}) \tag{5.63}$$

■

5.7 EXISTENCE AND REGULARITY THEORY IN $H^s(\Omega)$, $s<2m$

The major results of the last two sections hold for functions in $H^s(\Omega)$ whenever $s\geqslant 2m$. We now consider existence and regularity theory when the data are such that a solution may exist in spaces $H^s(\Omega)$ for which $s<2m$, even $s<0$. We run into some immediate problems as to whether $\mathcal{P}$ can still be bounded. If $u\in H^s(\Omega)$, then still $Au\in H^{s-2m}(\Omega)$ for $s<2m$; however, the trace theorem does not hold for the boundary operators B_k when $s<2m$. In other words, the mapping $B_k: H^s(\Omega)\to H^{s-q_k-1/2}(\partial\Omega), s<2m$, is not necessarily continuous, and the operator $\mathcal{P}: H^s(\Omega)\to J^{s-2m}_{s-q_k-1/2}(\bar{\Omega})$ therefore cannot be bounded for $s<2m$.

We circumvent this problem in the following developments by using a technique that is by now quite familiar. Redefine a new Hilbert space on which $\mathcal{P}$ acts, which has sufficiently weak topology that $\mathcal{P}$ can be continuous. In particular, we define a new space $\tilde{H}^{s,r}(\Omega), s<r\leqslant 2m$, and show that the operator $\mathcal{P}$ is bounded from $\tilde{H}^{s,r}(\Omega)$ to $H^{s-r}(\Omega)$, and we then can proceed to develop existence and regularity results.

From results established in Chapter 4, we know that for any real number s Sobolev spaces $H^s(\Omega)$ and $H^s(\partial\Omega)$ can be defined, the spaces for $s<0$ being associated with the duals of $H_0^{-s}(\Omega)$ and $H^{-s}(\partial\Omega)$ and endowed with the norm associated with linear functionals; e.g.,

$$\|u\|_{H^{-s}(\Omega)} = \sup_{v\in H_0^s(\Omega)} \frac{|(u,v)_{H^0(\Omega)}|}{\|v\|_{H^s(\Omega)}} \qquad s>0, \quad v\neq 0 \tag{5.64}$$

We use such duality properties repeatedly in the present analysis. In particular, we exploit the fact that, if A maps a Banach space $\mathcal{U}$ linearly and continuously into a Banach space $\mathcal{V}$, its adjoint A^* maps the dual $\mathcal{V}'$ into $\mathcal{U}'$ linearly and continuously.

Let $E(\bar{\Omega})$ denote the space of infinitely differentiable functions defined on Ω such that all their derivatives have continuous extensions to the boundary $\partial\Omega$. We now introduce a Hilbert space $\tilde{H}^{s,r}(\Omega)\subset H^s(\Omega)$, defined as the completion of $E(\bar{\Omega})$ in the norm

$$\|u\|^2_{\tilde{H}^{s,r}(\Omega)} = \|u\|^2_{H^s(\Omega)} + \sum_{k=0}^{r-1} \|D_n^k u\|^2_{H^{s-k-1/2}(\partial\Omega)} \tag{5.65}$$

where r is a positive integer, s is an integer, and $D_n^k \equiv \partial^k/\partial n^k$ is the kth normal derivative. We define $\tilde{H}^{s,r}(\Omega)$ for arbitrary real s by interpolation.

It is important that we be able to interpret the relation between $\tilde{H}^{s,r}(\Omega)$ and the spaces $H^s(\Omega)$ and $H^{s-k-1/2}(\partial\Omega)$ for various values of s and r. To accomplish this, we introduce a space $\tilde{K}^{s,r}(\Omega)$ defined as the product

$$\tilde{K}^{s,r}(\Omega) = H^s(\Omega)\times \prod_{k=0}^{r-1} H^{s-k-1/2}(\partial\Omega) \tag{5.66}$$

where s is any real number $\neq$integer$+\frac{1}{2}$, and r is a positive integer.

Now $E(\bar{\Omega})$ is, by definition, dense in $\tilde{H}^{s,r}(\Omega)$. Thus, if $u_k \equiv D_n^k u$ denotes the kth derivative of u normal to the boundary $\partial\Omega$, the mapping

$$S\colon \tilde{H}^{s,r}(\Omega)\to \tilde{K}^{s,r}(\Omega) \tag{5.67}$$

defined by

$$Su=(u;u_0,u_1,\dots,u_{r-1}) \qquad u_k=D_n^k u|_{\partial\Omega} \tag{5.68}$$

has meaning if $u\in E(\overline{\Omega})\subset \tilde{H}^{s,r}(\Omega)$. We can now define S (or its extension) for any $u\in \tilde{H}^{s,r}(\Omega)$ by continuity, since $E(\overline{\Omega})$ is dense in $\tilde{H}^{s,r}(\Omega)$.

Three distinct cases of (5.67) and (5.68) are of interest (cf. [5.6]):

(i) $s<\frac{1}{2}$. In this case S is a homeomorphism from $\tilde{H}^{s,r}(\Omega)$ onto $\tilde{K}^{s,r}(\Omega)$. That is, the two spaces are algebraically and topologically the same, and we write

$$\tilde{H}^{s,r}(\Omega)\approx\tilde{K}^{s,r}(\Omega)$$

(ii) $s+\frac{1}{2}>r$. Here $Su=(u;u|_{\partial\Omega},D_n u|_{\partial\Omega},\dots,D_n^{r-1}u|_{\partial\Omega}), u\in H^s(\Omega)$ so that $\tilde{H}^{s,r}(\Omega)$ can be identified with $H^s(\Omega)$.

(iii) $1<s+\frac{1}{2}<r$. In this case we introduce an integer q such that $q-\frac{1}{2}<s<q+\frac{1}{2}$. Then, for $u\in H^s(\Omega)$,

$$Su=(u;u_0,u_1,\dots,u_{r-1}) \qquad u_k=D_n^k u|_{\partial\Omega}, 0\leqslant k\leqslant q$$

and u_k is an arbitrary element in $H^{s-k-1/2}(\partial\Omega)$ for $q<k\leqslant r-1$.

The importance of the space $\tilde{H}^{s,r}(\Omega)$ is made clear in the following lemma (see Berezanskii [5.2, p. 233] and also [5.4–5.7]):

LEMMA 5.4. Let A be a differential operator of order $r\leqslant 2m$ with sufficiently smooth coefficients defined on a domain Ω with a smooth boundary $\partial\Omega$. Then there exists a constant $C_s>0$ such that

$$\|Au\|_{H^{s-r}(\Omega)}\leqslant C_s\|u\|_{\tilde{H}^{s,r}(\Omega)} \tag{5.69}$$

that is, $A:\tilde{H}^{s,r}(\Omega)\to H^{s-r}(\Omega)$ is bounded.

Moreover, if B is a differential boundary operator of order $r\leqslant 2m-1$ with sufficiently smooth coefficients on $\partial\Omega\in C^\infty$, there is a constant $\overline{C}_s>0$ such that

$$\|Bu\|_{H^{s-r-1/2}(\partial\Omega)}\leqslant \overline{C}_s\|u\|_{\tilde{H}^{s,r}(\Omega)} \tag{5.70}$$

PROOF. For $s\geqslant r$, inequality (5.69) follows from Lemma 5.2 and the trace theorem. Obviously, in this case $\|u\|_{\tilde{H}^{s,r}(\Omega)}\leqslant C\|u\|_{H^s(\Omega)}$, and $\tilde{H}^{s,r}(\Omega)$

can be identified with $H^s(\Omega)$. Thus we need consider only $s<r$.

First suppose that $0\leqslant s<r$, and use definition (5.64); we obtain for $u,v\in E(\bar{\Omega})$,

$$\|Au\|_{H^{s-r}(\Omega)}\leqslant \sup_{v\in H^{r-s}(\Omega)} \frac{|(Au,v)_{H^0(\Omega)}|}{\|v\|_{H^{r-s}(\Omega)}} \quad (v\neq 0) \tag{5.71}$$

(the equality holds when v is restricted to $H_0^{r-s}(\Omega)$). We next use integration by parts to transfer $r-s$ derivatives in the expression $(Au,v)_{H^0(\Omega)}$ from u to v so as to obtain

$$(Au,v)_{H^0(\Omega)}=(A_s u,A_{r-s}v)_{H^0(\Omega)}+\sum_{k=0}^{r-s-1}(D_n^{r-k-1}u,B_k v)_{H^0(\partial\Omega)} \tag{5.72}$$

where A_s, A_{r-s}, and B_k are differential operators whose orders are given by their subscripts. Taking absolute values and using trace inequalities (5.51), we obtain

$$\begin{aligned}
|(Au,v)_{H^0(\Omega)}| &\leqslant \|A_s u\|_{H^0(\Omega)}\|A_{r-s}v\|_{H^0(\Omega)}\\
&\quad+\sum_{k=0}^{r-s-1}\|D_n^{r-k-1}u\|_{H^{k+s+1/2-r}(\partial\Omega)}\|B_k v\|_{H^{r-s-k-1/2}(\partial\Omega)}\\
&\leqslant C_1\|u\|_{H^s(\Omega)}\|v\|_{H^{r-s}(\Omega)}\\
&\quad+C_2\sum_{k=0}^{r-s-1}\|D_n^{r-k-1}u\|_{H^{k+s+1/2-r}(\partial\Omega)}\|v\|_{H^{r-s}(\Omega)}\\
&=C_1\|u\|_{H^s(\Omega)}\|v\|_{H^{r-s}(\Omega)}\\
&\quad+C_2\sum_{k=s}^{r-1}\|D_n^{k}u\|_{H^{s-k-1/2}(\partial\Omega)}\|v\|_{H^{r-s}(\Omega)}\\
&\leqslant C_s\left[\|u\|^2_{H^s(\Omega)}+\sum_{k=0}^{r-1}\|D_n^k u\|^2_{H^{s-k-1/2}(\partial\Omega)}\right]^{1/2}\|v\|_{H^{r-s}(\Omega)}\\
&=C_s\|u\|_{\tilde{H}^{s,r}(\Omega)}\|v\|_{H^{r-s}(\Omega)}
\end{aligned}$$

where C_1, C_2, and C_s are positive constants. Hence

$$|(Au,v)_{H^0(\Omega)}|\leqslant C_s\|u\|_{\tilde{H}^{s,r}(\Omega)}\|v\|_{H^{r-s}(\Omega)} \tag{5.73}$$

Substituting (5.73) into (5.71), we obtain (5.69).

When $s<0$, we use the *method of transposition*, which is sufficiently important to warrant a few detailed comments. For negative s, we define the operator

$$Q: H^{-s+r}(\Omega) \to H^{-s}(\Omega) \times \prod_{k=0}^{r-1} H^{-s+r-k-1/2}(\partial\Omega)$$

by

$$Qu = \left(A^* u; \overline{B}_0 v, \overline{B}_1 v, \ldots, \overline{B}_{r-1} v\right)$$

where A^* is the adjoint of A, and the operators $\overline{B}_k$ are the boundary operators appearing in Green's formula

$$(Au, v)_{H^0(\Omega)} = (u, A^* v)_{H^0(\Omega)} + \sum_{k=0}^{r-1} \left(D_n^{r-k-1} u, \overline{B}_k v\right)_{H^0(\partial\Omega)}$$

The trace theorem shows that Q is a continuous linear operator. Hence its adjoint Q^* is also linear and continuous on the dual spaces:

$$Q^*: H^s(\Omega) \times \prod_{k=0}^{r-1} H^{s-r+k+1/2}(\partial\Omega) \to H^{s-r}(\Omega), \quad s<0$$

We recognize the domain of Q^* as the space $\tilde{K}^{s,r}(\Omega)$ defined earlier. Recall that, for $s<0, \tilde{K}^{s,r}(\Omega)$ can be identified with $\tilde{H}^{s,r}(\Omega)$ by means of the map $Su = (u; u|_{\partial\Omega}, \ldots, D_n^{r-1} u|_{\partial\Omega})$. Thus Q^* can be *regarded* as a bounded linear map of $\tilde{H}^{s,r}(\Omega)$ into $H^{s-r}(\Omega)$.

With this result in mind, take $u \in E(\overline{\Omega})$. Then we write

$$\begin{aligned} \langle Q^* u, v \rangle &= \langle u, Qv \rangle \\ &= (u, A^* v)_{H^0(\Omega)} + \sum_{k=0}^{r-1} \left(D_n^{r-1-k} u, \overline{B}_k v\right)_{H^0(\partial\Omega)} \\ &= (Au, v)_{H^0(\Omega)} \end{aligned}$$

Therefore

$$|(Au, v)_{H^0(\Omega)}| \leqslant C \|u\|_{\tilde{H}^{s,r}(\Omega)} \|v\|_{H^{r-s}(\Omega)}$$

Here Au is regarded as a linear functional, and it can be extended to a linear functional on $H^{r-s}(\Omega)$; i.e., $Au \in H^{s-r}(\Omega)$ and, for $u \in E(\overline{\Omega})$,

$$\|Au\|_{H^{s-r}(\Omega)} \leqslant C \|u\|_{\tilde{H}^{s,r}(\Omega)}$$

Equation (5.69) now follows from continuity and from the fact that $E(\overline{\Omega})$ is dense in $\tilde{H}^{s,r}(\Omega)$.

When s is not an integer, we can define A on $\tilde{H}^{s,r}(\Omega)$ by a Hilbert scale connecting two adjacent integral values of s. Then (5.69) is obtained by interpolation.

Inequality (5.70) can be established along similar lines. For $s \geqslant r$, it follows directly from (5.51). Now consider the case $s<r$. With the help of trace inequalities (5.51), we obtain

$$\|Bu\|_{H^{s-r-1/2}(\partial\Omega)} = \sup_{v \in H^{r-s+1/2}(\partial\Omega)} \frac{|(Bu,v)_{H^0(\partial\Omega)}|}{\|v\|_{H^{r-s+1/2}(\partial\Omega)}}$$

$$\leqslant C_1 \sup_{v \in H^{r-s+1}(\Omega)} \frac{|(Bu,v)_{H^0(\partial\Omega)}|}{\|v\|_{H^{r-s+1}(\Omega)}} \qquad (v \neq 0) \qquad (5.74)$$

In $(Bu,v)_{H^0(\partial\Omega)}$, we shift the tangential directional derivatives D_τ^k from u to v. We obtain an expression of the form

$$\int_{\partial\Omega} b_{ijk}(x) D_t^i D_n^j u D_\tau^k v \, ds = (Bu,v)_{H^0(\partial\Omega)}$$

where $j=0,1,\ldots,r \leqslant 2m-1, |i|+|k| \leqslant r-j$. For $j>s$, we shift the tangential derivatives such that $i=0$. Recalling that $v \in H^{r-s+1}(\Omega)$, we obtain

$$\left| \int_{\partial\Omega} b_{0jk}(\mathbf{x}) D_n^j u D_\tau^k v \, ds \right| \leqslant C_2 \|D_n^j u\|_{H^{s-j-1/2}(\partial\Omega)} \|D_\tau^k v\|_{H^{j-s+1/2}(\partial\Omega)}$$

$$\leqslant C_3 \|u\|_{\tilde{H}^{s,r}(\Omega)} \sum_{|\alpha| \leqslant r-j} \|D^\alpha v\|_{H^{j-s+1}(\Omega)}$$

$$\leqslant C_4 \|u\|_{\tilde{H}^{s,r}(\Omega)} \|v\|_{H^{r-s+1}(\Omega)}$$

Substitution of last expression into (5.74) leads to (5.70).

For $j \leqslant s$, the tangential derivative is shifted such that $|k| \leqslant r-s$, and the integral over $\partial\Omega$ is estimated as above. This completes the proof of the lemma. ■

Thus we have shown that the operator A is a bounded mapping from $\tilde{H}^{s,r}(\Omega)$ to $H^{s-r}(\Omega)$, and that boundary operators B are bounded mappings from $\tilde{H}^{s,r}(\Omega)$ to $H^{s-k-1/2}(\partial\Omega), k \leqslant r-1$. We now extend, with the help of the space $\tilde{H}^{s,r}(\Omega)$ and the above properties, the results of Sections 5.5 and 5.6 to cases in which $s<2m$.

Again we denote by A a properly elliptic operator of order $2m$, by $\{B_k\}_{k=0}^{m-1}$ the normal covering family of bounded boundary operators with orders $q_k, 0 \leqslant k \leqslant m-1$, and define the mapping

$$\mathcal{P} : \tilde{H}^{s,2m}(\Omega) \to J^{s-2m}_{s-q_k-1/2}(\bar{\Omega})$$

by

$$\mathcal{P} u = (Au; B_0 u, B_1 u, \ldots, B_{m-1} u) \qquad u \in \tilde{H}^{s,2m}(\Omega) \tag{5.75}$$

In view of Lemma 5.4, it is clear that $\mathcal{P}$ is bounded; i.e.,

$$\|\mathcal{P} u\|_{J^{s-2m}_{s-q_k-1/2}(\bar{\Omega})} \leqslant C \|u\|_{\tilde{H}^{s,2m}(\Omega)} \tag{5.76}$$

EXAMPLE 5.25. (Cf. [5.6].) Consider again the Dirichlet problem

$$-\nabla^2 u + u = f \text{ in } \Omega$$

$$u = g \text{ on } \partial\Omega$$

but now suppose that $f \in H^s(\Omega)$ and $g \in H^{s-1/2}(\partial\Omega)$, with $s<0$. In this case, $r=2m=2$ and S: $\tilde{H}^{s,2}(\Omega) \to H^{s-2}(\Omega)$, where $Su=(u; u_0, u_1)$. Thus a solution to this problem must be interpreted as a triplet of functions, $u \in H^{s+2}(\Omega)$, $u_0 \in H^{s+3/2}(\partial\Omega)$, and $u_1 \in H^{s+1/2}(\partial\Omega)$, such that, for $\phi \in E(\bar{\Omega})$,

$$-\langle u, \nabla^2\phi\rangle_\Omega + \langle u, \phi\rangle_\Omega - \langle u_0, \frac{\partial\phi}{\partial n}\rangle_{\partial\Omega} - \langle u_1, \phi\rangle_{\partial\Omega} = \langle f, \phi\rangle_\Omega$$

and $u_0 = g$, where $\langle\cdot,\cdot\rangle_\Omega$ and $\langle\cdot,\cdot\rangle_{\partial\Omega}$ denote duality pairings on restrictions to Ω and $\partial\Omega$, respectively. ■

Recall that the null space $\mathfrak{N}$ of $\mathcal{P}$ is a finite-dimensional subspace of $H^s(\Omega)$, $s \geqslant 2m$. It is easy to see that $\mathfrak{N}$ and $\mathfrak{N}^*$ are also subspaces of $\tilde{H}^{s,r}(\Omega) \subset H^s(\Omega)$, for any s.

We next cite a useful lemma on projections, the proof of which makes use of the finite dimensionality of $\mathfrak{N}$ and $\mathfrak{N}^*$ and of Riesz's theorem. For details, see [5.2, pp. 228–231].

LEMMA 5.5. Every element $u \in H^s(\Omega)$, for any s, can be written uniquely in the form

$$u = v + w, \qquad v \in \mathfrak{N}, \qquad w \in \mathfrak{N}^\perp \tag{5.77}$$

Moreover, the projection operators $\Pi u = v$, and $\Pi^\perp u = w$ are continuous on $H^s(\Omega)$. Similarly, every element $F=(f;g_0,g_1,\dots,g_{m-1})\in J^{s-2m}_{s-q_k-1/2}(\bar{\Omega})$ can be uniquely written in the form

$$F=F'+F''; \qquad F'=(f';0,0,\dots,0), \qquad f'\in\mathfrak{N}^*, \qquad F''\in J^{s-2m}_{s-q_k-1/2}(\bar{\Omega}) \tag{5.78}$$

where F'' satisfies the compatibility conditions (5.57). The projection operators $Q_* F=(f';0,0,\dots,0)$ and $Q_*^{\perp}F=F''$ are continuous on $J^{s-2m}_{s-q_k-1/2}(\bar{\Omega})$. ■

Lemmas similar to that above also hold with $\mathfrak{N}$ replaced by $\mathfrak{N}^*$ in (5.77) and Π by Π_*, and with $\mathfrak{N}^*$ replaced by $\mathfrak{N}$, $J^{s-2m}_{s-q_k-1/2}(\bar{\Omega})$ by $J^{s-2m}_{s-q_k'-1/2}(\bar{\Omega})$ in (5.78) ($q_k'=2m-1-p_k=$ order of C_k), and Q_* by Q. We now introduce the following set of spaces:

$$\hat{H}^{s,r}(\Omega)=\Pi^{\perp}(\tilde{H}^{s,r}(\Omega)); \qquad \hat{H}^{s,r}_*(\Omega)=\Pi^{\perp}_*(\tilde{H}^{s,r}(\Omega)) \tag{5.79}$$

$$\hat{J}^{s-2m}_{s-q_k-1/2}(\bar{\Omega})=Q^{\perp}_*\left(J^{s-2m}_{s-q_k-1/2}(\bar{\Omega})\right); \qquad \tilde{J}^{s-2m}_{s-q_k'-1/2}(\bar{\Omega})=Q^{\perp}\left(J^{s-2m}_{s-q_k-1/2}(\bar{\Omega})\right) \tag{5.80}$$

Clearly, for $s\geqslant 2m$, $\hat{H}^{s,r}(\Omega)=\mathfrak{N}^{\perp}$. We provide $\hat{H}^{s,r}(\Omega)$ and $\hat{J}^{s-2m}_{s-q_k-1/2}(\bar{\Omega})$ with the same norms as those for $\tilde{H}^{s,r}(\Omega)$ and $J^{s-2m}_{s-q_k-1/2}(\bar{\Omega})$, respectively. Note that $\hat{J}^{s-2m}_{s-q_k-1/2}(\bar{\Omega})$ is the space of elements that satisfy the compatibility conditions (5.57). Throughout the discussion to follow, we denote by $\mathcal{P}_\sigma$ ($\sigma\geqslant 0$) the restriction of $\mathcal{P}$ to $\hat{H}^{2m+\sigma,2m}(\Omega)$ ($=H^{2m+\sigma}(\bar{\Omega})\perp\mathfrak{N}$).

We now give a fundamental theorem on regularity of solutions to (5.36) (see [5.2, p. 235]).

THEOREM 5.7. Suppose that the problem (5.36) is regularly elliptic. Consider the pair of spaces $\hat{H}^{s,2m}(\Omega)$ and $\hat{J}^{s-2m}_{s-q_k-1/2}(\bar{\Omega})$ defined above for any s, and the operator $\mathcal{P}: H^{2m}(\Omega)\to J^{0}_{2m-q_k-1/2}(\bar{\Omega})$. If $s\geqslant 2m$, the restriction $\mathcal{P}_{s-2m}$ of $\mathcal{P}$ to $\hat{H}^{s,2m}(\Omega)$ establishes a homeomorphism between $\hat{H}^{s,2m}(\Omega)$ and $\hat{J}^{s-2m}_{s-q_k-1/2}(\bar{\Omega})$. If $s<2m$, such a homeomorphism is established by the closure of the operator $\mathcal{P}_0$: $\hat{H}^{s,2m}(\Omega)\to\hat{J}^{s-2m}_{s-q_k-1/2}(\bar{\Omega})$. Further $\mathcal{P}^{-1}$: $\hat{J}^{s-2m}_{s-q_k-1/2}(\bar{\Omega})\to\hat{H}^{s,2m}(\Omega)$ exists and is continuous, and the following estimate holds:

$$\|u\|^2_{\hat{H}^{s,2m}(\Omega)}\leqslant C\left(\|Au\|^2_{H^{s-2m}(\Omega)}+\sum_{k=0}^{m-1}\|B_k u\|^2_{H^{s-q_k-1/2}(\partial\Omega)}\right) \tag{5.81}$$

PROOF. For $s \geqslant 2m$, as we observed earlier $\tilde{H}^{s,r}(\Omega) = H^s(\Omega)$ and $\hat{H}^{s,r}(\Omega)$ coincides with $\mathfrak{N}^\perp$, and therefore the theorem reduces to Theorem 5.6.

We prove the theorem for $s < 2m$. The assertion of the theorem follows if we establish the inequality (5.81), since we have already established the continuity of $\mathcal{P}$.

Suppose that $0 \leqslant s \leqslant 2m$. The operator $\mathcal{P}_0$ establishes a homeomorphism from $\hat{H}^{2m,2m}(\Omega)$ onto $\hat{J}^0_{2m-q_k-1/2}(\bar{\Omega})$, and therefore $\mathcal{P}_0^{-1}$ exists and is continuous from $\hat{J}^0_{2m-q_k-1/2}(\bar{\Omega})$ to $\hat{H}^{2m,2m}(\Omega)$. We define an operator $\mathfrak{M}$ on $F \in \hat{J}^0_{2m-q_k-1/2}(\bar{\Omega})$ by

$$\mathfrak{M}F = \left(\mathcal{P}_0^{-1}F, S_0(\mathcal{P}_0^{-1}F), S_1(\mathcal{P}_0^{-1}F), \ldots, S_{m-1}(\mathcal{P}_0^{-1}F)\right) \quad (5.82)$$

where $\{S_k\}_{k=0}^{m-1}$ are the operators in Green's formula (5.24). Clearly, $\mathfrak{M}F \in J^{2m}_{2m-p_k-1/2}(\bar{\Omega})$, and $\mathfrak{M}$ is a bounded operator from $\hat{J}^0_{2m-q_k-1/2}(\bar{\Omega})$ to $\hat{J}^{2m}_{2m-p_k-1/2}(\bar{\Omega})$. Hence, $\mathfrak{M}Q_*^\perp$ is a bounded operator from $J^0_{2m-q_k-1/2}(\bar{\Omega})$ to $J^{2m}_{2m-p_k-1/2}(\bar{\Omega})$. Analogously, we can define an operator $\mathfrak{M}^*$ from $\tilde{J}^0_{2m-p_k'-1/2}(\bar{\Omega})$ to $J^{2m}_{2m-q_k'-1/2}(\bar{\Omega})$ by setting

$$\mathfrak{M}^*F^* = \left((\mathcal{P}_0^*)^{-1}F^*, T_0\left((\mathcal{P}_0^*)^{-1}F^*\right), \ldots, T_{m-1}\left((\mathcal{P}_0^*)^{-1}F^*\right)\right) \quad (5.83)$$

where F^* is the data of the adjoint problem, $\mathcal{P}_0^*$ is the restriction of $\mathcal{P}^*$ to $\hat{H}_*^{2m,2m}(\Omega)$, and $\{T_k\}_{k=0}^{m-1}$ are the operators in the Green's formula. Again $\mathfrak{M}^*Q^\perp$ is bounded from all of $J^0_{2m-p_k'-1/2}(\bar{\Omega})$ to $J^{2m}_{2m-q_k'-1/2}(\bar{\Omega})$.

Now using (5.82), (5.83), and Green's formula we can show that (see [5.2, p. 237]), for $F \in J^0_{2m-q_k-1/2}(\bar{\Omega})$ and $F^* \in J^0_{2m-p_k'-1/2}(\bar{\Omega})$,

$$\left[\mathfrak{M}Q_*^\perp F, F^*\right]_0 = \left[F, \mathfrak{M}^*Q^\perp F^*\right]_0 \quad (5.84)$$

where

$$[F, (v, T_0 v, T_1 v, \ldots, T_{m-1}v)]_0 = (f, v)_{H^0(\Omega)} + \sum_{k=0}^{m-1} (g_k, T_k v)_{H^0(\partial\Omega)}$$

We can construct the adjoint operator $(\mathfrak{M}^*Q^\perp)^*$ to $\mathfrak{M}^*Q^\perp$ with respect to $[\cdot,\cdot]_0$, which acts continuously from all of $J^{-2m}_{-(2m-q_k'-1/2)}(\bar{\Omega})$ to $J^0_{-(2m-p_k'-1/2)}(\bar{\Omega})$. It can be verified that $(\mathfrak{M}^*Q^\perp)^*$ is an extension of the operator $\mathfrak{M}Q_*^\perp$ and that it is continuous from $J^0_{2m-q_k-1/2}(\bar{\Omega})$ into $J^{2m}_{2m-p_k-1/2}(\bar{\Omega})$ and from $J^{-2m}_{-(2m-q_k'-1/2)}(\bar{\Omega})$ into $J^0_{-(2m-p_k'-1/2)}(\bar{\Omega})$. That is, for $0 \leqslant s \leqslant 2m$, $(\mathfrak{M}^*Q^\perp)^*$ is a bounded operator from $J^{s-2m}_{s-q_k-1/2}(\bar{\Omega})$ into $J^s_{s-p_k-1/2}(\bar{\Omega})$. Hence, we have the inequality,

$$\|(\mathfrak{M}^*Q^\perp)^*F\|_{J^s_{s-p_k-1/2}(\bar{\Omega})} \leqslant C_1\|F\|_{J^{s-2m}_{s-q_k}-1/2(\bar{\Omega})} \quad (5.85)$$

for $F \in J^{s-2m}_{s-q_k-1/2}(\bar{\Omega})$, $0 \leqslant s \leqslant 2m$. If $u \in \hat{H}^{2m,2m}(\Omega)$, then $F = \mathcal{P}u \in \bar{J}^0_{2m-q_k-1/2}(\bar{\Omega}) \subset J^{s-2m}_{s-q_k-1/2}(\bar{\Omega})$ and $F = Q_*^{\perp}F$. Substituting this into (5.85) and noting that $(\mathfrak{M}^* Q^{\perp})^* = \mathfrak{M} Q_*^{\perp}$, we obtain, for $u \in \hat{H}^{2m,2m}(\Omega)$, $0 \leqslant s \leqslant 2m$,

$$\|u\|^2_{H^s(\Omega)} + \sum_{k=0}^{m-1} \|S_k u\|^2_{H^{s-p_k-1/2}(\partial\Omega)}$$

$$\leqslant C_1^2 \left[\|Au\|^2_{H^{s-2m}(\Omega)} + \sum_{k=0}^{m-1} \|B_k u\|^2_{H^{s-q_k-1/2}(\partial\Omega)} \right] \tag{5.86}$$

Now consider the $2m$ boundary differential operators $(B_0, B_1, \ldots, B_{m-1}; S_0, S_1, \ldots, S_{m-1})$ that form a Dirichlet system of order $2m$. Denote $B_j = Q_j, S_j = Q_{j+m}$ for $j = 0, 1, \ldots, m-1$, where the operator Q_j is a boundary operator of order j. Recall from (5.15) that it is possible to represent D_n^j in the form

$$D_n^j = \frac{\partial^j}{\partial n^j} = \sum_{k=0}^{j} \tau_{jk}(\mathbf{x}, D_\tau) Q_k \qquad j = 0, \ldots, 2m-1; \mathbf{x} \in \partial\Omega$$

where $\tau_{jk}(\mathbf{x}, D_\tau)$ are differential expressions, containing only tangential derivatives D_τ, and of order $j-k$. Then, for $u \in H^{2m}(\Omega)$, $0 \leqslant s \leqslant 2m$,

$$\|D_n^j u\|_{H^{s-j-1/2}(\partial\Omega)} = \left\| \sum_{k=0}^{j} \tau_{jk}(\mathbf{x}, D_\tau) Q_k u \right\|_{H^{s-j-1/2}(\partial\Omega)}$$

$$\leqslant \sum_{k=0}^{j} \|\tau_{jk}(\mathbf{x}, D_\tau) Q_k u\|_{H^{s-j-1/2}(\partial\Omega)}$$

$$\leqslant C_2 \sum_{k=0}^{j} \|Q_k u\|_{H^{s-k-1/2}(\partial\Omega)}$$

and

$$\sum_{j=0}^{2m-1} \|D_n^j u\|^2_{H^{s-j-1/2}(\partial\Omega)} \leqslant C_2^2 \sum_{j=0}^{2m-1} \|Q_j u\|^2_{H^{s-j-1/2}(\partial\Omega)}$$

$$= C_2^2 \sum_{j=0}^{m-1} \Big(\|B_j u\|^2_{H^{s-q_j-1/2}(\partial\Omega)}$$

$$+ \|S_j u\|^2_{H^{s-p_j-1/2}(\partial\Omega)} \Big)$$

Therefore

$$\|u\|^2_{H^s(\Omega)}+\sum_{k=0}^{2m-1}\|D_n^k u\|^2_{H^{s-k-1/2}(\partial\Omega)}$$

$$\leqslant C_2^2\left[\|u\|^2_{H^s(\Omega)}+\sum_{k=0}^{m-1}\left(\|B_k u\|^2_{H^{s-q_k-1/2}(\partial\Omega)}+\|S_k u\|^2_{H^{s-p_k-1/2}(\partial\Omega)}\right)\right]$$

That is,

$$\|u\|_{\tilde{H}^{s,2m}(\Omega)}\leqslant C_2|||u|||_{H^s(\Omega)} \tag{5.87}$$

Inequalities (5.86) and (5.87) imply that the norms $|||u|||_{H^s(\Omega)}$ and $\|u\|_{\tilde{H}^{s,2m}(\Omega)}$ are equivalent. Hence, according to (5.81),

$$\|u\|_{\tilde{H}^{s,2m}(\Omega)}\leqslant C\left(\|Au\|_{H^{s-2m}(\Omega)}+\sum_{k=0}^{m-1}\|B_k u\|_{H^{s-q_k-1/2}(\partial\Omega)}\right)$$

which is the required inequality.

Now we sketch the proof for the case $s\leqslant 0$. The arguments are similar to the previous case, and use the idea of passage to the adjoint operators (however, we now do not need interpolation). Let $r=-s\geqslant 0$, and define the operator $\mathfrak{M}_r^*$ similar to (5.83), but with $\mathcal{P}_0^*$ replaced by $\mathcal{P}_r^*$ (the restriction of $\mathcal{P}^*$ to $\hat{H}^{r+2m,2m}(\Omega)$). The operator $\mathfrak{M}_r^*$ is bounded from $\tilde{J}^r_{r+p_k+1/2}(\overline{\Omega})$ to $\hat{H}_*^{r+2m,2m}(\Omega)$. The operator $\mathfrak{M}_r^*Q^\perp$ is bounded from all of $J^r_{r+p_k+1/2}(\overline{\Omega})$ to $J^{2m+r}_{r+q_k+1/2}(\overline{\Omega})$. The adjoint of $\mathfrak{M}_r^*Q^\perp$ (in the sense of $[\cdot,\cdot]_0$) is a bounded operator from all of $J^{-2m-r}_{-(r+q_k+1/2)}(\overline{\Omega})$ $(=J^{s-2m}_{s-q_k-1/2}(\overline{\Omega}))$ to $J^{-2m-r}_{-(r+q'+1/2)}(\overline{\Omega})(=J^s_{s-p_k-1/2}(\overline{\Omega}))$. As before, $(\mathfrak{M}_r^*Q^\perp)^*$ is an extension of the operator $\mathfrak{M}_r Q_*^\perp$. Thus the inequality (5.85) holds for $s\leqslant 0$. The remaining arguments are similar to those used previously. ■

COROLLARY 5.7.1. Let the conditions of Theorem 5.7 hold, with $s<2m$. Then we have the *a priori estimate*

$$\|u\|_{\tilde{H}^{s,2m}(\Omega)}\leqslant C\left(\|f\|_{H^{s-2m}(\Omega)}+\sum_{k=0}^{m-1}\|g_k\|_{H^{s-q_k-1/2}(\partial\Omega)}\right) \tag{5.88}$$

■

REFERENCES

5.1. Lions, J. L., and Magenes, E., *Non-Homogeneous Boundary-Value Problems and Applications*, Vol. I, Translated from a revision of the 1968 French edition by P. Kenneth, Springer-Verlag, New York, 1972.

5.2. Berezanskii, Ju., M., *Expansions in Eigenfunctions of Self-Adjoint Operators*, Translations of Mathematical Monographs, Vol. 17, American Mathematical Society, Providence, R.I., 1968.

5.3. Dunford, N. and Schwartz, J. T., *Linear Operators, Part I: General Theory*, Interscience, New York, 1958.

5.4. Volevič, L. R., "Solvability of Boundary Value Problems for General Elliptic Systems," *Amer. Math. Soc. Transl.*, Vol. 67, No. 2, pp. 182–225, 1968.

5.5. Roitberg, J. A. and Seftel, Z. G., "A Theorem on Homeomorphisms for Elliptic Systems and its Applications," *Math. U.S.S.R. Sb.*, Vol. 7, No. 3, pp. 439–465, 1969.

5.6. Kellogg, B., "Properties and Solutions of Elliptic Boundary-Value Problems," Chapter 3 of "Survey Lectures on the Mathematical Foundations of the Finite Element Method," by I. Babuška and A. K. Aziz, in Aziz, A. K., Ed., *The Mathematical Foundations of the Finite Element Method with Applications to Partial Differential Equations*, Academic Press, New York, 1972, pp. 47–82.

5.7. Nečas, J., *Les Méthodes Directes en Théorie des Équations Elliptiques*, Masson, Paris, 1967.

PART II

THE THEORY OF FINITE ELEMENTS

6
FINITE-ELEMENT INTERPOLATION

6.1 INTRODUCTION

The finite-element method is, first, a systematic and very powerful method of interpolation. Its most attractive features are threefold: first, the interpolation process is virtually independent of the geometry of the domain under consideration. Consequently, functions defined on very irregular, even multiply connected, domains can be quite easily interpolated. Second, conditions on given functions at points on the boundary can be matched in a systematic manner by the finite-element interpolant. Finally, since the concepts of interpolation and approximation are actually quite closely related, the finite-element concept leads to a very powerful method for the approximate solution of boundary-value problems. The properties of the technique as a method of approximation, however, cannot be fully appreciated until its properties as a method of interpolation are established, and it is toward this objective that the present chapter is directed.

Following this introduction, we describe so-called topological properties of finite-element models. The philosophy we adopt here is essentially that which has been described in [6.1] and [6.2], and it emphasizes the connectivity and local character of finite-element representations. Next we introduce the concepts of restrictions, prolongations, and projections, and then give a summary account of the theory of conjugate finite-element basis functions. Next we describe a variety of so-called families of finite elements and their properties, including Lagrange and Hermite families on simplexes and hypercubes, serendipity elements, isoparametric elements, and rational elements. We then go on to develop the fundamental theorems for finite-element interpolation of smooth functions and of functions in Sobolev spaces. Then we follow the penetrating work of Ciarlet and Raviart [6.3, 6.4] and establish a priori error estimates for finite-element interpolation in appropriate sup and Sobolev norms, in terms of certain mesh parameters and the degree of the polynomials used in the interpolation.

6.2 CONNECTIVITY OF FINITE-ELEMENT MODELS OF DOMAINS $\Omega \subset \mathsf{R}^n$

Consider a smooth function $u(\mathbf{x})$ whose domain $\bar{\Omega}$ is the closure of an open bounded region Ω in n-dimensional euclidean space R^n. As usual, $\mathbf{x} = (x_1, x_2, \ldots, x_n)$ is a point in $\bar{\Omega} = \Omega \cup \partial\Omega$, where $\partial\Omega$ is the boundary of Ω.

To construct a finite-element model of $u(\mathbf{x})$ [i.e., a finite-element interpolation of $u(\mathbf{x})$] we must first construct a finite-element model of $\bar{\Omega}$.

A *finite-element* model of the closed domain $\bar{\Omega}$ is a region $\tilde{\Omega} \subset \mathsf{R}^n$ which is the union of a finite number E of closed bounded subregions $\bar{\Omega}_e$ of R^n, each $\bar{\Omega}_e$ being the closure of an open region Ω_e:

$$\bar{\Omega}_e = \Omega_e \cup \partial\Omega_e \qquad e = 1, 2, \ldots, E \tag{6.1}$$

where $\partial\Omega_e$ is the boundary of Ω_e. The subregions $\bar{\Omega}_e$ are called *finite elements* of $\tilde{\Omega}$, and we obviously select $\tilde{\Omega}$ to coincide with (or, at least, closely approximate) $\bar{\Omega}$. The region $\tilde{\Omega}$ is called the *connected model* of $\bar{\Omega}$, and the open elements Ω_e are pairwise disjoint. We have

$$\tilde{\Omega} = \bigcup_{e=1}^{E} \bar{\Omega}_e \tag{6.2}$$

$$\Omega_e \cap \Omega_f = \varnothing \qquad e \neq f \tag{6.3}$$

In the connected model, we identify a finite number G of points, called *global nodes*, and we label them consecutively $\mathbf{x}^1, \mathbf{x}^2, \ldots, \mathbf{x}^G$. Likewise, we identify within each element $\bar{\Omega}_e$ a number N_e of points, called *local nodes*, and we label them consecutively $\mathbf{x}_e^1, \mathbf{x}_e^2, \ldots, \mathbf{x}_e^{N_e}; e = 1, 2, \ldots, E$. A correspondence must exist between points in $\bar{\Omega}_e$ and $\tilde{\Omega}$ and, in particular, between nodal points $\mathbf{x}_e^N$ in $\bar{\Omega}_e$ and nodes $\mathbf{x}^i$ in $\tilde{\Omega}$, if the elements are to fit together smoothly to form $\tilde{\Omega}$. That is, if a given model $\tilde{\Omega}$ is to correspond to a given class $\{\bar{\Omega}_e\}$ of finite elements, then certain compatibility conditions must hold.

Compatibility Conditions for Finite-Element Models

If we are *given* a region Ω, then it is often a simple matter to partition it into a collection of finite elements. A more common situation encountered in applications, however, is one in which a domain $\tilde{\Omega}$ is "built-up" by fitting a collection of disjoint elements. We refer to this process as *assembling* elements into a *connected model* $\tilde{\Omega}$. Conditions that must be met if such an assembly is to be possible are fairly obvious and are called *compatibility conditions.* To establish them, let $\bar{\Omega}_e$ and $\bar{\Omega}_f$ denote two distinct finite elements which we wish to make adjacent to each other in a model $\tilde{\Omega}$, and let $\partial\Omega_{ef}$ denote a subset of $\tilde{\Omega}$ such that the intersection of $\partial\Omega_{ef}$ with the boundaries $\partial\Omega_e$ and $\partial\Omega_f$ is nonempty. Denote the intersection $\partial\Omega_e \cap \partial\Omega_{ef}$ by $\partial\Omega'_e$ and $\partial\Omega_f \cap \partial\Omega_{ef}$ by $\partial\Omega'_f$, so that $\partial\Omega'_e$ and $\partial\Omega'_f$ represent the portions of $\partial\Omega_e$ and $\partial\Omega_f$ mapped into $\partial\Omega_{ef}$ on assembling elements. Further, let $\{\mathbf{x}_e^N(\partial\Omega'_e)\}_{N=1}^{N'_e}$, $\{\mathbf{x}_f^M(\partial\Omega'_f)\}_{M=1}^{N'_f}$, and $\{\mathbf{x}^i(\partial\Omega_{ef})\}_{i=1}^{G'}$ denote the sets of all nodal points in $\partial\Omega'_e$, $\partial\Omega'_f$, and $\partial\Omega_{ef}$, respectively. Then elements Ω_e and Ω_f are compatible with $\tilde{\Omega}$ if and only if

(i) There exists a one-to-one mapping Λ of $\partial\Omega'_e \times \partial\Omega'_f$ onto $\partial\Omega_{ef}$.

(ii) $N'_e = N'_f = G'$, and there exists a one-to-one mapping

$$\Lambda_{(ef)} \text{ of } \{\mathbf{x}_e^N(\partial\Omega'_e)\}_{N=1}^{N'_e} \times \{\mathbf{x}_f^M(\partial\Omega'_f)\}_{M=1}^{N'_f} \text{ onto } \{\mathbf{x}^i(\partial\Omega_{ef})\}_{i=1}^{G'}.$$

(iii) For every nodal point $\mathbf{x}_e^N$ and each finite element Ω_e, there is a corresponding node $\mathbf{x}^i \in \tilde{\Omega}$; i.e., there is a one-to-one mapping $\Lambda^{(e)}$ of local nodes $\mathbf{x}_e^N$ into global nodes $\mathbf{x}^i$ for each fixed e.

If these conditions are satisfied, then elements $\bar{\Omega}_e$ and $\bar{\Omega}_f$ can be fit smoothly together in $\tilde{\Omega}$, and we have

$$\partial\Omega_{ef} = \partial\Omega_e \cap \partial\Omega_f \tag{6.4}$$

The set $\partial\Omega_{ef}$ is then called the *interelement boundary* of adjacent elements $\bar{\Omega}_e$ and $\bar{\Omega}_f$. An example of compatible elements is shown in Fig. 6.1. There $N'_e = N'_f = G' = 3$.

Boolean Transformations

Unless noted otherwise, we henceforth assume that the compatibility conditions are satisfied and that therefore the elements fit continuously together to form $\tilde{\Omega}$. Also, without loss in generality, we take $\tilde{\Omega} = \bar{\Omega}$.

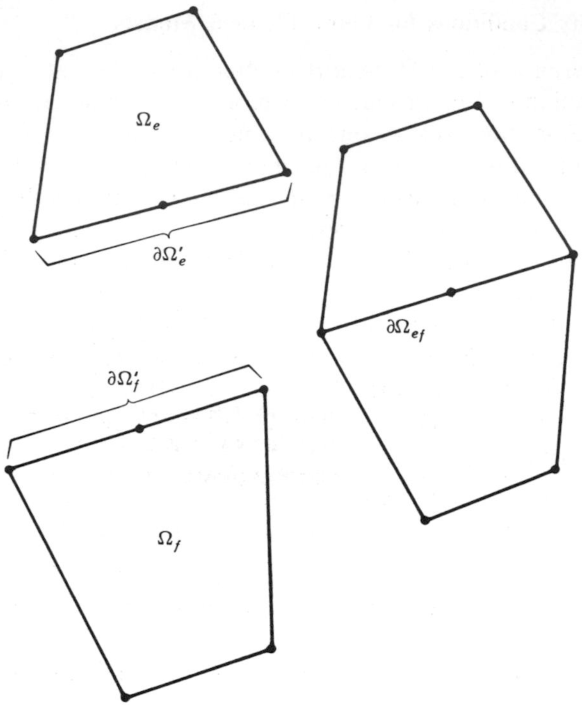

Figure 6.1 Compatible finite elements.

Recall that we have distinguished two types of nodal points: the global nodal points $\{\mathbf{x}^i\}_{i=1}^G$ and the local nodal points $\{\mathbf{x}_e^N\}_{N=1}^{N_e}; e=1,2,\ldots,E$. We remark that a dual meaning is to be assigned to the symbol $\mathbf{x}^i$(and $\mathbf{x}_e^N$). The notation $\mathbf{x}^i$ shall be used to indicate both the point $\mathbf{x}\in\tilde{\Omega}$ labeled i and the label i itself. The particular meaning intended is always clear from the context.

Now consider a collection of finite elements connected so as to form a region $\tilde{\Omega}$. An example of a collection of two-dimensional elements is shown if Fig. 6.2. Consider a typical element Ω_e of this collection, isolated from the rest as indicated in Fig. 6.2. Since the compatibility conditions are satisfied, we see that there must exist a simple correspondence between the labels $\mathbf{x}_e^N$ used to count nodal points locally in $\tilde{\Omega}_e$, and the labels $\mathbf{x}^i$ used to count nodes globally in $\tilde{\Omega}$. For example, in the model indicated in Fig. 6.2, node 2 of element Ω_e corresponds to node 6 of the connected model (or, in our choice of notation, $\mathbf{x}_e^2$ corresponds to $\mathbf{x}^6$). When such a correspondence exists, we say that node $\mathbf{x}_e^N$ of element Ω_e is *incident on* or *coincident with* node $\mathbf{x}^i$ of the connected model. Mathematically, we describe this corre-

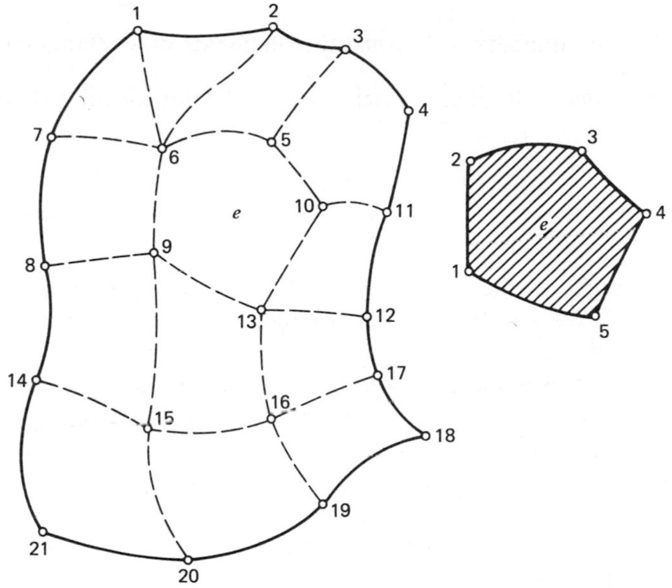

Figure 6.2 A global and local numbering scheme for nodes in a finite-element model.

spondence for each element label e as a mapping $\overset{(e)}{\Lambda}$ of local labels into global ones:

$$\overset{(e)}{\Lambda} : \{\mathbf{x}_e^N\}_{N=1}^{N_e} \rightarrow \{\mathbf{x}^i\}_{i=1}^{G} \tag{6.5}$$

Equivalently, since the correspondence is linear, we can effect it by the transformation

$$\mathbf{x}^i = \sum_{N=1}^{N_e} \overset{(e)}{\Lambda}{}^i_N \mathbf{x}_e^N \qquad e \text{ fixed} \tag{6.6}$$

where $\overset{(e)}{\Lambda}{}^i_N$ is an array of ones and zeroes defined by

$$\overset{(e)}{\Lambda}{}^i_N = \begin{cases} 1 & \text{if node } N \text{ of element } \Omega_e \text{ is} \\ & \text{incident on node } i \text{ of the} \\ & \text{connected model} \\ 0 & \text{if otherwise} \end{cases} \tag{6.7}$$

For each e, the numbers $\overset{(e)}{\Lambda}{}^{i}_{N}$ constitute elements of a retangular $(G \times N_e)$ matrix $\overset{(e)}{\mathbf{\Lambda}}$ called a Boolean matrix. The collection of all E mappings of this type,

$$\Lambda = \left\{ \overset{(1)}{\Lambda}, \overset{(2)}{\Lambda}, \ldots, \overset{(E)}{\Lambda} \right\} \tag{6.8}$$

effectively maps all the elements into their proper positions in the connected model. We say that $\mathbf{\Lambda}$ describes the *discrete connectivity* of the finite element model of $\overline{\Omega}$.

The reverse procedure of assembling elements into a global model is called *decomposition*. We also view the process of decomposing a connected collection of elements Ω into E isolated finite elements as a renumbering of the global nodal points associated with an element so that they correspond to the labels adopted for numbering local nodes. For example, $\mathbf{x}^5$ in Fig. 6.2 corresponds to $\mathbf{x}_e^3$. We describe this correspondence by a mapping

$$\overset{(e)}{\Omega} : \{\mathbf{x}^i\}_{i=1}^{G} \rightarrow \{\mathbf{x}_e^N\}_{N=1}^{N_e} \tag{6.9}$$

or, equivalently,

$$\mathbf{x}_e^N = \sum_{i=1}^{G} \overset{(e)}{\Omega}{}^{N}_{i} \mathbf{x}^i \qquad e = 1, 2, \ldots, E \tag{6.10}$$

where $\overset{(e)}{\Omega}{}^{N}_{i}$ is an array of ones and zeros defined by

$$\overset{(e)}{\Omega}{}^{N}_{i} = \begin{cases} 1 & \text{if node } i \text{ of the connected} \\ & \text{model is incident on node} \\ & N \text{ of element } \Omega_e \\ 0 & \text{if otherwise} \end{cases} \tag{6.11}$$

For each e, the numbers $\overset{(e)}{\Omega}{}^{N}_{i}$ constitute elements of a rectangular $(N_e \times G)$ matrix $\overset{(e)}{\mathbf{\Omega}}$ also called a Boolean matrix. The collection of all E mappings of this type,

$$\mathbf{\Omega} = \{\overset{(1)}{\mathbf{\Omega}}, \overset{(2)}{\mathbf{\Omega}}, \ldots, \overset{(E)}{\mathbf{\Omega}}\} \tag{6.12}$$

effectively maps the connected model into a collection of disjoint elements. We say that Ω describes the *discrete decomposition* of the finite-element model of the region Ω.

Some properties of the mappings $\overset{(e)}{\Lambda}$ and $\overset{(e)}{\Omega}$ are summarized in the following theorem.

THEOREM 6.1. Let $\mathcal{G}=\{\mathbf{x}^i\}_{i=1}^{G}$ and $\mathcal{L}_e=\{\mathbf{x}_e^N\}_{N=1}^{N_e}$ denote the set of global node labels and local node labels corresponding to element Ω_e, respectively and let $\mathcal{L}=\cup_{e=1}^{E}\mathcal{L}_e$. Let $\overset{(e)}{\Lambda}$ denote the mapping of $\mathcal{L}_e$ into $\mathcal{G}$ defined by (6.6) and let $\overset{(e)}{\Omega}$ denote the mapping of $\mathcal{G}$ into $\mathcal{L}$ defined by (6.10). Then,

(i) $\overset{(e)}{\Lambda}{}^{i}_{N}$ is the transpose of $\overset{(e)}{\Omega}{}^{N}_{i}$;

$$\overset{(e)}{\Lambda}{}^{i}_{N}=\left(\overset{(e)}{\Omega}{}^{N}_{i}\right)^{T} \tag{6.13}$$

(ii) For each fixed e, $\overset{(e)}{\Lambda}$ defines a right inverse of $\overset{(e)}{\Omega}$; i.e.,

$$\sum_{i=1}^{G}\overset{(e)}{\Omega}{}^{N}_{i}\overset{(e)}{\Lambda}{}^{i}_{M}=\delta^N_M \qquad M,N=1,2,\ldots,N_e \tag{6.14}$$

where δ^N_M is the Kronecker delta ($\delta^N_M=1$ if $M=N, \delta^N_M=0$ if $M\neq N$).

(iii) The restriction of $\overset{(e)}{\Omega}$ to the subset of $\mathcal{G}$ contained in $\overline{\Omega}_e$ defines a right inverse of $\overset{(e)}{\Lambda}$; i.e.,

$$\sum_{N=1}^{N_e}\overset{(e)}{\Lambda}{}^{i}_{N}\overset{(e)}{\Omega}{}^{N}_{j}=\begin{cases}\delta^i_j & \text{if } \mathbf{x}^i \text{ and } \mathbf{x}^j\in\overline{\Omega}_e\\ 0 & \text{if } \mathbf{x}^i \text{ and } \mathbf{x}^j\not\in\Omega_e\end{cases} \tag{6.15}$$

PROOF. Property (i) follows immediately from definitions (6.7) and (6.11). Property (ii) is also obvious; consider

$$\sum_{i=1}^{G}\overset{(e)}{\Omega}{}^{N}_{i}\overset{(e)}{\Lambda}{}^{i}_{M}=\overset{(e)}{\Omega}{}^{N}_{1}\overset{(e)}{\Lambda}{}^{1}_{M}+\overset{(e)}{\Omega}{}^{N}_{2}\overset{(e)}{\Lambda}{}^{2}_{M}+\cdots+\overset{(e)}{\Omega}{}^{N}_{G}\overset{(e)}{\Lambda}{}^{G}_{M}$$

All items in this sum vanish that do not correspond to global nodes contained in element e. For the remaining terms, when $N \neq M$, we always have a product of 1 with zero, hence the sum is zero. When $M=N$, suppose $\overset{(e)}{\Omega}{}_1^N=1$. Then $\overset{(e)}{\Lambda}{}_N^1=1$ by (i), but all other $\overset{(e)}{\Omega}{}_i^N=0$. Hence the sum equals unity and (6.14) holds. In other words, $\mathbf{\Lambda}$ of (6.8) is a surjective mapping of $\mathcal{L}$ onto $\mathcal{G}$, but it is not one-to-one; its restriction $\overset{(e)}{\Lambda}$ to $\mathcal{L}_e$ is one-to-one but not onto. Hence $\overset{(e)}{\Lambda}$ has an inverse defined on its range, and that inverse is precisely $\overset{(e)}{\Omega}$. Part (iii) is proved in a similar manner. We identify a subset of $\mathcal{G}$ which $\overset{(e)}{\Omega}$ maps one-to-one and onto $\{\mathbf{x}_e^N\}_{e=1}^{N_e}$. This subset is the set of global nodes contained in element $\overline{\Omega}_e$. Hence $\overset{(e)}{\Omega}$ has an inverse on this set, and a simple use of the definitions (6.7) and (6.11) shows that this inverse is $\overset{(e)}{\Lambda}$. ■

EXAMPLE 6.1. Consider the collection of three finite elements shown in Fig. 6.3. The node numbering scheme indicated inside each element

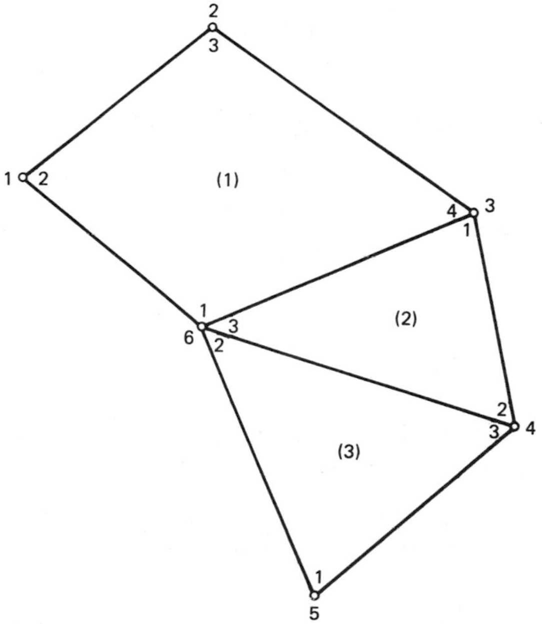

Figure 6.3 A collection of three finite elements.

describes the local nodes, whereas that indicated externally labels the global nodes. Here $G=6, N_1=4, N_2=3$, and $N_3=3$.

Observe that

$$\begin{bmatrix} \mathbf{x}^1_{(1)} \\ \mathbf{x}^2_{(1)} \\ \mathbf{x}^3_{(1)} \\ \mathbf{x}^4_{(1)} \end{bmatrix} = \begin{bmatrix} 0 & 0 & 0 & 0 & 0 & 1 \\ 1 & 0 & 0 & 0 & 0 & 0 \\ 0 & 1 & 0 & 0 & 0 & 0 \\ 0 & 0 & 1 & 0 & 0 & 0 \end{bmatrix} \begin{bmatrix} \mathbf{x}^1 \\ \mathbf{x}^2 \\ \mathbf{x}^3 \\ \mathbf{x}^4 \\ \mathbf{x}^5 \\ \mathbf{x}^6 \end{bmatrix} = \begin{bmatrix} \mathbf{x}^6 \\ \mathbf{x}^1 \\ \mathbf{x}^2 \\ \mathbf{x}^3 \end{bmatrix}$$

Hence, the Boolean matrices $\overset{(1)}{\boldsymbol{\Omega}}$ and $\overset{(1)}{\boldsymbol{\Lambda}}$ are given by

$$\overset{(1)}{\boldsymbol{\Omega}} = \overset{(1)}{\boldsymbol{\Lambda}}{}^T = \begin{bmatrix} 0 & 0 & 0 & 0 & 0 & 1 \\ 1 & 0 & 0 & 0 & 0 & 0 \\ 0 & 1 & 0 & 0 & 0 & 0 \\ 0 & 0 & 1 & 0 & 0 & 0 \end{bmatrix}$$

Likewise,

$$\overset{(2)}{\boldsymbol{\Omega}} = \overset{(2)}{\boldsymbol{\Lambda}}{}^T = \begin{bmatrix} 0 & 0 & 1 & 0 & 0 & 0 \\ 0 & 0 & 0 & 1 & 0 & 0 \\ 0 & 0 & 0 & 0 & 0 & 1 \end{bmatrix}$$

$$\overset{(3)}{\boldsymbol{\Omega}} = \overset{(3)}{\boldsymbol{\Lambda}}{}^T = \begin{bmatrix} 0 & 0 & 0 & 0 & 1 & 0 \\ 0 & 0 & 0 & 0 & 0 & 1 \\ 0 & 0 & 0 & 1 & 0 & 0 \end{bmatrix}$$

wherein rows in $\overset{(e)}{\boldsymbol{\Omega}}$ correspond to local node labels and columns correspond to global node labels.

Also note, for example, that

$$\overset{(1)}{\boldsymbol{\Omega}}\,\overset{(1)}{\boldsymbol{\Lambda}} = \begin{bmatrix} 0 & 0 & 0 & 0 & 0 & 1 \\ 1 & 0 & 0 & 0 & 0 & 0 \\ 0 & 1 & 0 & 0 & 0 & 0 \\ 0 & 0 & 1 & 0 & 0 & 0 \end{bmatrix} \begin{bmatrix} 0 & 1 & 0 & 0 \\ 0 & 0 & 1 & 0 \\ 0 & 0 & 0 & 1 \\ 0 & 0 & 0 & 0 \\ 0 & 0 & 0 & 0 \\ 1 & 0 & 0 & 0 \end{bmatrix} = \begin{bmatrix} 1 & 0 & 0 & 0 \\ 0 & 1 & 0 & 0 \\ 0 & 0 & 1 & 0 \\ 0 & 0 & 0 & 1 \end{bmatrix} = \mathbf{I}$$

and

$$\overset{(1)}{\Lambda}\,\overset{(1)}{\Omega} = \begin{bmatrix} 1 & 0 & 0 & 0 & 0 & 0 \\ 0 & 1 & 0 & 0 & 0 & 0 \\ 0 & 0 & 1 & 0 & 0 & 0 \\ 0 & 0 & 0 & 0 & 0 & 0 \\ 0 & 0 & 0 & 0 & 0 & 0 \\ 0 & 0 & 0 & 0 & 0 & 1 \end{bmatrix}$$

If we delete from $\overset{(1)}{\Lambda}\,\overset{(1)}{\Omega}$ those rows and columns corresponding to nodes not contained in element (1), i.e., rows and columns 4 and 5, as indicated by the dashed lines, we obtain a 4×4 identity matrix, in agreement with (6.15).

Likewise

$$\overset{(2)}{\Omega}\,\overset{(2)}{\Lambda} = \begin{bmatrix} 1 & 0 & 0 \\ 0 & 1 & 0 \\ 0 & 0 & 1 \end{bmatrix}$$

etc. ■

6.3 LOCAL AND GLOBAL REPRESENTATIONS OF FUNCTIONS

We are now ready to consider finite-element representations of a given smooth function $u(\mathbf{x})$, $\mathbf{x}\in\overline{\Omega}$. By a *representation* we mean an interpolant of $u(\mathbf{x})$ given in the form of a function $U(\mathbf{x})$, defined on the finite-element model $\tilde{\Omega}$ of $\overline{\Omega}$, which coincides with $u(\mathbf{x})$ and various derivatives of $u(\mathbf{x})$ at the nodal points $\mathbf{x}^i, i=1,2,\ldots,G$.

A key feature of such finite-element interpolants is that they are built up from so-called local interpolations. To appreciate what is meant by the term "local interpolation" consider a typical finite element Ω_e which, for the moment, is isolated from the model $\tilde{\Omega}$ of $\overline{\Omega}$. We next introduce a system of $\{\psi_N^{\alpha(e)}(\mathbf{x})\}_{N=1}^{N_e}$, $|\boldsymbol{\alpha}|\leqslant q$, called *local interpolation functions of order q for element* Ω_e, defined so as to have the following properties.

(i) $$\psi_N^{\alpha(e)}(\mathbf{x})\equiv 0 \qquad \text{if } \mathbf{x}\not\in\overline{\Omega}_e \tag{6.16}$$

(ii) $$D^{\beta}\psi_N^{\alpha(e)}(\mathbf{x}_e^M)=\delta_N^M\delta^{\beta\alpha} \tag{6.17}$$

where

$$\left.\begin{array}{c}\boldsymbol{\alpha},\boldsymbol{\beta}\in\hat{Z}\subset Z_{+}^{n}\qquad|\boldsymbol{\alpha}|,|\boldsymbol{\beta}|\leqslant q\\ M,N=1,2,\dots,N_{e}\\ e=1,2,\dots,E\\ \delta^{\boldsymbol{\beta}\boldsymbol{\alpha}}=\delta^{\beta_{1}\alpha_{1}}\delta^{\beta_{2}\alpha_{2}}\cdots\delta^{\beta_{n}\alpha_{n}}\end{array}\right\}\tag{6.18}$$

Thus the local interpolation functions $\psi_N^{\boldsymbol{\alpha}(e)}(\mathbf{x})$ have the property that they assume nonzero values only within element Ω_e, and the values of their derivatives of order $\leqslant q$ are either unity or zero at the nodal points of the element. The quantities $\boldsymbol{\alpha}$ and $\boldsymbol{\beta}$ are, as usual, multi-indices [e.g.,$\boldsymbol{\alpha}=(\alpha_1,\alpha_2,\dots,\alpha_n)\in Z_+^n, \alpha_i=$ integer $\geqslant 0$]. We confine the choices of $\boldsymbol{\alpha}$ and $\boldsymbol{\beta}$ to a subset $\hat{Z}$ of Z_+^n in order to maintain flexibility in our choices of local interpolation functions. For example, we may construct a set in which second partial derivatives of the functions $\psi_N^{\boldsymbol{\alpha}(e)}(\mathbf{x})$ assume unit or zero values at only a subset of nodal points in $\overline{\Omega}_e$, but only the values of the functions are one or zero at the remaining nodes. Later, when we discuss specific properties of certain families of finite elements, we adopt a more elaborate but more specific notation for describing interpolation functions of order $q>0$ [see (6.119) or, in particular, (6.123)].

A local finite-element representation of order q is any function $V_e(\mathbf{x})$ given as a linear combination of local interpolation functions of order q; i.e.,

$$V_e(\mathbf{x})=\sum_{N=1}^{N_e}\sum_{\substack{|\boldsymbol{\alpha}|\leqslant q\\ \boldsymbol{\alpha}\in\hat{Z}}} a_{\boldsymbol{\alpha}}^{N(e)}\psi_N^{\boldsymbol{\alpha}(e)}(\mathbf{x})\tag{6.19}$$

Notice that, because of (6.17), the coefficients $a_{\boldsymbol{\alpha}}^{N(e)}$ have a special property:

$$D^{\boldsymbol{\beta}}V_e(\mathbf{x}_e^N)=a_{\boldsymbol{\beta}}^{N(e)}\tag{6.20}$$

Now let $u(\mathbf{x})$ be a given function in $C^r(\overline{\Omega})$, where $r>q$, and suppose $\Omega_e\subset\Omega$. A *local interpolant of* $u(\mathbf{x})$ *of order q corresponding to element* Ω_e is the local representation of the form (6.19) obtained by the special choice of coefficients

$$a_{\boldsymbol{\alpha}}^{N(e)}=D^{\boldsymbol{\alpha}}u(\mathbf{x}_e^N)\tag{6.21}$$

The corresponding function

$$U_e(\mathbf{x}) = \sum_{N=1}^{N_e} \sum_{\substack{|\alpha| \leqslant q \\ \alpha \in \hat{Z}}} D^\alpha u(\mathbf{x}_e^N)\psi_N^{\alpha(e)}(\mathbf{x}) \tag{6.22}$$

interpolates $u(\mathbf{x})$ locally, at the nodes of element Ω_e; i.e., by construction,

$$D^\alpha U_e(\mathbf{x}_e^N) = D^\alpha u(\mathbf{x}_e^N) \qquad N = 1,2,\ldots,N_e; \quad |\boldsymbol{\alpha}| \leqslant q; \boldsymbol{\alpha} \in \hat{Z} \tag{6.23}$$

EXAMPLE 6.2. Perhaps the simplest local interpolation functions are the locally linear functions

$$\psi_1^{(e)}(x) = \frac{x^2 - x}{h} \qquad \psi_2^{(e)}(x) = \frac{x - x^1}{h} \tag{6.24}$$

where h is the length of the element between nodes x^1 and x^2. These are shown in Fig. 6.4*a*. These functions are of order $q=0$. The local interpolant of a function $u(x)$ is then the linear function shown in Fig. 6.4*b*; which coincides with $u(x)$ at the nodes. ■

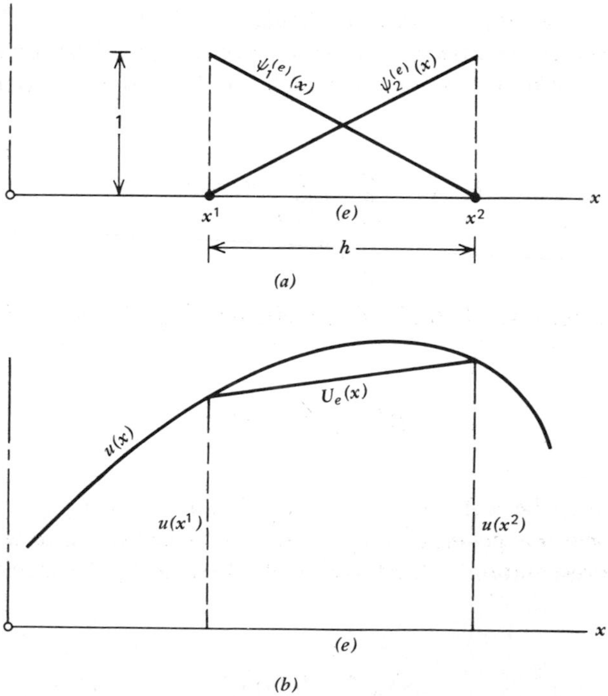

Figure 6.4 Linear, one-dimensional, local interpolation functions.

EXAMPLE 6.3. A two-dimensional version of Example 6.2 involves linear interpolation over triangles, as indicated in Fig. 6.5. To determine the functions $\psi_N^{(e)}(x,y)$, note that each local representation function is necessarily linear:

$$V(x,y) = a + bx + cy$$

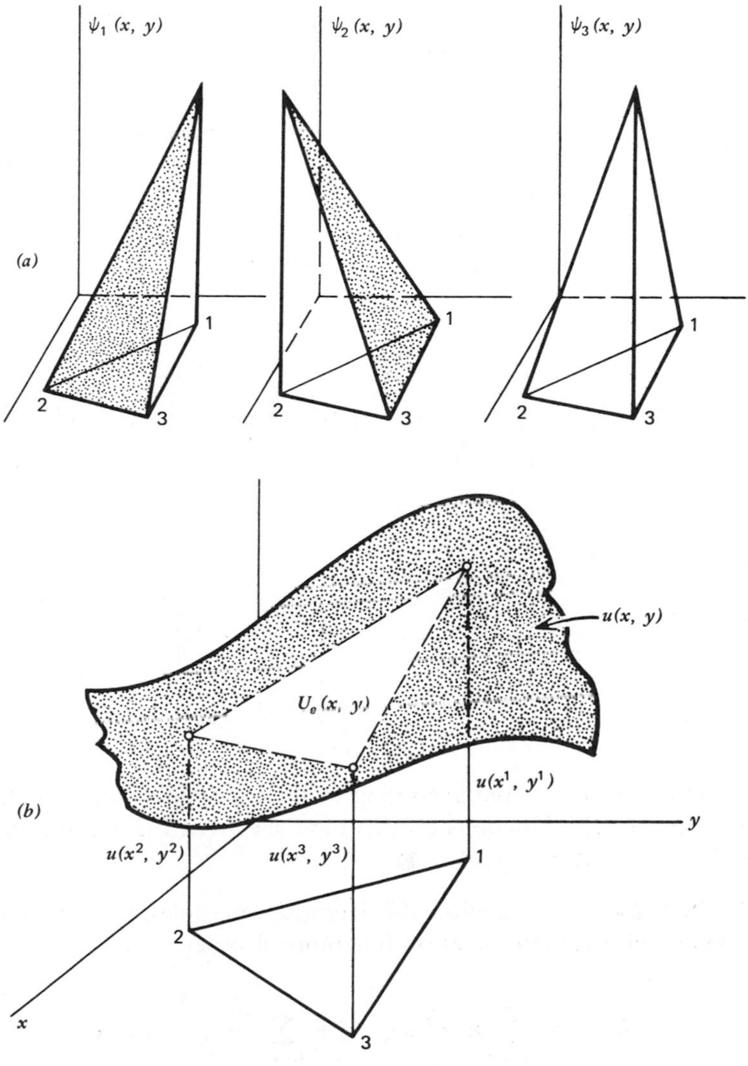

Figure 6.5 Linear interpolation functions defined on a triangular element.

Let V_i denote the value of $V(x,y)$ at node (x^i,y^i), $i=1,2,3$. Then

$$\begin{bmatrix} V_1 \\ V_2 \\ V_3 \end{bmatrix} = \begin{bmatrix} 1 & x^1 & y^1 \\ 1 & x^2 & y^2 \\ 1 & x^3 & y^3 \end{bmatrix} \begin{bmatrix} a \\ b \\ c \end{bmatrix} = \mathbf{Ca}$$

and therefore we can write

$$V(x,y) = [1,x,y]\mathbf{C}^{-1} \begin{Bmatrix} V_1 \\ V_2 \\ V_3 \end{Bmatrix}$$

We deduce that

$$\psi_N^{(e)}(x,y) = a_N^{(e)} + \sum_{\mu=1}^{2} b_{N\mu}^{(e)} x_\mu \qquad N=1,2,3; \quad x_1 = x, x_2 = y \tag{6.25}$$

where

$$\{a_N^{(e)}\} = \frac{1}{2A} \begin{bmatrix} x^2y^3 - x^3y^2 \\ x^3y^1 - x^1y^3 \\ x^1y^2 - x^2y^1 \end{bmatrix} \qquad \{b_{N\mu}^{(e)}\} = \frac{1}{2A} \begin{bmatrix} y^2 - y^3 & x^3 - x^2 \\ y^3 - y^1 & x^1 - x^3 \\ y^1 - y^2 & x^2 - x^1 \end{bmatrix}$$

(6.26*a*,*b*)

and

$$2A = \det \mathbf{C} = \det \begin{bmatrix} 1 & x^1 & y^1 \\ 1 & x^2 & y^2 \\ 1 & x^3 & y^3 \end{bmatrix} \tag{6.26c}$$

If the nodes are numbered according to the right-hand rule, A is the area of the triangle. The functions $\psi_N^{(e)}(x,y)$ are shown in Fig. 6.5 together with a local interpolant of $u(x,y)$. ■

EXAMPLE 6.4. We mention the Hermite interpolation polynomials as an example of local interpolation functions of order 1. Here

$$U_e(x) = \sum_{N=1}^{2} u(x^N)\psi_N^0(x) + \sum_{N=1}^{2} \frac{du(x^N)}{dx}\psi_N^1(x) \tag{6.27}$$

wherein $N = 1, 2$ and

$$\left.\begin{aligned}\psi_1^0(x) &= 1 - 3\xi^2 + 2\xi^3\\ \psi_2^0(x) &= 3\xi^2 - 2\xi^3\\ \psi_1^1(x) &= h(\xi - 2\xi^2 + \xi^3)\\ \psi_2^1(x) &= h(\xi^3 - \xi^2)\end{aligned}\right\} \tag{6.28}$$

Here

$$\xi = \frac{x - x^1}{h}$$

These functions are illustrated in Fig. 6.6. Notice that

$$\begin{aligned}U_e(x^1) &= u(x^1)\psi_1^0(x^1) + u(x^2)\psi_2^0(x^1) + u'(x^1)\psi_1^1(x^1)\\ &\quad + u'(x^2)\psi_2^1(x^1)\\ &= u(x^1)\cdot 1 + 0 + 0 + 0\\ &= u(x^1)\end{aligned}$$

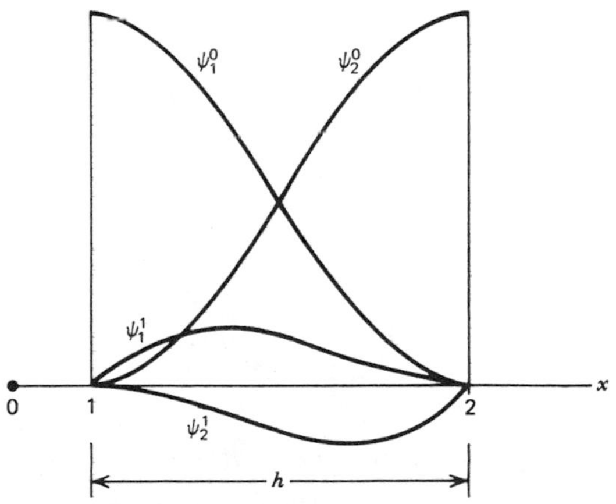

Figure 6.6 Hermite interpolation functions.

and

$$\frac{dU_e(x^1)}{dx} = u(x^1)\psi_1^{0\prime}(x^1) + u(x^2)\psi_2^{0\prime}(x^1) + u'(x^1)\psi_1^{1\prime}(x^1)$$

$$+ u'(x^2)\psi_2^{1\prime}(x^1)$$

$$= 0 + 0 + u'(x^1)\cdot 1 + 0$$

$$= \frac{du(x^1)}{dx}$$

etc. ■

We now turn to the global representation of functions. This is done by (1) "summing up" the local representations and (2) converting our labeling scheme from a local one to a global one. It is reasonable to require that all the global nodal points $\mathcal{G}$ are in $\overline{\Omega}$; i.e.,

$$\mathcal{G} \subset \overline{\Omega} \tag{6.29}$$

The *global* finite-element representation of order q is a function given by

$$V(\mathbf{x}) = \bigcup_{e=1}^{E} V_e(\mathbf{x}) \qquad \mathbf{x} \in \tilde{\Omega} \tag{6.30}$$

where $V_e(\mathbf{x})$ is defined in (6.19).

In keeping with (6.20), we use the values of the partial derivatives of order $\boldsymbol{\alpha}$ at node $\mathbf{x}^i$ of the connected model as coefficients in the global representations (6.30); i.e., if we denote

$$D^{\boldsymbol{\alpha}} V(\mathbf{x}^i) = A^i_{\boldsymbol{\alpha}} \tag{6.31}$$

then we wish to write $V(\mathbf{x})$ in the form

$$V(\mathbf{x}) = \sum_{i=1}^{G} \sum_{\substack{|\boldsymbol{\alpha}| \leq q \\ \boldsymbol{\alpha} \in \hat{Z}}} A^i_{\boldsymbol{\alpha}} \phi_i^{\boldsymbol{\alpha}}(\mathbf{x}) \tag{6.32}$$

where the $\phi_i^{\boldsymbol{\alpha}}(\mathbf{x})$ are *global interpolation functions*.

The next step is fundamental. What is the relationship between the global interpolation functions $\phi_i^{\boldsymbol{\alpha}}(\mathbf{x})$ and the local interpolation functions

$\psi_N^{\alpha(e)}(\mathbf{x})$? The answer is reached by noting that the coefficients $a_\alpha^{N(e)}$ and A_α^i represent precisely the same numbers for a given node, by virtue of (6.20) and (6.31). The difference is simply in the choice of numbering scheme. Recalling (6.10), we can write down this relationship formally:

$$a_\alpha^{N(e)} = \sum_{i=1}^{G} \overset{(e)}{\Omega}{}_i^N A_\alpha^i \tag{6.33}$$

Likewise,

$$A_\alpha^i = \sum_{N=1}^{N_e} \overset{(e)}{\Lambda}{}_N^i a_\alpha^{N(e)} \qquad e \text{ fixed} \tag{6.34}$$

Substituting (6.33) into (6.19) and using (6.30), we finally arrive at the basic formula

$$\phi_i^\alpha(\mathbf{x}) = \bigcup_{e=1}^{E} \sum_{N=1}^{N_e} \overset{(e)}{\Omega}{}_i^N \psi_N^{\alpha(e)}(\mathbf{x}) \tag{6.35}$$

Notice also that, in view of (6.17), the global interpolation functions have the property

$$D^\beta \phi_i^\alpha(\mathbf{x}^j) = \delta_i^j \delta^{\beta\alpha} \tag{6.36}$$

with $i,j = 1,2,\ldots,G; \boldsymbol{\alpha},\boldsymbol{\beta} \in \hat{Z} \subset Z_+^n; |\boldsymbol{\alpha}|, |\boldsymbol{\beta}| \leqslant q$.

A global interpolant of $u(\mathbf{x}) \in C^r(\bar{\Omega})$ of order $q, r > q$, is then the function

$$U(\mathbf{x}) = \bigcup_{e=1}^{E} U_e(\mathbf{x}) = \sum_{i=1}^{G} \sum_{\substack{|\alpha| \leqslant q \\ \alpha \in \hat{Z}}} D^\alpha u(\mathbf{x}^i) \phi_i^\alpha(\mathbf{x}) \tag{6.37}$$

where $\mathbf{x} \in \tilde{\Omega}$. In view of (6.36), $U(\mathbf{x})$ is a genuine interpolant of $u(\mathbf{x})$:

$$D^\alpha U(\mathbf{x}^i) = D^\alpha u(\mathbf{x}^i) \qquad |\alpha| \leqslant q; \alpha \in \hat{Z}; i = 1,2,\ldots,G \tag{6.38}$$

EXAMPLE 6.5. Global interpolation functions are obtained by simply fitting local interpolation functions of adjacent elements together appropriately at common nodal points. For example, the global functions corresponding to the linear interpolants of Fig. 6.4 are the pyramid functions shown in Fig. 6.7*a*. The global functions corresponding to the two-dimensional linear elements of Fig. 6.5 are pyramid functions of the type shown in Fig. 6.7*b*. The Hermite interpolation functions of Fig. 6.6 lead to piecewise cubic functions of the type shown in Fig. 6.7*c* and *d*. ■

REMARK. Observe that an intrinsic property of the global functions $\phi_i^\alpha(\mathbf{x})$ is that they have compact support in $\tilde{\Omega}$. Indeed, $\phi_i^\alpha(\mathbf{x})$ assumes

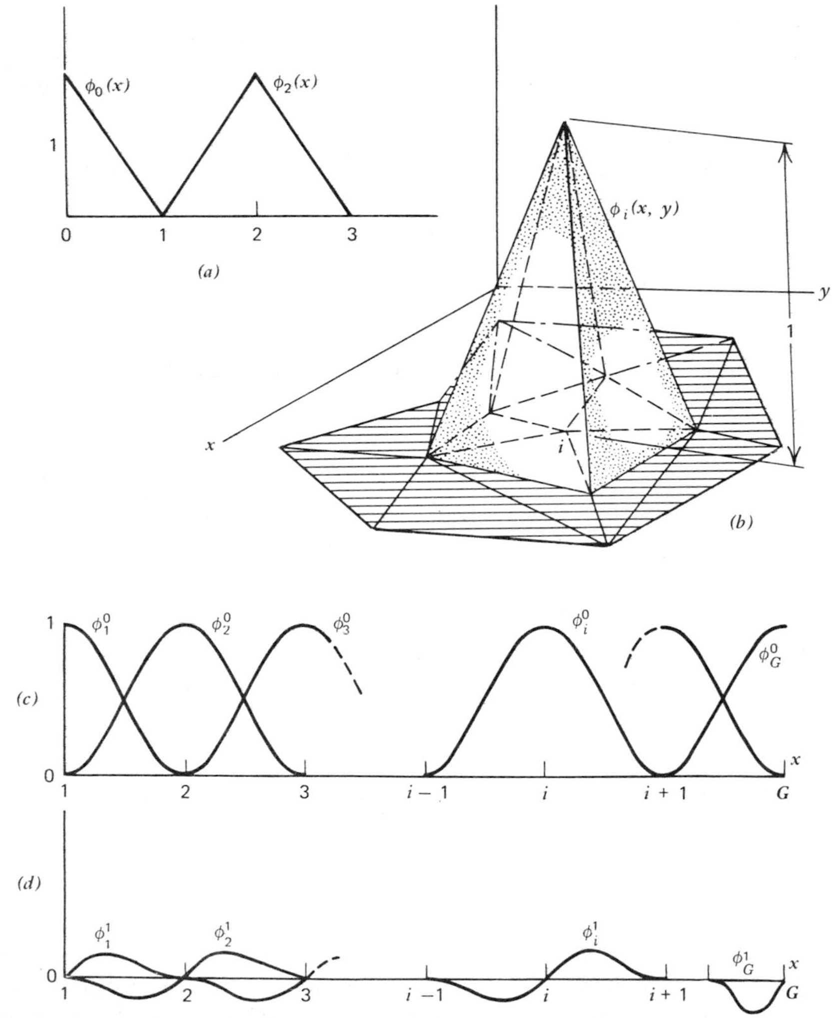

Figure 6.7 Some global finite-element interpolation functions.

nonzero values only within a collection of elements meeting at node $\mathbf{x}^i$. Moreover, their support is almost disjoint, since the intersection of the supports of any two distinct interpolation functions is either empty or, at most, a portion of an interelement boundary.

We also refer to a connected finite-element model as a finite-element *mesh*. To each mesh we assign a positive real number h called the *mesh parameter* which is generally designed to give a measure of the coarseness or fineness of the gradation of elements in $\tilde{\Omega}$. If the interpolation functions are properly chosen, then it is reasonable to expect that the more elements we use in $\tilde{\Omega}$ the closer the interpolant $U(\mathbf{x})$ will resemble $u(\mathbf{x})$. We would like to choose h so that the tendency of $U(\mathbf{x})$ to $u(\mathbf{x})$ as the mesh is refined (i.e., as more elements of smaller dimensions are used to model $\overline{\Omega}$) is represented by a tendency of h toward zero. For this purpose it is a standard to denote by h_e the *diameter* of element Ω_e,

$$h_e = \max_{\mathbf{x},\mathbf{y}\in\overline{\Omega}_e} |\mathbf{x}-\mathbf{y}|_{\mathbb{R}^n} = \max_{\mathbf{x},\mathbf{y}\in\overline{\Omega}_e} \left(\sum_{i=1}^{n} |x_i - y_i|^2 \right)^{1/2} \tag{6.39}$$

Then define h as the maximum diameter of all of the elements:

$$h = \max\{h_1, h_2, \dots, h_E\} \tag{6.40}$$

■

6.4 RESTRICTIONS, PROLONGATIONS, AND PROJECTIONS

One of the most primitive features of a finite-element interpolant $U(\mathbf{x})$ of a given function $u(\mathbf{x})$ is that, once a specific set of global functions $\phi_i^{\boldsymbol{\alpha}}(\mathbf{x})$ is defined, the interpolant is uniquely defined by a finite number of values of $u(\mathbf{x})$ and various of its derivatives at the nodal points of the finite-element mesh. In other words, the interpolation process involves the construction of a correspondence between $u(\mathbf{x})$ and a finite set of numbers $\{D^{\boldsymbol{\alpha}}u(\mathbf{x}^i)\}_{i=1}^{G}, |\boldsymbol{\alpha}| \leqslant q, \boldsymbol{\alpha} \in \hat{Z}$. Suppose that there are Q numbers in this set. Then we may describe this correspondence as a mapping R_h of $u(\mathbf{x})$ into $\mathbb{R}^Q$. The mapping R_h, which is linear and which depends on the particular mesh used to model $u(\mathbf{x})$ (as indicated by the subscript h), is called a *restriction* of $u(\mathbf{x})$ to $\mathbb{R}^Q$. We write

$$R_h u(\mathbf{x}) = (\mu_1, \mu_2, \dots, \mu_Q) \in \mathbb{R}^Q \tag{6.41}$$

The restriction R_h is thus defined by a system $\{l_i\}_{i=1}^{Q}$ of Q linear functionals on the space $\mathcal{H}$ to which $u(\mathbf{x})$ belongs:

$$l_i(u) = \mu_i \qquad i = 1, 2, \ldots, Q \tag{6.42}$$

There are many ways to construct such restrictions, but we emphasize here a class of restrictions suggested by finite-element concepts and the fact that the function $u(\mathbf{x})$ under consideration is generally an element of a Hilbert space.

For example, suppose that $u(\mathbf{x}) \in H^m(\Omega)$. Moreover, suppose we use finite-element methods to construct a set of Q linearly independent global interpolation functions $\{\phi_i^\alpha(\mathbf{x})\}$ which are contained in $H^m(\Omega)$. For simplicity in notation, let us transform this set into the more convenient form

$$\{\phi_1(\mathbf{x}), \phi_2(\mathbf{x}), \ldots, \phi_Q(\mathbf{x})\} \in H^m(\Omega) \tag{6.43}$$

Thus an interpolant $U(\mathbf{x})$ is written as the linear combination

$$U(\mathbf{x}) = \sum_{i=1}^{Q} a^i \phi_i(\mathbf{x}) \tag{6.44}$$

Since the set $\{\phi_i\}_{i=1}^{Q}$ is linearly independent, it forms the basis of a Q-dimensional subspace $S_h(\Omega)$ of $H^m(\Omega)$. Now the functions $\{\phi_i\}_{i=1}^{Q}$ are not, in general, orthonormal with respect to any inner product. However, a variety of *biorthogonal bases* can be constructed by using the following procedure.

(i) Let $(u, v)_s$ denote the inner product in $H^s(\Omega)$, $s \leqslant m$, i.e.,

$$(u, v)_s = \int_\Omega \sum_{|\alpha| \leqslant s} D^\alpha u \, \overline{D^\alpha v} \, dx \tag{6.45}$$

For a given basis $\{\phi_i\}$ of $S_h(\Omega) \subset H^m(\Omega)$, and a given $s \leqslant m$, we construct the Qth-order matrix

$$G_{ij}^{(s)} = (\phi_i, \phi_j)_s \qquad i, j = 1, 2, \ldots, Q \tag{6.46}$$

The matrix $G_{ij}^{(s)}$ is called the *Gram matrix of category s corresponding to the set* $\{\phi_i\}$. It is symmetric, invertible, and positive definite.

(ii) Next we invert $G_{ij}^{(s)}$. The inverse is denoted $G_{(s)}^{ij}$:

$$G_{(s)}^{ij} = \left(G_{ij}^{(s)}\right)^{-1} \qquad s \leqslant m \tag{6.47}$$

Then

$$\sum_{j=1}^{Q} G^{ij}_{(s)} G^{(s)}_{jm} = \delta^i_m \qquad i,m=1,2,\ldots,Q \tag{6.48}$$

(iii) Next, we construct the *conjugate* (or *dual*) *basis functions* $\{\phi^i_{(s)}\}_{i=1}^Q$ *of category* s by the sum

$$\phi^i_{(s)}(\mathbf{x}) = \sum_{j=1}^{Q} G^{ij}_{(s)} \phi_j(\mathbf{x}) \tag{6.49}$$

Since the $\phi^i_{(s)}(\mathbf{x})$ are linear combinations of the functions $\phi_i(\mathbf{x})$, they also belong to the subspace $S_h(\Omega)$; indeed, they are also linearly independent and therefore provide a (dual) basis for $S_h(\Omega)$. What is important about the functions defined in (6.49) is that they have the *biorthogonality property*

$$\left(\phi^i_{(s)}, \phi_j\right)_s = \delta^i_j \tag{6.50}$$

This follows immediately on taking the inner product $(\phi^i_{(s)}, \phi_j)_s$ of the function in (6.49) and making use of (6.46) and (6.48).

(iv) Finally, we use the property (6.50) to define the coefficients a^i of $U(\mathbf{x})$ in (6.44):

$$\left(U, \phi^i_{(s)}\right)_s = \sum_{j=1}^{Q} a^j \left(\phi_j, \phi^i_{(s)}\right)_s = \sum_{j=1}^{Q} a^j \delta^i_j = a^i$$

and, in view of (6.38), the a^i represent values of various derivatives of the given function $u(\mathbf{x})$ at the global nodal points.

Returning to the notion of restriction of $u(\mathbf{x}) \in H^m(\Omega)$ to $\mathbb{R}^Q$, we observe that one way of defining a restriction operator R_h is to use the conjugate basis functions $\{\phi^i_{(s)}\}_{i=1}^Q$ to generate Q linear functionals on $u(\mathbf{x})$ by means of the inner products $(\cdot, \cdot)_s$. We then have

$$R_h^{(s)} u = \left(u^1_{(s)}, u^2_{(s)}, \ldots, u^Q_{(s)}\right) \tag{6.51}$$

wherein

$$u^i_{(s)} = \left(u, \phi^i_{(s)}\right)_s \tag{6.52}$$

It is also possible to describe restrictions

$$R_h^{(s,r)} u = \left\{u^i_{(s,r)}\right\}_{i=1}^Q \qquad u^i_{(s,r)} = \left(u, \phi^i_{(s)}\right)_r \tag{6.53}$$

with $r, s \leqslant m$, but this rarely seems to be of interest.

Now every function $V(\mathbf{x})$ in the subspace $S_h(\Omega)\subset H^m(\Omega)$ is, by definition, of the form

$$V(\mathbf{x})=\sum_{i=1}^{Q} a^i\phi_i(\mathbf{x}) \tag{6.54}$$

What is important to note is that the ordered array of coefficients $\{a^i\}=(a^1,a^2,\ldots,a^Q)$ is an element in the discrete space $\mathbb{R}^Q$. Thus (6.55) defines a mapping P_h of $\mathbb{R}^Q$ onto $S_h(\Omega)$:

$$P_h\left(u^1_{(s)},u^2_{(s)},\ldots,u^Q_{(s)}\right)=\sum_{i=1}^{Q} u^i_{(s)}\phi_i(\mathbf{x})\in S_h(\Omega) \tag{6.55}$$

The operator P_h is called a *prolongation* of $\mathbb{R}^Q$ into $H^m(\Omega)$.

We are now ready to set down some basic properties of the restrictions $R_h^{(s)}$ and prolongations P_h.

THEOREM 6.2. Let $R_h^{(s)}$ be the restriction of $H^m(\Omega)$ to $\mathbb{R}^Q$ defined by (6.51) and let P_h be a prolongation of $\mathbb{R}^Q$ onto $S_h(\Omega)\subset H^m(\Omega)$, where $S_h(\Omega)$ is a Q-dimensional subspace of $H^m(\Omega)$ spanned by the basis functions $\{\phi_i\}_{i=1}^{Q}$. Then

(i) The composition

$$\Pi_h^{(s)}=P_hR_h^{(s)} \tag{6.56}$$

defines a projection of $H^m(\Omega)$ onto $S_h(\Omega)$.

(ii) The restriction $R_h^{(r)}u=\{u^i_{(r)}\}$ such that

$$\|u-P_hR_h^{(r)}u\|_{H^s(\Omega)}=\inf_{V\in S_h}\|u-V\|_{H^s(\Omega)} \tag{6.57}$$

is obtained when $r=s$; i.e.,

$$u^i_{(r)}=\left(u,\phi^i_{(s)}\right)_s=u^i_{(s)}$$

(iii) The function

$$E^{(s)}(\mathbf{x})=u(\mathbf{x})-\Pi_h^{(s)}u(\mathbf{x}) \tag{6.58}$$

where $\Pi_h^{(s)}$ is given by (6.56), is orthogonal to $S_h(\Omega)$ with respect to the inner product $(\cdot,\cdot)_s$; i.e., $\Pi_h^{(s)}$ is an orthogonal projection:

$$(E^{(s)},V)_s=0 \qquad \forall V\in S_h(\Omega) \tag{6.59}$$

(iv) If $u(\mathbf{x}) \in S_h(\Omega)$, all restrictions $\{u^i_{(r)}\}$ coincide; i.e., $u^i_{(r)} = u^i_{(s)} \forall r, s \leqslant m$.

PROOF. (i) Observe that

$$\begin{aligned}
\Pi_h^{(s)}\Pi_h^{(s)}u(\mathbf{x}) &= P_h R_h^{(s)} \sum_{j=1}^{Q} u^j_{(s)}\phi_j(\mathbf{x}) \\
&= P_h\left[\sum_{j=1}^{Q} u^j_{(s)}\left(\phi_j, \phi^1_{(s)}\right)_s, \sum_{j=1}^{Q} u^j_{(s)}\left(\phi_j, \phi^2_{(s)}\right)_s, \right. \\
&\qquad \left. \ldots, \sum_{j=1}^{Q} u^j_{(s)}\left(\phi_j, \phi^Q_{(s)}\right)_s\right] \\
&= P_h\left(u^1_{(s)}, u^2_{(s)}, \ldots, u^Q_{(s)}\right) \\
&= \sum_{j=1}^{Q} u^j_{(s)}\phi_j(\mathbf{x}) \\
&= \Pi_h^{(s)}u(\mathbf{x})
\end{aligned}$$

Therefore

$$(\Pi_h^{(s)})^2 = \Pi_h^{(s)}$$

(ii) Since each $V(\mathbf{x}) \in S_h(\Omega)$ is of the form (6.54), a simple calculation reveals that

$$\begin{aligned}
\|u - V\|^2_{H^s(\Omega)} &= \|u\|^2_{H^s(\Omega)} - 2\sum_{j=1}^{Q} a^j (u, \phi_j)_s \\
&\quad + \sum_{i=1}^{Q}\sum_{j=1}^{Q} a^i a^j (\phi_i, \phi_j)_s
\end{aligned}$$

This quadratic form in a^i assumes a minimum value whenever $\partial\|u - V\|^2_{H^s(\Omega)}/\partial a^k = 0, k = 1, 2, \ldots, Q$. Hence, using the notation introduced in (6.46) and (6.47), we have

$$\sum_{j=1}^{Q} a^j G^{(s)}_{ij} - (u, \phi_i)_s = 0 \qquad 1 \leqslant i \leqslant Q$$

and, for each $1 \leqslant j \leqslant Q$,

$$a^j = \sum_{i=1}^{Q} G^{ij}_{(s)}(u,\phi_i)_s$$

$$= \left(u, \sum_{i=1}^{Q} G^{ij}_{(s)}\phi_i\right)_s$$

$$= \left(u, \phi^j_{(s)}\right)_s$$

$$= u^j_{(s)}$$

Thus the $V(\mathbf{x})$ that provides the infimum in (6.58) is $\Sigma_{i=1}^{Q} u^j_{(s)}\phi_j(\mathbf{x}) = P_r R_h^{(s)} u$.

(iii) Since

$$\Pi_h^{(s)} u = \sum_{j=1}^{Q} u^j_{(s)}\phi_j(\mathbf{x})$$

we have, for arbitrary $V(\mathbf{x}) = \Sigma_{j=1}^{Q} b^j \phi_j(\mathbf{x}) \in S_h(\Omega)$, $1 \leqslant i,j,k \leqslant Q$,

$$(E^{(s)}, V)_s = \left(u - \sum_i u^i_{(s)}\phi_i, \sum_j b^j \phi_j\right)_s$$

$$= \sum_j b^j \left[(u,\phi_j)_s - \sum_i u^i_{(s)} G^{(s)}_{ij} \right]$$

$$= \sum_j b^j \left[\left(u, \sum_k G^{(s)}_{jk} \phi^k_{(s)}\right)_s - \sum_k u^k_{(s)} G^{(s)}_{kj} \right]$$

$$= \sum_j \sum_k b^j G^{(s)}_{jk} \left[u^k_{(s)} - u^k_{(s)} \right]$$

$$= 0$$

(iv) When $u(\mathbf{x}) \in S_h(\Omega)$, it is of the form $\Sigma_{i=1}^{Q} \hat{a}^i \phi_i(\mathbf{x})$, where $\hat{a}^i$ are specific coefficients. Clearly,

$$\left(\phi^j_{(s)}, u\right)_s = \sum_i \hat{a}^i \left(\phi^j_{(s)}, \phi_i\right)_s = \hat{a}^j = \sum_i \hat{a}^i \left(\phi^j_{(r)}, \phi_i\right)_r$$

$$= \left(u^j_{(r)}, \phi_i\right)_r \qquad r,s \leqslant m$$

This completes the proof of the theorem. ■

REMARKS. 1. The function $E^{(s)}(\mathbf{x})$ of (6.58) is called the *interpolation error* relative to $\Pi_h^{(s)}$. In view of (6.59), $\Pi_h^{(s)}$ is an orthogonal projection of $H^m(\Omega)$ onto $S_h(\Omega)$.

2. The interpolant $\Pi_h^{(s)}u$ is referred to as the *best approximation* to u in $S_h(\Omega)$ in the $\|\cdot\|_{H^s(\Omega)}$ norm. This terminology is used because $\Pi_h^{(s)}u$ is the unique function in $S_h(\Omega)$ which makes the norm of the interpolation error $\|E^{(s)}\|_{H^s(\Omega)}$ as small as possible for a given u.

3. For functions $u(\mathbf{x})$ in $H^m(\Omega)$, the notion of an interpolant must be modified slightly, because the values of certain derivatives of $u(\mathbf{x})$ may not be defined at points in Ω. For this reason, we refer to the projection $\Pi_h^{(s)}u(\mathbf{x}) = U(\mathbf{x})$ as an interpolant of $u(\mathbf{x})$ of category s. In most cases, we use the L_2 projection of $u(\mathbf{x})$ as its interpolant; i.e., $\Pi_h^{(0)}u(\mathbf{x})$. ■

6.5 CONJUGATE BASIS FUNCTIONS

In this article we summarize briefly several properties of the conjugate functions $\phi^i_{(s)}(\mathbf{x})$ described in the previous article and defined in (6.49). For emphasis, consider the class of zero-order finite element interpolations ($q=0$); i.e., we consider finite-element representations in which the value of a function is prescribed at nodal points and not its derivatives. Then the local interpolation functions corresponding to element Ω_e have the properties

$$\left.\begin{array}{lll}
\text{(i)} & \psi_N^{(e)}(\mathbf{x}) = 0 & \mathbf{x} \not\subset \bar{\Omega}_e \\
\text{(ii)} & \psi_N^{(e)}(\mathbf{x}^M) = \delta_N^M & M, N = 1, 2, \ldots, N_e \\
\text{(iii)} & \displaystyle\sum_{N=1}^{N_e} \psi_N^{(e)}(\mathbf{x}) = 1 & \mathbf{x} \in \bar{\Omega}_e
\end{array}\right\} \qquad (6.60)$$

and the global interpolation functions are such that

$$\left.\begin{array}{lll}
\text{(i)} & \phi_i(\mathbf{x}) = \displaystyle\bigcup_{e=1}^{E} \sum_{N=1}^{N_e} \overset{(e)}{\Omega}{}_i^N \psi_N^{(e)}(\mathbf{x}) & \\
\text{(ii)} & \phi_i(\mathbf{x}^j) = \delta_i^j & i, j = 1, 2, \ldots, G \\
\text{(iii)} & \displaystyle\sum_{i=1}^{G} \phi_i(\mathbf{x}) = 1 &
\end{array}\right\} \qquad (6.61)$$

Properties (i) and (ii) of (6.60) and (6.61) follow immediately from (6.16), (6.17), (6.35), and (6.36), and property (iii) expresses the fact that $V(\mathbf{x})=1$ belongs to the subspace $S_h(\Omega)$ spanned by $\{\phi_i\}_{i=1}^{G}$. Our present account follows [6.1], [6.5], and [6.6].

We begin at the local level. Let Ω_e be a typical finite element in a connected finite-element model $\tilde{\Omega}$, and let us denote by $(\cdot,\cdot)_0$ the inner product in $H^0(\Omega)$ [i.e., in $L_2(\Omega)$]. Then the *local* Gram matrix of category 0 is the $N_e \times N_e$ matrix

$$g_{NM}^{(e)}=(\psi_N^{(e)},\psi_M^{(e)})_0 \qquad N,M=1,2,\ldots,N_e \tag{6.62}$$

Likewise, the global Gram matrix of category 0 is given by

$$G_{ij}=(\phi_i,\phi_j)_0 \qquad i,j=1,2,\ldots,G \tag{6.63}$$

and, in view of (i) of (6.61),

$$G_{ij}=\left(\bigcup_{e=1}^{E}\sum_{N=1}^{N_e}\overset{(e)}{\Omega}{}_i^N\psi_N^{(e)},\bigcup_{f=1}^{E}\sum_{M=1}^{N_f}\overset{(f)}{\Omega}{}_j^M\psi_M^{(f)}\right)_0$$

$$=\sum_{e,f}\sum_{M,N}\overset{(e)}{\Omega}{}_i^N\overset{(f)}{\Omega}{}_j^M(\psi_N^{(e)},\psi_M^{(f)})_0$$

But

$$(\psi_N^{(e)},\psi_M^{(f)})_0=0 \qquad \text{if } e\neq f \tag{6.64}$$

Therefore

$$G_{ij}=\sum_{e=1}^{E}\sum_{N,M=1}^{N_e}\overset{(e)}{\Omega}{}_i^N g_{NM}^{(e)}\overset{(e)}{\Omega}{}_j^M \tag{6.65}$$

The global Gram matrix is thus built up by performing a series of transformations of local matrices associated with each element.

EXAMPLE 6.6. Consider the collection of finite elements shown in Fig. 6.8. Suppose that the local Gram matrices of elements $a,b,c,d,$ and e are

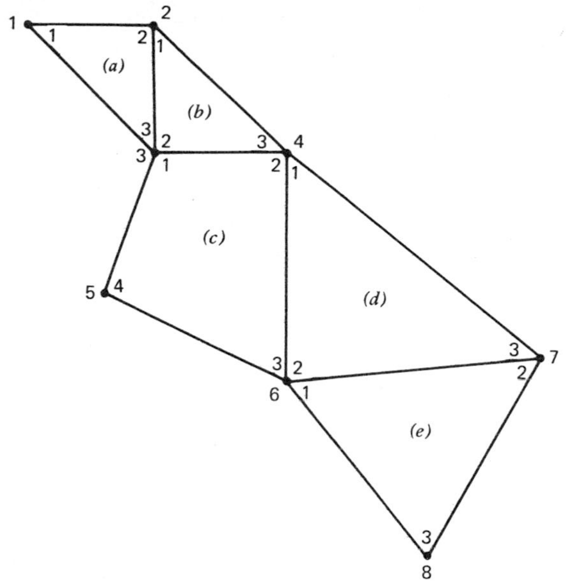

Figure 6.8 A collection of finite elements.

given by

$$\mathbf{g}^{(a)} = \begin{bmatrix} a_{11} & a_{12} & a_{13} \\ a_{21} & a_{22} & a_{23} \\ a_{31} & a_{32} & a_{33} \end{bmatrix} \qquad \mathbf{g}^{(b)} = \begin{bmatrix} b_{11} & b_{12} & b_{13} \\ b_{21} & b_{22} & b_{23} \\ b_{31} & b_{32} & b_{33} \end{bmatrix}$$

$$\mathbf{g}^{(c)} = \begin{bmatrix} c_{11} & c_{12} & c_{13} & c_{14} \\ c_{21} & c_{22} & c_{23} & c_{24} \\ c_{31} & c_{32} & c_{33} & c_{34} \\ c_{41} & c_{42} & c_{43} & c_{44} \end{bmatrix}$$

$$\mathbf{g}^{(d)} = \begin{bmatrix} d_{11} & d_{12} & d_{13} \\ d_{21} & d_{22} & d_{23} \\ d_{31} & d_{32} & d_{33} \end{bmatrix}$$

$$\mathbf{g}^{(e)} = \begin{bmatrix} e_{11} & e_{12} & e_{13} \\ e_{21} & e_{22} & e_{23} \\ e_{31} & e_{32} & e_{33} \end{bmatrix}$$

Since there are eight nodes in the connected model, the global Gram matrix is given by the 8×8 matrix

$$\mathbf{G}=\begin{bmatrix} a_{11} & a_{12} & a_{13} & 0 & 0 & 0 & 0 & 0 \\ a_{21} & a_{22}+b_{11} & a_{23}+b_{12} & b_{13} & 0 & 0 & 0 & 0 \\ a_{31} & a_{32}+b_{21} & a_{33}+b_{22}+c_{11} & b_{23}+c_{12} & c_{14} & c_{13} & 0 & 0 \\ 0 & b_{31} & b_{32}+c_{21} & b_{33}+c_{22}+d_{11} & c_{24} & c_{23}+d_{12} & d_{13} & 0 \\ 0 & 0 & c_{41} & c_{42} & c_{44} & c_{43} & 0 & 0 \\ 0 & 0 & c_{31} & c_{32}+d_{21} & c_{34} & c_{33}+d_{22}+e_{11} & d_{23}+e_{12} & e_{13} \\ 0 & 0 & 0 & d_{31} & 0 & d_{32}+e_{21} & d_{33}+e_{22} & e_{23} \\ 0 & 0 & 0 & 0 & 0 & e_{31} & e_{32} & e_{33} \end{bmatrix} \tag{6.66}$$

Note that $\mathbf{G}$ is sparce and symmetric. ■

REMARKS. 1. The Gram matrices of category 0 are called *mass matrices*, since their elements depict a distribution of mass in a physical body with a unit mass density; see [6.1].

2. The pattern by which the global matrix $\mathbf{G}$ is generated from the local ones is quite clear from the structure of the matrix in (6.66). In practice, the Boolean matrices $\Omega_i^{(e)N}$ are seldom actually used. Since they simply invoke a renumbering of local node labels, this can be done automatically by using the global numbering scheme for the identification of nodes in each element. This means that, instead of $g_{NM}^{(e)}$, we calculate an expanded local Gram matrix

$$G_{ij}^{(e)} = \sum_{N,M=1}^{N_e} \overset{(e)}{\Omega}{}_i^N g_{NM}^{(e)} \overset{(e)}{\Omega}{}_j^M \tag{6.67}$$

Then the global matrix is obtained by a simple matrix sum:

$$G_{ij} = \sum_{e=1}^{E} G_{ij}^{(e)} \tag{6.68}$$

For example, the matrix $\mathbf{g}^{(d)}$ in Example 6.6 would be replaced by

$$
\mathbf{G}^{(d)} = \begin{bmatrix}
0 & 0 & 0 & 0 & 0 & 0 & 0 & 0 \\
0 & 0 & 0 & 0 & 0 & 0 & 0 & 0 \\
0 & 0 & 0 & 0 & 0 & 0 & 0 & 0 \\
0 & 0 & 0 & d_{11} & 0 & d_{12} & d_{13} & 0 \\
0 & 0 & 0 & 0 & 0 & 0 & 0 & 0 \\
0 & 0 & 0 & d_{21} & 0 & d_{22} & d_{23} & 0 \\
0 & 0 & 0 & d_{31} & 0 & d_{32} & d_{33} & 0 \\
0 & 0 & 0 & 0 & 0 & 0 & 0 & 0
\end{bmatrix}
$$

The process (6.68) for superimposing local matrices is called the *direct stiffness method*.

3. It is an important feature of the finite-element method that the associated Gram matrices G_{ij} always have a structure typified by (6.66); that is, symmetric, positive definite, sparse (many zero elements), and banded. In other words, all the elements are grouped in a tight band around the main diagonal; for instance, matrices of the form

G =

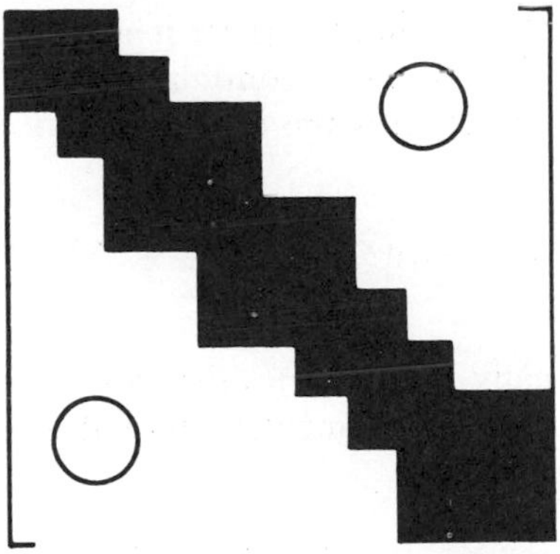

where the shaded overlapping portions indicate blocks of elements of individual local matrices. Computationally, this structure is a very attractive feature. ■

Once the global Gram matrix G_{ij} of (6.66) has been assembled, we can proceed directly to the calculation of the conjugate basis functions:

$$\phi^i(\mathbf{x}) = \sum_{j=1}^{G} G^{ij}\phi_j(\mathbf{x}) \qquad G^{ij} = (G_{ij})^{-1} \tag{6.69}$$

Then

$$(\phi^i, \phi_j)_0 = \delta^i_j \tag{6.70}$$

We now come to some important observations. While the basis functions $\phi_i(\mathbf{x})$ have compact support in $\tilde{\Omega}$ by the very nature of the finite-element construction, the dual basis functions $\phi^i(\mathbf{x})$ do not. Indeed, each $\phi^i(\mathbf{x})$ is, by virtue of (6.69), a linear combination of *all* the functions $\phi_i(\mathbf{x})$. While G_{ij} is sparse and banded, its inverse is fully populated; hence the value $\phi^i(\mathbf{x})$ at any point depends on each $\phi_i(\mathbf{x})$.

Moreover, the conjugate functions $\phi^i(\mathbf{x})$ cannot in general be generated from local functions in the same way as the functions $\phi_i(\mathbf{x})$. Usually we design the functions $\psi^{(e)}_N(\mathbf{x})$ so that, on assembly of the elements, $\phi_i(\mathbf{x})$ will be continuous. [In particular, if the function $u(\mathbf{x})$ is only in $H^1(\Omega)$, then so also must be $\phi_i(\mathbf{x})$. This means that the first derivatives of $\phi_i(\mathbf{x})$ are in $H^0(\Omega)$. We must, then, at least make the functions continuous across interelement boundaries.] However, there is no reason to expect functions such as $\Sigma_e\Sigma_N(g^{(e)}_{NM})^{-1}\psi^{(e)}_M(\mathbf{x})$ to be continuous on Ω. The closest thing to a local conjugate basis function is the function

$$\psi^N_{(e)}(\mathbf{x}) = \sum_{i=1}^{G} \overset{(e)}{\Omega}{}^N_i \phi^i(\mathbf{x}) \tag{6.71}$$

which, while not actually local, relates the global labeling scheme used for $\phi^i(\mathbf{x})$ to that used locally for element e. Here, if we write an element $V(\mathbf{x})$ in $S_h(\Omega)$ in the form

$$V(\mathbf{x}) = \sum_{i=1}^{G} a_i\phi^i(\mathbf{x}) = \bigcup_{e=1}^{E} V_e(\mathbf{x}) = \sum_{i=1}^{G} a^i\phi_i(\mathbf{x}) \tag{6.72}$$

then we can write

$$a_i = \sum_{e=1}^{E}\sum_{N=1}^{N_e} \overset{(e)}{\Omega}{}^N_i a^{(e)}_N \tag{6.73}$$

where

$$a_N^{(e)} = (V, \psi_N^{(e)})_0 \tag{6.74}$$

In this case

$$\psi_{(e)}^N(\mathbf{x}) = \bigcup_{e=1}^{E} \sum_{i,j=1}^{G} \sum_{M=1}^{N_e} \overset{(e)}{\Omega}{}_i^N G^{ij} \overset{(e)}{\Omega}{}_j^M \psi_M^{(e)}(\mathbf{x}) \tag{6.75}$$

The coefficients a^i and a_i are related to one another in a simple manner; e.g.,

$$\left.\begin{aligned} a^i &= (V, \phi^i)_0 = \sum_{j=1}^{G} G^{ij} a_j \\ a_i &= (V, \phi_i)_0 = \sum_{j=1}^{G} G_{ij} a^j \end{aligned}\right\} \tag{6.76}$$

Interestingly enough, the local coefficients can be shown to obey similar relations,

$$\left.\begin{aligned} a_{(e)}^N &= (V, \psi_{(e)}^N)_0 = \sum_{M=1}^{N_e} g_{(e)}^{NM} a_M^{(e)} \\ a_N^{(e)} &= (V, \psi_N^{(e)})_0 = \sum_{M=1}^{N_e} g_{NM}^{(e)} a_{(e)}^M \end{aligned}\right\} \tag{6.77}$$

where $(g_{NM}^{(e)})^{-1} = g_{(e)}^{NM}$; but the functions $\psi_N^{(e)}$ and $\psi_{(e)}^N$ are not biorthogonal. In fact,

$$(\psi_N^{(e)}, \psi_{(e)}^M)_0 = \sum_{i,j=1}^{G} \sum_{P=1}^{N_e} \overset{(e)}{\Omega}{}_i^M G^{ij} \overset{(e)}{\Omega}{}_j^P g_{NP}^{(e)} \tag{6.78}$$

The conjugate functions $\phi^i(\mathbf{x})$ admit to a simple but illustrative interpretation. Since they describe (generate) linear functionals on the set $\{\phi_i\}_{i=1}^G$, technically they belong to the dual space of $S_h(\Omega)$. Now recall that each $V(\mathbf{x}) \in S_h(\Omega)$ has the property

$$V(\mathbf{x}^i) = \sum_{j=1}^{G} a^j \phi_j(\mathbf{x}^i) = a^i \tag{6.79}$$

Hence

$$(V,\phi^i)_0 = V(\mathbf{x}^i) \tag{6.80}$$

Thus, within the subspace $S_h(\Omega)$, $\phi^i(\mathbf{x})$ behaves as a Dirac measure, $\delta(\mathbf{x}-\mathbf{x}^i) \approx \phi^i(\mathbf{x})$. We can express this property in another way by recalling that the projection Π_h of an element $u(\mathbf{x}) \in H^m(\Omega)$ into $S_h(\Omega)$ is of the form

$$\Pi_h u = \sum_{i=1}^{G} (u,\phi^i)_0 \phi_i(\mathbf{x}) = \sum_{i=1}^{G} (u,\phi_i)_0 \phi^i(\mathbf{x}) \tag{6.81}$$

Now *project* δ by replacing (u,ϕ^i) by the duality pairing $\langle \delta,\phi^i \rangle$. If $\mathbf{x}^k$ is a fixed global node, we have

$$\begin{aligned}\Delta(\mathbf{x}-\mathbf{x}^k) &= \Pi_h \delta(\mathbf{x}-\mathbf{x}^k)\\ &= \sum_{i=1}^{G} \langle \delta_k,\phi^i \rangle \phi_i(\mathbf{x})\\ &= \sum_{i=1}^{G} \phi^i(\mathbf{x}^k)\phi_i(\mathbf{x})\end{aligned} \tag{6.82}$$

Thus the values of $\phi^i(\mathbf{x})$ at a point $\mathbf{x}^k \in \Omega$ are the components of the projection of the Dirac distribution $\delta(\mathbf{x}-\mathbf{x}^k)$ into $S_h(\Omega)$. A similar interpretation applies to $\phi_i(\mathbf{x}^k)$, since also

$$\Delta(\mathbf{x}-\mathbf{x}^k) = \sum_{i=1}^{G} \phi_i(\mathbf{x}^k)\phi^i(\mathbf{x}) \tag{6.83}$$

EXAMPLE 6.7. Suppose we choose for the local interpolation functions $\psi_N^{(e)}(x)$, for purposes of interpolation of a function defined on a closed interval of the real line, the linear functions defined in (6.24) (see Fig. 6.4). Then each $\phi_i(x)$ assumes the form of the pyramid function in Fig. 6.7*a*. The corresponding conjugate functions are shown in Fig. 6.9*a*. Notice that they assume nonzero values almost everywhere on Ω. The conjugate function $\phi^5(x)$, for example, assumes its maximum value at node 5, and its magnitude decays rapidly as one moves away from node 5. Recall that $\phi^5(x)$ represents the discrete finite-element analog of the Dirac delta, $\delta(x-x^5)$. ■

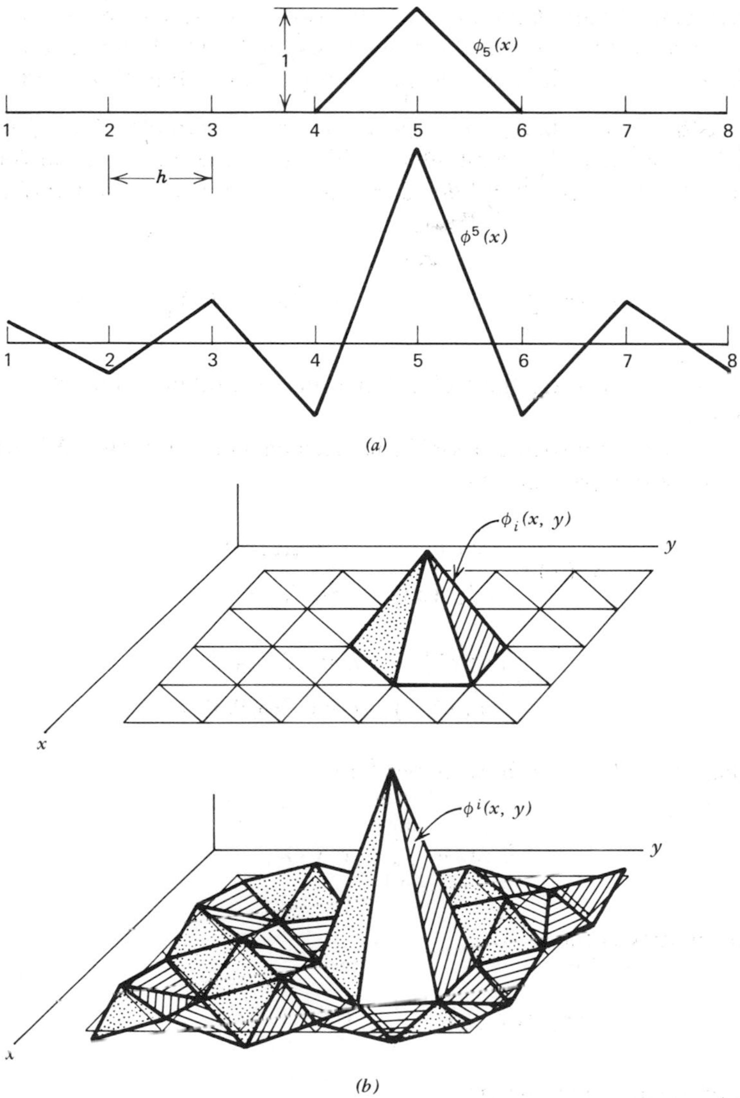

Figure 6.9 Conjugate functions for piecewise linear interpolation.

EXAMPLE 6.8. Figure 6.9*b* illustrates the two-dimensional version of the piecewise linear interpolants of the previous example. The function $\phi^i(x,y)$ shown there is dual to the pyramid function $\phi_i(x,y)$ of Fig. 6.7*b*. ■

EXAMPLE 6.9. (Cf. [6.6].) We now consider an example of conjugate functions associated with the piecewise cubic, Hermite interpolation functions of (6.27) and (6.28). On connecting elements, we obtain interpolants of the form

$$U(x) = \sum_{k=1}^{G} \left[u(x^k)\phi_k^0(x) + u'(x^k)\phi_k^1(x) \right]$$

where $\phi_k^0(x)$ and $\phi_k^1(x)$ are the global functions illustrated in Fig. 6.7*c* and *d*, respectively.

Consider a mesh consisting of only four elements of unit length. We next adopt the more compact notation

$$\chi_k(x) = \begin{cases} \phi_k^0(x) & k = 1,2,3,4,5 \\ \phi_{k-5}^1(x) & k = 6,7,8,9,10 \end{cases}$$

$$b^k = \begin{cases} u(x^k) & k = 1,2,3,4,5 \\ u'(x^{k-5}) & k = 6,7,8,9,10 \end{cases}$$

so that the interpolant can now be written

$$U(x) = \sum_{k=1}^{10} b^k \chi_k(x)$$

The Gram matrix is then

$$G_{ik} = (\chi_i, \chi_k)_0 = \int_0^4 \chi_i(x)\chi_k(x)\,dx$$

and the conjugate functions are

$$\chi^k(x) = \sum_{j=1}^{10} G^{kj}\chi_j(x) \qquad G^{kj} = (G_{kj})^{-1}$$

Figure 6.10 contains examples of the functions $\chi_k(x)$ and $\chi^k(x)$. Notice again that the $\chi^k(x)$ decay as one moves to points away from the node to which χ^k is associated. Notice also that their support is the entire domain $[0,4]$. ■

At this point a few additional definitions are in order. Notice that, according to (6.72), each function $V(\mathbf{x})$ in $S_h(\Omega)$ can be written in the form

$$V(\mathbf{x}) = \sum_{i=1}^{G} a^i \phi_i(\mathbf{x}) = \sum_{i=1}^{G} a_i \phi^i(\mathbf{x})$$

The coefficients a^i are determined by the inner product (V, ϕ^i) and are called the *contravariant* components of $V(\mathbf{x})$; the coefficients a_i are called *covariant* components.

We define contravariant and covariant *moments* by

$$M^i = (1, \phi^i)_0 = \int_\Omega \phi^i(\mathbf{x})\, dx \tag{6.84}$$

and

$$M_i = (1, \phi_i)_0 = \int_\Omega \phi_i(\mathbf{x})\, dx \tag{6.85}$$

respectively.

Several properties of the covariant and contravariant components are summarized in the following theorem.

THEOREM 6.3. Let properties (6.60) and (6.61) hold, and let M^i and M_i be the moments defined in (6.84) and (6.85). Then

(i)

$$\int_\Omega V(\mathbf{x})\, dx = \sum_{i=1}^{G} V^i M_i = \sum_{i=1}^{G} V_i M^i \tag{6.86}$$

where

$$V^i = a^i = (V, \phi^i)_0 \qquad \text{and} \qquad V_i = a_i = (V, \phi_i)_0$$

(ii) If $m_N^{(e)}$ denotes the local moment,

$$m_N^{(e)} = \int_{\Omega_e} \psi_N^{(e)}(\mathbf{x})\, dx \tag{6.87}$$

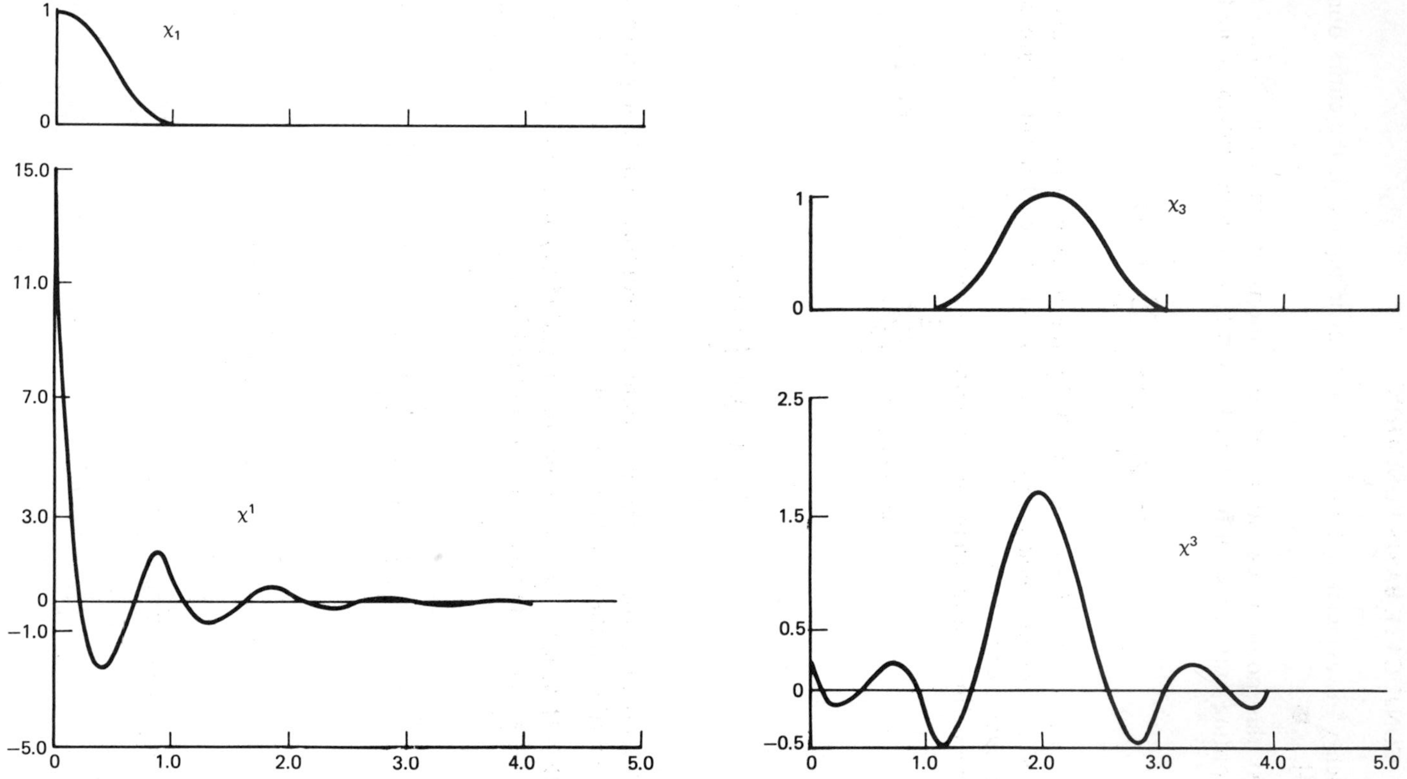
1
0
χ_1
15.0
11.0
7.0
3.0
0
−1.0
−5.0
χ^1
0
1.0
2.0
3.0
4.0
5.0
1
0
χ_3
2.5
1.5
0.5
0
−0.5
χ^3
0
1.0
2.0
3.0
4.0
5.0

(a) *(b)*

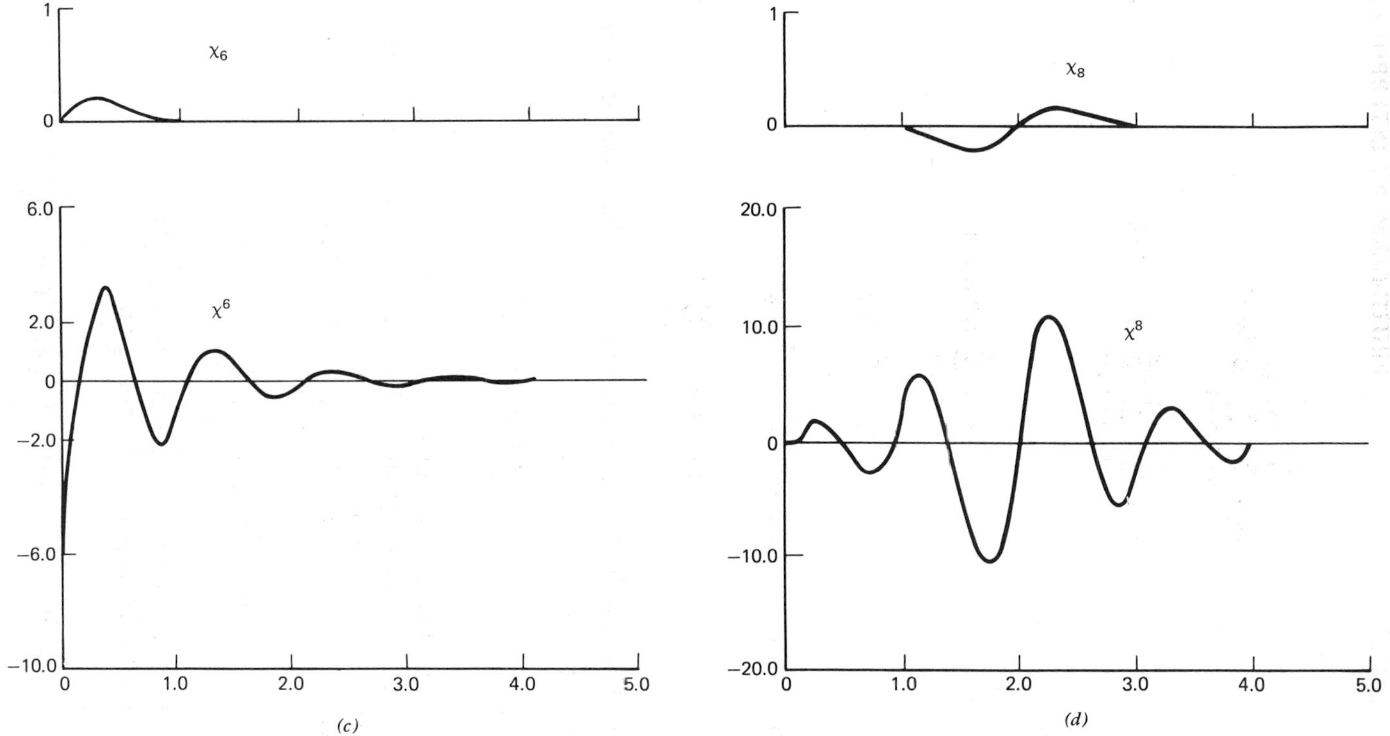

Figure 6.10 Piecewise cubic interpolation functions $\chi_k(x)$ and their conjugates. $\chi^k(x)$, (*a*) $\chi_1(x)$ and $\chi^1(x)$. (*b*) $\chi_3(x)$ and $\chi^3(x)$. (*c*) $\chi_6(x)$ and $\chi^6(x)$. (d) $\chi_8(x)$ and $\chi^8(x)$.

we have

$$M_i = \sum_{e=1}^{E} \sum_{N=1}^{N_e} \overset{(e)}{\Omega}{}_i^N m_N^{(e)} \tag{6.88}$$

$$M^i = \sum_{e=1}^{E} \sum_{N=1}^{N_e} \sum_{j=1}^{G} G^{ij} \overset{(e)}{\Omega}{}_j^N m_N^{(e)} \tag{6.89}$$

(iii)

$$M^i = \sum_{j=1}^{G} G^{ij} M_j = 1 \tag{6.90}$$

$$M_i = \sum_{j=1}^{G} G_{ij} M^j \tag{6.91}$$

(iv) The *volume* (measure) $|\tilde{\Omega}|$ of $\tilde{\Omega}$ is given by

$$|\tilde{\Omega}| = \int_{\tilde{\Omega}} dx = \sum_{i,j=1}^{G} G_{ij} \tag{6.92}$$

(v) If $V(\mathbf{x}) \in S_h(\Omega)$, its average value is $1/|\tilde{\Omega}|$ times the sum of its covariant components,

$$\int_{\Omega} V(\mathbf{x})\, dx = \sum_{i=1}^{G} V_i \tag{6.93}$$

PROOF. (i) This follows immediately on substituting $V(\mathbf{x})$ of (6.72) into $(1, V)_0$ and using (6.84) and (6.85). Part (ii) is obvious from the definitions and because of (*i*) of (6.61).

To verify (6.90), note that $\sum_i \phi_i(\mathbf{x}) = 1$. Therefore

$$M^i = \int_{\Omega} \phi^i \sum_j \phi_j \, dx = \sum_j (\phi^i, \phi_j)_0 = \sum_j \delta_j^i = 1$$

Then (6.91) follows from the fact that $M_i = \sum_j G_{ij} M^j$.

Equation (6.92) is obtained by noting that

$$\sum_i M_i = \sum_i \int_{\tilde{\Omega}} \phi_i(\mathbf{x})\, dx = \int_{\tilde{\Omega}} \sum_i \phi_i(\mathbf{x})\, dx = \int_{\tilde{\Omega}} 1\, dx = |\tilde{\Omega}|$$

and (6.93) follows from (6.90) and the fact that

$$\int_{\tilde{\Omega}} \sum_i V_i \phi^i(\mathbf{x})\,dx = \sum_i V_i M^i = \sum_i V_i \quad \blacksquare$$

Several additional examples of properties of the functions $\{\phi_i\}_{i=1}^G$ and $\{\phi^i\}_{i=1}^G$ can be found in [6.1].

6.6 FINITE-ELEMENT FAMILIES

There exists a remarkably large number of examples of effective local finite-element approximations, and large catalogs of the more popular elements in use in applications are available. See, e.g., [6.1, pp. 125–147], [6.7], and [6.8]. While we shall not attempt to reproduce an extensive list of such examples here, it is fitting to at least describe several of the more common finite-element families and some of their intrinsic properties.

Lagrange Families

We first dwell on the *Lagrange families* of finite elements, by which we mean a class of elements in which the local interpolant is uniquely determined by prescribing its values at a finite number N_e of nodal points. The interpolation functions $\psi_N^{(e)}(\mathbf{x})$ associated with elements in such families are of order $q=0$. These provide for the direct extension of classical Lagrange interpolation, i.e., the classical notion of curve fitting by interpolation of values at points, not derivatives at points.

There are two major families of Lagrange elements, the family defined on n-simplexes and the family defined on n-hypercubes. By an n-simplex, we mean a region $\Omega_e \subset \mathbf{R}^n$ determined by $n+1$ distinct points, not all of which lie in any $(n-1)$-dimensional hyperplane. A 2-simplex is a triangle, and a 3-simplex is a tetrahedron. These $n+1$ points define only the basic geometry of the element. We also want to introduce other nodes on the sides of the simplex so that $N_e \geqslant n+1$.

By an n-hypercube, we mean a region $\Omega_e \subset \mathbf{R}^n$ generated as a cartesian product of n intervals on the real line; that is, if $[a_i, b_i] = \{x: -\infty < a_i \leqslant x \leqslant b_i < \infty\}$,

$$\Omega = [a_1, b_1] \times [a_2, b_2] \times \cdots \times [a_n, b_n]$$

$$= \prod_{i=1}^{n} [a_i, b_i]$$

A 1-hypercube is a straight line connecting two points, a 2-hypercube is a rectangle in the plane, and a 3-hypercube is a right rectangular parallelepiped.

We take for granted here that the local interpolation function $\phi_N^{(e)}(\mathbf{x})$ shall, for simplicity, contain complete polynomials of degree $\leqslant k$, where k is an integer $\geqslant 0$. We use throughout the notation,

$$\mathcal{P}_k(\Omega) = \{\, p(\mathbf{x}) \colon p(\mathbf{x}) \text{ is a polynomial in } x_1, x_2, \ldots, x_n \\ \text{of degree} \leqslant k, \mathbf{x} \in \Omega \subset \mathbb{R}^n \} \tag{6.94}$$

For example, $\psi_N^{(e)}(\mathbf{x})$ may be a restriction of an element of $\mathcal{P}_k(\Omega)$ to Ω_e for some k or, frequently, $\psi_N^{(e)} \in \mathcal{P}_k(\Omega_e)$.

It is obvious from what has been said that, when $\psi_N^{(e)}(\mathbf{x}) \in \mathcal{P}_k(\Omega_e)$, the set of local nodal points of element Ω_e, $\mathcal{L}_e = \{\mathbf{x}_e^N\}_{N=1}^{N_e}$, must be such that, given any set of real numbers $\mu_N, N = 1, 2, \ldots, N_e$, there exists a *unique* polynomial $p(\mathbf{x}) \in \mathcal{P}_k(\Omega_e)$ such that

$$p(\mathbf{x}_e^N) = \mu_N \qquad 1 \leqslant N \leqslant N_e \tag{6.95}$$

Only then can local interpolation functions $\psi_N^{(e)}(\mathbf{x})$ exist in $\mathcal{P}_k(\Omega_e)$. When condition (6.95) is satisfied, we say that the nodal set $\mathcal{L}_e = \{\mathbf{x}_e^N\}_{N=1}^{N_e}$ is *k-unisolvent*, i.e. specifying N_e values of $p(\mathbf{x}) \in \mathcal{P}_k(\Omega_e)$ at the points $\mathbf{x}_e^N$ uniquely determines the polynomial $p(\mathbf{x})$. A set $M \subset \mathbb{R}^n$ is termed $\mathcal{L}_e$-*admissible* if, whenever it contains a point $\mathbf{x}$, it also contains the closed segment joining $\mathbf{x}$ to $\mathbf{x}_e^N$, for all $N = 1, 2, \ldots, N_e$; i.e., M is star-shaped with respect to each node. The intersection of all convex sets containing $\mathcal{L}_e$ is called the *convex hull* of $\mathcal{L}_e$. This terminology is used by Ciarlet and Raviart [6.3]. It is clear that the convex hull of $\mathcal{L}_e$ is always $\mathcal{L}_e$-admissible, but it is easy to construct examples of $\mathcal{L}_e$-admissible sets that are not convex.

In many cases it is possible to write explicitly a smooth function $u(\mathbf{x})$ to be interpolated in terms of its local Lagrange interpolant and some error term. Such expressions follow easily from Taylor-type expansions in $\mathbb{R}^n$, which make use of the concept of Fréchet derivatives (for an exhaustive treatment of this subject, see Nashed [6.9]). Recall that the *Fréchet differential* $\delta\mathcal{Q}(u,\eta)$ of an operator $\mathcal{Q}$ from a Banach space $\mathcal{U}$ into a Banach space $\mathcal{V}$ is defined as the form $\delta\mathcal{Q}(u,\eta)$, linear in η, where η is an arbitrary element in $\mathcal{U}$, such that

$$\lim_{\|\eta\|_{\mathcal{U}} \to 0} \frac{1}{\|\eta\|_{\mathcal{U}}} \|\mathcal{Q}(u+\eta) - \mathcal{Q}(u) - \delta\mathcal{Q}(u,\eta)\|_{\mathcal{V}} = 0 \tag{6.96}$$

Then $\delta \mathcal{Q}(u, \cdot)$, regarded as a linear operator on η, is the *Fréchet derivative* of $\mathcal{Q}$ at u. By repeated use of (6.96), we can define Fréchet derivatives of all orders.

In the case in which $\mathcal{Q}$ is merely a function $u(\mathbf{x})$ from R^n into R, the form of $\delta \mathcal{Q}(u,\eta)$ is particularly simple. Then the kth Fréchet differential of $u(\mathbf{x})$ at a point $\mathbf{x} \in \mathsf{R}^n$ is a symmetric k-linear mapping of $\mathsf{R}^n \times \mathsf{R}^n \times \cdots \times \mathsf{R}^n$ (k times) into R denoted $\mathcal{D}^k u(\mathbf{x})$ with the property that

$$\mathcal{D}^k u(\mathbf{x})\cdot(\xi_1,\xi_2,\ldots,\xi_k) = \mathcal{D}^k u(\mathbf{x})\cdot(\xi_{i_1},\xi_{i_2},\ldots,\xi_{i_k}) \tag{6.97}$$

Here $(\xi_1,\xi_2,\ldots,\xi_k)$ is an arbitrary set of k vectors in R^n and $(\xi_{i_1},\xi_{i_2},\ldots,\xi_{i_k})$ denotes a rearrangement of these vectors for any permutation of the integers from 1 to k. We also use the notation

$$\mathcal{D}^k u(\mathbf{x})\cdot(\xi_1,\xi,\xi,\ldots,\xi) = \mathcal{D}^k u(\mathbf{x})\cdot(\xi_1,\xi^{k-1}) \text{ and}$$

$$\mathcal{D}^k u(\mathbf{x})\cdot(\xi,\xi,\ldots,\xi) = \mathcal{D}^k u(\mathbf{x})\cdot\xi^k.$$

The usual partial derivatives of $u(\mathbf{x})$ are easily recovered from $\mathcal{D}^m u$, $0 \leqslant m \leqslant k$, as *directional derivatives*. For example, if we endow R^n with the orthonormal basis $(\mathbf{e}_1,\mathbf{e}_2,\ldots,\mathbf{e}_n)$, where

$$\mathbf{e}_i = \Big(0,0,\ldots,0,\underset{\substack{\uparrow\\ i}}{1},0,\ldots,0\Big)$$

$$\mathbf{e}_i\cdot\mathbf{e}_j = \delta_{ij} \qquad i,j = 1,2,\ldots,n$$

then

$$\frac{\partial u(\mathbf{x})}{\partial x_i} = \mathcal{D} u(\mathbf{x})\cdot\mathbf{e}_i$$

$$\frac{\partial^2 u(\mathbf{x})}{\partial x_i\,\partial x_j} = \mathcal{D}^2 u(\mathbf{x})\cdot(\mathbf{e}_i,\mathbf{e}_j)$$

and

$$D^\alpha u(\mathbf{x}) = \mathcal{D}^{|\alpha|} u(\mathbf{x})\cdot(\mathbf{e}_{\alpha_1},\mathbf{e}_{\alpha_2},\ldots,\mathbf{e}_{\alpha_n}) \tag{6.98}$$

Also, we define

$$\|\mathcal{D}^k u(\mathbf{x})\| = \sup\{|\mathcal{D}^k u(\mathbf{x})\cdot(\xi_1,\xi_2,\ldots,\xi_n)|\ ,\quad |\xi_i| \leqslant 1;\ i=1,2,\ldots,n\} \tag{6.99}$$

where $|\xi_i|$ denotes the usual euclidean norm of ξ_i. This definition, together with (6.98), also leads to the useful inequalities,

$$\left|\frac{\partial u(\mathbf{x})}{\partial x_i}\right| \leqslant \|\mathfrak{D}u(\mathbf{x})\| \leqslant C_1 \max_i \left|\frac{\partial u(\mathbf{x})}{\partial x_i}\right| \qquad i=1,2,\ldots,n$$

$$\left|\frac{\partial^2 u(\mathbf{x})}{\partial x_i\,\partial x_j}\right| \leqslant \|\mathfrak{D}^2 u(\mathbf{x})\| \leqslant C_2 \max_{i,j} \left|\frac{\partial^2 u(\mathbf{x})}{\partial x_i\,\partial x_j}\right| \qquad i,j=1,2,\ldots,n \tag{6.100}$$

etc., where C_1, C_2 depend only on n.

A key to some of our interpolation results is that we assume that the given function $u(\mathbf{x})$, which is to be interpolated, can be represented by a Taylor formula of the type

$$u(\mathbf{x}_e^N) = \sum_{r=0}^{k} \frac{1}{r!}\mathfrak{D}^r u(\mathbf{x})\cdot(\mathbf{x}_e^N - \mathbf{x})^r$$

$$+\frac{1}{(k+1)!}\mathfrak{D}^{k+1}u(\eta_N(\mathbf{x}))\cdot(\mathbf{x}_e^N-\mathbf{x})^{k+1}$$

$$\eta_N(\mathbf{x}) = \theta_N\mathbf{x} + (1-\theta_N)\mathbf{x}_e^N \qquad 0<\theta_N<1;\ 1\leqslant N \leqslant N_e \tag{6.101}$$

for each $\mathbf{x}_e^N \in \mathfrak{L}_e$ and for each $\mathbf{x}$ belonging to some set M containing $\mathfrak{L}_e$. It is clear that (6.101) will hold whenever M is a convex set containing $\mathfrak{L}_e$, $u \in C^k(M)$, and $\mathfrak{D}^{k+1}u(\mathbf{x})$ exists at every $\mathbf{x}\in M$. More generally, M need not actually be convex, but it must be $\mathfrak{L}_e$-admissible, and it is sufficient to demand that (6.101) hold on the convex hull of $\mathfrak{L}_e$.

The above ideas are also useful in the construction of Taylor-type expressions for the Lagrange interpolant over Ω_e. The principal results are summarized in the following theorem due to Ciarlet and Raviart [6.3]. Similar results are given in [6.1, pp. 106–114] and were originally used in finite-element error analysis in [6.10]; see also [6.11].

THEOREM 6.4. Let $\mathfrak{L}_e = \{\mathbf{x}_e^N\}_{N=1}^{N_e}$ be a k-unisolvent set, where k is a fixed integer $\geqslant 1$. Let M be an $\mathfrak{L}_e$-admissible set in R^n and let $u(\mathbf{x})$ be a function defined on M such that

(i) $u(\mathbf{x}) \in C^k(M)$.

(ii) $\mathfrak{D}^{k+1}u(\mathbf{x})$ exists, $\forall \mathbf{x}\in M$, where $\mathfrak{D}^{k+1}u(\mathbf{x})$ is the $(k+1)$st Fréchet derivative of $u(\mathbf{x})$. Then at any point $\mathbf{x} \in M$ and for any integer $0 \leqslant m \leqslant k$,

$$\mathfrak{D}^m U_e(\mathbf{x}) = \mathfrak{D}^m u(\mathbf{x}) + \frac{1}{(k+1)!} \sum_{N=1}^{N_e} \left[\mathfrak{D}^{k+1} u(\eta_N(\mathbf{x})) \cdot (\mathbf{x}_e^N - \mathbf{x})^{k+1} \right] \cdot \mathfrak{D}^m \psi_N^{(e)}(\mathbf{x}) \tag{6.102}$$

where $(\mathbf{x}_e^N - \mathbf{x})^{k+1} = (\mathbf{x}_e^N - \mathbf{x}, \ldots, \mathbf{x}_e^N - \mathbf{x})$, $k+1$ times, $U_e(\mathbf{x})$ is the local finite-element interpolant of $u(\mathbf{x})$, i.e.,

$$U_e(\mathbf{x}_e^N) = u(\mathbf{x}_e^N) \qquad N = 1, 2, \ldots, N_e \tag{6.103}$$

the functions $\psi_N^{(e)}(\mathbf{x})$ are uniquely determined polynomials of degree $\leqslant k$ such that

$$\psi_N^{(e)}(\mathbf{x}_e^M) = \delta_N^M \qquad M, N = 1, 2, \ldots, N_e$$

and $\eta_N(\mathbf{x})$ are functions of the form

$$\eta_N(\mathbf{x}) = \theta_N \mathbf{x} + (1 - \theta_N)\mathbf{x}_e^N \qquad 0 < \theta_N < 1 \tag{6.104}$$

PROOF. We begin the proof with two observations. First, conditions (i) and (ii) allow us to write $u(\mathbf{x})$ in a Taylor expansion of the form in (6.101). Second, we know that, for $\mathbf{x} \in \Omega_e \subset M$,

$$U_e(\mathbf{x}) = \sum_{N=1}^{N_e} u(\mathbf{x}_e^N) \psi_N^{(e)}(\mathbf{x})$$

and therefore that

$$\mathfrak{D}^m U_e(\mathbf{x}) = \sum_{N=1}^{N_e} u(\mathbf{x}_e^N)\, \mathfrak{D}^m \psi_N^{(e)}(\mathbf{x})$$

Combining these facts, we have immediately

$$\mathfrak{D}^m U_e(\mathbf{x}) = \sum_{r=0}^{k} F_r^m(u; \mathbf{x}) + \frac{1}{(k+1)!} \sum_{N=1}^{N_e} \left[\mathfrak{D}^{k+1} u(\eta_N(\mathbf{x})) \cdot (\mathbf{x}_e^N - \mathbf{x})^{k+1} \right] \cdot \mathfrak{D}^m \psi_N^{(e)}(\mathbf{x}) \tag{6.105}$$

where, for compactness, we have denoted

$$F_r^m(u;\mathbf{x}) = \frac{1}{r!} \sum_{N=1}^{N_e} \left[\mathfrak{D}^r u(\mathbf{x}) \cdot (\mathbf{x}_e^N - \mathbf{x})^r \right] \mathfrak{D}^m \psi_N^{(e)}(\mathbf{x}) \tag{6.106}$$

The theorem will be proved if we can show that $F_r^m(u;\mathbf{x})$ equals $\mathfrak{D}^m u(\mathbf{x})$. We show in fact that

$$F_r^m(u;\mathbf{x}) = \begin{cases} 0 & \text{for } 0 \leqslant r \leqslant m-1 \\ \mathfrak{D}^m u(\mathbf{x}) & \text{for } r = m \\ 0 & \text{for } m+1 \leqslant r \leqslant k \end{cases} \tag{6.107}$$

Consider first the case in which $0 \leqslant r \leqslant m-1$ and $u(\mathbf{x})$ is a constant C_0. Since (6.105) must also hold in this case, we have $F_0^m(u;\mathbf{x}) = 0$. The fact that $F_r^m(u;\mathbf{x}) = 0$ for any $r \leqslant m-1$ then follows from induction: If $F_r^m(u;\mathbf{x}) = 0$ for $r = 0, 1, \ldots, r_0, r_0 \leqslant m-2$, then, for some r_0-linear mapping $A_{r_0} \in \mathfrak{L}(\Pi^{r_0}\mathbf{R}^n, \mathbf{R})$

$$0 = \sum_{N=1}^{N_e} \left[A_{r_0} \cdot (\mathbf{x}_e^N - \mathbf{x})^{r_0} \right] \mathfrak{D}^m \psi_N^{(e)}(\mathbf{x})$$

This follows from (6.105) and the fact that we can associate with any such operator A_{r_0} a polynomial $q(\mathbf{x})$ in $\mathfrak{P}_{r_0}(M)$ such that $\mathfrak{D}^{r_0} q(\mathbf{x}) = A_{r_0}$. By setting $A_{r_0} = \mathfrak{D}^{r_0} u(\mathbf{x})$, we obtain

$$\sum_{N=1}^{N_e} \left[\mathfrak{D}^{r_0} u(\mathbf{x}) \cdot (\mathbf{x}_e^N - \mathbf{x})^{r_0} \right] \mathfrak{D}^m \psi_N^{(e)}(\mathbf{x}) = 0$$

Therefore $F_r^m(u;\mathbf{x}) = 0$ for $r \leqslant m-1$. The case $m+1 \leqslant r \leqslant k$ is proved in an identical manner.

When $r = m$, note that $U_e(\mathbf{x}) = u(\mathbf{x})$ if $u(\mathbf{x}) \in \mathfrak{P}_m(M)$. Each element $q(\mathbf{x})$ in $\mathfrak{P}_m(M)$ is of the form $q(\mathbf{x}) = \Sigma_{|\alpha| \leqslant m} A_\alpha \mathbf{x}^\alpha$ where the A_α's are constants. Then $\mathfrak{D}^m q(\mathbf{x})$ is some constant array, say A_m. Setting $u(\mathbf{x}) = U_e(\mathbf{x}) = q(\mathbf{x})$ in (6.105) for the special case in which $u(\mathbf{x}) \in \mathfrak{P}_m(M)$, and recalling that the terms of $r \neq m$ vanish, we see that

$$A_m = \frac{1}{m!} \sum_{N=1}^{N_e} \left[A_m \cdot (\mathbf{x}_e^N - \mathbf{x})^m \right] \mathfrak{D}^m \psi_N^{(e)}(\mathbf{x}) \tag{6.108}$$

This must hold for every A_m in $\mathcal{L}(\Pi^m \mathsf{R}^n, \mathsf{R})$ so that

$$\mathfrak{D}^m u(\mathbf{x}) = \frac{1}{m!} \sum_{N=1}^{N_e} \left[\mathfrak{D}^m u(\mathbf{x}) \cdot (\mathbf{x}_e^N - \mathbf{x})^m \right] \mathfrak{D}^m \psi_N^{(e)}(\mathbf{x})$$

We have thus proved (6.107) and have therefore completed the proof of the theorem. ■

EXAMPLE 6.10. The arguments of the above theorem are particularly easy to visualize if we consider a specific one-dimensional example. Suppose $k=4$, $m=2$, and the function $u(x)$ is defined on a one-dimensional domain. Thus (6.105) leads to the expansion

$$D^2 U(x) = \sum_{r=0}^{4} \frac{1}{r!} D^r u(x) \sum_{N=1}^{N_e} (x^N - x)^r D^2 \psi_N(x)$$

$$+ \frac{1}{5!} \sum_{N=1}^{N_e} D^5 u(\eta_N(x))(x^N - x)^5 D^2 \psi_N(x)$$

where $U(x)$ is in a space of finite-element interpolants, spanned by $\{\psi_N\}_{N=1}^{N_e}$, which contains $\mathcal{P}_4(\Omega)$.

The next step is fundamental. Since the above expansion holds for all $u(x) \in C^4(\Omega)$, with fifth derivatives existent for all $x \in \Omega_e$, it certainly holds for a $u(x) \in \mathcal{P}_4(\Omega)$. Thus choose $u(x)$ to be a polynomial of degree $\leqslant 4$. Then, the remainder term vanishes and $U(x)$ and $u(x)$ coincide. We then have, for $u \in \mathcal{P}_4(\Omega)$,

$$D^2 U(x) = D^2 u(x) = u(x) \sum_{N=1}^{N_e} D^2 \psi_N(x) + Du(x) \sum_{N=1}^{N_e} (x^N - x) D^2 \psi_N(x)$$

$$+ \frac{1}{2} D^2 u(x) \sum_{N=1}^{N_e} (x^N - x)^2 D^2 \psi_N(x)$$

$$+ \sum_{r=3}^{N_e} \frac{1}{r!} D^r u(x) \sum_{N=1}^{N_e} (x^N - x)^r D^2 \psi_N(x)$$

Next consider the three choices of polynomials $u = C_0 = \text{constant}$, $u(x) = C_0 + C_1 x$, and $u(x) = C_0 + C_1 x + C_2 x^2$. We have

$$D^2 C_0 = 0 = C_0 \sum_{N=1}^{N_e} D^2 \psi_N$$

thus

$$\sum_{N=1}^{N_e} D^2\psi_N = 0$$

$$D^2(C_0 + C_1 x) = 0 = C_1 \sum_{N=1}^{N_e} (x^N - x) D^2\psi_N$$

thus

$$\sum_{N=1}^{N_e} (x^N - x) D^2\psi_N = 0$$

and

$$D^2(C_0 + C_1 x + C_2 x^2) = 2C_2 = 2C_2 \cdot \frac{1}{2} \sum_{N=1}^{N_e} (x^N - x)^2 D^2\psi_N$$

which shows that

$$\frac{1}{2} \sum_{N=1}^{N_e} (x^N - x)^2 D^2\psi_N = 1$$

Summing up, we have shown that, for $u \in \mathcal{P}_4(\Omega)$,

$$D^2U(x) = D^2u(x) + \sum_{r=3}^{4} \frac{1}{r!} D^r u(x) \sum_{N=1}^{N_e} (x^N - x)^r D^2\psi_N(x)$$

But for this choice $U(x) = u(x)$, so that also

$$\sum_{N=1}^{N_e} (x^N - x)^r D^2\psi_N = 0 \qquad r = 3, 4$$

It follows that, for arbitrary (but sufficiently smooth) u,

$$D^2U(x) = D^2u(x) + \frac{1}{5!} \sum_{N=1}^{N_e} D^5u(\eta_N)(x^N - x)^5 D^2\psi_N(x)$$

in agreement with (6.102). ■

Lagrange Simplexes

We now pass on to some specific examples of Lagrange simplex finite-element families. The most popular among these are best illustrated by the Pascal triangle shown in Fig. 6.11. For two-dimensional elements, all triangles in this case, various members of the family are represented by various tiers of the triangle. Each nodal point of an element corresponds to an entry in the Pascal triangle. The first element has a single node at its centroid. The specification of the value of a polynomial of zero degree (i.e., a constant) uniquely determines this polynomial. The point is thus 0-unisolvent. The second element has three nodes; specification of the values of a linear function at these three points uniquely defines it. Therefore this set of nodes is 1-unisolvent. Likewise, the next tier is 2-unisolvent, and a unique quadratic is determined by the six nodal values; etc.

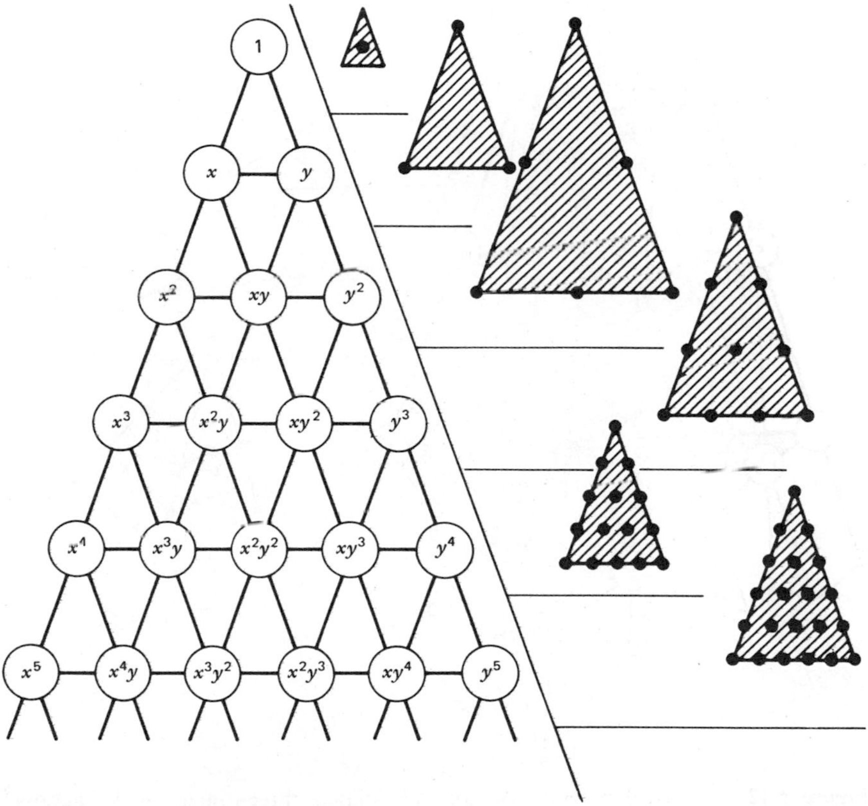

Figure 6.11 The Pascal triangle and the associated Lagrange simplex elements.

The idea is easily extended to three dimensions by using a "Pascal tetrahedron" as indicated in Fig. 6.12. Again, each tier defines a tetrahedral finite element with one nodal point allotted to each monomial entry in the Pascal tetrahedron. Tier r of the tetrahedron corresponds to a complete polynomial of degree r in x, y, and z and defines an r-unisolvent set of nodal points.

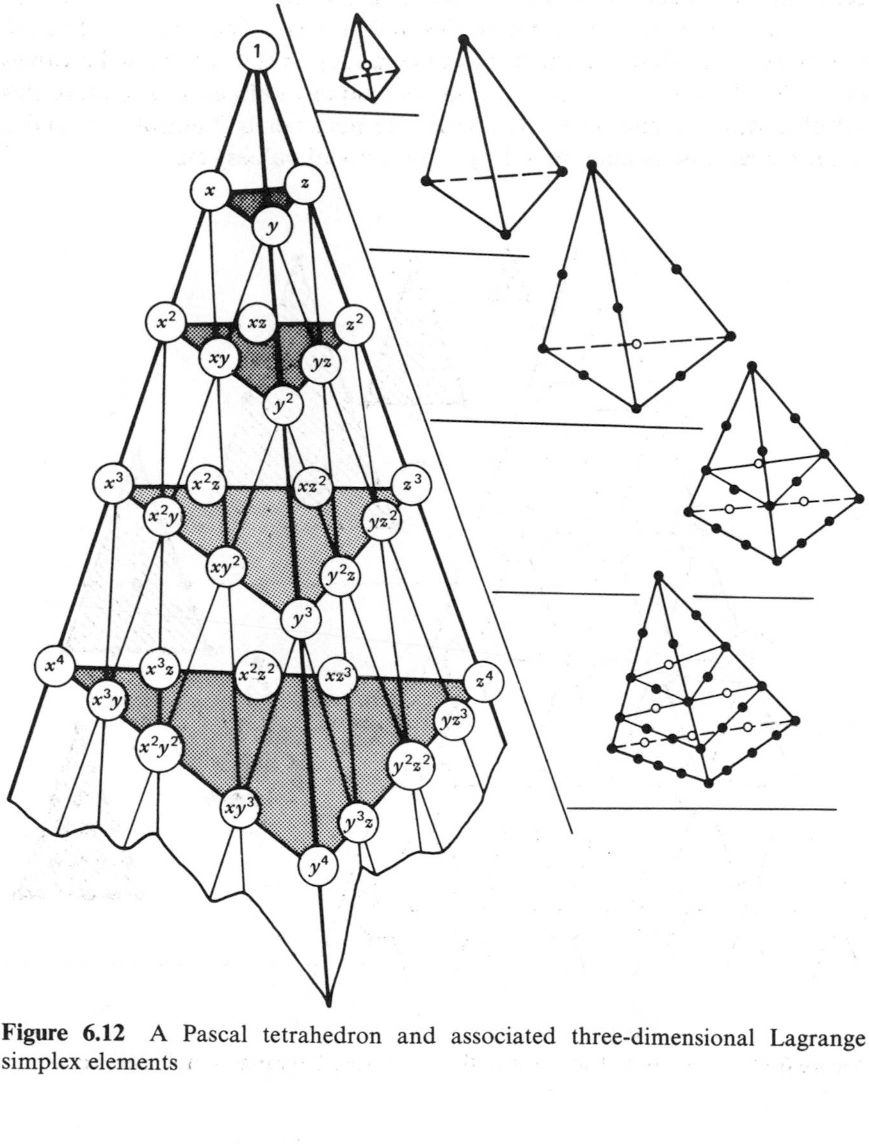

Figure 6.12 A Pascal tetrahedron and associated three-dimensional Lagrange simplex elements

In the n-dimensional case, a similar but more abstract situation is encountered. There, if Ω_e is a nondegenerate n-simplex in $\mathbb{R}^n$, the set of nodes

$$\mathfrak{L}_e^{(I)} = \{\mathbf{x}_e^N\}_{N=1}^{N_e} \tag{6.109}$$

is 1-unisolvent. If $\mathbf{x}_e^{NM}$ denotes the node at the midpoint of the edge connecting $\mathbf{x}_e^N$ and $\mathbf{x}_e^M$, then

$$\mathfrak{L}_e^{(II)} = \{\mathbf{x}_e^N\}_{N=1}^{N_e} \cup \{\mathbf{x}_e^{NM}\} \qquad 1 \leqslant N < M \leqslant N_e + 1 \tag{6.110}$$

is 2-unisolvent; etc. (cf. [6.3]).

Lagrange Hypercubes

The simplest examples of the Lagrange hypercube finite-element family are furnished again by the Pascal triangle. Again, each node of an element of this family corresponds to an entry of a monomial in the triangle, but now the triangle is extended in each case to form a square, as indicated in Fig. 6.13. A three-dimensional version is indicated in Fig. 6.14, where a skewed right Pascal tetrahedron can be visualized.

The key feature of Lagrange families of this type is that they can be generated systematically from classical one-dimensional Lagrange interpolation polynomials. For instance, the function

$$L_i^{(m)}(x) = \frac{(x-x^0)(x-x^1)\cdots(x-x^{i-1})(x-x^{i+1})\cdots(x-x^m)}{(x^i-x^0)(x^i-x^1)\cdots(x^i-x^{i-1})(x^i-x^{i+1})\cdots(x^i-x^m)}$$

$$= \frac{\prod\limits_{\substack{i\neq j\\ j=0}}^{m} (x-x^j)}{\prod\limits_{\substack{i\neq j\\ j=0}}^{m} (x^i-x^j)} \qquad i=0,1,2,\ldots,m \tag{6.111}$$

is a polynomial of degree $m+1$, which assumes a unit value at node x^i and a value of zero at all other nodes, $x^j, j\neq i, 0\leqslant i,j\leqslant m$. The classical Lagrange interpolant of a function $u(x)$ on $[x^0,x^m]$ is then

$$\Pi_x u = U(x) = \sum_{N=0}^{m} u(x^N) L_N^{(m)}(x) \tag{6.112}$$

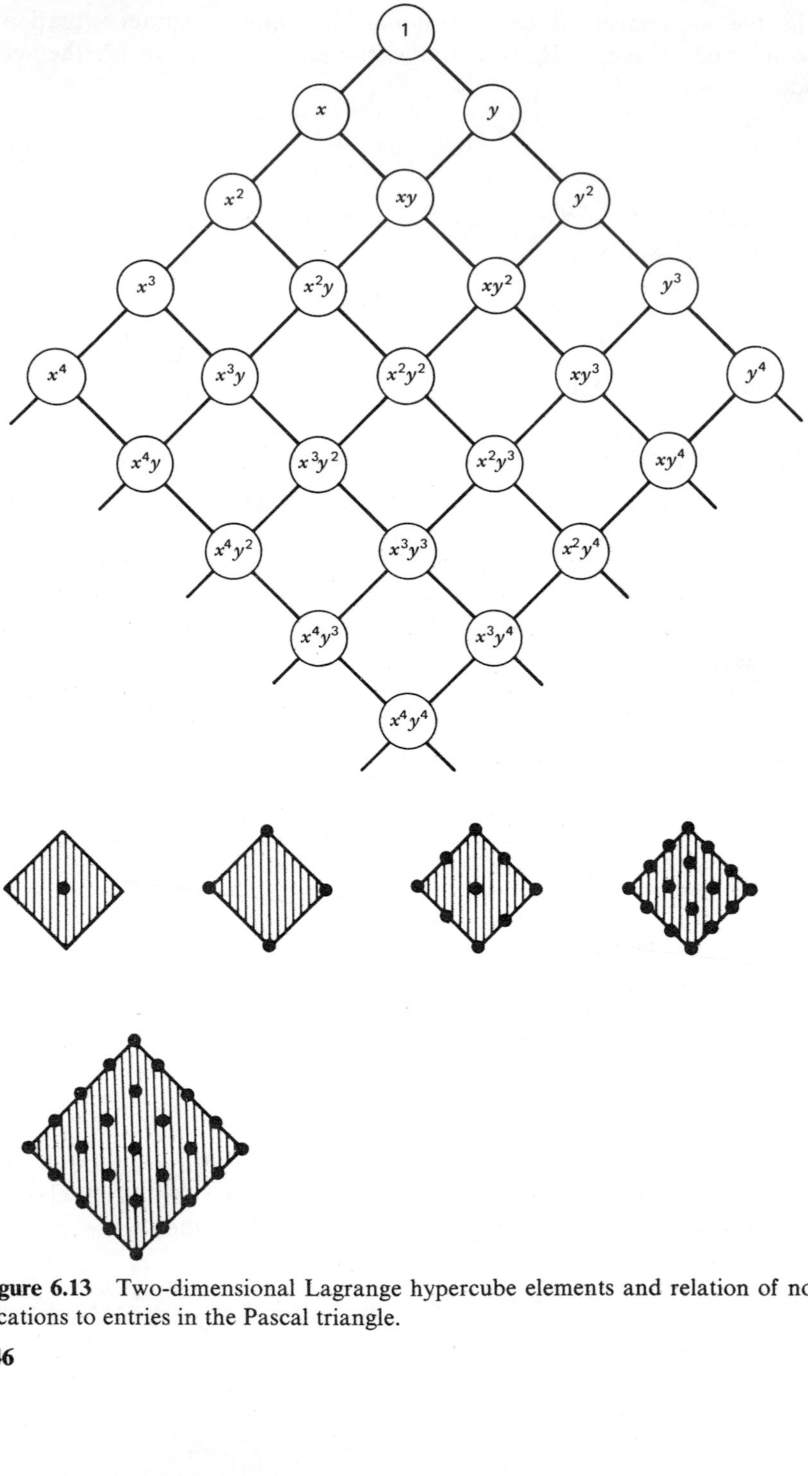

Figure 6.13 Two-dimensional Lagrange hypercube elements and relation of node locations to entries in the Pascal triangle.

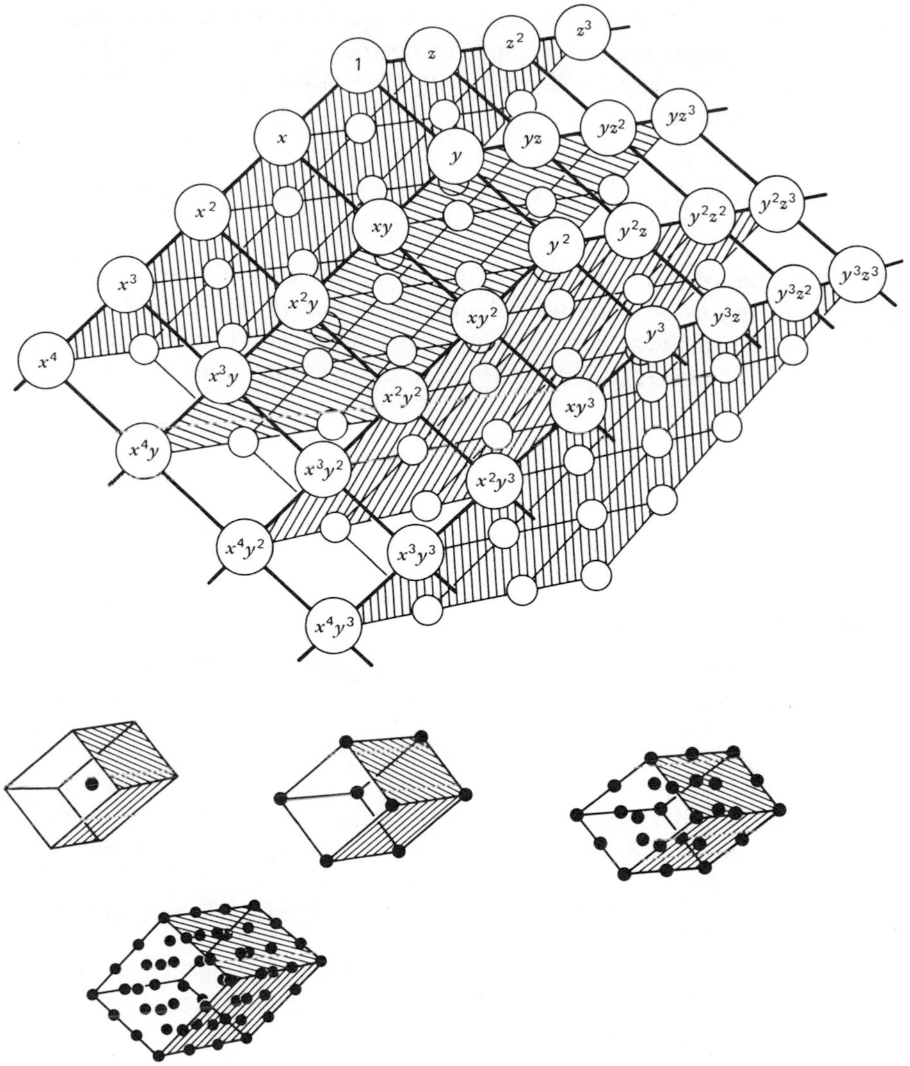

Figure 6.14 Three-dimensional Lagrange hypercube elements.

where Π_x is the projection operator mapping $u(x)$ onto the $(m+1)$-dimesional space spanned by the functions $L_N^{(m)}(x)$. Then two-dimensional Lagrange elements of the type in Fig. 6.13 are described by the composition

$$\Pi_y \cdot \Pi_x u(x,y) = \sum_{M=0}^{m} \sum_{R=0}^{r} u(x^M, y^R) L_M^{(m)}(x) L_R^{(r)}(y) \tag{6.113}$$

Indeed, when $m \neq r$, we can expand on the class of elements indicated in Fig. 6.13 by considering polynomials of degree $m+1$ in x and $r+1$ in y. This leads to a rectangular element with $(m+1)\times(r+1)$ nodes. In three dimensions we have

$$\Pi_z \cdot \Pi_y \cdot \Pi_x u = \sum_{M=0}^{m} \sum_{R=0}^{r} \sum_{S=0}^{s} u(x^M, y^R, z^S) L_M^{(m)}(x) L_R^{(r)}(y) L_S^{(s)}(z) \quad (6.114)$$

etc. Functions of the type in (6.113) and (6.114) are regarded as tensor products of spaces of Lagrange polynomials in one variable. For example, if Π_x produces a linear combination of the functions $(1,x,x^2)$ and Π_y a combination of $(1,y,y^2)$, then $\Pi_x \cdot \Pi_y$ produces combinations of

$$(1,x,x^2)\otimes(1,y,y^2) \equiv \begin{bmatrix} 1 \\ x \\ x^2 \end{bmatrix} \begin{bmatrix} 1, y, y^2 \end{bmatrix} = \begin{bmatrix} 1 & y & y^2 \\ x & xy & xy^2 \\ x^2 & x^2y & x^2y^2 \end{bmatrix}$$

Boolean Sums

There is another way to compound the projection operators Π_x and Π_y called the *Boolean Sum,* denoted

$$\Pi_x \oplus \Pi_y = \Pi_x + \Pi_y - \Pi_x \cdot \Pi_y \quad (6.115)$$

Gordon [6.12] and Gordon and Hall [6.13] have made extensive use of such compositions in bivariate and multivariate interpolation. Since

$$\Pi_x \oplus \Pi_y u(x^M, y) = u(x^M, y) \qquad \text{and} \qquad \Pi_x \oplus \Pi_y u(x, y^R) = u(x, y^R)$$

$M = 1,2,\ldots,m$; $R = 1,2,\ldots,r$, the operation $\Pi_x \oplus \Pi_y$ is seen to interpolate $u(x,y)$ along the *lines* $x = x^M$, $y = y^R$. This has prompted Gordon to use the term *transfinite interpolation.*

The most important aspect of such transfinite interpolation schemes is their increased accuracy over tensor product schemes such as (6.113). This can be seen by introducing *remainder projectors:*

$$R_x = I - \Pi_x \qquad \text{and} \qquad R_y = I - \Pi_y \quad (6.116)$$

Then

$$I - \Pi_x \cdot \Pi_y = R_x + R_y - R_x \cdot R_y = R_x \oplus R_y \quad (6.117)$$

but

$$I - \Pi_x \oplus \Pi_y = R_x \cdot R_y \qquad (6.118)$$

Now R_x and R_y effectively give the truncation errors in the interpolants $\Pi_x u$ and $\Pi_y u$. Thus in the tensor schemes the overall truncation error is a linear combination of the individual errors produced by Π_x and Π_y taken independently, minus a product term. By contrast, the Boolean sum leads to a product of R_x and R_y and, consequently, a generally higher-order asymptotic rate-of-convergence.

The Serendipity Elements

Closely related to the Lagrange families of finite elements are the *serendipity families* of Zienkiewicz [6.8, pp. 107–110], which are obtained by direct construction of independent interpolation functions which assume a unit value at one node of an element and zero values at the remaining nodes. In essence, most of these elements can be obtained directly from the Lagrange families by imposing constraints on the values of interpolants at points between boundary nodes. For example, if we require that the value of the interpolant at the center node of the Lagrange element in Fig. 6.15*a* be the average of the values at opposite nodes at the midpoint of parallel sides, then the central node can be eliminated. The result is a serendipity element of the type shown. The local interpolation functions for this element are

$$\psi_N(x,y) = \tfrac{1}{4}(1 + x^N x)(1 + y^N y)(x^N x + y^N y - 1) \qquad N = 1,2,3,4$$

$$\psi_N(x,y) = \tfrac{1}{2}(1 - x^2)(1 + y^N y) \qquad N = 5,7$$

$$\psi_N(x,y) = \tfrac{1}{2}(1 + x^N x)(1 - y^2) \qquad N = 6,8$$

Examples of three-dimensional serendipity elements are given in Fig. 6.15*b*.

Conforming Elements

Consider any specific example of a Lagrange finite element described previously, and note that on fitting any two of such elements together at nodal points along a common interelement boundary we uniquely define a polynomial of a certain degree through these boundary nodes. For example, by fitting two triangles together, over each of which the local interpolant is piecewise linear and uniquely determined by its values at the

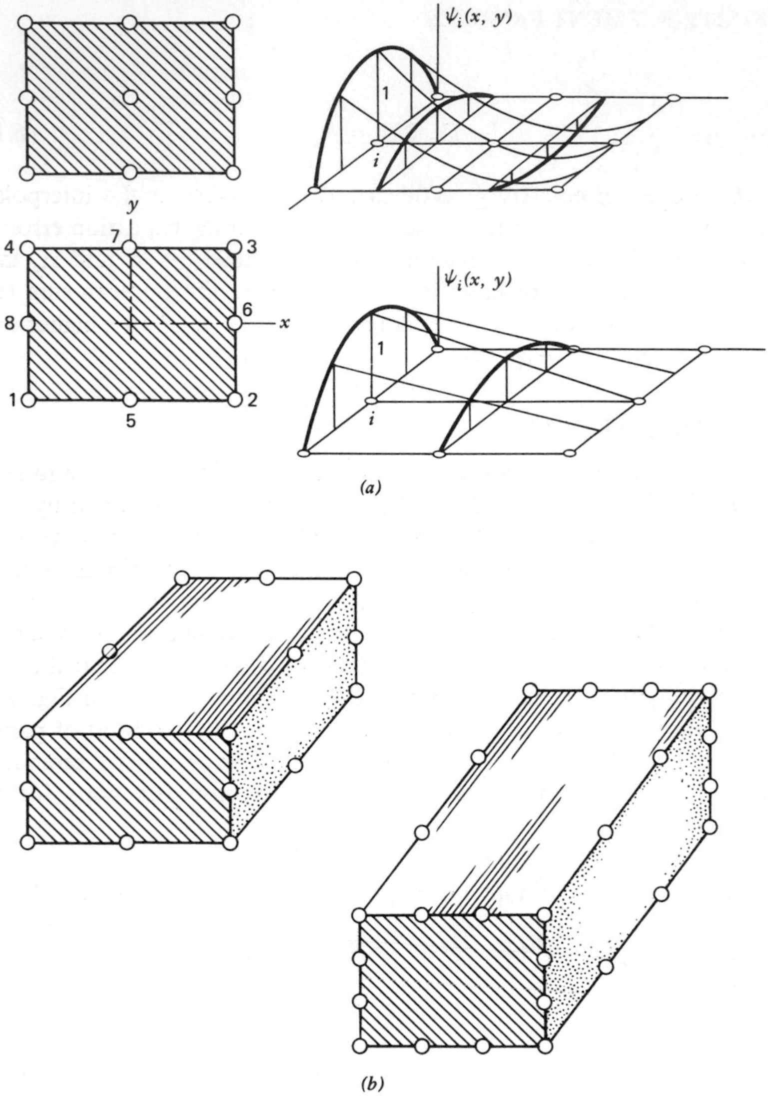

Figure 6.15 Serendipity elements.

vertices, we match values of the local interpolants at two common boundary nodes. Two such values uniquely determine a linear function along this straight interelement boundary. Likewise, requiring that two quadratics coincide at three points on a line uniquely determines a quadratic polynomial through these three values.

It follows that in all cases of this type the global interpolant $U(\mathbf{x})$ is *continuous across interelement boundaries.* Indeed, if $\tilde{\Omega}$ denotes the connected model, then $U(\mathbf{x}) \in C^0(\tilde{\Omega})$.

When the choice of nodal points, element geometry, and local interpolating functions leads to a global interpolant that is continuous across interelement boundaries in the connected finite element model, then the individual finite elements are said to be *conforming.* The Lagrange families of finite elements contain C^0*-conforming elements.* As a rule, to develop smoother conforming elements, e.g., C^q elements, $q > 0$, it is necessary to also prescribe values of derivatives of the interpolant at element nodes. We consider some examples of such elements below.

Hermite Families

We now come to another large class of finite elements characterized by interpolants for which the values of various derivatives, as well as the values of the function itself, are specified at nodal points. The elementary Hermite interpolation polynomials of Fig. 6.6 belong to this class, so we call specific subclasses of such elements Hermite families. Hermite families involving both n-simplexes and hypercubes can be constructed.

The first problem we face in describing Hermite families of elements is that of unisolvence: What criteria can we lay down to guarantee that the specification of derivatives of various orders at selected nodal points uniquely determines polynomials of a specified degree k? This question was elegantly answered by Ciarlet and Raviart [6.3], and we outline their method here.

Again, let $\mathcal{P}_k(\Omega)$ denote the space of polynomials of degree $\leqslant k$ defined on $\Omega \subset \mathbf{R}^n$, and let k be a fixed integer $\geqslant 1$. Consider a finite element $\Omega_e \subset \tilde{\Omega} \subset \mathbf{R}^n$, and divide the set $\mathcal{L}_e = \{\mathbf{x}_e^N\}_{N=1}^{N_e}$ of local nodes into the $\nu+1$ subsets

$$\Sigma^{(0)} = \left\{{}_{(0)}\mathbf{x}^N\right\}_{N=1}^{N_0} \qquad \Sigma^{(1)} = \left\{{}_{(1)}\mathbf{x}^N\right\}_{N=1}^{N_1} \qquad \cdots \qquad \Sigma^{(\nu)} = \left\{{}_{(\nu)}\mathbf{x}^N\right\}_{N=1}^{N_\nu} \tag{6.119}$$

where, for clarity, we omit writing the element identification label e. The designation of these $\nu+1$ sets allows us to distinguish subsets of nodal points at which derivatives of a specific order are prescribed. Clearly,

$$\mathcal{L}_e = \bigcup_{\mu=0}^{\nu} \Sigma^{(\mu)}$$

Continuing, let $(\mathsf{R}^n)^r$ denote the product space $\Pi^r \mathsf{R}^n$, and denote by $\mathcal{L}_r(\mathsf{R}^n, \mathsf{R})$ the space of continuous r-linear mappings from $(\mathsf{R}^n)^r$ into R. We assume in the Hermite interpolation problem that to each point ${}_{(\mu)}\mathbf{x}^N$, $0 \leqslant \mu \leqslant \nu$, in $\mathcal{L}_e$ there is associated a subset χ_μ^N of $(\mathsf{R}^n)^\mu$.

Collecting all these preliminary ideas, we now define the set $\mathcal{L}_e$ of (6.119) to be k-unisolvent if and only if, for any N_μ μ-linear arrays $A_\mu^N \in \mathcal{L}_\mu(\mathsf{R}^n, \mathsf{R}), 1 \leqslant N \leqslant N_\mu, 0 \leqslant \mu \leqslant \nu$, there is a unique polynomial $p(\mathbf{x}) \in \mathcal{P}_k(\Omega_e)$ such that

$$\mathfrak{D}^\mu p\left({}_{(\mu)}\mathbf{x}^N\right) \cdot \Xi = A_\mu^N \cdot \Xi \tag{6.120}$$

for all sets Ξ of μ n-vectors $\Xi = (\xi_1, \xi_2, \ldots, \xi_\mu) \in \chi_\mu^N$, and for all $1 \leqslant N \leqslant N_\mu$, $0 \leqslant \mu \leqslant \nu$. It is to be understood that $p({}_{(0)}\mathbf{x}^N) = A_0^N$ when $\mu = 0$.

When a given function $u(\mathbf{x})$ is such that $\mathfrak{D}^\mu u({}_{(\mu)}\mathbf{x}^N)$ is defined for all $1 \leqslant N \leqslant N_\mu, 0 \leqslant \mu \leqslant \nu$, the corresponding polynomial $U_e(\mathbf{x})$ is the local finite-element interpolating polynomial of $u(\mathbf{x})$ if it is the unique polynomial of degree $\leqslant k$ in $\mathcal{P}_k(\Omega_e)$ such that

$$\mathfrak{D}^\mu U_e\left({}_{(\mu)}\mathbf{x}^N\right) = \mathfrak{D}^\mu u\left({}_{(\mu)}\mathbf{x}^N\right) \tag{6.121}$$

for every $1 \leqslant N \leqslant N_\mu, 0 \leqslant \mu \leqslant \nu$.

We reproduce the following theorem from Ciarlet and Raviart [6.3, p. 188] which is an extension of Theorem 6.4 to Hermite interpolation for the case $\nu = 1$. Results for $\nu \geqslant 2$ are obtained in an analogous manner.

THEOREM 6.5. Let

$$\mathcal{L}_e = \left\{{}_{(0)}\mathbf{x}^N\right\}_{N=1}^{N_0} \cup \left\{{}_{(1)}\mathbf{x}^N\right\}_{N=1}^{N_1}$$

be a k-unisolvent set such that to each node ${}_{(1)}\mathbf{x}^N$ there corresponds a set $\chi_1^N \subset \mathsf{R}^n$ spanned by the linearly independent vectors $(\mathbf{e}_1^N, \mathbf{e}_2^N, \ldots, \mathbf{e}_\gamma^N)$. Let $M \subset \mathsf{R}^n$ be an $\mathcal{L}_e$-admissible set and let $u(\mathbf{x})$ be a function defined on M such that

(i) $u(\mathbf{x}) \in C^k(M)$.

(ii) $\mathfrak{D}^{k+1}u(\mathbf{x})$ exists, $\forall \mathbf{x} \in M$, where $\mathfrak{D}^{k+1}u(\mathbf{x})$ is the $(k+1)$st Fréchet derivative of $u(\mathbf{x})$.

Then at any point $\mathbf{x}\in M$ and for any integer m such that $0\leqslant m\leqslant k$,

$$\mathfrak{D}^m U_e(\mathbf{x})=\mathfrak{D}^m u(\mathbf{x})+\frac{1}{(k+1)!}\sum_{N=1}^{N_0}\left[\mathfrak{D}^{k+1}u(\eta_N^0(\mathbf{x}))\cdot\left({}_{(0)}\mathbf{x}^N-\mathbf{x}\right)^{k+1}\right]\mathfrak{D}^m\psi_N^0(\mathbf{x})$$

$$+\frac{1}{k!}\sum_{N=1}^{N_1}\sum_{q=1}^{\gamma}\left[\mathfrak{D}^{k+1}u\left(\eta_{Nq}^1(\mathbf{x})\right)\cdot\left(\mathbf{e}_q^N,\left({}_{(1)}\mathbf{x}^N-\mathbf{x}\right)^k\right)\right]\mathfrak{D}^m\psi_N^{1,q}(\mathbf{x}) \tag{6.122}$$

Here $U_e(\mathbf{x})$ is the local finite-element interpolant of $u(\mathbf{x})$; $\psi_N^0(\mathbf{x})$ and $\psi_N^{1,q}(\mathbf{x})$ are unique polynomials of degree $\leqslant k$ such that

$$\left.\begin{aligned}
&\psi_N^0\left({}_{(0)}\mathbf{x}^M\right)=\delta_N^M && M,N=1,2,\ldots,N_0\\
&\mathfrak{D}\psi_N^0\left({}_{(1)}\mathbf{x}^M\right)\cdot\mathbf{e}_q^M=0 && N=1,2,\ldots,N_0\\
& && M=1,2,\ldots,N_1\\
& && q=1,2,\ldots,\gamma\\
&\psi_N^{1,q}\left({}_{(0)}\mathbf{x}^M\right)=0 && N=1,2,\ldots,N_1\\
& && M=1,2,\ldots,N_0\\
& && q=1,2,\ldots,\gamma\\
&\mathfrak{D}\psi_N^{1,q}\left({}_{(1)}\mathbf{x}^M\right)\cdot\mathbf{e}_p^M=\delta_N^M\delta_p^q && M,N=1,2,\ldots,N_1\\
& && q,p=1,2,\ldots,\gamma
\end{aligned}\right\} \tag{6.123}$$

and

$$\left.\begin{aligned}
\eta_N^0(\mathbf{x})&=\theta_N^0\mathbf{x}+\left(1-\theta_N^0\right){}_{(0)}\mathbf{x}^N\\
\eta_{Nq}^1(\mathbf{x})&=\theta_{Nq}^1\mathbf{x}+\left(1-\theta_{Nq}^1\right){}_{(1)}\mathbf{x}^N
\end{aligned}\right\} \tag{6.124}$$

for $0<\theta_N^0,\ \theta_{Nq}^1<1$.

PROOF. We first observe that

$$U_e(\mathbf{x})=\sum_{N=1}^{N_0}u\left({}_{(0)}\mathbf{x}^N\right)\psi_N^0(\mathbf{x})$$

$$+\sum_{N=1}^{N_1}\sum_{q=1}^{\gamma}\left[\mathfrak{D}u\left({}_{(1)}\mathbf{x}^N\right)\cdot\mathbf{e}_q^N\right]\psi_N^{1,q}(\mathbf{x}) \tag{6.125}$$

and

$$\mathfrak{D}^m U_e(\mathbf{x}) = \sum_{N=1}^{N_0} u\left({}_{(0)}\mathbf{x}^N\right) \mathfrak{D}^m \psi_N^0(\mathbf{x})$$

$$+ \sum_{N=1}^{N_1} \sum_{q=1}^{\gamma} \left[\mathfrak{D} u\left({}_{(1)}\mathbf{x}^N\right) \cdot \mathbf{e}_q^N \right] \mathfrak{D}^m \psi_N^{1,q}(\mathbf{x}) \tag{6.126}$$

Using (6.101) with $\mathbf{x}^N$ replaced by ${}_{(0)}\mathbf{x}^N$ and noting that

$$\mathfrak{D} u\left({}_{(1)}\mathbf{x}^N\right) \cdot \mathbf{e}_q^N = \mathfrak{D} u(\mathbf{x}) \cdot \mathbf{e}_q^N + \mathfrak{D}^2 u(\mathbf{x}) \cdot \left(\mathbf{e}_q^N, {}_{(1)}\mathbf{x}^N - \mathbf{x}\right)$$
$$+ \cdots + \frac{1}{(k-1)!} \mathfrak{D}^k u(\mathbf{x}) \cdot \left(\mathbf{e}_q^N, \left({}_{(1)}\mathbf{x}^N - \mathbf{x}\right)^{k-1}\right)$$
$$+ \frac{1}{k!} \mathfrak{D}^{k+1} u\left(\eta_{Nq}^1(\mathbf{x})\right) \cdot \left(\mathbf{e}_q^N, \left({}_{(1)}\mathbf{x}^N - \mathbf{x}\right)^k\right)$$

we arrive at the lengthy expansion,

$$\mathfrak{D}^m U_e(\mathbf{x}) = \sum_{r=0}^{k} \frac{1}{r!} \sum_{N=1}^{N_0} \left[\mathfrak{D}^r u(\mathbf{x}) \cdot \left({}_{(0)}\mathbf{x}^N - \mathbf{x}\right)^r \right] \mathfrak{D}^m \psi_N^0(\mathbf{x})$$
$$+ \frac{1}{(k+1)!} \sum_{N=1}^{N_0} \left[\mathfrak{D}^{k+1} u\left(\eta_N^0(\mathbf{x})\right) \cdot \left({}_{(0)}\mathbf{x}^N - \mathbf{x}\right)^{k+1} \right] \mathfrak{D}^m \psi_N^0(\mathbf{x})$$
$$+ \sum_{r=0}^{k} \frac{1}{(r-1)!} \sum_{N=1}^{N_1} \sum_{q=1}^{\gamma} \left[\mathfrak{D}^r u(\mathbf{x}) \cdot \left(\mathbf{e}_q^N, \left({}_{(1)}\mathbf{x}^N - \mathbf{x}\right)^{r-1}\right) \right] \mathfrak{D}^m \psi_N^{1,q}(\mathbf{x})$$
$$+ \frac{1}{k!} \sum_{N=1}^{N_1} \sum_{q=1}^{\gamma} \left[\mathfrak{D}^{k+1} u\left(\eta_{Nq}^1(\mathbf{x})\right) \cdot \left(\mathbf{e}_q^N, \left({}_{(1)}\mathbf{x}^N - \mathbf{x}\right)^k\right) \right] \mathfrak{D}^m \psi_N^{1,q}(\mathbf{x})$$

We next follow precisely the strategy used in the proof of Theorem 6.4. Using the fact that $U_e(\mathbf{x}) = u(\mathbf{x})$ whenever $u(\mathbf{x}) \in \mathcal{P}_k(\Omega)$, we deduce that

$$\sum_{r=0}^{k} \left\{ \frac{1}{r!} \sum_{N=1}^{N_0} \left[\mathfrak{D}^r u(\mathbf{x}) \cdot \left({}_{(0)}\mathbf{x}^N - \mathbf{x}\right)^r \right] \mathfrak{D}^m \psi_N^0(\mathbf{x}) \right.$$
$$\left. + \frac{1}{(r-1)!} \sum_{N=1}^{N_1} \sum_{q=1}^{\gamma} \left[\mathfrak{D}^r u(\mathbf{x}) \cdot \left(\mathbf{e}_q^N, \left({}_{(1)}\mathbf{x}^N - \mathbf{x}\right)^{r-1}\right) \mathfrak{D}^m \psi_N^{1,q}(\mathbf{x}) \right\}$$
$$= \mathfrak{D}^m u(\mathbf{x}) \tag{6.127}$$

This completes the proof of the theorem. ■

Hermite Simplexes

We cite a few examples of Hermite elements defined on simplexes, particularly triangles and tetrahedra.

Among the most popular triangular element families of higher order are the Hermite triangles investigated by Bramble and Zlamal [6.14], who make the following observations.

A polynomial $p(x,y)$ in two variables of degree $4m+1$ has exactly $(2m+1)(4m+3)$ coefficients.

If $\Omega \subset \mathbf{R}^2$ is a triangle with vertices P_1, P_2, P_3, centroid P_0, and sides l_1, l_2, and l_3, and if $u(x,y)$ is any sufficiently smooth function, specification of the following values of $u(x,y)$ and its derivatives involves exactly $(2m+1)$ $(4m+3)$ independent conditions:

$$D^{(\alpha_1,\alpha_2)}u(P_0) \qquad \alpha_1+\alpha_2 \leqslant m-2 \quad (m>1)$$

$$D^{(\alpha_1,\alpha_2)}u(P_N) \qquad N=1,2,3, \quad \alpha_1+\alpha_2 \leqslant 2m$$

$$\frac{\partial^r}{\partial n_j^r}u\left(Q_j^{(p,r)}\right) \qquad \begin{aligned} &j=1,2,3\\ &p=1,2,\ldots,r\\ &r=1,2,\ldots,m \end{aligned}$$

Here $\partial^r/\partial n_j^r$ denotes the normal derivative of order r on side l_j, and $Q_j^{(p,r)}$ are points separating $r+1$ equal segments along side l_j. Consequently, there is a unique polynomial $p(x,y)$ of degree $4m+1$ which assumes the above values. The most popular element of this class is the quintic C^1-conforming element proposed by Bosshard [6.16], Bell [6.15], and others. In this case $m=1$, and the 21 terms in a quintic are determined by prescribing all derivatives of u of order $\leqslant 2$ at each node (i.e., u, u_x, u_y, u_{xx}, u_{xy}, and u_{yy}) and the normal derivative u_n at the midpoint of each side. This element is illustrated in Fig. 6.16*a*.

We also mention that some of the TUBA elements of Argyris et al. [6.17] fall into the category of Hermite triangles, and their TET elements provide examples of members of tetrahedral Hermite families. Examples of such elements are given in Fig. 6.16*b–d*, with the TUBA 13 element in Fig. 6.16*b*, TUBA 15 in Fig. 6.16*c*, and TEA 8 in Fig. 6.16*d*; the element in Fig. 6.16*e* is described in [6.8, p.201].

Hermite Hypercubes

The most popular Hermite families of the hypercube type pertain to bivariate interpolation. We cite a few examples.

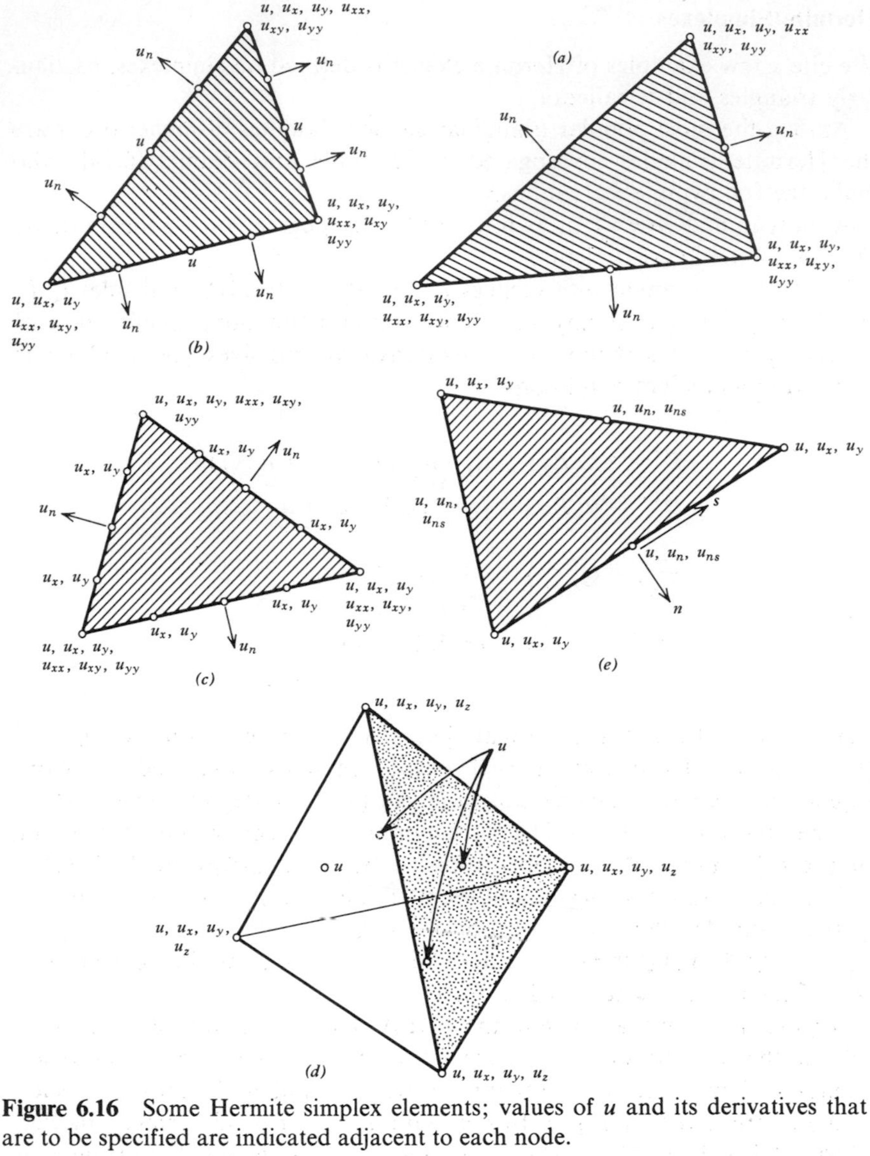

Figure 6.16 Some Hermite simplex elements; values of u and its derivatives that are to be specified are indicated adjacent to each node.

Consider once again the extended Pascal triangle (square) shown in Fig. 6.17. We generate an important family of two-dimensional Hermite rectangles in the following way: (i) Identify those square blocks of monomials whose sides are formed by terms in complete polynomials in x and y of odd degree. In Fig. 6.17, these correspond to blocks $1,3,5,\ldots,m,\ldots$ and contain $4,16,36,\ldots,\ (m+1)^2,\ldots$ independent monomials. (ii) Let the degrees of each monomial term in a block of order m correspond to the order of the derivatives to be prescribed at the four corners of an element whose local interpolants contain all the terms in the block of order $2m+1$. For each element in this family, this process leads to a C^m-conforming element.

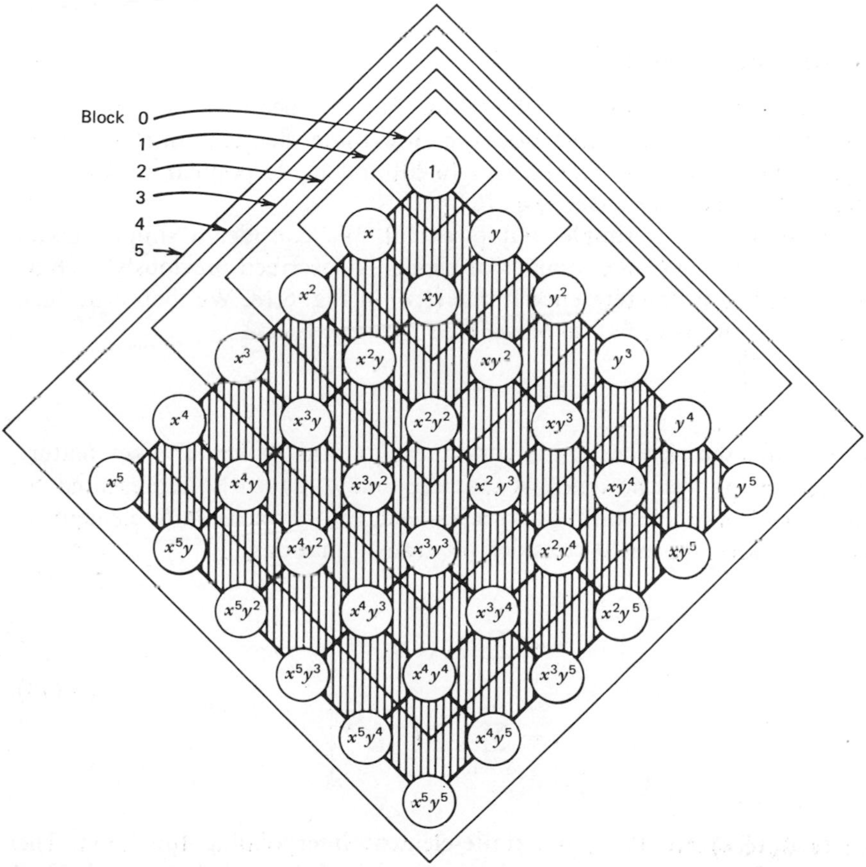

Figure 6.17 Scheme for generating a Hermite family of rectangular finite elements.

For example, when $m=0$, we prescribe only the value of the function $u(x,y)$ at each node. Since $2m+1=1$, all terms in block 1, i.e., 1, x, y, and xy, are used in the local interpolation. This gives the familiar bilinear element and is a degenerate Hermite element; i.e., it is precisely the Lagrange element described earlier. It is a C^0-conforming element. For $m=1$, we have 16 terms in block $2m+1=3$, and we prescribe derivatives at each node corresponding to the terms (1, x, y, xy) in block m; i.e., we prescribe (u, u_x, u_y, and u_{xy}) at each of the four nodes. This is a well-known C^1-finite element, introduced by Bogner, Fox, and Schmidt [6.18] and Birkhoff and de Boor [6.19]. The next element in this family, $m=2$, involves the prescription of block-2 derivatives $(u, u_x, u_y, u_{xx}, u_{xy}, u_{yy}, u_{xxy}, u_{xyy}, u_{xxyy})$ at each of the four nodes and is an example of a C^2-finite element.

Isoparametric Elements

All the finite elements described thus far can be used as a basis for constructing the curvilinear isoparametric finite elements introduced by Ergatoudis, Irons, and Zienkiewicz [6.20] and made popular by Zienkiewicz and his associates (see, e.g., [6.8]).

The basic idea is simple, but powerful. We consider a simple model element from one of the families of elements described previously such as that described in the reference frame (ξ,η) in Fig. 6.18. We then introduce a coordinate transformation of the type

$$x=x(\xi,\eta) \qquad y=y(\xi,\eta)$$

which maps the element into a curvilinear form as shown. The key feature of these mappings, however, is that they may in turn be represented by polynomial interpolation functions. Indeed, for a typical finite element Ω_e with N_e nodes, we might consider the transformations

$$\left.\begin{aligned} x &= \sum_{N=1}^{N_i} \psi_N(\xi,\eta)\, x^N \\ y &= \sum_{N=1}^{N_j} \psi_N(\xi,\eta)\, y^N \end{aligned}\right\} \tag{6.128}$$

where $\psi_N(\xi,\eta)$ are the usual finite-element interpolating functions. The number of parameters used to define the transformation law may be less than, the same as, or greater than that used to define the local interpolants

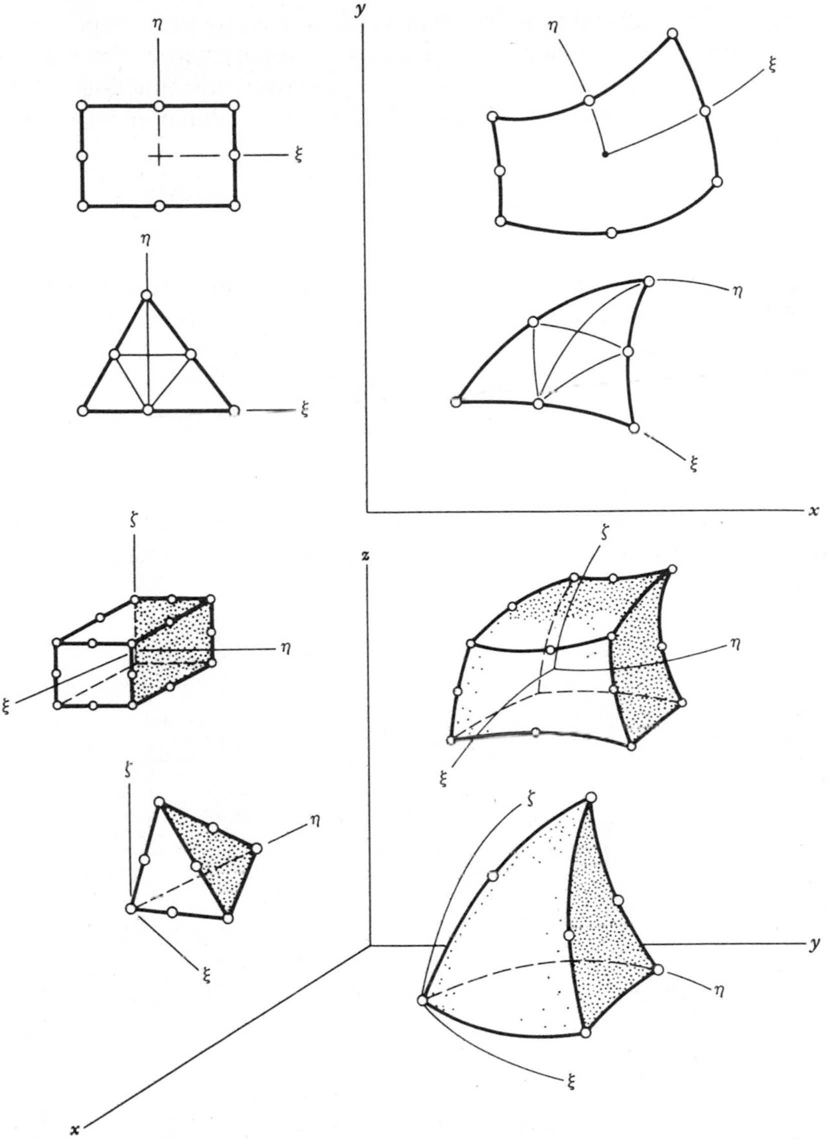

Figure 6.18 Examples of isoparametric finite elements.

over Ω_e. Accordingly, the curved elements produced by such mappings are termed subparametric, isoparametric, and superparametric elements, respectively. Thus, if $\{\psi_N^{(e)}\}_{N=1}^{N_e}$ are the local interpolation functions for Ω_e, an isoparametric element is obtained through transformations of the type

$$x = \sum_{N=1}^{N_e} \psi_N(\xi,\eta)x^N \qquad y = \sum_{N=1}^{N_e} \psi_N(\xi,\eta)y^N \tag{6.129}$$

Some examples of commonly used isoparametric elements are shown in Fig. 6.18.

We remark that, when

$$\sum_{N=1}^{N_e} \psi_N^{(e)}(x,y) = 1$$

which is true for Lagrange families by virtue of (iii) of (6.60), then (6.129) guarantees that completeness of the polynomial interpolant is preserved under the transformation. By this we mean that, if for example U_e is linear in x and y,

$$U_e(\xi,\eta) = \sum_{N=1}^{N_e} u(x^N,y^N)\psi_N^{(e)}(\xi,\eta) = a + bx + cy$$

then $u(x^N,y^N) = a + bx^N + cy^N, \quad N = 1,2,\dots,N_e$, so that

$$a + bx + cy = a\sum_{N=1}^{N_e} \psi_N^{(e)} + b\sum_{N=1}^{N_e} x^N\psi_N^{(e)} + c\sum_{N=1}^{N_e} y^N\psi_N^{(e)}$$

from which we recover

$$\sum_{N=1}^{N_e} \psi_N^{(e)}(\xi,\eta) = 1 \qquad \sum_{N=1}^{N_e} x^N\psi_N^{(e)}(\xi,\eta) = x \qquad \sum_{N=1}^{N_e} y^N\psi_N^{(e)}(\xi,\eta) = y \tag{6.130}$$

Rational Finite-Element Interpolation

The use of rational functions for finite-element interpolation was pioneered by Wachspress [6.21,6.22, 6.23] and represents an elegant extension of the concept of finite elements to quite general shapes.

For example, it is possible to construct two-dimensional Lagrange elements in the form of a polygon with n straight sides such as that shown in

Fig. 6.19*a*. The functions

$$f_i(x,y)=a_i x+b_i y+c_i$$

are constructed so that the equation of the line forming side i of the polygon is

$$f_i(x,y)=0 \qquad 1\leqslant i\leqslant n$$

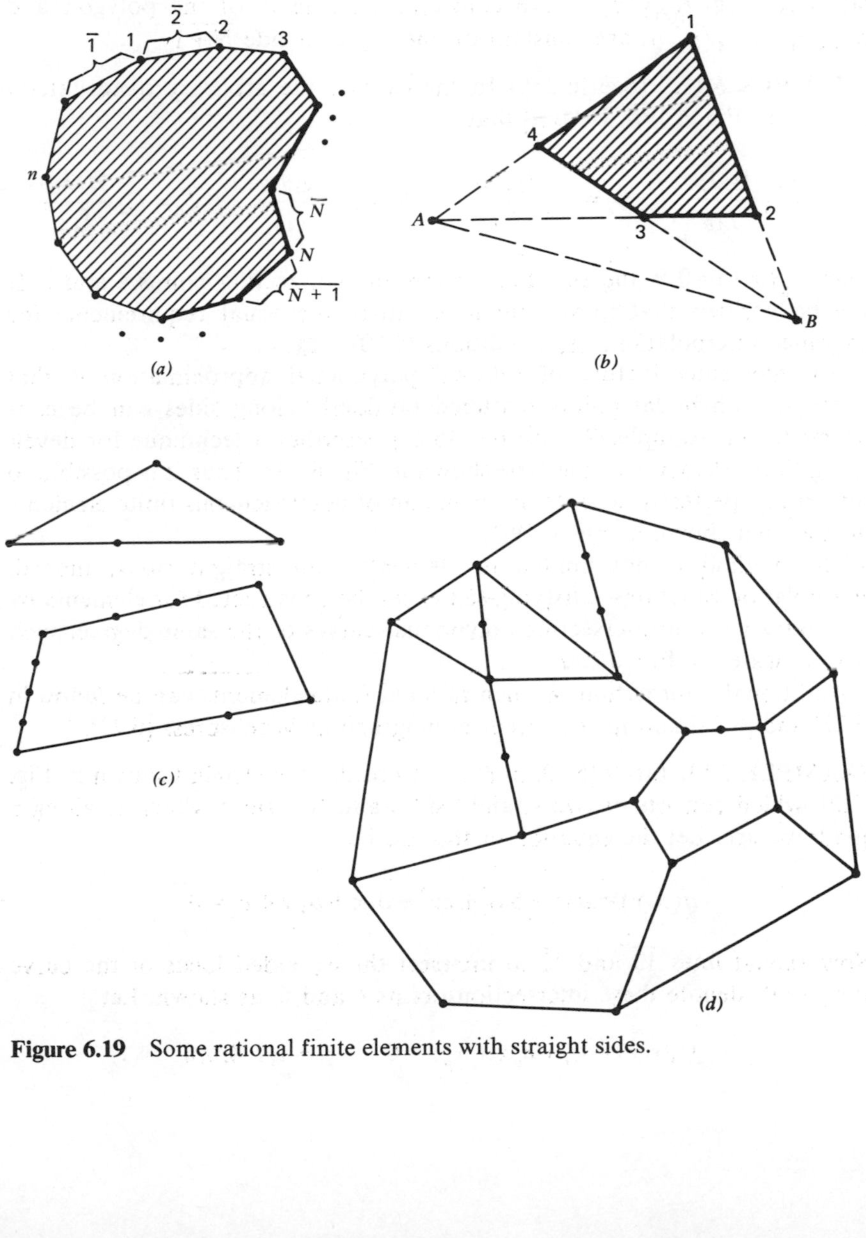

Figure 6.19 Some rational finite elements with straight sides.

Wachspress [6.21] shows that the general form of the Lagrange interpolation function corresponding to node N is

$$\psi_N(x,y)=C_N\frac{f_{N+2}(x,y)f_{N+3}(x,y)\cdots f_{N-1}(x,y)}{g_1(x,y)g_2(x,y)\cdots g_{n-3}(x,y)} \tag{6.131}$$

where the C_N are constants selected so that $\psi_N(x^M,y^M)=\delta_N^M$, $1\leqslant M,N\leqslant n+1$, and the functions $g_i(x,y)$ are linear functions selected so that the ratios $f_{N+2}/g_1, f_{N+3}/g_2,\ldots$ are constants on side $\overline{N}$ of the polygon and $f_{N+3}/g_1, f_{N+4}/g_2,\ldots$ are constant on the adjacent side $\overline{N+1}$

EXAMPLE 6.11. (See [6.21].) In the case of the arbitrary quadrilateral element in Fig. 6.19*b*, we find that

$$\psi_1=C_1\frac{f_{23}f_{34}}{f_{AB}}\qquad \psi_2=C_2\frac{f_{34}f_{41}}{f_{AB}}\qquad \psi_3=C_3\frac{f_{41}f_{12}}{f_{AB}}\qquad \psi_4=C_4\frac{f_{12}f_{23}}{f_{AB}}$$

where $f_{ij}(x,y)=0$ is the equation of the line connecting points i and j. It can be verified that these functions satisfy the usual requirements for Lagrange interpolation, i.e., conditions (6.60). ■

An interesting feature of rational polynomial approximation is that elements with nodal points scattered randomly along sides can be constructed. For example, Wachspress [6.21] describes a technique for developing finite elements of the type shown in Fig. 6.19*c*. Thus it is possible to construct a perfectly acceptable collection of heterogeneous finite elements such as that shown in Fig. 6.19*d*.

The method is not limited to elements with straight sides; indeed, interpolation functions satisfying (6.61) can be constructed for elements by an arbitrary set of intersecting polynomial curves of the same degree, such as is indicated in Fig. 6.20*a*.

Additional information on such rational finite elements can be found in [6.21] and [6.22] and in the recent monograph of Wachspress [6.23].

EXAMPLE 6.12. (after [6.22, p. 88]). Consider the element shown in Fig. 6.20*b* which consists of two straight sides and two sides which lie along a quadratic arc. Let the equation of this arc be

$$q(x,y)=ax^2+bxy+cy^2+a_1x+b_1y+c_1=0$$

Now extend lines $\overline{13}$ and $\overline{12}$ to intersect the extended locus of the curve $q(x,y)=0$; denote these intersection points 6 and 5, as shown. Let

$$f_\mu(x,y)=a_\mu+b_\mu x+x_\mu y=0\qquad \mu=1,2,4,5,6$$

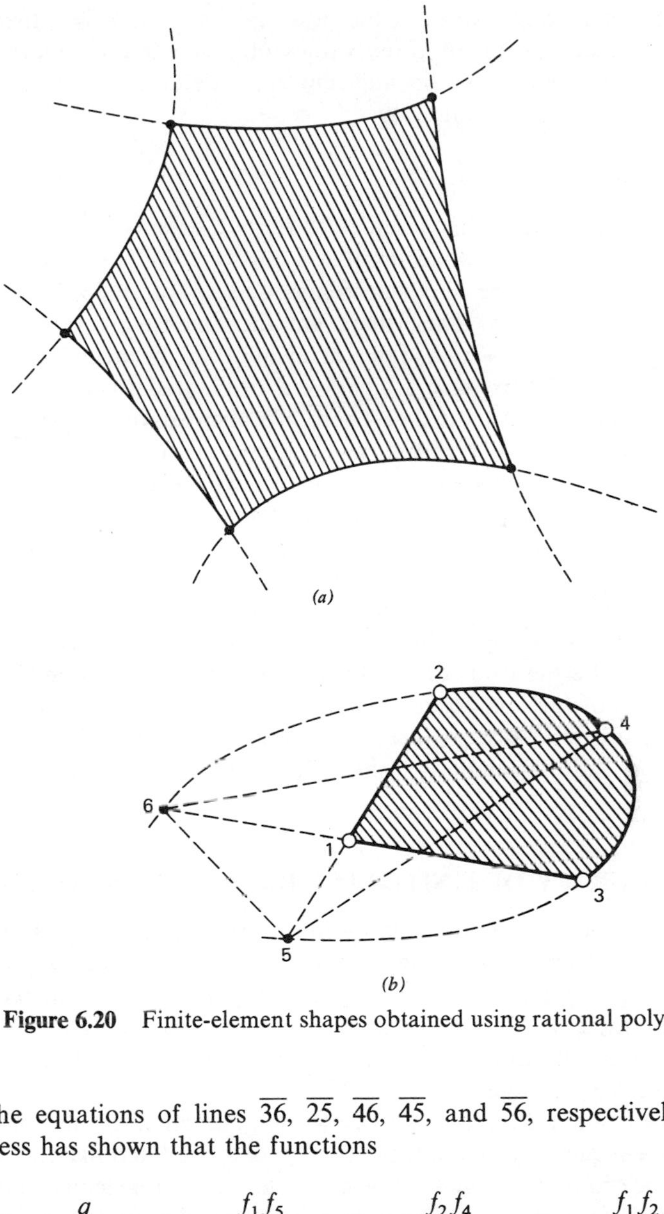

Figure 6.20 Finite-element shapes obtained using rational polynomials.

denote the equations of lines $\overline{36}$, $\overline{25}$, $\overline{46}$, $\overline{45}$, and $\overline{56}$, respectively. Then Wachspress has shown that the functions

$$\psi_1 = c_1 \frac{q}{f_6} \qquad \psi_2 = c_2 \frac{f_1 f_5}{f_6} \qquad \psi_3 = c_3 \frac{f_2 f_4}{f_6} \qquad \psi_4 = c_4 \frac{f_1 f_2}{f_6}$$

where the constants c_i are chosen so that $\psi_N(x^M, y^M) = \delta_N^M$, satisfy exactly the usual requirements of local interpolants; i.e., ψ_N assumes a unit value at node N and zero values at the other nodes, (6.130) is satisfied and, in this case, specification of three values of a linear combination of these functions at three nodes on adjacent sides determines uniquely a linear function over the element. ■

Other Finite Elements

There are numerous other families of finite elements that could be added to the relatively small list accumulated thus far. We have not, for example, mentioned any of the shell elements used to produce C^1 elements on noneuclidian surfaces. Then there are the extensive families of discrete Kirchhoff elements which fall somewhere in between two-and three-dimensional elements, and which were originated in an early paper on plate and shell analysis [6.24]. There are also families of nonconforming elements which have received wide attention, and there are numerous examples of so-called composite elements which are developed by fitting together in various ways certain subelements.

Perhaps the most widely used element of this latter type is the C^1-conforming Clough-Tocher triangle [6.25] which consists of three complete cubics defined over three interlocked triangles. Through a systematic elimination process, only nine parameters are left to be specified, and these are taken to be the value of the local interpolant and its first derivatives at the vertices of the composite triangle.

We again refer the reader to standard sources for more complete catalogs of finite elements. In particular, see [6.1], [6.7], [6.8], and [6.17].

6.7 ACCURACY OF FINITE-ELEMENT INTERPOLATIONS

We now come to a fundamentally important aspect of the finite-element method—determination of the accuracy of finite-element interpolations by establishing a priori estimates of the interpolation error in terms of the mesh parameter h. Our investigation is based on the analysis of Ciarlet and Raviart [6.3,6.4] and culminates in the basic interpolation theorems, Theorems 6.6, through 6.8.

The key to our approach is to view the generation of any finite-element mesh as a sequence of mappings $\{T_e\}_{e=1}^{E}$ of some fixed master element Ω_M into E locations in the mesh. For example, the finite-element mesh shown in Fig. 6.21 is obtained by repeatedly mapping the shaded master element shown into each separate location in the mesh. When the mesh is *refined* (i.e., when element dimensions are decreased and the number of elements is increased), the same master element is still used, but a new sequence of

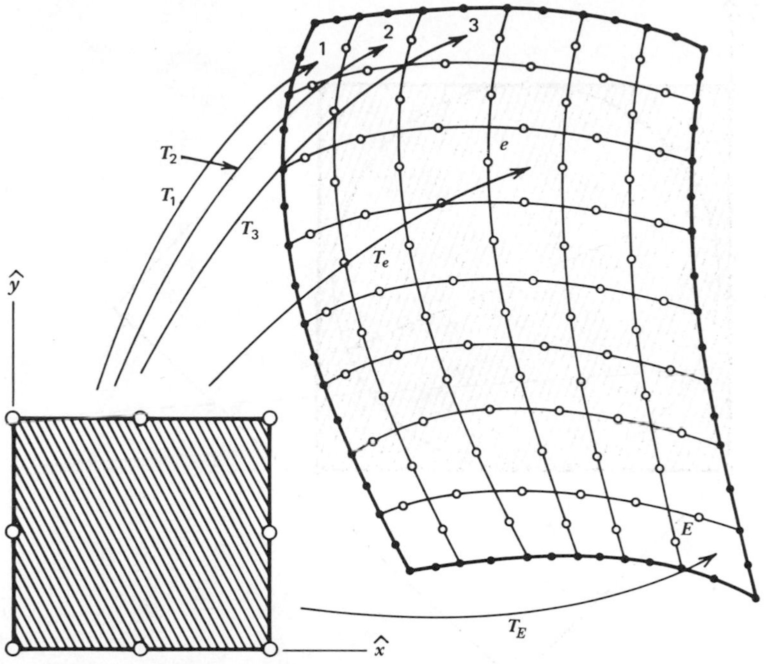

Figure 6.21 Generation of a finite element mesh by sequence $\{T_e\}_e^E=1$ of invertible mappings of a fixed master element.

mappings $\{T_e\}_{e=1}^{E'}$, $E'>E$, is introduced. All these ideas are reminiscent of the notion of isoparametric elements discussed previously.

For clarity in exposition, we begin by considering only linear transformations. Then a master element with straight sides is transformed into a skewed element with straight sides, as indicated in Fig. 6.22. The basic geometric features are determined by the parameters indicated in the figure. We denote by $\hat{\mathcal{L}}_M$ the nodal points in the master element, denoted Ω_M, and by $\mathcal{L}_e$ the local nodes associated with the transformed element Ω_e.

$$\hat{\mathcal{L}}_M=\{\hat{\mathbf{x}}^N\}_{N=1}^{N_e} \qquad \mathcal{L}_e=\{\mathbf{x}_e^N\}_{N=1}^{N_e} \tag{6.132}$$

the two sets are of course related by a linear transformation

$$\mathbf{x}_e^N=\mathbf{T}_e\hat{\mathbf{x}}^N+\mathbf{b}_e \tag{6.133}$$

where $\mathbf{T}_e$ is a square $N_e\times N_e$ matrix, and $\mathbf{b}_e$ is the translation vector giving

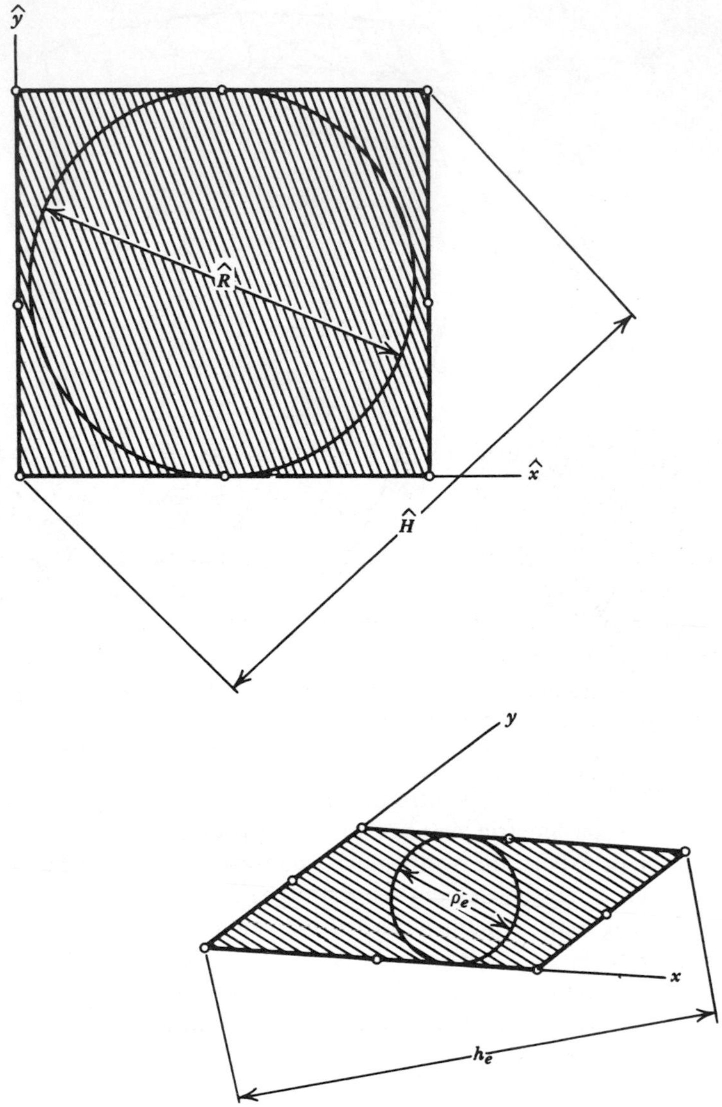

Figure 6.22 A linear transformation of a master element and associated geometric parameters.

the origin of the local element coordinate system relative to that of the fixed $\mathbf{x}$ frame. We assume throughout that the sets $\hat{\mathfrak{L}}_M$ and $\mathfrak{L}_e$ are *equivalent*, which simply means that each $\mathbf{T}_e$ in (6.133) is invertible. It is easily verified that, whenever $\hat{\mathfrak{L}}_M$ is k-unisolvent, then any equivalent set $\mathfrak{L}_e$ is also k-unisolvent.

We use the notation indicated in Fig. 6.22; i.e.,

$$\left.\begin{aligned}
&\hat{H} = \text{diameter of } \bar{\Omega}_M = \max_{\mathbf{x},\mathbf{y}\in\bar{\Omega}_M} |\mathbf{x}-\mathbf{y}| \equiv \bar{\hat{\Omega}} \\
&\hat{R} = \sup\{\text{diameters of all spheres contained in } \Omega_M\} \\
&h_e = \text{diameter of } \Omega_e \; [\text{see (6.39)}] \\
&\rho_e = \sup\{\text{diameter of all spheres contained in } \Omega_e\}
\end{aligned}\right\} \quad (6.134)$$

Also, we add to the notation of (6.40) the following:

$$h = \max_{1\leqslant e\leqslant E}\{h_e\} \quad \text{and} \quad \rho = \min_{1\leqslant e\leqslant E}\{\rho_e\} \tag{6.135}$$

We now establish several simple but useful identities connected with linear mappings (6.133).

LEMMA 6.1. Let $\mathbf{T}$ be any square invertible matrix of finite order n and let $\mathbf{b}$ be a translation vector of dimension $n\times 1$. Consider the coordinate transformation

$$\mathbf{x} = \mathbf{T}\hat{\mathbf{x}} + \mathbf{b} \tag{6.136}$$

where $\mathbf{x}$ and $\hat{\mathbf{x}}$ are n-vectors. Then

(i)
$$\|\mathbf{T}\| = \frac{1}{\hat{R}} \sup_{|\hat{\mathbf{x}}|=\hat{R}} \|\mathbf{T}\hat{\mathbf{x}}\| \tag{6.137}$$

where $\hat{R}$ is any positive constant, and $\|\cdot\|$ indicates the matrix norm induced by the usual euclidean norm on $\mathbf{R}^n$. Then

(ii)
$$dx = \det(\mathbf{T})\, d\hat{x} \tag{6.138}$$

(iii)
$$|D^\alpha u(\hat{\mathbf{x}})| \leqslant \|\mathbf{T}\|^s |D^\alpha u(\mathbf{x})| \qquad |\alpha| = s \tag{6.139}$$

PROOF. (i) Let $\mathbf{z}=\hat{R}\hat{\mathbf{x}}/|\hat{\mathbf{x}}|$. Then $|\mathbf{z}|=\hat{R}$. From the usual definition of norms of bounded linear operators

$$\|\mathbf{T}\| = \sup_{|\hat{\mathbf{x}}|\neq 0} \frac{|\mathbf{T}\hat{\mathbf{x}}|}{|\hat{\mathbf{x}}|}$$

$$= \sup \frac{|(|\hat{\mathbf{x}}|/\hat{R})\mathbf{T}\mathbf{z}|}{|\hat{\mathbf{x}}|}$$

$$= \frac{1}{\hat{R}} \sup_{|\mathbf{z}|=\hat{R}} |\mathbf{T}\mathbf{z}|$$

(ii) This is a well-known elementary identity which follows immediately from the fact that

$$dx = dx_1\, dx_2 \cdots dx_n = \sum_{i_1, i_2, \cdots, i_n = 1}^{n} \epsilon_{i_1 i_2 \cdots i_n} T_{i_1 1} T_{i_2 2} \cdots T_{i_n n}\, d\hat{x}_1\, d\hat{x}_2 \cdots d\hat{x}_n$$

where $\epsilon_{i_1 i_2 \cdots i_n}$ is the n-dimensional permutation symbol.

(iii) Observe that

$$D_{\hat{x}_i} u(\hat{\mathbf{x}}) = \sum_{j=1}^{n} T_{ji} D_{x_j} u(\mathbf{x})$$

$$D_{\hat{x}_i}^2 u(\hat{\mathbf{x}}) = \sum_{j,k=1}^{n} T_{ji} T_{jk} D_{x_k}^2 u(\mathbf{x})$$

$$\vdots$$

$$D_{\hat{x}_i}^{\alpha_i} u(\hat{\mathbf{x}}) = \sum_{j_1, j_2, \cdots, j_{\alpha_i} = 1}^{n} T_{j_1 i} T_{j_1 j_2} \cdots T_{j_{\alpha_i - 1} j_{\alpha_i}} D_{x_{j_{\alpha_i}}}^{\alpha_i} u(\mathbf{x})$$

Thus

$$|D_{\hat{x}_i}^{\alpha_i} u(\hat{\mathbf{x}})| \leqslant \underbrace{\|T\|\,\|T\| \cdots \|T\|}_{\alpha_i \text{ times}} |D_{x_{j_{\alpha_i}}}^{\alpha_i} u(\mathbf{x})|$$

Repeating this process for each component in $\boldsymbol{\alpha}=(\alpha_1, \alpha_2 \ldots, \alpha_n)$, we get (6.139). ■

We next add to Lemma 6.1 the following simple inequalities.

LEMMA 6.2. Let **T** be a linear transformation of the type described in Lemma 6.1. Then, in the notation of (6.134) and (6.135),

$$\|\mathbf{T}\| \leqslant \frac{h}{\hat{R}} \quad \text{and} \quad \|\mathbf{T}^{-1}\| \leqslant \frac{\hat{H}}{\rho} \tag{6.140}$$

PROOF. Pick two points $\hat{\mathbf{y}}$ and $\hat{\mathbf{z}}$ in Ω_M on the sphere of diameter $\hat{R}$. Then

$$|\hat{\mathbf{y}} - \hat{\mathbf{z}}| = \hat{R}$$

and, from (6.137),

$$\begin{aligned}
\|\mathbf{T}\| &= \frac{1}{\hat{R}} \sup|\mathbf{T}(\hat{\mathbf{y}} - \hat{\mathbf{z}})| \\
&= \frac{1}{\hat{R}} \sup|\mathbf{T}(\hat{\mathbf{y}}) + \mathbf{b} - \mathbf{T}(\hat{\mathbf{z}}) - \mathbf{b}| \\
&= \frac{1}{\hat{R}} \sup|\mathbf{y} - \mathbf{z}| \\
&\leqslant \frac{h}{\hat{R}}
\end{aligned}$$

An identical type of argument leads to the second inequality. ■

All the groundwork is now laid for the fundamental theorem:

THEOREM 6.6 (The Finite-Element Interpolation Theorem for C^k Functions). Let $u(\mathbf{x})$ denote any function with the properties

(i) $\quad u(\mathbf{x}) \subset C^k(M) \qquad M \subset \mathbf{R}^n$

(ii) $\quad \mathfrak{D}^{k+1}u(\mathbf{x})$ exists $\qquad \forall \mathbf{x} \in M$

where k is a fixed integer $\geqslant 0$, and M is an $\mathfrak{L}_e$-admissible set for some collection of nodal points $\mathfrak{L}_e$ in $\mathbf{R}^n$. Let

$$C_{k+1} = \sup_{\mathbf{x} \in \Omega_e} \|\mathfrak{D}^{k+1}u(\mathbf{x})\| < \infty \tag{6.141}$$

where $\|\mathfrak{D}^{k+1}u\|$ is the operator norm described in (6.99). In addition, let either of the following two sets of conditions hold:

(I) (Lagrange finite elements)

$$\mathfrak{L}_e = \left\{ \mathbf{x}_e^N \right\}_{N=1}^{N_e}$$

is a k-unisolvent set of nodal points of a finite element $\Omega_e \subset \mathsf{R}^n$, and $U_e(\mathbf{x})$ is the unique interpolating polynomial of degree $\leqslant k$ of $u(\mathbf{x})$.

(II) (Hermite finite elements)

$$\mathfrak{L}_e = \bigcup_{\mu=0}^{\nu} \Sigma^{(\mu)} \qquad \Sigma^{(\mu)} = \left\{ {}_{(\mu)}\mathbf{x}_e^N \right\}_{N=1}^{N_\mu}$$

is a set of k-unisolvent nodes of an $\mathfrak{L}_e$-admissible finite element Ω_e such that each ${}_{(\mu)}\mathbf{x}_e^N$ is associated with a subset χ_μ^N of $\Pi^\mu \mathsf{R}^n \equiv (\mathsf{R}^n)^\mu$, $1 \leqslant N \leqslant N_\mu$, $0 \leqslant \mu \leqslant \nu$, and $U_e(\mathbf{x})$ is the unique interpolating polynomial in $\mathcal{P}_k(M)$ such that

$$\left.\begin{aligned} &U_e\left({}_{(0)}\mathbf{x}^N\right) = u\left({}_{(0)}\mathbf{x}^N\right) \qquad 1 \leqslant N \leqslant N_0 \\ &\mathfrak{D}^\mu U_e\left({}_{(\mu)}\mathbf{x}^N\right) \cdot (\xi_1, \xi_2, \ldots, \xi_\mu) = \mathfrak{D}^\mu u\left({}_{(\mu)}\mathbf{x}^N\right) \cdot (\xi_1, \xi_2, \ldots, \xi_\mu) \end{aligned}\right\} \tag{6.142}$$

$$\forall (\xi_1, \xi_2, \ldots, \xi_\mu) \in \chi_\mu^N, \quad 1 \leqslant N \leqslant N_\mu, 0 \leqslant \mu \leqslant \nu.$$

Then there exist positive constants $C = C(n, k, m, \hat{\mathfrak{L}}_M)$ depending on n, k, m, and $\hat{\mathfrak{L}}_M$ but not on u, h, or ρ such that, for any integer m, $0 \leqslant m \leqslant k$, we have

$$\sup_{\mathbf{x} \in \Omega_e} \| \mathfrak{D}^m u(\mathbf{x}) - \mathfrak{D}^m U_e(\mathbf{x}) \| \leqslant C C_{k+1} \frac{h^{k+1}}{\rho^m} \tag{6.143}$$

PROOF. We begin by assuming that conditions (I) hold. Then the conclusions of Theorem 6.4 are valid, and we use (6.102) to write down immediately

$$\| \mathfrak{D}^m (u - U_e) \| \leqslant \frac{1}{(k+1)!} \sum_{N=1}^{N_e} \left| \mathfrak{D}^{k+1} u(\eta_N(\mathbf{x})) \cdot (\mathbf{x}^N - \mathbf{x})^{k+1} \right|$$

$$\cdot \| \mathfrak{D}^m \psi_N^{(e)}(\mathbf{x}) \| \tag{6.144}$$

Clearly,

$$\left| \mathfrak{D}^{k+1} u(\eta_N(\mathbf{x})) \cdot (\mathbf{x}^N - \mathbf{x})^{k+1} \right| \leqslant C_{k+1} h^{k+1}$$

wherein we also use (6.141).

We next transform the element from the set $\mathcal{L}_e$ to an equivalent set $\hat{\mathcal{L}}_M$. We have

$$\psi_N^{(e)}(\mathbf{x})=\hat{\psi}_N^{(e)}(\mathbf{T}_e^{-1}(\mathbf{x}-\mathbf{b}_e))$$

and

$$\mathfrak{D}^m\psi_N^{(e)}(\mathbf{x})\cdot(\xi_1,\xi_2,\dots,\xi_m)=\mathfrak{D}^m\hat{\psi}_N^{(e)}(\mathbf{T}_e^{-1}(\mathbf{x}-\mathbf{b}_e))$$
$$\cdot(\mathbf{T}_e^{-1}\xi_1,\mathbf{T}_e^{-1}\xi_2,\dots,\mathbf{T}_e^{-1}\xi_m)$$

Therefore, with $\Omega_M\equiv\hat{\Omega}$,

$$\sup_{\mathbf{x}\in\Omega_e}\|\mathfrak{D}^m\psi_N^{(e)}(\mathbf{x})\|\leqslant\|T_e^{-1}\|^m\sup_{\hat{\mathbf{x}}\in\hat{\Omega}}\|\mathfrak{D}^m\hat{\psi}_N^{(e)}(\hat{\mathbf{x}})\|$$
$$\leqslant\frac{\hat{H}^m}{\rho^m}\sup_{\hat{\mathbf{x}}\in\hat{\Omega}}\|\mathfrak{D}^m\hat{\psi}_N^{(e)}(\hat{\mathbf{x}})\| \tag{6.145}$$

Introducing this result, together with (6.145), into (6.144) and denoting

$$C=\frac{\hat{H}^m}{(k+1)!}\sum_{N=1}^{N_e}\sup_{\hat{\mathbf{x}}\in\hat{\Omega}}\|\mathfrak{D}^m\hat{\psi}_N^{(e)}(\hat{\mathbf{x}})\| \tag{6.146}$$

gives (6.143)

The proof that (II) implies (6.143) follows exactly the same lines, hence we sketch only the essential features. We consider, for simplicity, the case in which $\nu=1$. This time we use Theorem 6.5, particularly (6.122), to establish that

$$\|\mathfrak{D}^m(u-U_e)\|\leqslant\frac{1}{(k+1)!}\sum_{N=1}^{N_0}\underline{\left|\mathfrak{D}^{k+1}u\left(\eta_N^0(\mathbf{x})\right)\cdot\left({}_{(0)}\mathbf{x}^N-\mathbf{x}\right)^{k+1}\right|}$$
$$\cdot\left\|\mathfrak{D}^m\psi_N^{0,(e)}(\mathbf{x})\right\|+\frac{1}{k!}\sum_{N=1}^{N_1}\sum_{q=1}^{\gamma}\underline{\left|\mathfrak{D}^{k+1}u\left(\eta_{Nq}^1(\mathbf{x})\right)\right.}$$
$$\underline{\left.\cdot\left(\mathbf{e}_q^N,\left({}_{(1)}\mathbf{x}^N-\mathbf{x}\right)^k\right)\right|}\left\|\mathfrak{D}^m\psi_N^{1,q,(e)}(\mathbf{x})\right\| \tag{6.147}$$

The underlined terms in (6.147) are seen to be bounded by $C_{k+1}h^{k+1}$ and $C_{k+1}\|e_q^N\|h^k$, respectively, and we easily argue that $\|e_q^N\|\leqslant h\|\hat{e}_q^N\|/\hat{R}$.

Likewise, we use the transformation properties of Lemma 6.2 to show that

$$\sup_{\mathbf{x}\in\Omega_e} \|\mathfrak{D}^m\psi_N^{0(e)}(\mathbf{x})\| \leqslant \frac{\hat{H}^m}{\rho^m} \sup_{\hat{\mathbf{x}}\in\hat{\Omega}} \|\mathfrak{D}^m\hat{\psi}_N^{0(e)}(\hat{\mathbf{x}})\|$$

and

$$\sup_{\mathbf{x}\in\Omega_e} \|\mathfrak{D}^m\psi_N^{1,q(e)}(\mathbf{x})\| \leqslant \frac{\hat{H}^m}{\rho^m} \sup_{\hat{\mathbf{x}}\in\hat{\Omega}} \|\mathfrak{D}^m\hat{\psi}_N^{1,q,(e)}(\hat{\mathbf{x}})\|$$

Finally, substituting these results into (6.147) and denoting

$$C = \frac{\hat{H}^m}{(k+1)!} \sum_{N=1}^{N_0} \sup_{\hat{\mathbf{x}}\in\hat{\Omega}} \|\mathfrak{D}^m\hat{\psi}_N^{0(e)}(\hat{\mathbf{x}})\|$$

$$+ \frac{1}{k!}\frac{\hat{H}^m}{\hat{R}} \sum_{N=1}^{N_1} \sum_{q=1}^{\gamma} \|\hat{e}_q^N\| \sup_{\hat{\mathbf{x}}\in\hat{\Omega}} \|\mathfrak{D}^m\psi_N^{1,q,(e)}(\hat{\mathbf{x}})\| \tag{6.148}$$

We obtain (6.143). ■

REMARKS. 1. The appearence of two mesh parameters, h and ρ, in these estimates provides a great deal of flexibility in interpreting the effects of a refinement of a mesh. For example, if we first interpolate a given $u(x,y)$ using the finite-element mesh in Fig. 6.23a, and then, to improve accuracy, choose the mesh in Fig. 6.23b, we must take into account that both mesh parameters h and ρ have changed in this refinement. Such a refinement is called *irregular* [6.1, 6.3].

2. It is customary to introduce the real numbers

$$\left.\begin{aligned} \sigma &= \frac{h}{\rho} = \frac{\max_{1\leqslant e\leqslant E}\{h_e\}}{\min_{1\leqslant e\leqslant E}\{\rho_e\}} \\ &\text{or} \\ \gamma &= \frac{h}{\hat{h}} = \frac{h}{\min_{1\leqslant e\leqslant E}\{h_e\}} \end{aligned}\right\} \tag{6.149}$$

When $\gamma = 1$, the mesh is *uniform* and, whenever there exists a fixed number $\sigma_0 > 0$ such that, for a sequence of refinements $\sigma_0 \geqslant \sigma$, the meshes are *quasi-uniform*.

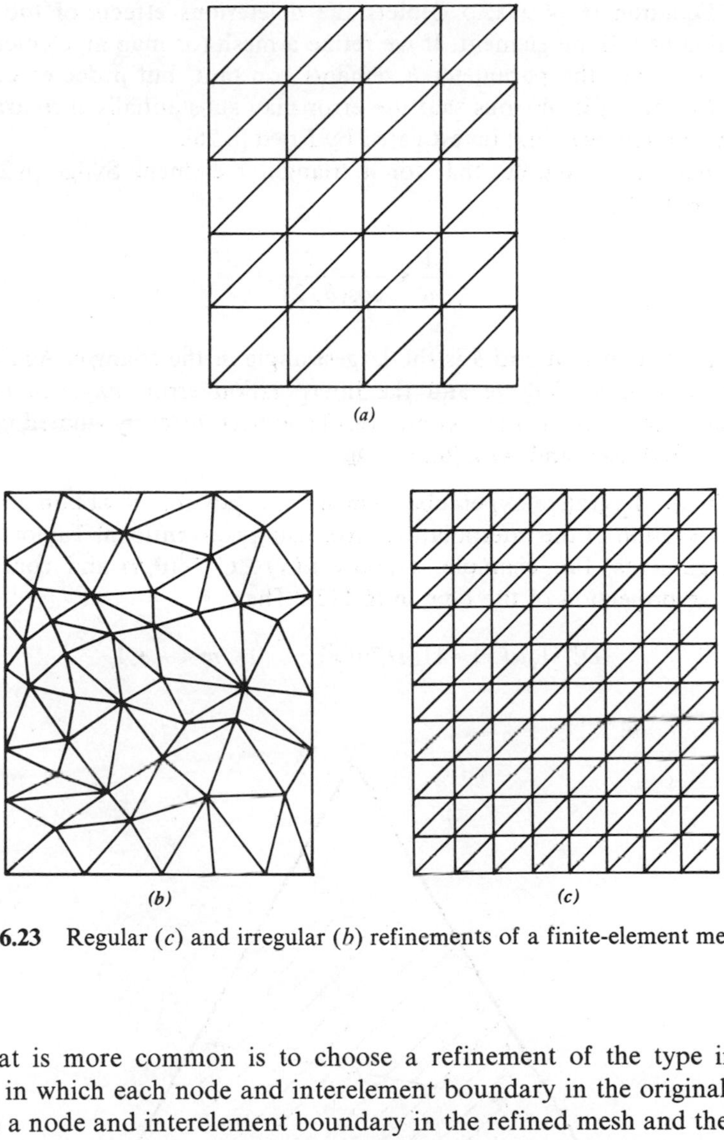

Figure 6.23 Regular (*c*) and irregular (*b*) refinements of a finite-element mesh (*a*).

What is more common is to choose a refinement of the type in Fig. 6.23*c*, in which each node and interelement boundary in the original mesh is also a node and interelement boundary in the refined mesh and the basic geometric pattern is preserved. In such cases, $h_e \leqslant \mu\rho_e$, $1 \leqslant e \leqslant E$, where μ is a constant. We call such refinements *regular*. Clearly, when the conditions of Theorem 6.6 hold for regular or quasi-uniform refinements,

$$\sup_{\mathbf{x}\in\Omega_e} \|\mathfrak{D}^m[u(\mathbf{x}) - U_e(\mathbf{x})]\| \leqslant CC_{k+1}h^{k+1-m} \tag{6.150}$$

3. Equation (6.142) also depicts the deleterious effects of too much distortion of a finite element. If we refine a mesh (or map an element into a mesh) so that the parameter h remains constant, but ρ decreases as in Fig. 6.24, then it is obvious that the error may substantially increase. This distortion effect was first investigated by Fried [6.26].

We note, for example, that for a triangular element Synge [6.27] has shown that

$$\frac{1}{\rho} \leqslant \frac{c}{\cos(\theta/2)} \tag{6.151}$$

where c is a constant and θ is the largest angle in the triangle. Again, as θ increases, so also does ρ, and the interpolation error may become unacceptably large. Such angle conditions have been recently studied in some detail by Babuška and Aziz [6.28]. ■

For many Lagrange hypercube elements it is possible to obtain a slightly finer resolution of the interpolation error using conventional Taylor expansion arguments. For example, suppose $u(x) \in C^{k+1}(0,h)$ and consider a Lagrange projection of the type in (6.112). Then

$$D_x^m \Pi_x u(x) = \Pi_x D_x^m u(x) \qquad 0 \leqslant m \leqslant k+1$$

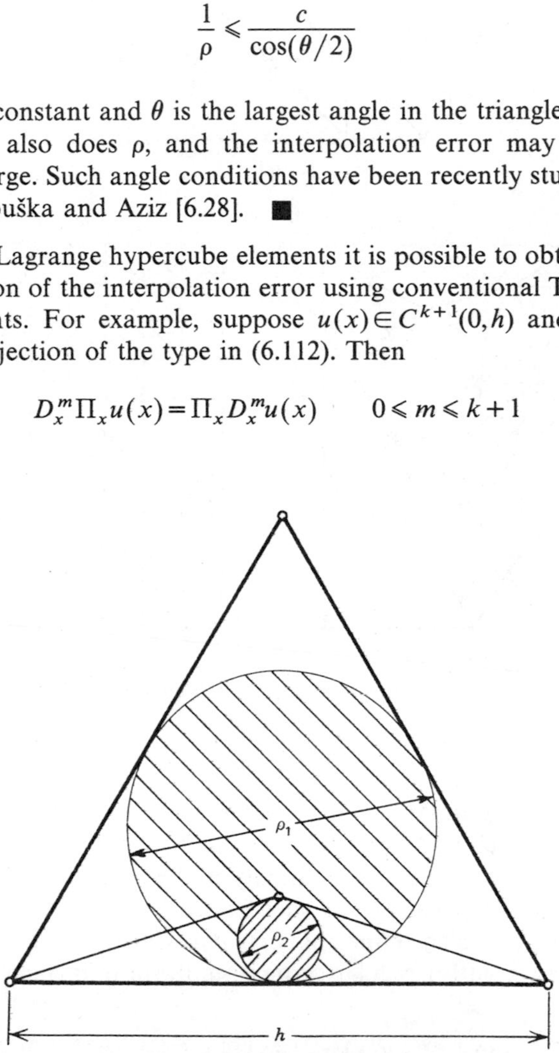

Figure 6.24 Distortion of a triangular element in which h is held constant.

and

$$D_x^m R_x u(x) \equiv D_x^m (u(x) - \Pi_x u(x)) = R_x D_x^m u(x)$$

Denoting $\Pi_x u(x) = U_e(x)$, it follows that, for regular refinements,

$$\|R_x D_x^m u\|_\infty \equiv \sup_{x \in [0,h]} |D_x^m (D_x^m (u(x) - U(x)))| \leqslant c(m,k) \|D_x^{k+1} u\|_\infty h^{k+1-m} \tag{6.152}$$

where $c(m,k)$ is a positive number.

Likewise, for functions $u(x,y) \in C^{(k+1,p+1)}(\Omega)$ defined on two-dimensional domains,

$$\begin{aligned} \|D^{(m,l)} R_x u(x,y)\|_\infty &\leqslant c_1(m,k) \|D^{(k+1,l)} u\|_\infty h^{k+1-m} \\ \|D^{(m,l)} R_y u(x,y)\|_\infty &\leqslant c_2(l,p) \|D^{(m,p+1)} u\|_\infty h^{p+1-l} \end{aligned} \tag{6.153}$$

and, consequently,

$$\begin{aligned} \|D^{(m,l)} R_x R_y u\|_\infty &\leqslant c_1(m,k) \|D^{(k+1,l)} R_y u\|_\infty h^{k+1-m} \\ &\leqslant c_1(m,k) c_2(l,p) \|D^{(k+1,p+1)} u\|_\infty \\ &\quad \cdot h^{k+1-m} \cdot h^{p+1-l} \end{aligned} \tag{6.154}$$

Summarizing, we have the following.

THEOREM 6.7. Let $u(x,y) \in C^{(k+1,p+1)}(\Omega)$, $\Omega \subset \mathsf{R}^2$. Then, for $0 \leqslant m \leqslant k$, $0 \leqslant l \leqslant p$,

$$\begin{aligned} \|D^{(m,l)}(u - U_e)\|_\infty &\leqslant c_1(m,k) \|D^{(k+1,l)} u\|_\infty h^{k+1-m} \\ &\quad + c_2(l,p) \|D^{(m,p+1)} u\|_\infty h^{p+1-l} \\ &\quad + c_1(m,k) c_2(l,p) \|D^{(k+1,p+1)} u\|_\infty h^{k+p+2-m-l} \end{aligned} \tag{6.155}$$

where $U_e = \Pi_x \cdot \Pi_y u(x,y)$, and Π_x, Π_y are Lagrange projectors defined in (6.112) such that $\Pi_x u$ and $\Pi_y u$ are polynomials of degree k and p respectively. ■

What is the actual *asymptotic* rate of convergence of the interpolation (6.154); i.e., as $h \to 0$, how fast does $D^{(m,l)} U_e$ converge to $D^{(m,l)} u$ in the sup norm? Obviously, for small h, $h^{\alpha} > h^{\beta}$ if $\alpha < \beta$; $\alpha, \beta > 0$. Thus the last term on the right side of (6.154) is of higher order than the first two, and we write

$$\left\|D^{(m,l)}(u - U_e)\right\|_{\infty} = O(h^{\mu}) \qquad \mu = \min\{k+1-m, p+1-l\} \tag{6.156}$$

The symbolism $O(h^{\mu})$ of course represents the fact that $\|D^{(m,l)}(u - U_e)\|_{\infty}$ is of *order* h^{μ}; i.e., for $h < 1$, we can find a $K > 0$ such that $[\|D^{(m,l)}(u - U_e)\|_{\infty}/h^{\mu}] \leqslant K$.

Gordon and Hall [6.13] point out that, if the Boolean sum (6.115) is used to define the finite-element interpolant, then a much more accurate scheme can be obtained. Indeed, in view of (6.118) and (6.155) we then have

$$\left\|D^{(m,l)}(u - \Pi_x \oplus \Pi_y u)\right\|_{\infty} \leqslant c\|D^{(k+1,p+1)}u\|_{\infty} h^{k+p+2-m-l} \tag{6.157}$$

Extensions of these results to n dimensions are straightforward.

Interpolation in Sobolev Spaces

We now come to one of the most important results of this chapter, the accuracy of finite-element interpolations of functions u in Sobolev spaces $H^{k+1}(\Omega)$. Recall that by an interpolant we now mean as an appropriately defined projection $\Pi_h u$ of u into a finite-dimensional subspace $S_h(\Omega) \subset H^{k+1}(\Omega)$ spanned by global interpolation functions. Our analysis once again follows the plan of Ciarlet and Raviart [6.3]. To make the arguments simple, we reach the basic theorem through several relatively simple lemmas.

We begin by establishing a version of a fundamental lemma due to Bramble and Hilbert [6.30]; see also [6.29].

LEMMA 6.3. (The Bramble-Hilbert Lemma). Let Ω be an open bounded domain in $\mathbf{R}^n$ satisfying the cone condition. Let f denote an element in the dual space $(H^{k+1}(\Omega))'$ of $H^{k+1}(\Omega)$ such that

$$\langle f, u \rangle = 0 \qquad \forall u \in \mathcal{P}_k(\Omega) \tag{6.158}$$

where $\langle \cdot, \cdot \rangle$ denotes the duality pairing between $(H^{k+1}(\Omega))'$ and $H^{k+1}(\Omega)$, and $\mathcal{P}_k(\Omega)$ is the space of polynomials on Ω of degree $\leqslant k$, k being a fixed integer, $k \geqslant 0$. Then there exists a positive constant c, depending on n, k, and Ω but not on u, such that

$$|\langle f, u \rangle| \leqslant c\|f\|_{k+1}^{*}|u|_{k+1} \qquad \forall u \in H^{k+1}(\Omega) \tag{6.159}$$

where $\|\cdot\|^*_{k+1}$ is the norm associated with the dual space, and $|u|_{k+1}$ is the seminorm,

$$|u|^2_{k+1} = \int_\Omega \sum_{|\alpha|=k+1} |D^\alpha u|^2\,dx \tag{6.160}$$

PROOF. Let u be an arbitrary element in $H^{k+1}(\Omega)$ and let $v \in \mathcal{P}_k(\Omega)$. Then $\langle f,u\rangle = \langle f,u+v\rangle$. According to (4.21),

$$\|f\|^*_{k+1} = \sup \frac{|\langle f,u+v\rangle|}{\|u+v\|_{k+1}} \qquad u+v \in H^{k+1}(\Omega)$$

where $\|u+v\|_{k+1} \equiv \|u+v\|_{H^{k+1}(\Omega)}$. Thus

$$|\langle f,u\rangle| \leqslant \|f\|^*_{k+1} \inf_{v\in\mathcal{P}_k} \|u+v\|_{k+1}$$

According to Theorem 3.14,

$$\inf_{v\in\mathcal{P}_k} \|u+v\|_{k+1} \leqslant c|u|_{k+1}$$

Thus (6.159) follows. ■

We next prove a lemma constructed in [6.3].

LEMMA 6.4. Let Π be any linear operator from $H^{k+1}(\Omega)$ into $H^m(\Omega)$, $0 \leqslant m \leqslant k+1$, such that

$$\Pi u = u \qquad \forall u \in \mathcal{P}_k(\Omega) \tag{6.161}$$

where, again, $\mathcal{P}_k(\Omega)$ is the space of polynomials on Ω of degree $\leqslant k$. Then, $\forall u \in H^{k+1}(\Omega)$, there is a constant $c>0$ such that

$$\|u-\Pi u\|_m \leqslant c\|I-\Pi\|_{\mathcal{L}(H^{k+1},H^m)}|u|_{k+1} \tag{6.162}$$

PROOF. Consider a continuous linear functional g on $H^m(\Omega)$. Then $g \in (H^m(\Omega))'$, and $\langle g,u\rangle$ denotes the usual duality pairing. If $u \in H^{k+1}(\Omega)$, $k+1 \geqslant m$, then $u - \Pi u \in H^m(\Omega)$ and $(H^m(\Omega))' \subset (H^{k+1}(\Omega))'$. Therefore we can construct a linear functional f on $H^{k+1}(\Omega)$ such that

$$\langle f,u\rangle = \langle g,u-\Pi u\rangle \qquad \forall u \in H^{k+1}(\Omega)$$

As in the previous lemma, let $\|f\|_{k+1}^{*}$ denote the norm of f in $(H^{k+1}(\Omega))'$. Then

$$\|f\|_{k+1}^{*}=\sup_{u}\frac{|\langle f,u\rangle|}{\|u\|_{k+1}} \qquad u\neq 0$$

$$=\sup_{u}\frac{|\langle g,u-\Pi u\rangle|}{\|u\|_{k+1}} \qquad u\neq 0 \tag{6.163}$$

Now, because the set of elements $u-\Pi u$ is contained in $H^{m}(\Omega)$,

$$\|g\|_{m}^{*}\geqslant \sup\frac{|\langle g,u-\Pi u\rangle|}{\|u-\Pi u\|_{m}} \qquad u-\Pi u\neq 0$$

Thus

$$|\langle g,u-\Pi u\rangle|\leqslant \|g\|_{m}^{*}\|I-\Pi\|_{\mathcal{L}(H^{k+1},H^{m})}\|u\|_{k+1}$$

Substituting this result into (6.163) gives

$$\|f\|_{k+1}^{*}\leqslant \|g\|_{m}^{*}\|I-\Pi\|_{\mathcal{L}(H^{k+1},H^{m})} \tag{6.164}$$

Recalling (4.29), and Lemma 6.3, we have, for $g\neq 0$,

$$\|u-\Pi u\|_{m}=\sup_{g}\frac{|\langle g,u-\Pi u\rangle|}{\|g\|_{m}^{*}}$$

$$=\sup_{g}\frac{|\langle f,u\rangle|}{\|g\|_{m}^{*}}$$

$$\leqslant c\sup_{g}\frac{\|f\|_{k+1}^{*}}{\|g\|_{m}^{*}}|u|_{k+1} \qquad g\neq 0$$

where the supremum is taken over $(H^{m}(\Omega))'$.

Finally, introducing (6.164), we have

$$\|u-\Pi u\|_{m}\leqslant c\|I-\Pi\|_{\mathcal{L}(H^{k+1},H^{m})}|u|_{k+1}$$

which completes the proof. ■

Collecting all our results, we are finally ready to establish the basic theorem, after [6.3].

THEOREM 6.8. (The Interpolation Theorem for Finite Elements). Let Ω be an open bounded domain in R^n satisfying the cone condition. Let k be a fixed integer and m an integer such that $0 \leqslant m \leqslant k+1$. Let $\Pi \in \mathcal{L}(H^{k+1}(\Omega), H^m(\Omega))$ be such that

$$\Pi u = u \qquad \forall u \in \mathcal{P}_k(\Omega) \tag{6.165}$$

Then, for any $u \in H^{k+1}(\Omega)$ and for sufficiently small h, there exist positive constants, C, independent of u and h, such that

$$\|u - \Pi u\|_{H^m(\Omega)} \leqslant C \frac{h^{k+1}}{\rho^m} |u|_{H^{k+1}(\Omega)} \tag{6.166}$$

where $|u|_{H^{k+1}(\Omega)}$ is the seminorm.

PROOF. From (6.139) we have

$$\int_{\hat{\Omega}} |D^\alpha u(\hat{\mathbf{x}})|^2 \, d\hat{x} \leqslant \|\mathbf{T}\|^{2s} \int_{\Omega} |D^\alpha u(\mathbf{x})|^2 (\det \mathbf{T})^{-1} \, dx$$

for $|\boldsymbol{\alpha}| = s$. Thus

$$|\hat{u}|_{H^s(\hat{\Omega})} \leqslant \|\mathbf{T}\|^s |\det \mathbf{T}|^{-1/2} |u|_{H^s(\Omega)} \tag{6.167}$$

and

$$|u|_{H^s(\Omega)} \leqslant \|\mathbf{T}^{-1}\|^s |\det \mathbf{T}|^{1/2} |\hat{u}|_{H^s(\hat{\Omega})} \tag{6.168}$$

where $\hat{u} = u(\hat{\mathbf{x}})$.

We can take $\|\mathbf{T}^{-1}\| \geqslant 1$ because our choices of $\hat{H}$ and $\hat{R}$ are arbitrary. Thus

$$\|u\|_{H^m(\Omega)} = \left(\sum_{|\alpha| \leqslant m} |u|^2_{H^{|\alpha|}(\Omega)} \right)^{1/2}$$

$$\leqslant |\det \mathbf{T}|^{1/2} \left(\sum_{|\alpha| \leqslant m} \|\mathbf{T}^{-1}\|^{2|\alpha|} |\hat{u}|^2_{H^{|\alpha|}(\hat{\Omega})} \right)^{1/2}$$

$$\leqslant |\det \mathbf{T}|^{1/2} \|\mathbf{T}^{-1}\|^m \|\hat{u}\|_{H^m(\hat{\Omega})} \tag{6.169}$$

Likewise,

$$\|u-\Pi u\|_{H^m(\Omega)} \leqslant |\det \mathbf{T}|^{1/2}\|\mathbf{T}^{-1}\|^m \|\hat{u}-\hat{\Pi}\hat{u}\|_{H^m(\hat{\Omega})} \tag{6.170}$$

since $\widehat{\Pi u}=\hat{\Pi}\hat{u}$.

From Lemma 6.4,

$$\|\hat{u}-\hat{\Pi}\hat{u}\|_{H^m(\hat{\Omega})} \leqslant \hat{c}\|\hat{I}-\hat{\Pi}\|_{\mathcal{L}(H^{k+1}(\hat{\Omega}),H^m(\hat{\Omega}))}|\hat{u}|_{H^{k+1}(\hat{\Omega})}$$

Therefore, in view of (6.167),

$$\|\hat{u}-\hat{\Pi}\hat{u}\|_{H^m(\hat{\Omega})} \leqslant \hat{c}\|\hat{I}-\hat{\Pi}\|_{\mathcal{L}(H^{k+1}(\hat{\Omega}),H^m(\hat{\Omega}))} \cdot \|\mathbf{T}\|^{k+1}|\det \mathbf{T}|^{-1/2}|u|_{H^{k+1}(\Omega)}$$

Substituting this inequality into (6.170), we have

$$\|u-\Pi u\|_{H^m(\Omega)} \leqslant |\det \mathbf{T}|^{1/2}\|\mathbf{T}^{-1}\|^m \hat{c}\|\hat{I}-\hat{\Pi}\|_{\mathcal{L}(H^{k+1}(\hat{\Omega}),H^m(\hat{\Omega}))} \cdot \|\mathbf{T}\|^{k+1}|\det \mathbf{T}|^{-1/2}|u|_{H^{k+1}(\Omega)}$$

Recall from (6.140) that

$$\|\mathbf{T}^{-1}\|^m\|\mathbf{T}\|^{k+1} \leqslant \frac{\hat{H}^m}{\hat{R}^{k+1}} \cdot \frac{h^{k+1}}{\rho^m}$$

Thus, if we denote by C the constant

$$C=C(n,k,\hat{\Omega},\hat{\Pi})=\hat{c}\frac{\hat{H}^m}{\hat{R}^{k+1}}\|\hat{I}-\hat{\Pi}\|_{\mathcal{L}(H^{k+1}(\hat{\Omega}),H^m(\hat{\Omega}))} \tag{6.171}$$

we then have

$$\|u-\Pi u\|_{H^m(\Omega)} \leqslant C\frac{h^{k+1}}{\rho^m}|u|_{H^{k+1}(\Omega)}$$

which was to be proved. ■

Curved Elements

The extension of the theory developed thus far to curved elements was also developed by Ciarlet and Raviart [6.4] (see also [6.31]), and we survey briefly the essential features of their analysis. The principal feature of the extended theory is that, instead of the linear transformations (6.136), we now have nonlinear mappings of the form

$$\mathbf{x}=\mathbf{F}(\hat{\mathbf{x}}) \tag{6.172}$$

where $\mathbf{F}(\hat{\mathbf{x}})$ is an invertible and sufficiently differentiable vector-valued function of $\hat{\mathbf{x}}$. Now the master element $\hat{\Omega}$ can be mapped consecutively into curvilinear elements, as indicated in Fig. 6.25.

The key to the analysis of such curved elements is to view the curved element as a perturbation of an element with straight sides. In other words, the mapping $\mathbf{F}(\hat{\mathbf{x}})$ is viewed as a deviation from a linear mapping $\tilde{\mathbf{F}}(\hat{\mathbf{x}})$, for which the results in Theorem 6.8 hold. The idea is illustrated in Fig. 6.25; $\tilde{\Omega}_e$ is the "linear" image of $\hat{\Omega}$ ($\tilde{\Omega}_e=\tilde{\mathbf{F}}(\hat{\Omega})$) and the final finite element Ω_e is then obtained as a perturbation of $\tilde{\Omega}_e$.

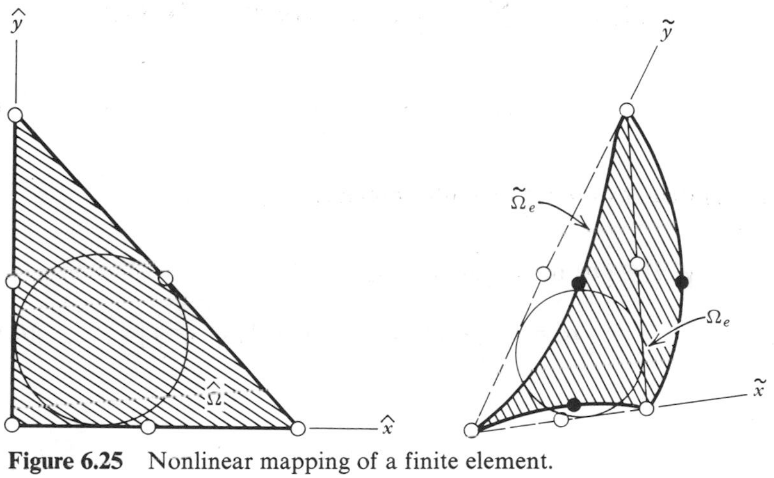

Figure 6.25 Nonlinear mapping of a finite element.

Noting that, if $\hat{u}$ is $u(\mathbf{x})$ in terms of the "master" coordinates $\hat{\mathbf{x}}$, it is obvious that

$$\frac{\partial \hat{u}}{\partial \hat{x}_k}=\sum_m \frac{\partial u}{\partial x_m}\cdot\frac{\partial F_m}{\partial \hat{x}_k}$$

$$\frac{\partial^2 \hat{u}}{\partial \hat{x}_k\,\partial \hat{x}_r}=\sum_{s,m}\left(\frac{\partial^2 u}{\partial x_s\,\partial x_m}\cdot\frac{\partial F_m}{\partial \hat{x}_k}\cdot\frac{\partial F_s}{\partial \hat{x}_r}+\frac{\partial u}{\partial x_m}\cdot\frac{\partial^2 F_m}{\partial \hat{x}_k\,\partial \hat{x}_r}\right)$$

etc., where

$$\frac{\partial F_m}{\partial \hat{x}_k} = \mathfrak{D} F_m \cdot \hat{\mathbf{e}}_k, \qquad \frac{\partial^2 F_m}{\partial \hat{x}_k \, \partial \hat{x}_r} = \mathfrak{D}^2 F_m \cdot (\hat{\mathbf{e}}_k, \hat{\mathbf{e}}_r) \qquad \ldots$$

and $\mathfrak{D}$ again denotes the Fréchet derivative and $\hat{\mathbf{e}}_i$ denotes an orthogonal basis in $\hat{\Omega} \subset \mathbf{R}^n$. With a little algebra (see [6.4]), we can show that these relations lead to the inequality

$$|D^{\beta}\hat{u}| \leqslant \sum_{|\boldsymbol{\alpha}|=1}^{k+1} |D^{\boldsymbol{\alpha}} u| \, |||\mathbf{F}|||_{k+1} \tag{6.173}$$

where

$$|||\mathbf{F}|||_{k+1}^2 = \sum_{j \in I(k+1,m)} \sup_{\hat{\mathbf{x}} \in \hat{\Omega}} \left\{ \|\mathfrak{D}\,\mathbf{F}(\hat{\mathbf{x}})\|^{j_1}, \|\mathfrak{D}^2 \mathbf{F}(\hat{\mathbf{x}})\|^{j_2}, \ldots, \|\mathfrak{D}^{k+1}\mathbf{F}(\hat{\mathbf{x}})\|^{j_{k+1}} \right\} \tag{6.174}$$

and $I(i,j)$ denotes the set of multi-integers,

$$I(i,j) = \left\{ \boldsymbol{\alpha}\colon \boldsymbol{\alpha} = (\alpha_1, \alpha_2, \ldots, \alpha_{j-1}) \in Z_+^{j-1};\; |\boldsymbol{\alpha}| = i;\; \alpha_1 + 2\alpha_2 + \cdots + (j-1)\alpha_{j-1} = j \right\} \tag{6.175}$$

Since

$$d\hat{x} = J(\hat{\mathbf{x}})\, dx$$

where $J(\hat{\mathbf{x}})$ is the jacobian,

$$J(\mathbf{x}) = \det \frac{\partial F_i(\hat{\mathbf{x}})}{\partial \hat{x}_j} = \det \mathfrak{D}\,\mathbf{F} \tag{6.176}$$

we integrate (6.173) and use the definition of the seminorm to obtain the inequality

$$|\hat{u}|_{k+1} \leqslant c \frac{|||\mathbf{F}|||_{k+1}}{\inf_{\hat{\mathbf{x}} \in \hat{\Omega}} |J(\hat{\mathbf{x}})|} \|u\|_{k+1} \tag{6.177}$$

Conversely,

$$|u|_m \leqslant c \sup_{\hat{\mathbf{x}} \in \hat{\Omega}} |J(\hat{\mathbf{x}})| \, |||\mathbf{F}^{-1}|||_m |\hat{u}|_m \tag{6.178}$$

Collecting these results, using the procedure developed in the proof of Theorem 6.8, and noting that use of the perturbation idea leads to the inequalities

$$\left.\begin{aligned} &\sup_{\hat{\mathbf{x}} \in \hat{\Omega}} \max_{|\alpha|=r} |D^{\alpha}\mathbf{F}(\hat{\mathbf{x}})| \leqslant c_r h^r \qquad 1 \leqslant r \leqslant k+1 \\ &\sup_{\hat{\mathbf{x}} \in \hat{\Omega}} \max_{|\alpha|=1} |D\mathbf{F}^{-1}(\mathbf{x})| \leqslant c_0 h^{-1} \end{aligned}\right\} \tag{6.179}$$

we obtain the new error estimate for regular refinements,

$$\|u - \Pi u\|_m \leqslant c \frac{\sup\{|J(\hat{\mathbf{x}})|; \hat{\mathbf{x}} \in \hat{\Omega}\}}{\inf\{|J(\hat{\mathbf{x}})|; \hat{\mathbf{x}} \in \hat{\Omega}\}} h^{k+1-m} \|u\|_{k+1} \tag{6.180}$$

Thus, if the curved element is not distorted too much from an associated linear one, the same order of accuracy possible in the linear theory (Theorem 6.8) is obtained.

REFERENCES

6.1. Oden, J. T., *Finite Elements of Nonlinear Continua*, McGraw-Hill, New York, 1972.

6.2. Oden, J. T., "Mathematical Aspects of Finite Element Approximation in Continuum Mechanics," in Nemmet-Nasser, S. Ed., *Mechanics Today*, Vol. 2, Pergamon Press, Oxford, 1975, pp. 159–250.

6.3. Ciarlet, P. G. and Raviart, P. A., "General Lagrange and Hermite Interpolation in $\mathbf{R}^n$ with Application to the Finite Element Method," *Arch. Ration Mech. Anal.*, Vol. 46, pp. 177–199, 1972.

6.4. Ciarlet, P. G. and Raviart, P. A., "Interpolation Theory over Curved Elements," *Comput. Meth. Appl. Mech. Eng.*, Vol. 1, pp. 217–249, 1972.

6.5. Brauchli, H. J. and Oden, J. T., "Conjugate Approximation Functions in Finite Element Analysis," *Quar. Appl. Math.*, Vol. 29, No. 3, pp. 65–90, 1971.

6.6. Oden, J. T., "Theory of Conjugate Projections in Finite Element Analysis," in *Lectures on Finite Element Methods in Continuum Mechanics,* Oden, J. T. and Oliveira, E. R. A., Eds., UAH Press, Huntsville, Ala., 1973, pp. 41–75.

6.7. Argyris, J. H., "Continua and Discontinua," *Proceedings, Conference on Matrix Methods in Structural Analysis,* AFFDL-TR-66-80, Wright-Patterson AFB, Ohio, 1965, pp. 11–190.

6.8. Zienkiewicz, O. C., *The Finite Element Method in Engineering Science,* McGraw-Hill, New York, 1971.

6.9. Nashed, M. Z., "Differentiability and Related Properties of Nonlinear Operators: Some

Aspects of the Role of Differentials in Functional Analysis," in Rall, L. B., Ed., *Nonlinear Functional Analysis and Applications,* Academic Press, New York, 1971, pp. 103–310.

6.10. Oliveira, E. R. A., "Theoretical Foundations of the Finite Element Method," *Int. J. Solids Struct.,* Vol. 4, pp. 929–952, 1968.

6.11. Ciarlet, P. G. and Wagschal, C., "Multipoint Taylor Formulas and Applications to the Finite Element Method," *Numer. Math.,* Vol. 17, pp. 84–100, 1971.

6.12. Gordon, W. J., " 'Blending-function' Methods of Bivariate and Multivariate Interpolation," *SIAM J. Numer. Anal.,* Vol. 8, pp. 158–177, 1971.

6.13. Gordon, W. J. and Hall, C. A., "Transfinite Element Methods: Blending Function Interpolation over Aribitrary Curved Element Domains," *Numer. Math.,* Vol. 21, pp. 109–129, 1973.

6.14. Bramble, J. H. and Zlamal, M., "Triangular Elements in the Finite Element Method," *Math. Comput.,* Vol. 24, No. 112, pp. 809–820, 1970.

6.15. Bell, K., "A Refined Triangular Plate Bending Element," *Int. J. Numer. Meth. Eng.,* Vol. 1, No. 1, pp. 101–122, 1969.

6.16. Bosshard, W., "Ein Neues Vollverträgliches Endliches Element für Plattenbiegung," Publication 1, *Assoc. Bridge Struct. Eng.*, Vol. 28, Part 1, 1968, pp. 27–40.

6.17. Argyris, J. H., Buck, K. E., Fried, I., Hilber, H. M. Mareczek, G., and Scharpf, D. W., "Some New Elements for the Matrix Displacement Method," *Proceedings, Second Conference on Matrix Methods in Structural Analysis*, AFFDL-TR-68-150, Wright Patterson AFB, Ohio, 1968, pp. 333–399.

6.18. Bogner, F. K., Fox, R. L., and Schmit, L. A. Jr., "The Generation of Interelement, Compatible, Stiffness and Mass Matrices by Use of Interpolation Formulas," *Proceedings, Conference on Matrix Methods in Structural Mechanics*, AFFDL-TR-66-80, Wright-Patterson AFB, Ohio, 1966, pp. 397–443.

6.19. Birkhoff, G. and de Boor, C., "Piecewise Polynomial Interpolation and Approximation," in Garabedian, H. L., Ed., *Approximation of Functions,* Elsevier, Amsterdam, 1965, pp. 164–190.

6.20. Ergatoudis, I., Irons, B. M., and Ziekiewicz, O. C., "Curved Isoparametric 'Quadrilateral' Elements for Finite Element Analysis," *Int. J. Solids Struct.,* Vol. 4, pp. 31–42, 1968.

6.21. Wachspress, E. L., "A Rational Basis for Function Approximation," *Lecture Notes in Mathematics* (Proceedings of the Conference in Application of Numerical Analysis), Vol. 228, Springer-Verlag, Berlin, 1971, pp. 223–252.

6.22. Wachspress, E. L., "A Rational Basis for Function Approximation, Part II: Curved Sides," *J. Inst. Math. Its Appl.,* Vol. 11, No. 1, pp. 83–104, 1973.

6.23. Wachspress, E. L., *A Rational Finite Element Basis*, Academic Press, New York, 1975.

6.24. Wempner, G. A., Oden, J. T., and Kross, D. A., "Finite Element Analysis of Thin Shells," *J. Eng. Mech. Div., ASCE,* Vol. 94, No. EM6, pp. 1273–1294, 1968.

6.25. Clough, R. W. and Tocher, J. L., "Finite Element Stiffness Matrices for Analysis of Plate Bending," *Proceedings, Conference on Matrix Methods in Structural Mechanics,* AFFDL-TR-66-80, Wright-Patterson AFB, Ohio, 1966, pp. 515–546.

6.26. Fried, I., "Discretization and Round-off Errors in the Finite Element Analysis of Elliptic Boundary-Value Problems and Eigenvalue Problems," *Ph. D. Dissertation*, Massachussetts Institute of Technology, Cambridge, 1971.

6.27. Synge, J. L., *The Hypercircle in Mathematical Physics,* Cambridge University Press, Cambridge, 1957.

6.28. Babuška, I. and Aziz, A. K., "On the Angle Condition in the Finite Element Method," *Technical Note*, No. BN-808, Institute for Fluid Dynamics and Applied Mathematics, University of Maryland, November, 1974.

6.29. Bramble, J. H. and Hilbert, S. R., "Estimation of Linear Functionals with Application to Fourier Transforms and Spline Analysis," *SIAM J. Numer. Anal.,* Vol. 7, pp. 112–124, 1970.

6.30. Bramble, J. H. and Hilbert, S. R., "Bounds for a Class of Linear Functionals with Applications to Hermite Interpolation," *Numer. Math.,* Vol. 16, pp. 362–369, 1971.

6.31. Ciarlet, P. G. and Raviart, P. A., "The Combined Effect of Curved Boundaries and Numerical Integration in Isoparametric Finite Element Methods," in Aziz, A. K., Ed., *The Mathematical Foundations of the Finite Element Method with Applications to Partial Differential Equations*, Academic Press, New York, 1972, pp. 409–474.

7

VARIATIONAL BOUNDARY-VALUE PROBLEMS

7.1 INTRODUCTION

In this chapter we bring together several of the ideas developed in previous chapters and use them to develop a framework for a theory of finite-element approximations of elliptic boundary-value problems. The key step in the construction of such a theory is to recast the elliptic problem in a variational setting firmly rooted in the Hilbert space theory described in Part I. Then the intrinsic relation between variational methods of approximation and all the results of Chapters 4 through 6 provide, in a very natural way, the framework for a fairly deep approximation theory. The main objective of the present chapter is to examine the structure of such variational settings for boundary-value problems. A study of how they can be used in conjunction with finite-element concepts to formulate approximate methods is taken up in Chapter 8.

To fix ideas, consider the simple second-order differential equation in two dimensions:

$$Au \equiv -\frac{\partial}{\partial x}\left[a_{11}(x,y)\frac{\partial u(x,y)}{\partial x}\right] - \frac{\partial}{\partial y}\left[a_{22}(x,y)\frac{\partial u(x,y)}{\partial y}\right]$$

$$+ a_0(x,y)u(x,y) = f(x,y), \qquad (x,y)\in\Omega\subset\mathbf{R}^2; \qquad u(x,y)=0, (x,y)\in\partial\Omega \tag{7.1}$$

Suppose that the coefficients a_{11}, a_{22}, a_0 are in $C^\infty(\overline{\Omega})$ and that they are such that (7.1) is a regularly elliptic boundary-value problem of order 2. Let us consider the case in which $f\in H^0(\Omega)$. Then, we know from the theory of Chapter 5 that a solution u of (7.1) exists and that it is in $H^2(\Omega)$

or, more specifically, a subspace of $H^2(\Omega)$ containing functions v such that $Av \in H^0(\Omega)$.

Next, consider an alternate boundary-value problem that closely resembles (7.1); find u such that $\forall\ v \in H_0^1(\Omega)$:

$$\int_\Omega \left(a_{11} \frac{\partial u}{\partial x} \frac{\partial v}{\partial x} + a_{22} \frac{\partial u}{\partial y} \frac{\partial v}{\partial y} + a_0 uv \right) dx\, dy = \int_\Omega fv\, dx\, dy \tag{7.2}$$

We refer to (7.2) as a *variational boundary-value problem*. Obviously, it suggests that weaker conditions may be required of the solution than that suggested by (7.1); second derivatives appear in (7.1), whereas only integrals of first derivatives appear in (7.2). A little thought leads us to the following observations:

(i) Every solution of (7.1) is also a solution of (7.2). This can be seen by multiplying both sides of (7.1) by a test function $\phi \in \mathfrak{D}(\Omega)$ and integrating by parts. Since $\mathfrak{D}(\Omega)$ is dense in $H_0^1(\Omega)$, the conclusion follows by continuity.

(ii) Judging from the smoothness of u needed in order that (7.2) make sense, the solution of (7.2) (assuming one exists) need be only in $H_0^1(\Omega)$. However, we are quite familiar with the fact that $H^2(\Omega)$ is densely embedded in $H^1(\Omega)$, so it is possible that these two solutions, one to (7.1) and one to (7.2), are the same. We shall see that this depends on the choice of v and f in (7.2), among other things, and that for the particular choices indicated the solutions do indeed coincide (almost everywhere). Thus (7.1) and (7.2) are *equivalent*, and it makes sense to say that the solution to (7.2) is an element $u \in H^2(\Omega) \cap H_0^1(\Omega)$ [or, more specifically, an element in the intersection of $H_0^1(\Omega)$ and the space of functions u such that $Au \in H^0(\Omega)$].

(iii) If A is the operator in (7.1),

$$A = -\frac{\partial}{\partial x}\left(a_{11} \frac{\partial}{\partial x} \right) - \frac{\partial}{\partial y}\left(a_{22} \frac{\partial}{\partial y} \right) + a_0$$

then, for $f \in H^0(\Omega)$, the domain of A can be regarded as a subset of $H^2(\Omega)$. The operator $\tilde{A}$ suggested in (7.2) is the unique self-adjoint extension of A to an operator on $H_0^1(\Omega)$; indeed $\tilde{A}$ maps $H_0^1(\Omega)$ continuously into $H^{-1}(\Omega) = (H_0^1(\Omega))'$. Recall that $H^1(\Omega)$ can be interpreted as the completion of $H^2(\Omega)$ in the $\|\cdot\|_{H^1(\Omega)}$ norm.

(iv) By specifying that $v \in H_0^1(\Omega)$ in (7.2), all the essential boundary conditions (see Section 7.2) are automatically provided for in the single statement of the variational boundary-value problem. Classically, $H_0^1(\Omega)$ corresponds to the *space of admissible variations,* as explained in the next observation.

(v) The term "variational boundary-value problem" arises from the fact that (7.2) is precisely the condition for the vanishing of the first variation of a quadratic functional which corresponds to (7.1) in the sense of the classical calculus of variations. Indeed, let $B(u,v)$ denote the *symmetric* bilinear form

$$B(u,v)=\int_{\Omega}\left(a_{11}\frac{\partial u}{\partial x}\frac{\partial v}{\partial x}+a_{22}\frac{\partial u}{\partial y}\frac{\partial v}{\partial y}+a_0uv\right)dx\,dy \tag{7.3}$$

and consider the quadratic functional

$$I(v)=B(v,v)-2\int_{\Omega}fv\,dx\,dy=B(v,v)-2(f,v)_0 \tag{7.4}$$

where $v\in H_0^1(\Omega)$. Then, for $\alpha\in\mathbf{R}$ and arbitrary $v\in H_0^1(\Omega)$,

$$I(u+\alpha v)=I(u)+2\alpha\left[B(u,v)-(f,v)_0\right]+\alpha^2B(v,v)$$

Thus, the *first variation* $\delta I(u,v)$ of $I(v)$ vanishes at the point $u\in H_0^1(\Omega)$ for which

$$\delta I(u,v)=\lim_{\alpha\to 0}\frac{1}{\alpha}\left[I(u+\alpha v)-I(u)\right]=2\left[B(u,v)-(f,v)_0\right]=0$$

That is, $\delta I(u,v)$ vanishes at that $u\in H_0^1(\Omega)$ for which

$$B(u,v)=(f,v)_0 \qquad \forall v\in H_0^1(\Omega) \tag{7.5}$$

The function u is a *critical point* of $I(v)$. We recognize that (7.5) is identical to (7.2); hence the term "variational" boundary-value problem.

(vi) It is clear that (7.5) provides a basis for a general class of boundary-value problems. Indeed, suppose $\mathfrak{U}$ is a Hilbert space and B is a continuous bilinear form from $\mathfrak{U}\times\mathfrak{U}$ into $\mathbf{R}$. Then the problem of finding $u\in\mathfrak{U}$ such that

$$B(u,v)=l(v) \qquad \forall v\in\mathfrak{U} \tag{7.6}$$

where l is a continuous linear functional on $\mathfrak{U}$, is an abstract variational boundary-value problem (we further generalize these ideas in subsequent sections). When B is symmetric, then (7.6) arises from the minimization of a quadratic functional similar to that in (7.4); when B is not symmetric, the minimization of $I(v)$ is not equivalent to (7.5), but we still refer to (7.6) as a variational boundary-value problem.

(vii) Suppose $f \in H^0(\Omega)$ in (7.1). Then

$$(Au, v)_{H^0(\Omega)} = (f, v)_{H^0(\Omega)}$$

and

$$B(u,v) = (f,v)_{H^0(\Omega)} \qquad \forall v \in H_0^1(\Omega)$$

Now, for fixed u, $B(u,v)$ is a continuous linear functional on $H_0^1(\Omega)$; i.e.,

$$B(u,v) = \tilde{A}_u(v) \equiv \langle \tilde{A}u, v \rangle \tag{7.7}$$

where $\tilde{A}u \in H^{-1}(\Omega)$, and therefore $\tilde{A}$ is a linear map from $H_0^1(\Omega)$ into $H^{-1}(\Omega)$. Since $H^0(\Omega) \subset H^{-1}(\Omega)$, it makes sense to define a set $\mathfrak{D} \subset H^1(\Omega)$ such that $\tilde{A}u \in H^0(\Omega) \forall u \in \mathfrak{D}$; i.e., $\mathfrak{D} = \tilde{A}^{-1}(H^0(\Omega))$. Then, if $\tilde{A}_{\mathfrak{D}}$ is the restriction of $\tilde{A}$ to $\mathfrak{D}$, we have

$$(f,v)_0 = (Au, v)_{H^0(\Omega)} = (\tilde{A}_{\mathfrak{D}} u, v)_{H^0(\Omega)}$$

In other words, $A = \tilde{A}_{\mathfrak{D}}$. *It follows that* $\tilde{A}$ *of* (7.7) *is the* (*unique*) *extension of* A *to* $H^1(\Omega)$.

It is also clear that A maps the space $\mathfrak{D}(\Omega)$ of test functions into itself and its transpose maps $(\mathfrak{D}(\Omega))'$ into itself. Likewise, $B(u,v)$ is a continuous-bilinear form on $(\mathfrak{D}(\Omega))' \times \mathfrak{D}(\Omega)$ *and* on $H^{-1}(\Omega) \times H_0^1(\Omega)$.

7.2 FORMULATION OF VARIATIONAL BOUNDARY-VALUE PROBLEMS

We continue our study of variational boundary-value problems by considering next a class of problems quite a bit more general than (7.1) or (7.2). Now suppose Ω is a smooth, open, bounded subset of $\mathbb{R}^n$ with a smooth boundary $\partial\Omega$, and consider the boundary-value problem

$$\begin{aligned} Au &= f \qquad \text{in } \Omega \\ B_k u &= 0 \qquad \text{on } \partial\Omega \qquad 0 \leqslant k \leqslant m-1 \end{aligned} \tag{7.8}$$

Here A is a $2m$th-order differential operator of the form

$$Au = \sum_{|\alpha|,|\beta| \leqslant m} (-1)^{|\alpha|} D^\alpha \left(a_{\alpha\beta}(\mathbf{x}) D^\beta u \right) \tag{7.9}$$

with infinitely differentiable coefficients $a_{\alpha\beta}(\mathbf{x})$. The boundary operators $\{B_k\}_{k=0}^{m-1}$ form a normal covering of A on $\partial\Omega$. We assume, in other words, that the boundary-value problem (7.8) is a regularly elliptic problem of order $2m$. We next associate with the operator A (or its extension) a bilinear form $B(u,v)$ defined by

$$B(u,v)=\sum_{|\alpha|,|\beta|\leqslant m}\int_{\Omega}a_{\alpha\beta}(\mathbf{x})D^{\alpha}vD^{\beta}u\,dx \tag{7.10}$$

where, for the moment, u and v are assumed to be in $H^m(\Omega)$.

We next choose among the boundary operators B_k those of order $q_k < m$. Let there be r of them and relabel them so as to form a system $\{B_k\}_{k=0}^{r-1}$ of operators of order $q_k < m$. Now we can always construct a system of boundary operators $\{B_k'\}_{k=r}^{m-1}$, $0 \leqslant r \leqslant m-1$, such that the collection $\{B_0, B_1, \ldots, B_{r-1}, B_r', B_{r+1}', \ldots, B_{m-1}'\}$ forms a Dirichlet system of order m on $\partial\Omega$. Following [7.1], let us denote this system of operators by

$$\{F_k\}_{k=0}^{m-1}=\{B_0, B_1, \ldots, B_{r-1}, B_r', \ldots, B_{m-1}'\}$$

It is not difficult to show (e.g., see [7.1, p. 120]) that there exists a complementary system of boundary operators $\{\Phi_k\}_{k=0}^{m-1}$, where the order of Φ_k is $2m-1-(\text{order of } F_k)$, such that the following variant of Green's formula holds for every u, $v \in C^{\infty}(\overline{\Omega})$:

$$B(u,v)=\int_{\Omega}Auv\,dx-\sum_{k=0}^{m-1}\oint_{\partial\Omega}\Phi_k u F_k v\,ds \tag{7.11}$$

Let $\mathcal{V}$ be a Hilbert space of functions v which satisfy

$$B_k v=0 \qquad \text{for } 0\leqslant k\leqslant r-1 \tag{7.12}$$

When $C^{\infty}(\overline{\Omega})$ is dense in $\mathcal{V}$, we extend (7.11) to functions in $\mathcal{V}$ by continuity arguments now quite familiar. Then (7.11) can be written

$$B(u,v)=\langle Au,v\rangle-\sum_{k=r}^{m-1}\langle\Phi_k u, B_k' v\rangle_{\partial\Omega} \tag{7.13}$$

Obviously, $\langle u,v\rangle$ and $\langle u,v\rangle_{\partial\Omega}$ denote the duality pairings extending $\int_{\Omega}uv\,dx$ and $\oint_{\partial\Omega}uv\,ds$, respectively. Now suppose u is such that

$$B(u,v)=\int_{\Omega}fv\,dx \qquad \forall v\in\mathcal{V} \tag{7.14}$$

Then it is clear from (7.13) that, simultaneously,

$$\langle Au-f,v\rangle=0 \qquad \text{and} \qquad \sum_{k=r}^{m-1}\langle \Phi_k u, B_k' v\rangle_{\partial\Omega}=0$$

for arbitrary $v\in\mathcal{V}$. This of course suggests that

$$Au=f \text{ in } \Omega \qquad \text{and} \qquad \Phi_k u=0 \text{ on } \partial\Omega \qquad r\leqslant k\leqslant m-1 \tag{7.15}$$

Obviously, if u satisfies (7.15), then it certainly satisfies (7.14).

We see that the system $\{B_k\}_{k=0}^{m-1}$ of boundary operators is naturally divided into two distinct classes of boundary operators when formulating the variational problem corresponding to (7.8). The first group $\{B_k\}_{k=0}^{r-1}$ consists of operators of order $q_k < m$, and the associated boundary conditions are called *essential* or *stable* boundary conditions. The second group consists of operators $\{B_k\}_{k=r}^{m-1}$ of order $q_k\geqslant m$, and the associated boundary conditions are called *natural* boundary conditions.

It is clear from these results that problem (7.8) can formally be reduced to an equivalent variational problem, *provided* the operators $B_r',\dots,B_{m-1}'$ can be chosen so that $\Phi_k=B_k$ for $r\leqslant k\leqslant m-1$. This latter condition imposes restrictions on the order of the natural boundary conditions. Indeed, in Green's formula (7.11) (A is assumed to be elliptic), recall that the order of Φ_k is $2m-1-$(order of F_k). If $p_r,\dots,p_{m-1}$ are numbers between 0 and $m-1$ so that $q_0,q_1,\dots,q_{r-1},p_r,\dots,p_{m-1}$ gives, in some arbitrary order, all the numbers $0,1,\dots,m-1$, p_k is the order of $B_k', r\leqslant k\leqslant m-1$, and the order of B_k is

$$q_k=2m-1-p_k \qquad k=r,r+1,\dots,m-1 \tag{7.16}$$

This condition is not necessarily satisfied by every regularly elliptic problem.

We now state a fairly general version of the variational formulation of an elliptic boundary-value problem. Let $\mathcal{F}$ be the Hilbert space in which the data f lie, and $\mathcal{V}$ be the closed subspace of $H^m(\Omega)$ such that $H_0^m(\Omega)\subset\mathcal{V}\subset\mathcal{F}\subset\mathcal{V}'$ with continuous injection. Consider the following Neumann problem: Find $u\in\mathcal{V}$ such that

$$\left.\begin{array}{lll} Au=f & \text{in } \Omega & f\in\mathcal{F} \\ B_k u=g_k & \text{on } \partial\Omega & g_k\in H^{m-q_k-1/2}(\partial\Omega) \\ \multicolumn{3}{c}{m\leqslant q_k\leqslant 2m-1;\ 0\leqslant k\leqslant m-1} \end{array}\right\} \tag{7.17}$$

where A is given by (7.9) The associated variational problem is to find $u \in \mathcal{V}$ such that

$$B(u,v) = l(v) \qquad \forall v \in \mathcal{V} \tag{7.18}$$

where $B(u,v)$ is given by (7.10) and $l(v)$ is a continuous linear functional on $\mathcal{V}$ given by

$$l(v) = \langle f, v \rangle - \sum_{k=0}^{m-1} \langle g_k, B_k' v \rangle_{\partial\Omega} \tag{7.19}$$

and $\langle \cdot, \cdot \rangle$ denotes the duality pairing between $\mathcal{V}'$ and $\mathcal{V}$. If we take $v \in \mathcal{D}(\mathbf{R}^n)$, it is clear that (7.18) implies (7.17) *provided the differentiations are interpreted in the sense of distributions*. For the corresponding Dirichlet problem, set $g_k = 0$ in (7.17), let $B_k u = 0$ on $\partial\Omega, 0 \leqslant q_k \leqslant m-1$. Then $B(u,v)$ is given by (7.10) and $l(v) = \langle f, v \rangle$. With these alterations, (7.18) and (7.17) are equivalent. With these conventions understood, we have:

THEOREM 7.1. The boundary-value problem (7.17) and the variational problem (7.18) are equivalent. ■

REMARK. The difference between essential (stable) and natural boundary conditions for variational boundary-value problems deserves some additional comments. Suppose we are given $f \in L_2(0,1)$ and u is a function in $H^2(0,1)$ which satisfies the distributional equations, $-D^2 u + u = f$ in $(0,1)$ and $u(0) = 0, Du(0) = g$. It is pointed out in [7.2] that if $\{\phi_n(x)\}$ is a complete infinite set of orthonormal eigenfunctions of the operator $-D^2 + 1$ *satisfying the boundary conditions*, the sequence $\{(f, \phi_n)_0 \phi_n\}$ converges to f in $L_2(0,1)$, even though f does *not* satisfy the boundary conditions, whereas $\{(f, \phi_n)_0 \phi_n / \lambda_n\}$, λ_n being the eigenvalue corresponding to ϕ_n, converges to the proper solution u in the $\|\cdot\|_{H^2(0,1)}$ norm. The essence of the problem is that the boundary conditions on f are unstable in the L_2-norm, whereas $\{(f, \phi_n)_0 \phi_n / \lambda_n\}$ converges to the correct solution u in the stronger norm $\|\cdot\|_{H^2(0,1)}$. In general, if $u \in H^m(0,1)$ we cannot impose conditions on derivatives of order m or higher. Such conditions are *unstable*. Thus, when the solution of a variational boundary-value problem is in $H^m(0,1)$ only those conditions involving derivatives of order $\leqslant m-1$ are stable (hence essential). ■

In this regard, it is important to again note that the variational problem (7.18) contains the natural boundary conditions explicilty, and that the essential boundary conditions are contained in the requirement $u \in \mathcal{V}$ and the definition of $\mathcal{V}$. For example, if $\mathcal{V}=H_0^m(\Omega)$, all the boundary conditions are essential ($r=m$), and (7.17) becomes the *Dirichlet problem.*

$$\left.\begin{array}{lll} Au=f & \text{in } \Omega & f\in H^{-m}(\Omega) \\ B_k u=0 & \text{on } \partial\Omega & 0\leqslant k\leqslant m-1 \end{array}\right\} \tag{7.20}$$

the order of B_k being less than m. If $\mathcal{V}=H^m(\Omega)$, all the boundary conditions are natural ($r=0$),

$$\Phi_k u=0 \qquad 0\leqslant k\leqslant m-1 \tag{7.21}$$

and the associated boundary-value problem is the *Neumann problem* with respect to Green's formula (7.11). If we define $\mathcal{V}$ to be the space

$$\mathcal{V}=\{v\colon v\in H^m(\Omega);\ F_k v=0 \text{ on } \partial\Omega_k;\ 0\leqslant k\leqslant m-1\} \tag{7.22}$$

where $\partial\Omega_k$ are open subsets of $\partial\Omega$, then (7.18) corresponds to the *mixed problem*

$$Au=f \text{ in } \Omega$$

$$F_k u=0 \text{ on } \partial\Omega_k \qquad \Phi_k u=0 \text{ on } \partial\Omega-\partial\Omega_k$$

for $k=0,\ldots,m-1$.

EXAMPLE 7.1. Consider the problem

$$\left.\begin{array}{ll} Au \doteq -\dfrac{\partial}{\partial x}\left(a_{11}\dfrac{\partial u}{\partial x}\right)-\dfrac{\partial}{\partial y}\left(a_{22}\dfrac{\partial u}{\partial y}\right)+a_0u=f & \text{in } \Omega \\ u=0 \qquad \text{on } \partial\Omega & \end{array}\right\} \tag{7.23}$$

Let $\partial\Omega\in C^\infty$, $0<a_0, a_{11}, a_{22}<M$, and $a_{ij}\in C^\infty(\Omega)$, for every $\mathbf{x}=(x,y)\in\Omega$, and let $f\in H^0(\Omega)$. Here $m=1$, $r=1$, and $\mathcal{V}=H_0^1(\Omega)$. The variational formulation assumes the form of (7.18); i.e., find $u\in H_0^1(\Omega)$ such that

$$B(u,v)=(f,v)_0 \qquad \forall v\in H_0^1(\Omega) \tag{7.24}$$

where $B(u,v)$ is given by (7.3). In this case Green's formula (7.11) becomes

$$B(u,v)=(Au,v)_{H^0(\Omega)} \tag{7.25}$$

for $u\in H^1(\Omega)$ and $v\in H_0^1(\Omega)$. Thus Green's formula makes sense whenever

$v \in H_0^1(\Omega)$ and u belongs to the space

$$H^1(\Omega, A) = \{u: u \in H^1(\Omega);\ Au \in L_2(\Omega)\} \tag{7.26}$$

Note that $H^1(\Omega, A)$ can be assigned the norm

$$\|u\|^2_{H^1(\Omega,A)} = \|u\|^2_{H^1(\Omega)} + \|Au\|^2_{L_2(\Omega)} \tag{7.27}$$

We can now show that the problem of finding $u \in H^1(\Omega, A)$ such that (7.23) holds for $f \in L_2(\Omega)$ is equivalent to the variational boundary-value problem (7.24). Clearly such a u is in $H_0^1(\Omega)$, because $H^1(\Omega, A) \subset H^1(\Omega)$ and $u|_{\partial\Omega} = \gamma_0 u = 0$. Multiplying the first member of (7.23) by an arbitrary $v \in H_0^1(\Omega)$ and using Green's formula gives (7.24).

Conversely, if u is a solution of (7.24) and $\phi \in \mathfrak{D}(\Omega)$ is a test function, (7.24) holds for $v = \phi$, because $\mathfrak{D}(\Omega)$ is dense in $H_0^1(\Omega)$.Thus (7.23) holds in the sense of distributions. Now $u \in H_0^1(\Omega)$ and $Au = f$ is in $L_2(\Omega)$. Thus $u \in H_0^1(\Omega, A)$, from which it follows that (7.23) and (7.24) are equivalent. ■

REMARKS. 1. The previous example brings to light an important consideration that must be made when we attempt to identify which type of boundary-value problem is equivalent to a given variational boundary-value problem. Namely, in precisely what spaces does Green's formula (7.11) or (7.13) make sense for given data f and g_k? For example, suppose

$$A = -\Delta + 1$$

Then Green's formula for sufficiently smooth u and v is

$$\int_\Omega \left(\sum_{i=1}^{n} \frac{\partial u}{\partial x_i} \frac{\partial v}{\partial x_i} + uv \right) dx = \int_\Omega Auv\, dx + \oint_{\partial\Omega} \frac{\partial u}{\partial n} v\, ds$$

and we would like to *extend* this by continuity to the weaker formula

$$B(u, v) = \langle Au, v \rangle + \langle \gamma_1 u, \gamma_0 v \rangle_{\partial\Omega} \tag{7.28}$$

where $\langle \cdot, \cdot \rangle$ is the duality pairing on $(\mathfrak{D}(\Omega))' \times \mathfrak{D}(\Omega)$ extending $\int_\Omega Auv\, dx$, $\langle \cdot, \cdot \rangle_{\partial\Omega}$ is the duality pairing extending $\oint_{\partial\Omega} uv\, ds$, and γ_0 and γ_1 are the trace operators studied in Chapter 4. The problem is that γ_1 *cannot be extended* to a continuous operator on $H^1(\Omega)$—a smaller space is needed. It so happens that this smaller space is precisely the space $H^1(\Omega, A)$ defined in (7.26) with $A = -\Delta + 1$, i.e., γ_1 *can* be extended to a continuous operator mapping $H^1(\Omega, A)$ into $H^{-1/2}(\partial\Omega)$. Then, when $u \in H^1(\Omega, A)$ and $v \in$

$H^1(\Omega)$, formula (7.28) has meaning. For additional details, see [7.3, pp. 164–172].

2. The data f and g_k play a crucial role in the equivalence question. Indeed, whenever the data are such that a classical solution exists in the sense described in Chapter 5, then the solution of the variational problem coincides with the classical solution. Of course, in many cases the data are such that the variational solution must be interpreted in a distribution sense if it is to be associated with a differential equation. ■

EXAMPLE 7.2. Consider again the operator A defined in Example 7.1, with $f \in H^0(\Omega)$, but now suppose we have the boundary condition

$$B_0 u \equiv \frac{\partial u}{\partial n} = 0 \text{ on } \partial\Omega \tag{7.29}$$

Here $m=1, r=0$ ($r=0$ means that all B_k's are of order $\geqslant m$), and the order of B_0 satisfies the restriction $q_0 = 2m-1-p_0 = 0$. The variational problem is to find $u \in H^1(\Omega)$ such that

$$B(u,v) = l(v) \equiv \langle f, v \rangle \tag{7.30}$$

for all $v \in H^1(\Omega)$. In (7.29), $v|_{\partial\Omega} \in H^{1/2}(\Omega)$ when $v \in H^1(\Omega)$. In this case Green's formula assumes the form

$$B(u,v) = \langle Au, v \rangle + \langle B_0 u, \gamma_0 v \rangle_{\partial\Omega} \tag{7.31}$$

with $\gamma_0 = F_0$ mapping $H^1(\Omega)$ onto $H^{1/2}(\partial\Omega)$ (recall the trace theorem, Theorem 4.17). However, notice that, since $B_0 u = 0$, we can again write

$$B(u,v) = \langle Au, v \rangle \tag{7.32}$$

Clearly, the problem of finding $u \in H^1(\Omega, A)$ such that $Au = f$ and (7.29) holds is equivalent to problem (7.30). Conversely, (7.30) holds $\forall v \in H^1(\Omega)$, and we deduce from (7.31) that

$$(Au, v)_{H^0(\Omega)} + \langle B_0 u, \gamma_0 v \rangle_{\partial\Omega} = (f, v)_{H^0(\Omega)}$$

It follows that $Au = f$ almost everywhere in Ω, and

$$\langle B_0 u, \gamma_0 v \rangle_{\partial\Omega} = 0 \qquad \text{for any } v \in H^1(\Omega)$$

Since γ_0 maps $H^1(\Omega)$ onto $H^{1/2}(\partial\Omega)$, we have (7.29). Last, since $Au = f \in H^0(\Omega)$, $u \in H^2(\Omega)$ or, more precisely, $u \in H^1(\Omega, A)$. ■

REMARK. Notice that when boundary condition (7.29) is homogeneous, the form of the variational boundary-value problem is precisely the same for the Dirichlet problem (7.23) as for the Neumann problem of Example 7.2. [compare (7.25) and (7.32)]. The only difference is that (7.24) holds for every v in $H_0^1(\Omega)$, whereas (7.30) [with B_0u set to zero in (7.31)] holds for every $v \in H^1(\Omega)$. It is also interesting to note that, if we merely add terms such as $-a_{12}\partial^2 u/\partial x\,\partial y$, $a_{12} \in C^\infty(\overline{\Omega})$, to the definition of A, then the space $H^1(\Omega, A)$ is equivalent to $H^2(\Omega)$. ■

EXAMPLE 7.3. Now consider the mixed boundary-value problem

$$\left.\begin{aligned} &Au = -\frac{\partial^2 u}{\partial x^2} - \frac{\partial^2 u}{\partial y^2} + u = f \\ &u = 0 \text{ on } \partial\Omega_1 \qquad \frac{\partial u}{\partial n} = 0 \text{ on } \partial\Omega_2 \equiv \partial\Omega - \partial\Omega_1 \end{aligned}\right\} \tag{7.33}$$

where $\partial\Omega_1$ is a subset of $\partial\Omega$. Let $H_*^1(\Omega)$ be the closure in $H^1(\Omega)$ of the functions u which vanish on $\partial\Omega_1$. Then the variational problem consists of finding $u \in H_*^1(\Omega)$ such that

$$B(u,v) = (f,v)_{H^0(\Omega)} \qquad \forall v \in H_*^1(\Omega)$$

From Green's formula we have

$$B(u,v) = \langle Au, v\rangle = \int_\Omega \left(\frac{\partial u}{\partial x}\frac{\partial v}{\partial x} + \frac{\partial u}{\partial y}\frac{\partial v}{\partial y} + uv \right) dx + \oint_{\partial\Omega_2} \frac{\partial u}{\partial n} v\, ds$$

which implies that

$$\oint_{\partial\Omega_2} \frac{\partial u}{\partial n} v\, ds = 0 \qquad \text{for all } v \in H_*^1(\Omega)$$

and $Au = f$ in Ω, in a distributional sense. Then $u \in H_*^1(\Omega)$ implies $u = 0$ on $\partial\Omega_1$, and the above condition implies $\partial u/\partial n = 0$ on $\partial\Omega_2$. ■

Nonhomogeneous Boundary Conditions

Note that in all the preceding examples we have considered only homogeneous boundary conditions. When this is not the case, it is always possible to reduce a given linear boundary-value problem with nonhomogeneous boundary conditions to one with homogeneous boundary conditions.

Let ϕ be a function such that $B_k\phi = g_k, 0 \leqslant k \leqslant m-1$, and $A\phi \in H^0(\Omega)$. Suppose that u is the solution of (7.17). Then $u - \phi = w$ satisfies

$$Aw = f - A\phi \text{ in } \Omega \qquad B_k w = 0 \text{ on } \partial\Omega \qquad 0 \leqslant k \leqslant m-1$$

Thus we are formally led back to problem (7.20). In the case of Dirichlet, Neumann, and mixed boundary-value problems of second order, the boundary conditions become, respectively,

(i) $u = \phi$ on $\partial\Omega$

(ii) $\partial u/\partial n = \partial\phi/\partial n$ on $\partial\Omega$

(iii) $u = \phi$ on $\partial\Omega_1$ and $\partial u/\partial n = \partial\phi/\partial n$ on $\partial\Omega - \partial\Omega_1$.

Thus, if we wish to prescribe u and $\partial u/\partial n$ on $\partial\Omega$, we need only to find a function $\phi \in H^0(\Omega)$ for which $A\phi \in H^0(\Omega)$ and $u = \phi$ or $\partial u/\partial n = \partial\phi/\partial n$ on the appropriate parts of $\partial\Omega$.

EXAMPLE 7.4. We now consider an example discussed in one form or another by many authors (see, e.g., Mikhlin [7.4] or Babuška and Aziz [7.5]).We refer to a general second-order boundary-value problem in $\mathbf{R}^2$ with nonhomogeneous boundary conditions of the form

$$\left.\begin{aligned} Au &= f \quad \text{in } \Omega \\ B_0 u &= g \quad \text{on } \partial\Omega \end{aligned}\right\} \tag{7.34}$$

with

$$\left.\begin{aligned} Au &= -\sum_{i,j=1}^{2} \frac{\partial}{\partial x_i}\left(a_{ij}\frac{\partial u}{\partial x_j}\right) + \sum_{i=1}^{2} b_i \frac{\partial u}{\partial x_i} + c_0 u \\ B_0 u &= \frac{\partial u}{\partial n} + \alpha\frac{\partial u}{\partial s} + \beta u \end{aligned}\right\} \tag{7.35}$$

Here n and s denote the exterior normal and tangent to the boundary. Next we consider the associated bilinear and linear forms

$$B(u,v) = \int_\Omega \left(\sum_{i,j=1}^{2} a_{ij}\frac{\partial u}{\partial x_i}\frac{\partial v}{\partial x_j} + \sum_{i=1}^{2} b_i \frac{\partial u}{\partial x_i} v + c_0 uv \right) dx$$

$$+ \oint_{\partial\Omega} \left[\alpha_1(s)\frac{\partial u}{\partial s} + \beta_1(s) u \right] v\, ds \tag{7.36}$$

with $u, v \in H^1(\Omega)$, and

$$l(v) = \int_\Omega fv\,dx + \oint_{\partial\Omega} gv\,ds \qquad v \in H^1(\Omega) \tag{7.37}$$

where $\alpha_1 = p\alpha - q, \beta_1 = p\beta, p \neq 0$, and

$$p = \sum_{i,j=1}^{2} a_{ij} \cos n_i \cos n_j \qquad q = -\sum_{i,j=1}^{2} a_{ij} \cos n_i \sin n_j$$

Here $\cos n_i$ and $\sin n_i$ are the cosine and sine of the angle between the outer normal to $\partial\Omega$ and x_i. The associated variational problem is this: Find $u \in H^1(\Omega)$ such that

$$B(u,v) = l(v) \qquad \forall v \in H^1(\Omega) \tag{7.38}$$

The equivalence of (7.34) and (7.38) can be verified by integrating $B(u,v)$ by parts and noting that

$$\sum_{i,j=1}^{2} a_{ij} \frac{\partial u}{\partial x_j} \cos n_i = p \frac{\partial u}{\partial n} + q \frac{\partial u}{\partial s}$$

We find

$$\begin{aligned} B(u,v) &= \int_\Omega Auv\,dx + \oint_{\partial\Omega} \left[p \frac{\partial u}{\partial n} + (q + \alpha_1) \frac{\partial u}{\partial s} + \beta_1 u \right] v\,ds \\ &= \int_\Omega fv\,dx + \oint_{\partial\Omega} p \left(\frac{\partial u}{\partial n} + \alpha \frac{\partial u}{\partial s} + \beta u \right) v\,ds \\ &= l(v) \end{aligned}$$

as asserted. We return to this example later. ■

EXAMPLE 7.5. We now give an example of a regularly elliptic boundary-value problem that *does not* admit to a variational formulation (without some modifications). Consider the boundary-value problem

$$\left.\begin{aligned} &Au \equiv \Delta\Delta u + u = f \text{ in } \Omega \\ &B_0 u \equiv u = g_0 \qquad B_1 u \equiv \frac{\partial}{\partial n}(\nabla^2 u) = g_1 \qquad \text{on } \partial\Omega \end{aligned}\right\} \tag{7.39}$$

where $\Delta = \sum_{i=1}^{n} \partial^2 / \partial x_i^2$ is the Laplace operator. Here we have $m = 2$, $r = 1$, and the order of B_1 is 3, which does not satisfy (7.16). Therefore the boundary-value problem does not admit variational formulation in the sense we have used it previously. ■

The character of the variational formulation obviously depends quite strongly on the choice of the bilinear form $B(u,v)$ and, if we choose $B(u,v)=(Au,v)_{H^0(\Omega)}$(or $=\langle Au,v\rangle$) for homogeneous boundary conditions, it is clear from the above example that the variational approach may be quite limited. However, we now show that a relatively straightforward extension of the definition of $B(u,v)$ is possible, which will considerably expand the range of applicability of the variational theory. This is accomplished by using, instead of $(Au,v)_{H^0(\Omega)}$, the definition

$$B(u,v)=(Au,Tv)_{H^0(\Omega)} \qquad v\in\mathcal{V} \tag{7.40}$$

where T is a linear operator, the form of which may be suggested by the problem at hand. The space $\mathcal{V}$ in fact depends on the choice of T, and therefore problems that do not admit a variational formulation for our earlier choice of $T=I$ (the identity operator) may now admit a variational formulation for some specific choice of T. The following example, adapted from Lions and Magenes [7.1, p. 214], illustrates this assertion.

EXAMPLE 7.6. We return to (7.39) and now take $\mathcal{V}$ to be

$$\mathcal{V}=\left\{u:\ u\in H_0^1(\Omega);\ \frac{\partial}{\partial x_i}(\Delta u)\in L_2(\Omega);\ 1\leqslant i\leqslant n\right\}$$

with the norm

$$\|u\|_{\mathcal{V}}=\left[\sum_{i=1}^{n}\left(\left\|\frac{\partial u}{\partial x_i}\right\|_{L_2(\Omega)}^2+\left\|\frac{\partial \Delta u}{\partial x_i}\right\|_{L_2(\Omega)}^2\right)\right]^{1/2}$$

In this case we define the bilinear form (7.40) by

$$B(u,v)=\sum_{i=1}^{n}\int_\Omega\left[\frac{\partial}{\partial x_i}(\Delta u)\,\overline{\frac{\partial}{\partial x_i}(\Delta v)}+\frac{\partial u}{\partial x_i}\,\overline{\frac{\partial v}{\partial x_i}}\right]dx$$

wherein we have taken $T=\Delta$. The associated variational problem is to find, given $f\in H^{-1}(\Omega)$, a $u\in\mathcal{V}$ such that

$$B(u,v)=\sum_{i=1}^{n}\int_\Omega\frac{\partial f}{\partial x_i}\,\overline{\frac{\partial v}{\partial x_i}}\,dx \qquad \forall v\in\mathcal{V} \tag{7.41}$$

Recall that the operator Δ is an isomorphism of $H_0^1(\Omega)$ onto $(H_0^1(\Omega))' = H^{-1}(\Omega)$.

We show that (7.41) is equivalent to (7.39) (the converse is obvious). We can write (7.41) in the form

$$\sum_{i=1}^{n} \int_\Omega \frac{\partial}{\partial x_i}(\Delta u)\, \overline{\frac{\partial}{\partial x_i}(\Delta v)}\, dx = \int_\Omega (u-f)\, \overline{\Delta v}\, dx \qquad \forall v \in \mathcal{V}$$

Let $\mathcal{W}$ be the space of functions of the form Δu, where $u \in \mathcal{V}$. Clearly, $H^1(\Omega) \supset \mathcal{W}$. Therefore $w = \Delta u \in \mathcal{W}$ is a solution of

$$\sum_{i=1}^{n} \int_\Omega \frac{\partial w}{\partial x_i} \overline{\frac{\partial z}{\partial x_i}}\, dx = \int_\Omega (u-f)\bar{z}\, dx \qquad \forall z \in H^1(\Omega)$$

This implies that w is a variational solution of

$$-\Delta w = u - f \text{ in } \Omega \qquad \frac{\partial w}{\partial n} = 0 \text{ on } \partial\Omega$$

Then, for $u \in \mathcal{V}$, we have

$$\Delta\Delta u + u = f \text{ in } \Omega$$

$$\frac{\partial \nabla^2 u}{\partial n} = 0 \qquad u = 0 \text{ on } \partial\Omega$$

which is precisely (7.39). The derivatives here are to be interpreted in the sense of distributions. Thus (7.39), which does not admit variational formulation with $T = I$, admits variational formulation for $T = \Delta$. ■

7.3 COERCIVE BILINEAR FORMS

Consider a linear operator A mapping a given Hilbert space $\mathcal{U}$ continuously into its dual $\mathcal{U}'$; i.e., $A \in \mathcal{L}(\mathcal{U}, \mathcal{U}')$. It is clear that we can use A and the duality pairing $\langle \cdot, \cdot \rangle$ on $\mathcal{U}' \times \mathcal{U}$ to construct the following bilinear form:

$$B(u,v) = \langle Au, v \rangle \tag{7.42}$$

Conversely, a continuous bilinear form $B(u,v)$ on $\mathcal{U} \times \mathcal{U}$ defines a continuous linear mapping A from $\mathcal{U}$ into $\mathcal{U}'$. Such an operator $A \in \mathcal{L}(\mathcal{U}, \mathcal{U}')$ is said to be $\mathcal{U}$-*elliptic* if there exists a positive constant μ_0 such that

$$B(u,u) \geqslant \mu_0 \|u\|_{\mathcal{U}}^2 \qquad \forall u \in \mathcal{U} \tag{7.43}$$

We use the term $\mathcal{U}$*-coercive* to describe the associated bilinear form $B(u,u)$ whenever (7.43) holds.

Now suppose that $\mathcal{U}$ is a dense and continuously embedded subspace of a pivot space $\mathcal{H}$; i.e., $\mathcal{U} \subset \mathcal{H} = \mathcal{H}' \subset \mathcal{U}'$. Then the bilinear form $B(u,v)$ [or the operator $A \in \mathcal{L}(\mathcal{U}, \mathcal{U}')$] is said to be $(\mathcal{U}, \mathcal{H})$*-coercive* if there exists a real number $\lambda_0 \in \mathbf{R}$ and a positive constant μ_0 such that

$$B(u,u) \geqslant \mu_0 \|u\|_{\mathcal{U}}^2 - \lambda_0 \|u\|_{\mathcal{H}}^2 \tag{7.44}$$

for all $u \in \mathcal{U}$. Obviously, the "coerciveness" depends on the spaces $\mathcal{U}$ and $\mathcal{H}$. For $\lambda_0 = 0$, (7.44) reduces to the definition of $\mathcal{U}$-coerciveness. Further, if $A + \lambda_0$ is $\mathcal{U}$-elliptic, then

$$\langle Au + \lambda_0 u, u \rangle \geqslant \mu_0 \|u\|_{\mathcal{U}}^2$$

which means that A is then $(\mathcal{U}, \mathcal{H})$-coercive if λ_0 is sufficiently large. A $\mathcal{U}$-elliptic operator $A + \lambda_0$ is clearly bounded below, because

$$\mu_0 \|u\|_{\mathcal{U}}^2 \leqslant \langle (A + \lambda_0)u, u \rangle \leqslant \|(A + \lambda_0)u\|_{\mathcal{U}'} \|u\|_{\mathcal{U}}$$

which means that

$$\|(A + \lambda_0)u\|_{\mathcal{U}'} \geqslant \mu_0 \|u\|_{\mathcal{U}}$$

Thus, according to the bounded inverse theorem, every linear operator bounded below (i.e., every operator $A : \mathcal{U} \to \mathcal{V}$ such that $\|Au\|_{\mathcal{V}} \geqslant M \|u\|_{\mathcal{U}}$) has a continuous inverse defined on its range. Consequently, $(A + \lambda_0 I)^{-1}$ exists and is continuous. Since $\mathcal{U}$ is dense in $\mathcal{H}$, it follows that $(A + \lambda_0 I)^{-1}$ is a compact operator, and therefore A can be expressed as a weighted sum of projections for λ_0 in the resolvent set of A.

The following theorem establishes some properties of such coercive operators.

THEOREM 7.2. Let A be a continuous linear operator from a Hilbert space $\mathcal{U}$ into $\mathcal{U}'$. If A is $\mathcal{U}$-elliptic, then A is an isomorphism from $\mathcal{U}$ into $\mathcal{U}'$. Further, if the injection from $\mathcal{U}$ into the pivot space $\mathcal{H}$ is dense and compact and if A is $(\mathcal{U}, \mathcal{H})$-coercive, the operator $A + \lambda_0$ is an isomorphism from $\mathcal{U}$ onto $\mathcal{U}'$, provided λ_0 belongs to the resolvent set of A.

PROOF. The operator A is one-to-one since, if $Au=0$, we have

$$0=\langle Au,u\rangle_{\mathcal{H}}\equiv B(u,u)\geqslant \mu_0\|u\|^2_{\mathcal{U}} \qquad \forall u\in\mathcal{U}$$

implying $u=0$.

Now we show that $A(\mathcal{U})$ is a closed subspace of $\mathcal{U}'$. Let $\{Au_n\}$ be a Cauchy sequence in $\mathcal{U}'$. Then $\{u_n\}$ is a Cauchy sequence in $\mathcal{U}$, since

$$\|u_n-u_m\|^2\leqslant\frac{1}{\mu_0}B(u_n-u_m,u_n-u_m)=\frac{1}{\mu_0}(Au_n-Au_m,u_n-u_m)_{\mathcal{H}}$$

$$\leqslant\frac{1}{\mu_0}\|Au_n-Au_m\|_{\mathcal{U}'}\|u_n-u_m\|_{\mathcal{U}}$$

Therefore u_n converges to $u\in\mathcal{U}$, and Au_n converges to $Au\in\mathcal{U}'$, which means that $A(\mathcal{U})$ is closed.

It remains to be shown that $A(\mathcal{U})=\mathcal{U}'$. Suppose that this is not the case. Since $A(\mathcal{U})$ is a closed subspace of $\mathcal{U}'$, there exists a nonzero element $v_0\in\mathcal{U}=(\mathcal{U}')'$ (v_0 is a linear functional on $\mathcal{U}'$) such that

$$\langle v,v_0\rangle=0 \qquad \text{for all } v\in A(\mathcal{U})$$

That is, $v\in A(\mathcal{U})^\perp$. But then

$$\langle Au,v_0\rangle=0 \qquad \text{for all } v\in\mathcal{U}$$

By taking $u=v_0$, we see that $B(v_0,v_0)=0$, which contradicts our hypothesis. Hence $A(\mathcal{U})=\mathcal{U}'$, and the first part of the theorem is proved.

Now consider a $(\mathcal{U},\mathcal{H})$-coercive operator A, $A+\lambda_0$ being an isomorphism from $\mathcal{U}$ into $\mathcal{U}'$. The canonical injection of $\mathcal{U}$ into $\mathcal{U}'$ is compact, because the injection of $\mathcal{U}$ into $\mathcal{H}$ is compact and $\mathcal{U}\subset\mathcal{H}\subset\mathcal{U}'$. Thus the remainder of the proof follows immediately from the Riesz-Fredholm property: If B: $\mathcal{U}\to\mathcal{V}$ is compact, $B+\lambda$ is an isomorphism from $\mathcal{U}$ onto $\mathcal{V}$, provided λ is not an eigenvalue of B (see, e.g., Akhiezer and Glazman [7.6, p. 119]). ■

EXAMPLE 7.7. Consider the second-order differential operator A defined by

$$Au=-\sum_{i,j=1}^{n}\frac{\partial}{\partial x_i}\left(a_{ij}\frac{\partial u}{\partial x_j}\right)+c_0u$$

Let us assume the coefficients a_{ij} are such that A is strongly elliptic. Define the associated bilinear form $B(u,v)$ by

$$B(u,v)=\sum_{i,j}^{n}\int_{\Omega}a_{ij}(\mathbf{x})\frac{\partial u}{\partial x_i}\frac{\partial v}{\partial x_j}\,dx+\int_{\Omega}c_0(\mathbf{x})uv\,dx$$

The strong ellipticity of A implies that there is some $\mu_0>0$ such that

$$\sum_{i,j=1}^{n}a_{ij}\xi_i\xi_j\geqslant\mu_0|\xi|^2\qquad\forall\xi\neq\mathbf{0}\tag{7.45}$$

where $|\cdot|$ denotes the euclidean norm. The type of coerciveness of $B(\cdot,\cdot)$ depends on the coefficient c_0. First let us assume that, for some positive constant M,

$$M\leqslant c_0(\mathbf{x})\qquad\forall\mathbf{x}\in\Omega$$

Then $B(\cdot,\cdot)$ is $H^1(\Omega)$-coercive; i.e.,

$$B(u,u)\geqslant C\left[\int_{\Omega}\left(\sum_{i=1}^{n}\left|\frac{\partial u}{\partial x_i}\right|^2+|u|^2\right)dx\right]=C\|u\|^2_{H^1(\Omega)}$$

where $C=\min(\mu_0,M)$.

Now suppose we assume only that $c_0(\mathbf{x})\geqslant 0$. Then, for $u\in H_0^1(\Omega)$, we use (7.45) and the Friedrichs' inequality (3.50) to establish that $\|u\|_{H^0(\Omega)}\leqslant C_0\|D^{\alpha}u\|_{H^0(\Omega)}$, $|\alpha|\leqslant 1$. Thus we obtain

$$B(u,u)\geqslant c\left(\sum_{i}^{n}\int_{\Omega}\left|\frac{\partial u}{\partial x_i}\right|^2dx\right)=c\|\nabla u\|^2_{H^0(\Omega)}\geqslant c\|u\|^2_{H^1(\Omega)}$$

Hence $B(\cdot,\cdot)$ is $H_0^1(\Omega)$-coercive. ■

EXAMPLE 7.8. Consider the bilinear form associated with the operator $\Delta\Delta+\lambda$, with $\lambda>0$:

$$B(u,v)=\int_{\Omega}(\Delta u\,\Delta v+\lambda uv)\,dx\qquad u,v\in H^2(\Omega)$$

Then $B(u,v)$ is *not* $H^2(\Omega)$-coercive, because the right-hand side cannot be bounded below by the $H^2(\Omega)$ norm of u. However, if $u\in H_0^2(\Omega)$, using

Poincare's inequality we have

$$B(u,u) \geqslant c\int_\Omega \left[\sum_{i=1}^n \left|\frac{\partial^2 u}{\partial x_i^2}\right|^2 + |u|^2\right] dx$$

$$\geqslant c_0\int_\Omega \left[\left(\sum_{i=1}^n \left|\frac{\partial u}{\partial x_i}\right|^2\right) + |u|^2\right] dx = c_0\|u\|^2_{H^1(\Omega)}$$

Hence, $B(u,v)$ is $H^1(\Omega)$-coercive (when $\lambda > 0$). ■

If A of (7.35) is strongly elliptic, the associated bilinear form

$$B(u,v) = \int_\Omega \left(\sum_{i,j=1}^n a_{ij}\frac{\partial u}{\partial x_i}\frac{\partial v}{\partial x_j} + \sum_{i=1}^n b_i\frac{\partial u}{\partial x_i}v + c_0 uv\right) dx$$

is $[H^1(\Omega), H^0(\Omega)]$-coercive, owing to the fact that the lower-order terms can be estimated in terms of $\|u\|^2_{H^1(\Omega)}$ and $\|u\|^2_{H^0(\Omega)}$ (see Theorem 7.4). However, this is not the case in higher-order problems ($m > 2$); strong ellipticity of A does not necessarily imply $[H^m(\Omega), H^0(\Omega)]$-coercivity. We have the following result due to Gårding on higher-order problems.

THEOREM 7.3. (Cf. [7.7].) Let Ω be a bounded set in $\mathbf{R}^n$ and A be given by (7.9) with $a_{\alpha\beta} \in C^\infty(\overline{\Omega})$ for $|\boldsymbol{\alpha}| = |\boldsymbol{\beta}| = m$. Assume that A is strongly elliptic in $\overline{\Omega}$ in the sense of (5.8). Then the bilinear form $B(u,v)$ of (7.10) associated with A is $[H_0^m(\Omega), H^0(\Omega)]$-coercive.

PROOF. We prove this theorem for the case of constant coefficients. For more general cases see, e.g., Friedman [7.8] or Agmon [7.9].

First let $u \in C_0^\infty(\Omega)$. We can extend u outside Ω to $\mathbf{R}^n$ by zero so that we can use Fourier transform (2.61). From Plancherel's equality (2.83) we have, for some $\varepsilon_0 > 0$,

$$\|D^{\boldsymbol{\alpha}}u\|^2_{H^0(\Omega)} = \|\widehat{D^{\boldsymbol{\alpha}}u}\|^2_{H^0(\Omega)} = \int_{\mathbf{R}^n} |\mathbf{y}^{\boldsymbol{\alpha}}\hat{u}(\mathbf{y})|^2\,dy$$

$$\leqslant \int_{|\mathbf{y}|^2 \leqslant \varepsilon_0} |\mathbf{y}|^{2\boldsymbol{\alpha}}|\hat{u}(\mathbf{y})|^2\,dy + \int_{|\mathbf{y}|^2 > \varepsilon_0} |\mathbf{y}|^{2\boldsymbol{\alpha}}|\hat{u}(\mathbf{y})|^2\,dy$$

$$\leqslant \varepsilon_0^j\int_{\mathbf{R}^n} |\hat{u}(\mathbf{y})|^2\,dy + \varepsilon_0^{j-m}\int_{|\mathbf{y}|^2 > \varepsilon_0} (1 + |\mathbf{y}|^2)^m |\hat{u}(\mathbf{y})|^2\,dy$$

where $|\mathbf{y}|^2 = \sum_{i=1}^n y_i^2$, and $|\boldsymbol{\alpha}| = j \leqslant m$. Here we have used the fact that

$\varepsilon_0^j = \varepsilon_0^m \varepsilon_0^{j-m} \leqslant \varepsilon_0^{j-m} |\mathbf{y}|^{2m} \leqslant \varepsilon_0^{j-m}(1+|\mathbf{y}|^2)^m$ for $|\mathbf{y}|^2 > \varepsilon_0$. Then we have

$$\sum_{|\alpha| \leqslant m-1} \|D^\alpha u\|^2_{H^0(\Omega)} \leqslant \sum_{|\alpha| \leqslant m-1} \varepsilon_0^{|\alpha|} \|u\|^2_{H^0(\Omega)} + \sum_{|\alpha| \leqslant m-1} \varepsilon_0^{|\alpha|-m} \|u\|^2_{H^m(\Omega)}$$

That is, for arbitrary $\varepsilon > 0$,

$$\|u\|^2_{H^{m-1}(\Omega)} \leqslant c(\varepsilon)\|u\|^2_{H^0(\Omega)} + \varepsilon \|u\|^2_{H^m(\Omega)} \tag{7.46}$$

where $c(\varepsilon)$ is constant >0 independent of u but dependent on ε.

Next, using property (2.81) of Fourier transforms and the ellipticity condition (5.8), we have

$$\begin{aligned}
\sum_{|\alpha|=|\beta|=m} \left(D^\alpha u, a_{\alpha\beta} D^\beta u\right)_{H^0(\Omega)} &= \int_{\mathbf{R}^n} \sum_{|\alpha|=|\beta|=m} a_{\alpha\beta} \mathbf{y}^\alpha \mathbf{y}^\beta |\hat{u}(\mathbf{y})|^2 \, dy \\
&\geqslant \mu_0 \int_{\mathbf{R}^n} |\mathbf{y}|^{2m} |\hat{u}(\mathbf{y})|^2 \, dy \\
&= \mu_0 \sum_{|\alpha|=m} (D^\alpha u, D^\alpha u)_{H^0(\Omega)} \\
&= \mu_0 |u|^2_{H^m(\Omega)}
\end{aligned}$$

We easily verify that the bilincar form $B(u,v)$ satisfies

$$\begin{aligned}
B(u,u) = &\sum_{|\alpha|=|\beta|=m} \left(D^\alpha u, a_{\alpha\beta} D^\beta u\right)_{H^0(\Omega)} \\
&+ \sum_{\substack{|\alpha|,|\beta| \leqslant m \\ |\alpha|+|\beta|<2m}} \left(D^\alpha u, a_{\alpha\beta} D^\beta u\right)_{H^0(\Omega)}
\end{aligned}$$

Now if $|\alpha| \leqslant m$ and $|\beta| \leqslant m-1$, it follows immediately from the Schwarz inequality that

$$\begin{aligned}
\left| \int_\Omega D^\alpha u D^\beta v \, dx \right| &\leqslant \|D^\alpha u\|_{H^0(\Omega)} \|D^\beta v\|_{H^0(\Omega)} \\
&\leqslant \|u\|_{H^m(\Omega)} \|v\|_{H^{m-1}(\Omega)} \qquad \forall u, v \in C_0^\infty(\Omega)
\end{aligned}$$

Thus

$$\left| \sum_{\substack{|\alpha|,|\beta| \leqslant m \\ |\alpha|+|\beta|<2m}} \left(D^\alpha u, a_{\alpha\beta} D^\beta u\right)_{H^0(\Omega)} \right| \leqslant K \|u\|_{H^m(\Omega)} \|u\|_{H^{m-1}(\Omega)}$$

where K is a positive constant depending on m, Ω, and $a_{\alpha\beta}$ but not u. Therefore, for $u \in C_0^\infty(\Omega)$,

$$B(u,u) \geqslant \sum_{|\alpha|=|\beta|=m} \left(D^\alpha u, a_{\alpha\beta} D^\beta u\right)_{H^0(\Omega)} - K\|u\|_{H^m(\Omega)}\|u\|_{H^{m-1}(\Omega)}$$

Hence

$$B(u,u) \geqslant \mu_0 |u|^2_{H^m(\Omega)} - K\|u\|_{H^m(\Omega)}\|u\|_{H^{m-1}(\Omega)}$$

Observing that $\|u\|^2_{H^m(\Omega)} = \|u\|^2_{H^{m-1}(\Omega)} + |u|^2_{H^m(\Omega)}$ and using the elementary inequality

$$ab \leqslant \frac{1}{4\varepsilon} a^2 + \varepsilon b^2 \qquad \forall \varepsilon > 0 \qquad a, b \in \mathbf{R} \tag{7.47}$$

we have, for some $\varepsilon_1 > 0$,

$$B(u,u) \geqslant \left(\mu_0 - \frac{K}{4\varepsilon_1}\right)\|u\|^2_{H^m(\Omega)} - (\mu_0 + K\varepsilon_1)\|u\|^2_{H^{m-1}(\Omega)}$$

Finally, introducing inequality (7.46) into this last result, we obtain

$$B(u,u) \geqslant \mu_1 \|u\|^2_{H^m(\Omega)} - \lambda_0 \|u\|^2_{H^0(\Omega)} \tag{7.48}$$

wherein $\mu_1 = \mu_0 - K/4\varepsilon_1 - \mu_0\varepsilon - K\varepsilon_1\varepsilon$ and $\lambda_0 = c(\varepsilon)(\mu_0 + K\varepsilon_1)$ and we choose ε and ε_1 so that $\mu_1, \lambda_0 > 0$.

This completes the proof for the case $u \in C_0^\infty(\Omega)$.

The proof now follows from continuity: e.g., let $u \in H_0^m(\Omega)$. Since $C_0^\infty(\Omega)$ is dense in $H_0^m(\Omega)$, there exists a sequence $\{u_k\}$ in $C_0^\infty(\Omega)$ such that $\|u - u_k\|_{H^m(\Omega)} \to 0$ as $k \to \infty, \forall u \in H_0^m(\Omega)$. Since

$$\left|\|u_k\|_{H^0(\Omega)} - \|u\|_{H^0(\Omega)}\right| \leqslant \|u_k - u\|_{H^0(\Omega)} \leqslant \|u_k - u\|_{H^m(\Omega)}$$

and

$$\left|\|u_k\|_{H^m(\Omega)} - \|u\|_{H^m(\Omega)}\right| \leqslant \|u_k - u\|_{H^m(\Omega)}$$

it follows that $\|u_k\|_{H^m(\Omega)} \to \|u\|_{H^m(\Omega)}$ and $\|u_k\|_{H^0(\Omega)} \to \|u\|_{H^0(\Omega)}$ as $k \to \infty$. Moreover, $B(u,v)$ is continuous on $H^m(\Omega)$; i.e. there is an $M > 0$ such that $B(u,v) \leqslant M\|u\|_{H^m(\Omega)}\|v\|_{H^m(\Omega)}$. Thus,

$$\begin{aligned} |B(u_k,u_k) - B(u,u)| &= |B(u_k - u, u_k) + B(u_k - u, u)| \\ &\leqslant M(\|u_k\|_{H^m(\Omega)} + \|u\|_{H^m(\Omega)})\|u_k - u\|_{H^m(\Omega)} \end{aligned}$$

Since the quantity in parentheses is bounded, $B(u_k, u_k) \to B(u,u)$ as $k \to \infty$. Hence (7.48) holds for $u \in H_0^m(\Omega)$. ■

Results similar to those in Theorem 7.3 can be obtained when we enlarge the definition of $B(u,v)$ to include boundary conditions. The following theorem provides an example of such extensions for second-order problems (cf. [7.5]).

THEOREM 7.4. Consider the bilinear form $B(u,v)$ of (7.36) which is associated with the operators A and B_0 of (7.35), the coefficients of which are infinitely differentiable. Let A be strongly elliptic; i.e., let there exist a constant $\mu_0 > 0$ such that

$$\sum_{i,j=1}^{2} a_{ij}\xi_i\xi_j \geqslant \mu_0|\xi|^2 \tag{7.49}$$

Then $B(u,v)$ is $[H^1(\Omega), H^0(\Omega)]$-coercive, i.e., there exists a constant $\omega_0 > 0$ such that

$$Q(u,u) \geqslant c\|u\|^2_{H^1(\Omega)} \tag{7.50}$$

where $c > 0$ and

$$\begin{aligned} Q(u,u) &= \int_\Omega \left(\sum_{i,j=1}^{2} a_{ij} \frac{\partial u}{\partial x_i}\frac{\partial u}{\partial x_j} + \sum_{i=1}^{2} b_i \frac{\partial u}{\partial x_i} u + \omega_0 u^2 \right) dx \\ &\quad + \oint_{\partial\Omega} \left(\alpha_1 \frac{\partial u}{\partial s} u + \beta_1 u^2 \right) ds \\ &= B(u,u) + \omega_0\|u\|^2_{H^0(\Omega)} - \int_\Omega c_0 u^2\, dx \end{aligned} \tag{7.51}$$

PROOF. First note that

$$\begin{aligned} \sum_{i=1}^{2} \int_\Omega b_i \frac{\partial u}{\partial x_i} u\, dx &\leqslant \left(\sum_{i=1}^{2} \left\| b_i \frac{\partial u}{\partial x_i} \right\|_{H^0(\Omega)} \right) \|u\|_{H^0(\Omega)} \\ &\leqslant c_1 \|u\|_{H^1(\Omega)} \|u\|_{H^0(\Omega)} \\ &\leqslant \frac{c_1}{4\varepsilon} \|u\|^2_{H^1(\Omega)} + c_1\varepsilon \|u\|^2_{H^0(\Omega)} \end{aligned}$$

where c_1 and ε are > 0, ε being arbitrary. In the last step we made use of the elementary inequality (7.47).

In addition,

$$\left|\oint_{\partial\Omega}\left(\alpha_1 \frac{\partial u}{\partial s} u+\beta_1 u^2\right) ds\right|=\left|\oint_{\partial\Omega}\left(\beta_1-\frac{1}{2} \frac{d\alpha_1}{ds}\right) u^2 ds\right| \leqslant c_2\|u\|_{H^0(\partial\Omega)}^2$$

Using the Sobolev embedding theorem and the trace theorem, we have

$$\begin{aligned}\left|\int_{\partial\Omega}\left(\alpha_1 \frac{\partial u}{\partial s} u+\beta_1 u^2\right) ds\right| &\leqslant c_2 c_3\|u\|_{H^{1/4}(\partial\Omega)}^2\\ &\leqslant c_2 c_3 c_4\|u\|_{H^{3/4}(\Omega)}^2\\ &\leqslant c_5(\varepsilon)\|u\|_{H^0(\Omega)}^2+c_6(\varepsilon)\|u\|_{H^1(\Omega)}^2\end{aligned}$$

where the magnitude of $c_5(\varepsilon)$, $c_6(\varepsilon)$ depends on the choice of an arbitrary $\varepsilon>0$, and in the last step we used a well-known interpolation property of Hilbert spaces (see, e.g. Lions [7.10]).

It follows that

$$\begin{aligned}Q(u,u) \geqslant &\int_{\Omega}\left(\sum_{i,j} a_{ij} \frac{\partial u}{\partial x_i} \frac{\partial u}{\partial x_j}+\omega_0 u^2\right) dx-\left[\frac{c_1}{4\varepsilon}+c_6(\varepsilon)\right]\|u\|_{H^1(\Omega)}^2\\ &-\left[c_1\varepsilon+c_5(\varepsilon)\right]\|u\|_{H^0(\Omega)}^2\\ \geqslant &\left[\mu_0-\frac{c_1}{4\varepsilon}-c_6(\varepsilon)\right]\|u\|_{H^1(\Omega)}^2+\left[\omega_0-\mu_0-c_1\varepsilon-c_5(\varepsilon)\right]\|u\|_{H^0(\Omega)}^2\end{aligned}$$

Now choose ε so that $\mu_0-c_1/4\varepsilon-c_6(\varepsilon)=c>0$ and, for this choice of ε, take $\omega_0-\mu_0-c_1\varepsilon-c_5(\varepsilon)\geqslant 0$. Then (7.50) immediately follows. ■

While strong ellipticity is sufficient for $[H_0^m(\Omega), H^0(\Omega)]$-coercivity for higher-order problems, we do not have, in general, $[H^m(\Omega), H^0(\Omega)]$-coerciveness. The next example illustrates this fact.

EXAMPLE 7.9. Consider the biharmonic operator in two dimensions:

$$\Delta^2=\Delta\Delta=\sum_{i,j}^{2} \frac{\partial^2}{\partial x_i^2}\left(\frac{\partial^2}{\partial x_j^2}\right)$$

We choose the natural bilinear form (there are infinitely many bilinear forms associated with the operator Δ^2) given by

$$B_1(u,v)=\int_{\Omega} \sum_{i,j}^{2} \frac{\partial^2 u}{\partial x_i^2} \frac{\partial^2 u}{\partial x_j^2}\, dx \tag{7.52}$$

which comes from Green's formula

$$\int_{\Omega}\sum_{i,j}^{2}\frac{\partial^4 u}{\partial x_i^2\,\partial x_j^2}v\,dx=\int_{\Omega}\sum_{i,j}^{2}\frac{\partial^2 u}{\partial x_i^2}\,\frac{\partial^2 v}{\partial x_j^2}\,dx+\oint_{\partial\Omega}\left[v\frac{\partial}{\partial n}(\Delta u)-\Delta u\frac{\partial v}{\partial n}\right]ds$$

The associated homogeneous boundary conditions are, clearly,

$$\Delta u=0 \qquad \frac{\partial}{\partial n}(\Delta u)=0 \text{ on } \partial\Omega$$

The boundary-value problem $\Delta^2 u=f$, with the above boundary conditions, is not regularly elliptic. The bilinear form $B_1(u,v)$ is not $[H^2(\Omega),H^0(\Omega)]$-coercive, since $B_1(u,u)$ vanishes for any function harmonic in Ω. ■

A fairly large literature exists on various conditions for coerciveness of bilinear forms associated with higher-order problems. As representative, we mention some of the results of Aronszajn [7.11] and Smith [7.12] on coerciveness. The first result follows from the use of the Gårding inequality.

THEOREM 7.5. ([7.11].) Let Ω be a bounded domain of $\mathbf{R}^n$, let $P_1(\xi),P_2(\xi),\dots,P_N(\xi)$ be homogeneous polynomials of degree m having constant coefficients, and define

$$B(u,v)=\sum_{k=1}^{N}\int_{\Omega}P_k(\mathbf{x},D)uP_k(\mathbf{x},D)v\,dx \tag{7.53}$$

for $u,v\in H^m(\Omega)$, $P_k(\mathbf{x},D)$ being differential operators of order $k\leqslant m$. Then a necessary and sufficient condition for $B(u,v)$ to be $[H_0^m(\Omega),H^0(\Omega)]$-coercive is that there be no common nonzero real root of $P_1(\xi),\dots,P_N(\xi)$. Furthermore, if Ω satisfies the cone condition, a necessary condition that $B(u,v)$ be $[H^m(\Omega),H^0(\Omega)]$-coercive is that there is no common nonzero complex root of $P_1(\xi),\dots,P_N(\xi)$. ■

THEOREM 7.6. (Cf. [7.12].) Let $P_k(\xi),k=1,2,\dots,N,$ be homogeneous polynomials of degree m having constant coefficients and having no common nonzero complex root. Then the bilinear form (7.53) is $[H^m(\Omega),H^0(\Omega)]$-coercive. Moreover, there exists a $c>0$ such that

$$\|u\|_{H^m(\Omega)}\leqslant c\left[\sum_{k=1}^{N}\|P_k(\mathbf{x},D)u\|_{H^0(\Omega)}+\|u\|_{H^0(\Omega)}\right]^{1/2} \tag{7.54}$$

$u\in H^m(\Omega)$, where Ω is a open bounded domain satisfying the cone condition. ■

EXAMPLE 7.10. Consider the biharmonic operator of Example 7.9 with the bilinear form

$$B_2(u,v)=\int_\Omega \hat{\nabla}^2 u \hat{\nabla}^2 v\, dx + 4\int_\Omega \frac{\partial^2 u}{\partial x_1 \partial x_2}\cdot\frac{\partial^2 v}{\partial x_1 \partial x_2}\, dx$$

where $\hat{\nabla}^2=(\partial^2/\partial x_1^2)-(\partial^2/\partial x_2^2)$. It can be shown that $B_2(u,v)$ leads to a regularly elliptic boundary-value problem. We have seen that the bilinear form $B_1(u,v)$ in Example 7.9 is is not $[H^2(\Omega), H^0(\Omega)]$-coercive. However, the bilinear form $B_2(u,v)$ *is* $[H^2(\Omega), H^0(\Omega)]$-coercive, since the polynomials $\xi_1^2-\xi_2^2$ and $2\xi_1\xi_2$ have no common complex roots, except $\xi_1=\xi_2\neq 0$.

In general, it can be shown (see [7.13]) that the bilinear form

$$B(u,v)=\theta B_1(u,v)+(1-\theta)B_2(u,v) \qquad u,v\in H^2(\Omega)$$

where θ is a real number, is $[H^2(\Omega), H^0(\Omega)]$-coercive if and only if $-3<\theta<1$. ■

7.4 WEAK COERCIVENESS

Important extensions of the theory of variational boundary-value problems can be made by introducing a notion of coerciveness more general than those defined in (7.43) and (7.44). Let $\mathcal{U}$ and $\mathcal{V}$ denote two Hilbert spaces and let $B(u,v), u\in\mathcal{U}$ and $v\in\mathcal{V}$, denote a bilinear form defining a mapping of $\mathcal{U}\times\mathcal{V}$ into $\mathbf{R}$. Then we refer to $B(\cdot,\cdot)$ as *weakly coercive* if there exists a constant γ such that

$$\inf_{\|u\|_{\mathcal{U}}=1}\ \sup_{\|v\|_{\mathcal{V}}\leqslant 1} |B(u,v)| \geqslant \gamma > 0 \tag{7.55}$$

$\forall u\in\mathcal{U}$ and $\forall v\in\mathcal{V}$, and if

$$\sup_{u\in\mathcal{U}} |B(u,v)|>0 \qquad v\neq 0 \tag{7.56}$$

where $\|\cdot\|_{\mathcal{U}}$ and $\|\cdot\|_{\mathcal{V}}$ are the norms on $\mathcal{U}$ and $\mathcal{V}$, respectively. Bilinear forms with properties (7.55) and (7.56) are mentioned in Nečas [7.14], but their use in broad classes of variational boundary-value problems was demonstrated by Babuška [7.15,7.16] and Babuška and Aziz [7.5].

REMARKS. 1. It is easily verified that, whenever the stronger coerciveness (7.43) holds, for the symmetric forms, then so also do (7.55) and (7.56); i.e., (7.43) also implies weak coerciveness. To see this, simply set

$\mathcal{U} = \mathcal{V} = \mathcal{W}$ and assume that (7.43) holds. Then

$$\sup_{\|v\|_{\mathcal{W}} \leqslant 1} |B(u,v)| \geqslant \sup_{v \in \mathcal{W}} \frac{|B(u,v)|}{\|v\|_{\mathcal{W}}} \geqslant \frac{|B(u,u)|}{\|u\|_{\mathcal{W}}} \geqslant \gamma \|u\|_{\mathcal{W}}$$

By taking the infimum over those u such that $\|u\|_{\mathcal{U}} = 1$ we obtain (7.55) for the case $\mathcal{U} = \mathcal{V} = \mathcal{W}$. Inequality (7.56) follows from (7.55) and the assumed symmetry of $B(u,v)$.

2. Notice also that, whenever u is an arbitrary element of $\mathcal{U}$, the element $z = u/\|u\|_{\mathcal{U}}$ has a unit norm. For weakly coercive forms, we have

$$0 < \gamma \leqslant \inf_{\|u\|_{\mathcal{U}} = 1} \sup_{\|v\|_{\mathcal{V}} \leqslant 1} |B(u,v)| \leqslant \sup_{\|v\|_{\mathcal{V}} \leqslant 1} \frac{|B(u,v)|}{\|u\|_{\mathcal{U}}}$$

Thus, when (7.55) holds,

$$\sup_{\|v\|_{\mathcal{V}} \leqslant 1} |B(u,v)| \geqslant \gamma \|u\|_{\mathcal{U}} \qquad \forall u \in \mathcal{U} \tag{7.57}$$

3. We also remark that (7.56) can be replaced by the dual of (7.55); i.e., whenever (7.55) and (7.56) hold, there is a $C > 0$ such that

$$\inf_{\|v\|_{\mathcal{V}} = 1} \sup_{\|u\|_{\mathcal{U}} \leqslant 1} |B(u,v)| \geqslant C > 0 \tag{7.58}$$

(see Babuška and Aziz [7.5, p. 117]). ■

An important example of a weakly coercive form is that in (7.36) corresponding to the general second-order elliptic problem (7.34). In this case $\mathcal{U} = \mathcal{V} = H^1(\Omega)$. According to Theorem 7.4, $B(u,v)$ is strongly coercive [i.e., coercive in the sense of (7.43)] only for special choices of the coefficient $c_0(\mathbf{x})$. We now show by means of a theorem of Babuška and Aziz [7.5] that $B(u,v)$ is weakly coercive, assuming only that c_0 is sufficiently smooth.

THEOREM 7.7. Let the conditions of Theorem 7.4 hold. Then the bilinear form $B(u,v)$ of (7.36) satisfies conditions (7.55) and (7.56) for $\mathcal{U} = \mathcal{V} = H^1(\Omega)$.

PROOF. Following [7.5], the plan is to construct a special element $w \in H^1(\Omega)$ such that

$$B(u,v) \geqslant c\|u\|_{H^1(\Omega)} \|v\|_{H^1(\Omega)}$$

for some $c > 0$, where $v = u + w$.

In view of (7.51), for any $v = u + w$,

$$B(u,v) = B(u,w) + Q(u,u) + \int_\Omega (c_0 - \omega_0)u^2\,dx$$

where $Q(u,u)$ is defined in (7.51). We next choose w to be the solution of the special boundary-value problem

$$A^* w = f^* \qquad f^* \equiv -(c_0 - \omega_0)u \qquad \text{in } \Omega;\ B_0^* u = 0 \text{ on } \partial\Omega$$

where (A^*, B_0^*) corresponds to the adjoint problem to (7.34). This problem has a solution w_0, and in fact

$$\|w_0\|_{H^3(\Omega)} \leqslant \hat{c}\|(c_0 - \omega_0)u\|_{H^1(\Omega)} \leqslant c'\|u\|_{H^1(\Omega)}$$

We find

$$B(u,w) = -\int_\Omega (c_0 - \omega_0)u^2\,dx$$

Thus, for $v = u + w_0$,

$$B(u,v) = Q(u,u) \geqslant c\|u\|^2_{H^1(\Omega)}$$

From this inequality and the fact that

$$\|v\|_{H^1(\Omega)} = \|u + w_0\|_{H^1(\Omega)} \leqslant (1 + c')\|u\|_{H^1(\Omega)}$$

it follows that (7.55) holds. Inequality (7.56) then follows by reversing the roles of u and v and using similar arguments. ■

Babuška's Theory of Lagrange Multipliers

Another interesting example of a variational boundary-value problem involving weakly coercive bilinear forms was developed by Babuška [7.15] and involves the treatment of nonhomogeneous boundary conditions as constraints through the introduction of Lagrange multipliers. We summarize some of Babuška's theory for the model nonhomogeneous boundary-value problem

$$\left.\begin{aligned} -\Delta u + u &= f \quad \text{in } \Omega \\ u &= g \quad \text{on } \partial\Omega \end{aligned}\right\} \tag{7.59}$$

where $f \in H^0(\Omega)$ and $g \in H^{1/2}(\partial\Omega)$. Now in conventional variational methods associated with (7.59), we seek the critical point u of functionals such as

$$I(v) = \int_\Omega (\nabla v \cdot \nabla v + v^2 - 2fv)\, dx$$

from the class (linear manifold) of functions satisfying the boundary condition. Such restrictions on the variations of $I(v)$ are impractical when we come to questions of approximation.

A natural way to overcome such restrictions is to treat the essential boundary conditions as constraints. This means that, instead of $I(v)$, we use the functional

$$I(v,\mu) = \int_\Omega (\nabla v \cdot \nabla v + v^2 - 2fv)\, dx - \oint_{\partial\Omega} 2\mu(v-g)\, ds$$

A simple calculation reveals that μ can be associated with $\partial v / \partial n$. The associated bilinear form is therefore

$$B((u,\lambda),(v,\mu)) \equiv \int_\Omega (\nabla u \cdot \nabla v + uv)\, dx - \oint_{\partial\Omega} (\lambda v + \mu u)\, ds \tag{7.60}$$

In other words, $B(\cdot,\cdot)$ maps pairs of ordered pairs (v,μ) into $\mathbb{R}$, with $v \in H^1(\Omega)$; and since $\gamma_0 v \in H^{1/2}(\partial\Omega)$, $\mu \in H^{-1/2}(\partial\Omega)$. It follows that in this particular example $B(\cdot,\cdot)$ is a symmetric bilinear form on the space

$$\mathcal{W} = H^1(\Omega) \times H^{-1/2}(\partial\Omega) \tag{7.61}$$

We introduce the following norm on $\mathcal{W}$:

$$\|(v,\mu)\|^2_{\mathcal{W}} = \|v\|^2_{H^1(\Omega)} + \|\mu\|^2_{H^{-1/2}(\partial\Omega)} \tag{7.62}$$

The variational problem is to find $(u,\lambda) \in \mathcal{W}$ such that

$$B((u,\lambda),(v,\mu)) = \int_\Omega fv\, dx - \oint_{\partial\Omega} g\mu\, ds \qquad \forall (v,\mu) \in \mathcal{W} \tag{7.63}$$

We now show that $B(\cdot,\cdot)$ is weakly coercive, i.e., (7.55) and (7.56) are satisfied.

Now let $(u,\lambda) \in \mathcal{W}$ be given, and let $w \in H^1(\Omega)$ be the solution of the Neumann problem

$$-\nabla^2 w + w = 0 \text{ in } \Omega, \qquad \frac{\partial w}{\partial n} = \lambda \qquad \text{on } \partial\Omega$$

The solution $w \in H^1(\Omega)$ is uniquely characterized by the variational problem

$$\int_\Omega (\nabla w \cdot \nabla v + wv)\, dx = \oint_{\partial\Omega} \lambda v\, ds \qquad \forall v \in H^1(\Omega) \tag{7.64}$$

We have from (7.64) with $v = w$,

$$\|w\|^2_{H^1(\Omega)} = \oint_{\partial\Omega} \lambda w\, ds \tag{7.65}$$

According to Corollary 4.17.1, there exists an inverse mapping ϕ of $H^{1/2}(\partial\Omega)$ into $H^1(\Omega)$ such that $\phi(w) = w$ on $\partial\Omega$ and

$$\|\phi(w)\|_{H^1(\Omega)} \leqslant c\|w\|_{H^{1/2}(\partial\Omega)}$$

Thus, if we replace v by $\phi(v)$, $v \in H^{1/2}(\partial\Omega)$, in (7.64) we obtain

$$(w, \phi(v))_{H^1(\Omega)} = \int_{\partial\Omega} \frac{\partial w}{\partial n} \phi(v)\, ds = \oint_{\partial\Omega} \frac{\partial w}{\partial n} v\, ds$$

Clearly,

$$|(w, \phi(v))_{H^1(\Omega)}| \leqslant \|w\|_{H^1(\Omega)} \|\phi(v)\|_{H^1(\Omega)}$$

or

$$\left| \oint_{\partial\Omega} \frac{\partial w}{\partial n} v\, ds \right| \leqslant c\|w\|_{H^1(\Omega)} \|v\|_{H^{1/2}(\partial\Omega)}$$

Hence

$$\begin{aligned} \|\lambda\|_{H^{-1/2}(\partial\Omega)} &= \sup_{v \in H^{1/2}(\partial\Omega)} \frac{\left| \oint_{\partial\Omega} \lambda v\, ds \right|}{\|v\|_{H^{1/2}(\partial\Omega)}} \\ &= \sup_{v \in H^{1/2}(\partial\Omega)} \frac{|(w, \phi(v))_{H^1(\Omega)}|}{\|v\|_{H^{1/2}(\partial\Omega)}} \\ &= \sup_{v \in H^{1/2}(\partial\Omega)} \frac{\left| \oint_{\partial\Omega} (\partial w / \partial n) v\, ds \right|}{\|v\|_{H^{1/2}(\partial\Omega)}} \qquad v \neq 0 \end{aligned}$$

or

$$\|\lambda\|_{H^{-1/2}(\partial\Omega)} \leqslant c\|w\|_{H^1(\Omega)} \tag{7.66}$$

Equation (7.64) then leads to the inequality

$$\|\lambda\|^2_{H^{-1/2}(\partial\Omega)} \leqslant c' \oint_{\partial\Omega} \lambda w\, ds \tag{7.67}$$

Let $v = u - w$, and $\mu = -2\lambda$. Then

$$\begin{aligned} \|(v,\mu)\|^2_{\mathscr{W}} &= \|u-w\|^2_{H^1(\Omega)} + \|-2\lambda\|^2_{H^{-1/2}(\partial\Omega)} \\ &\leqslant 2\|u\|^2_{H^1(\Omega)} + 2\|w\|^2_{H^1(\Omega)} + 4\|\lambda\|^2_{H^{-1/2}(\partial\Omega)} \\ &\leqslant c''\|(u,\lambda)\|^2_{\mathscr{W}} \end{aligned} \tag{7.68}$$

where we have used inequality (7.66). Recalling (7.62) and setting $v = u - w$ and $\mu = -2\lambda$ in (7.60) gives

$$B((u,\lambda),(u-v,-2\lambda)) = \|u\|^2_{H^1(\Omega)} + \oint_{\partial\Omega} \lambda w\, ds$$

Then using (7.67), we have

$$\begin{aligned} B((u,\lambda),(v,\mu)) &\geqslant \|u\|^2_{H^1(\Omega)} + \frac{1}{c'}\|\lambda\|^2_{H^{-1/2}(\partial\Omega)} \\ &\geqslant C\|(u,\lambda)\|^2_{\mathscr{W}} \end{aligned}$$

and, in view of (7.67), for this special choice of (v,μ),

$$|B((u,\lambda),(v,\mu))| \geqslant (C/c'')\|(u,\lambda)\|_{\mathscr{W}}\|(v,\mu)\|_{\mathscr{W}}$$

Thus

$$\sup_{\substack{(v,\mu)\in\mathscr{W} \\ \|(v,\mu)\|_{\mathscr{W}}\leqslant 1}} |B((u,\lambda),(v,\mu))| \geqslant (C/c'')\|(u,\lambda)\|_{\mathscr{W}}$$

from which (7.55) immediately follows. Condition (7.56) then follows from the symmetry of $B(u,v)$.

7.5 EXISTENCE AND UNIQUENESS OF SOLUTIONS

In the preceding sections we established several properties of bilinear forms that appear in a variety of elliptic varational boundary-value problems. It is natural to inquire whether or not solutions to such variational problems exist and, if so, whether they are unique? More specifically, what conditions on the associated bilinear forms are sufficient to guarantee the existence of solutions? These questions were first addressed by Lax and Milgram for strongly coercive bilinear forms, and their work led to the well-known Lax-Milgram theorem [7.17]. An extension of this theorem to weakly coercive forms was presented by Babuška [7.15] (see also [7.5]). In the present section we present a detailed proof of the extended theorem following [7.5], and we cite several examples.

THEOREM 7.8 (The Generalized Lax-Milgram Theorem [7.5]**).** Let $\mathcal{U}$ and $\mathcal{V}$ be two real Hilbert spaces with the inner produces $(\cdot,\cdot)_{\mathcal{U}}$ and $(\cdot,\cdot)_{\mathcal{V}}$, respectively, and let $B(u,v)$ $(u\in\mathcal{U}, v\in\mathcal{V})$ be a continuous, weakly coercive, bilinear form on $\mathcal{U}\times\mathcal{V}$; i.e., let $B(u,v)$ be such that the following conditions hold:

$$\text{(i)} \qquad |B(u,v)| \leqslant M\|u\|_{\mathcal{U}}\|v\|_{\mathcal{V}} \qquad \forall u\in\mathcal{U} \text{ and } \forall v\in\mathcal{V} \tag{7.69}$$

$$\text{(ii)} \qquad \inf_{\substack{u\in\mathcal{U}\\ \|u\|_{\mathcal{U}}=1}} \sup_{\substack{v\in\mathcal{V}\\ \|v\|_{\mathcal{V}}\leqslant 1}} |B(u,v)| \geqslant \gamma > 0 \tag{7.70}$$

$$\text{(iii)} \qquad \sup_{u\in\mathcal{U}} |B(u,v)| > 0 \qquad v\neq 0 \tag{7.71}$$

where M and γ are finite positive constants. Then, if l is a continuous linear functional on $\mathcal{V}$ (i.e., $l\in\mathcal{V}'$), there exists a unique element $u^*\in\mathcal{U}$ such that

$$B(u^*,v) = l(v) \qquad \forall v\in\mathcal{V} \tag{7.72}$$

and

$$\|u^*\|_{\mathcal{U}} \leqslant \frac{1}{\gamma}\|l\|_{\mathcal{V}}^* \tag{7.73}$$

Here $\|\cdot\|_{\mathcal{U}}$ and $\|\cdot\|_{\mathcal{V}}$ denote the norms in $\mathcal{U}$ and $\mathcal{V}$, respectively, and $\|\cdot\|_{\mathcal{V}}^*$ denotes the norm in the dual space $\mathcal{V}'$.

PROOF. By (i) and the Riesz representation theorem, we can, for each fixed $u\in\mathcal{U}$, use $B(u,v)$ to define a continuous linear functional F_u on $\mathcal{V}$ such that

$$F_u(v) = B(u,v)$$

and

$$\|F_u\|^*_{\mathcal{V}} = \sup_{\substack{v\in\mathcal{V}\\ \|v\|_{\mathcal{V}}\leqslant 1}} |B(u,v)| \leqslant M\|u\|_{\mathcal{U}} \tag{7.74}$$

Again, by the Riesz representation theorem, there exists a unique element $z\in\mathcal{V}$ such that $F_u(v)=(z(u),v)_{\mathcal{V}}$ for every $v\in\mathcal{V}$. Now this element z depends linearly on the choice of u, so that we can write $z=\Lambda u$, where Λ is a linear map from $\mathcal{U}$ into $\mathcal{V}$. Summarizing, we have

$$(\Lambda u,v)_{\mathcal{V}} = F_u(v) = B(u,v) \tag{7.75}$$

Clearly, the operator so defined is linear and continuous. Indeed, the continuity follows from (7.69) by choosing $v=\Lambda u$:

$$\|\Lambda u\|^2_{\mathcal{V}} = B(u,\Lambda u) \leqslant M\|u\|_{\mathcal{U}}\|\Lambda u\|_{\mathcal{V}}$$

or

$$\|\Lambda u\|_{\mathcal{V}} \leqslant M\|u\|_{\mathcal{U}} \tag{7.76}$$

To show that Λ has a continuous inverse on $\mathcal{V}$ we first show that Λ is bounded below, i.e.,

$$\|\Lambda u\|_{\mathcal{V}} \geqslant \gamma\|u\|_{\mathcal{U}} \tag{7.77}$$

and, second, that the range of Λ is dense in $\mathcal{V}$; in fact,

$$\mathcal{R}(\Lambda) = \mathcal{V} \tag{7.78}$$

Indeed, that Λ is bounded below follows from (7.75) and (7.70), when $v\in\mathcal{V}$,

$$\|\Lambda u\|_{\mathcal{V}} = \|F_u\|^*_{\mathcal{V}} = \sup_{\|v\|_{\mathcal{V}}\leqslant 1} |(\Lambda u,v)_{\mathcal{V}}| = \sup_{\|v\|_{\mathcal{V}}\leqslant 1} |B(u,v)| \geqslant \gamma\|u\|_{\mathcal{U}}$$

Thus (7.77) holds.

To establish (7.78), we first prove that $\mathcal{R}(\Lambda)$ is a closed set in $\mathcal{V}$. Let $\{\Lambda u_n\}$ be a Cauchy sequence in $\mathcal{V}$. Then we have from (7.77) and the linearity of Λ that

$$\begin{aligned} 0 &= \lim_{m,n\to\infty} \|\Lambda u_n - \Lambda u_m\|_{\mathcal{V}} = \lim_{m,n\to\infty} \|\Lambda(u_n-u_m)\|_{\mathcal{V}} \\ &\geqslant \gamma \lim_{m,n\to\infty} \|u_n-u_m\|_{\mathcal{U}} \end{aligned}$$

Hence $\{u_n\}$ is a Cauchy sequence, and since Λ is continuous, $\mathcal{R}(\Lambda)$ is closed.

Now suppose that $\mathcal{R}(\Lambda) \neq \mathcal{V}$. Since $\mathcal{R}(\Lambda)$ is a closed subspace of $\mathcal{V}$, there could then exist a nonzero element $v^* \in \mathcal{V}, v^* \in \mathcal{R}(\Lambda)^{\perp}$ such that

$$(v, v^*)_{\mathcal{V}} = 0 \qquad \text{for all } v \in \mathcal{R}(\Lambda)$$

or

$$(\Lambda u, v^*)_{\mathcal{V}} = 0 \qquad \text{for all } u \in \mathcal{U}$$

In view of (7.75), this implies that $B(u, v^*) = 0$ for all $u \in \mathcal{U}$. However, this is a contradiction of (7.71). Hence $\mathcal{R}(\Lambda) = \mathcal{V}$.

Thus Λ^{-1} exists, and it is linear and continuous:

$$\|\Lambda^{-1}u\|_{\mathcal{V}} \leqslant \frac{1}{\gamma}\|u\|_{\mathcal{U}} \qquad \text{and} \qquad \|\Lambda^{-1}\|_{\mathcal{L}(\mathcal{U},\mathcal{V})} \leqslant \frac{1}{\gamma} \tag{7.79}$$

Again using the Riesz representation theorem, let v^* be the unique element in $\mathcal{V}$ corresponding to the given linear functional $l(v)$ in (7.72). Then

$$l(v) = (v^*, v)_{\mathcal{V}} \qquad \text{and } \|l\|_{\mathcal{V}}^* = \|v^*\|_{\mathcal{V}}$$

Introducing this identity together with (7.75) into (7.72) gives

$$(\Lambda u - v^*, v)_{\mathcal{V}} = 0 \quad \forall v \in \mathcal{V}$$

Hence a solution to (7.72) does exist, and in fact is precisely

$$u^* = \Lambda^{-1}v^* \tag{7.80}$$

This completes the existence part of the theorem.

Uniqueness follows from (7.80), but it is further verified if we assume u^* is not unique, and that $u_0 \neq u^*$ is also a solution of (7.72). This implies that

$$B(u^* - u_0, v) = (\Lambda(u^* - u_0), v)_{\mathcal{V}} = 0 \qquad \forall v \in \mathcal{V}$$

By taking $v = \Lambda(u^* - u_0)$, we obtain $\|\Lambda(u^* - u_0)\|_{\mathcal{V}}^2 = 0$ or $\Lambda u^* = \Lambda u_0$. Thus $u^* = \Lambda^{-1}\Lambda u_0 = u_0$, hence u^* is unique.

Finally, we show that the solution u of (7.72) depends continuously on the data l. Simply note that

$$\|u^*\|_{\mathcal{U}} = \|\Lambda^{-1}v^*\|_{\mathcal{U}} \leqslant \|\Lambda^{-1}\|_{\mathcal{L}(\mathcal{U},\mathcal{V})}\|v^*\|_{\mathcal{V}}$$

which, in view of (7.79), reduces to (7.73). This completes the proof of the theorem. ■

REMARK. The conventional Lax-Milgram theorem applies to symmetric strongly coercive bilinear forms $[B(u,v)=B(v,u)]$ and follows from Theorem 7.8 by setting $\mathcal{U}=\mathcal{V}=\mathcal{W}$. Then conditions (7.69) through (7.71) reduce to the two conditions,

$$|B(u,v)| \leqslant M\|u\|_{\mathcal{W}}\|v\|_{\mathcal{W}} \tag{7.81}$$

$$B(u,u) \geqslant \gamma\|u\|^2_{\mathcal{W}} \tag{7.82}$$

■

EXAMPLE 7.11. Consider the problem of finding $u \in H_0^1(\Omega)$ such that

$$\int_\Omega (\nabla u \cdot \nabla v + c_0 uv)\,dx = \int_\Omega fv\,dx \qquad \forall v \in H_0^1(\Omega) \tag{7.83}$$

where $c_0=\text{constant}>0$, $f \in H^0(\Omega)$. In this case $\mathcal{U}=\mathcal{V}=H_0^1(\Omega)$. This of course is equivalent to the Dirichlet problem for the operator $-\Delta+c_0$.

Problem (7.83) has a unique solution. The bilinear form in (7.83) is strongly coercive, as can be seen from Example 7.7. We need only show that it is continuous. This is not difficult. Observe that

$$|B(u,v)|^2 = \left|\int_\Omega (\nabla u \cdot \nabla v + c_0 uv)\,dx\right|^2$$

$$\leqslant \left|\sum_{i=1}^n \left\|\frac{\partial u}{\partial x_i}\right\|_{H^0(\Omega)}\left\|\frac{\partial v}{\partial x_i}\right\|_{H^0(\Omega)} + c_0\|u\|_{H^0(\Omega)}\|v\|_{H^0(\Omega)}\right|^2 \tag{7.84}$$

We next recall the well-known version of the Schwarz inequality for $\mathbf{R}^n$:

$$\left(\sum_{i=1}^n a_i b_i\right)^2 \leqslant \left(\sum_{i=1}^n a_i^2\right)\left(\sum_{i=1}^n b_i^2\right) \tag{7.85}$$

Applying this inequality to (7.84), we have

$$|B(u,v)|^2 \leqslant \left[c_0 \|u\|^2_{H^0(\Omega)} + \sum_{i=1}^{n} \left\| \frac{\partial u}{\partial x_i} \right\|^2_{H^0(\Omega)} \right]$$

$$\times \left[c_0 \|v\|^2_{H^0(\Omega)} + \sum_{i=1}^{n} \left\| \frac{\partial v}{\partial x_i} \right\|^2_{H^0(\Omega)} \right]$$

$$\leqslant [\max(1,c_0)]^2 \|u\|^2_{H^1(\Omega)} \|v\|^2_{H^1(\Omega)}$$

Thus (7.69) holds with $\mathcal{U} = \mathcal{V} = H^1(\Omega)$, and $M = \max(1, c_0)$. ■

EXAMPLE 7.12. Now consider the Neumann problem:

$$\int_\Omega \nabla u \cdot \nabla v \, dx = \int_\Omega fv \, dx \qquad \forall v \in H^1(\Omega) \tag{7.86}$$

where $f \in H^0(\Omega)$ satisfies the compatibility condition (5.56). The conditions (7.82) and (7.69) are *not* satisfied, hence a unique solution to (7.86) may not exist. However, we can still use the results of Theorem 7.8 by regarding u as an element of the quotient space $L_2^1(\Omega) = H^1(\Omega)/\mathcal{P}_0(\Omega)$ (see Theorem 3.14), the norm on which is $|u|_{H^1(\Omega)}$ [see (3.63)]. It is then obvious that the form in (7.86) is continuous and coercive relative to $|u|_{H^1(\Omega)}$, and that therefore (7.86) possesses a solution that is unique to within an arbitrary constant. ■

EXAMPLE 7.13. The variational problem (7.38), with $B(u,v)$ given by (7.36) and subject to the conditions of Theorem 7.4, has a unique solution. It is weakly coercive on $\mathcal{U} = \mathcal{V} = H^1(\Omega)$ by virtue of Theorem 7.4. It is also continuous, since (7.69) holds with $M = \max\{\max_{ij}(\sup_{\mathbf{x}}|a_{ij}|), \max_i(\sup_{\mathbf{x}}|b_i|), \sup_{\mathbf{x}}|c_0|\}$. ■

EXAMPLE 7.14. The variational problem (7.63) with $B((u,\lambda),(v,\mu))$ given by (7.60) and $\mathcal{U} = \mathcal{V} = \mathcal{W} \equiv H^1(\Omega) \times H^{-1/2}(\partial\Omega)$ has a unique solution. We established earlier that $B(\cdot,\cdot)$ is weakly coercive on $\mathcal{W} \times \mathcal{W}$; i.e., (7.70) and (7.71) hold. To establish that (7.69) holds, note that

$$|B((u,\lambda),(v,\mu))|^2 \leqslant \left[\sum_{i=1}^{n} \left\| \frac{\partial u}{\partial x_i} \right\|_{H^0(\Omega)} \left\| \frac{\partial v}{\partial x_i} \right\|_{H^0(\Omega)} \right.$$

$$+ \|u\|_{H^0(\Omega)} \|v\|_{H^0(\Omega)} + \|v\|_{H^{1/2}(\partial\Omega)} \|\lambda\|_{H^{-1/2}(\partial\Omega)}$$

$$\left. + \|u\|_{H^{1/2}(\partial\Omega)} \|\mu\|_{H^{-1/2}(\partial\Omega)} \right]^2$$

We next use the trace theorem to write $\|v\|_{H^{1/2}(\partial\Omega)} \leqslant c\|v\|_{H^1(\Omega)}$, etc. Then expanding the right side of this inequality and using (7.85), we obtain

$$|B((u,\lambda),(v,\mu))|^2 \leqslant M^2\left(\|u\|^2_{H^0(\Omega)} + \sum_i \left\|\frac{\partial u}{\partial x_i}\right\|^2_{H^0(\Omega)} + \|\lambda\|^2_{H^{-1/2}(\partial\Omega)}\right)$$

$$\left(\|v\|^2_{H^0(\Omega)} + \sum_i \left\|\frac{\partial v}{\partial x_i}\right\|^2_{H^0(\Omega)} + \|\mu\|^2_{H^{-1/2}(\partial\Omega)}\right)$$

$$\leqslant M^2\|(u,\lambda)\|^2_{\mathcal{W}}\|(v,\mu)\|^2_{\mathcal{W}}$$

wherein $\|(\cdot,\cdot)\|_{\mathcal{W}}$ is defined in (7.62). ■

REFERENCES

7.1. Lions, J. L. and Magenes, E., *Non-Homogeneous Boundary Value Problems and Applications*, Vol. I. Translated from a revision of the 1968 French edition by P. Kenneth, Springer-Verlag, New York, 1972.

7.2. Strang, G. and Fix, G., *An Analysis of the Finite Element Method*, Prentice-Hall, Englewood Cliffs, N. J., 1973.

7.3. Aubin, J. P., *Approximation of Elliptic Boundary-Value Problems*, Wiley-Interscience, New York, 1972.

7.4. Mikhlin, S. G., *Variational Methods in Mathematical Physics*, translated from the 1957 Russian edition by T. Boddington, Pergamon Press, Oxford, 1964.

7.5. Babuška, I. and Aziz, A. K., "Survey Lectures on the Mathematical Foundations of the Finite Element Method," in Aziz, A. K., Ed., *The Mathematical Foundations of the Finite Element Method with Applications to Partial Differential Equations*, Academic Press, New York, 1972, pp. 5–359.

7.6. Akhiezer, N. I. and Glazman, I. M., *Theory of Linear Operators in Hilbert Space*, Vol. I, Translated from the Russian edition by M. Nestell, Frederick Ungar, New York, 1961.

7.7. Yoshida, K., *Functional Analysis*, Academic Press, New York, 1965.

7.8. Friedman, A., *Partial Differential Equations*, Holt, New York, 1969.

7.9. Agmon, S., *Lectures on Elliptic Boundary Value Problems*, D. Van Nostrand, Princeton, N. J., 1965.

7.10. Lions, J. L., *Lectures on Elliptic Partial Differential Equations*, Tata Institute of Fundamental Research, Bombay, 1957.

7.11. Aronszajn, N., "On Coercive Integro-Differential Quadratic Forms," *Technical Report*, No. 14, University of Kansas, pp. 94–106, 1955.

7.12. Smith, K. T., "Inequalities for Formally Positive Integro-Differential Forms," *Bull. Amer. Math. Soc.*, Vol. 67, pp. 368–370, 1961.

7.13. Agmon, S., *Proceedings of the International Symposium on Linear Spaces*, Pergamon Press, Oxford, 1961, pp. 1–13.

7.14. Nečas, J., *Les Méthodes Directes en Théorie des Équations Elliptiques*, Masson, Paris, 1967.

7.15. Babuška, I., "Error-Bounds for the Finite Element Method," *Numer. Math.*, Vol. 16, pp. 322–333, 1971.

7.16. Babuška, I., "The Finite Element Method with Lagrange Multipliers," *Numer. Math.*, Vol. 20, pp. 179–192, 1973.

7.17. Lax, P. D. and Milgram, N., "Parabolic Equations, Contributions to the Theory of Partial Differential Equations," *Ann. Math. Studies, Princeton*, No. 33, pp. 167–190, 1954.

8 FINITE-ELEMENT APPROXIMATIONS OF ELLIPTIC BOUNDARY-VALUE PROBLEMS

8.1 INTRODUCTION

In this chapter we come to grips with a central issue of this book, the theory of finite-element approximations of elliptic boundary-value problems. We begin with a description of Galerkin methods, and we go on to show how the finite-element concept provides a systematic way to implement Galerkin approximations of variational boundary-value problems. We then proceed to investigate the mathematical structure of finite-element approximations, with particular emphasis on convergence questions and on the construction of a priori error estimates.

8.2 GALERKIN APPROXIMATIONS

In 1915 B. G. Galerkin [8.1] proposed an elegant technique for the approximate solution of boundary-value problems, generalizations of which have since proved to be fundamentally important, not only to approximation theory but also to existence theory for partial differential equations. Galerkin's method amounts to a direct and very obvious discretization of the variational formulation of elliptic problems.

To be specific, suppose that we are given the abstract variational boundary-value problem of finding $u \in \mathcal{U}$ such that

$$B(u,v) = l(v) \qquad \forall v \in \mathcal{V} \tag{8.1}$$

where $B(u,v)$ is a continuous bilinear form on $\mathcal{U} \times \mathcal{V}$, $\mathcal{U}$ and $\mathcal{V}$ being Hilbert spaces, and l is a linear functional on $\mathcal{V}$. Galerkin's method is to seek solutions to (8.1) when the variational problem involves only functions in certain finite-dimensional subspaces of $\mathcal{U}$ and $\mathcal{V}$. Indeed, if

$$\mathcal{U}_h \subset \mathcal{U} \qquad \text{and} \qquad \mathcal{V}_h \subset \mathcal{V}$$

are two finite-dimensional subspaces of $\mathcal{U}$ and $\mathcal{V}$, respectively, then the Galerkin approximation of the solution u of (8.1) is the function $U(\mathbf{x}) \in \mathcal{U}_h$ such that

$$B(U,V) = l(V) \qquad \forall V \in \mathcal{V}_h \tag{8.2}$$

We assign a subscript h to these subspaces to imply that their properties generally depend on some real parameter h (such as mesh size) such that, as h decreases, the dimensions of $\mathcal{U}_h$ and $\mathcal{V}_h$ increase and each tends to "fill up" $\mathcal{U}$ and $\mathcal{V}$. The *quality* of the approximation U of u (indeed, the very existence of U) depends on the properties of the subspaces $\mathcal{U}_h$ and $\mathcal{V}_h$. We see in the next section that the finite-element concept provides a systematic method for constructing such subspaces which have several desirable properties.

We assume that the subspaces $\mathcal{U}_h$ and $\mathcal{V}_h$ have the same dimension G. Let $\{\phi_k\}_{k=1}^{G}$ be a basis of $\mathcal{U}_h$ and $\{\psi_k\}_{k=1}^{G}$ be a basis of $\mathcal{V}_h$. Then a solution $U \in \mathcal{U}_h$ can be represented as the linear combination

$$U = \sum_{k=1}^{G} a^k \phi_k \tag{8.3}$$

where the a^k are real numbers. Likewise, each $V \in \mathcal{V}_h$ is of the form

$$V = \sum_{k=1}^{G} b^k \psi_k \tag{8.4}$$

where the b^k are arbitrary. Problem (8.2) then becomes one of finding the specific collection of coefficients a^k such that

$$B\left(\sum_{k=1}^{G} a^k \phi_k, \sum_{j=1}^{G} b^j \psi_j \right) = l\left(\sum_{j=1}^{G} b^j \psi_j \right)$$

or

$$\sum_{j=1}^{G} b^j \left[\sum_{k=1}^{G} a^k B(\phi_k, \psi_j) - l(\psi_j) \right] = 0$$

Since the b^j are arbitrary, we are led to the system of linear equations

$$\sum_{k=1}^{G} a^k K_{kj} = f_j \qquad j = 1, 2, \ldots, G \tag{8.5}$$

where K_{kj} is the *stiffness matrix*

$$K_{kj} = B(\phi_k, \psi_j) \tag{8.6}$$

and f_j is the *generalized force*

$$f_j = l(\psi_j) \tag{8.7}$$

The stiffness matrix is seen to be a bilinear form on $\mathcal{U}_h \times \mathcal{V}_h$, and the generalized force represents a linear functional on $\mathcal{V}_h$. If K_{kj} is invertible, we have

$$a^k = \sum_{j=1}^{G} K_{kj}^{-1} f_j \tag{8.8}$$

and the Galerkin approximant (8.3) assumes the specific form

$$U = \sum_{k,j=1}^{G} K_{kj}^{-1} f_j \phi_k \tag{8.9}$$

The subspace $\mathcal{U}_h$ is often referred to as the *space of trial functions*, and the subspace $\mathcal{V}_h$ as the *space of test functions*. When $\mathcal{U} = \mathcal{V}$, we usually take $\mathcal{U}_h = \mathcal{V}_h$, in which case K_{kj} is symmetric for symmetric $B(\cdot, \cdot)$.

REMARKS. 1. (*Orthogonality of the Error*). Since (8.1) holds for any $v \in \mathcal{V}$, it holds in particular for $V \in \mathcal{V}_h \subset \mathcal{V}$. Therefore

$$B(u, V) = l(V) \qquad \forall V \in \mathcal{V}_h \tag{8.10}$$

Subtracting (8.2) from (8.10), we obtain

$$B(u - U, V) = 0 \qquad \forall V \in \mathcal{V}_h \tag{8.11}$$

We may interpret the results as an *orthogonality condition*; the error

$e = u - U$ of the Galerkin approximation U of the solution of (8.1) is "orthogonal" to the subspace $\mathcal{V}_h$ in the sense of (8.11), i.e., with respect to the bilinear map $B(\cdot,\cdot)$.

2. (*Best Approximation*). Now suppose $\mathcal{U} = \mathcal{V}$ and $\mathcal{U}_h = \mathcal{V}_h = S_h$. Further, let $B(u,v)$ be symmetric and positive definite (i.e., $B(v,v) \geqslant 0, = 0$ if and only if $v = 0$). Then, for arbitrary $V \in S_h$,

$$B(u-V, u-V) = B(e+(U-V), e+(U-V))$$
$$= B(e,e) + B(U-V, U-V) \tag{8.12}$$

where we used (8.11). In (8.12), u is the exact solution of (8.1), and U is its Galerkin approximation. Clearly, of all the functions $V \in S_h$, that closest to the actual solution u in the sense of the metric $[B(u-v, u-v)]^{1/2}$ is the Galerkin approximation U. For this reason, U is called the *best approximation* to u in S_h. ■

8.3 EXISTENCE AND UNIQUENESS OF GALERKIN APPROXIMATIONS

Since the subspaces $\mathcal{U}_h$ and $\mathcal{V}_h$ described previously are closed, each is a complete inner-product space, with the same inner products $(\cdot,\cdot)_{\mathcal{U}}$ and $(\cdot,\cdot)_{\mathcal{V}}$ as $\mathcal{U}$ and $\mathcal{V}$, respectively. Thus Theorem 7.8 also applies to variational problems posed on these spaces, and the conditions listed in that theorem apply immediately to the Galerkin problem (8.2).

THEOREM 8.1. Let $\mathcal{U}_h$ and $\mathcal{V}_h$ be closed subspaces of Hilbert spaces $\mathcal{U}$ and $\mathcal{V}$, respectively, and let $B(U,V)$ denote a continuous weakly coercive form on $\mathcal{U}_h \times \mathcal{V}_h$; i.e., let there exist constants M_0, $\Gamma > 0$, such that the following conditions hold:

$$|B(U,V)| \leqslant M_0 \|U\|_{\mathcal{U}} \|V\|_{\mathcal{V}} \qquad \forall U \in \mathcal{U}_h \qquad V \in \mathcal{V}_h \tag{8.13}$$

$$\inf_{\substack{U \in \mathcal{U}_h \\ \|U\|_{\mathcal{U}} = 1}} \sup_{\substack{V \in \mathcal{V}_h \\ \|V\|_{\mathcal{V}} \leqslant 1}} |B(U,V)| \geqslant \Gamma > 0 \tag{8.14}$$

$$\sup_{U \in \mathcal{U}_h} |B(U,V)| > 0 \qquad \forall V \in \mathcal{V}_h;\ V \neq 0 \tag{8.15}$$

Then there exists a unique element $U^* \in \mathcal{U}_h$ that satisfies the Galerkin problem (8.2); i.e.,

$$B(U^*, V) = l(V) \qquad \forall V \in \mathcal{V}_h \tag{8.16}$$

Moreover,

$$\|U^*\|_{\mathcal{U}} \leqslant \frac{1}{\Gamma}\|l\|^*_{\mathcal{V}} \tag{8.17}$$

where we have employed the notation of Theorem 7.8. ■

It is important to note that the Galerkin problem (8.16) may possess a unique solution even though the actual variational problem (8.1) may not. Indeed, in the approximate problem we only require that $B(U,V)$ be continuous and weakly coercive on certain closed subspaces, and not the full spaces $\mathcal{U}$ and $\mathcal{V}$.

However, it is not difficult to determine sufficient conditions for the existence of a unique solution to (8.16) to guarantee the existence of a unique solution to (8.1). Suppose $\mathcal{U}_0$ and $\mathcal{V}_0$ are pivot spaces such that

$$\mathcal{U} \subset \mathcal{U}_0 = \mathcal{U}_0' \subset \mathcal{U}' \qquad \mathcal{V} \subset \mathcal{V}_0 = \mathcal{V}_0' \subset \mathcal{V}'$$

the inclusions being dense and continuous. In addition, let $\{\mathcal{U}_h\}$ and $\{\mathcal{V}_h\}$, $0<h\leqslant 1$, be families of subspaces of $\mathcal{U}$ and $\mathcal{V}$, respectively, each member of which is spanned by a finite number of linearly independent functions of the type $\{\phi_k\}_{k=1}^G$ or $\{\psi_k\}_{k=1}^G$ described previously [see (8.3) and (8.4)]. Then, for each $\mathcal{U}_h$ and $\mathcal{V}_h$, there are uniquely defined orthogonal projections Π_h and P_h of $\mathcal{U}$ and $\mathcal{V}$ onto $\mathcal{U}_h$ and $\mathcal{V}_h$, respectively, each projection being defined with respect to the inner product on $\mathcal{U}_0$ or $\mathcal{V}_0$, respectively (i.e., Π_h: $\mathcal{U} \to \mathcal{U}_h$, P_h: $\mathcal{V} \to \mathcal{V}_h$).

Next, let us use members of the families $\{\mathcal{U}_h\}$ and $\{\mathcal{V}_h\}$ (or, equivalently, $\{\Pi_h\}$ and $\{P_h\}$) to form Galerkin approximations of problem (8.1). The real number

$$C_h(u,v) = B(\Pi_h u, P_h v) - B(u,v) \tag{8.18}$$

is referred to as the *lack of consistency*, and the families $\{\mathcal{U}_h\}$ and $\{\mathcal{V}_h\}$ are said to provide *consistent* Galerkin approximations to (8.2) if and only if

$$\lim_{h\to 0} C_h(u,v) = 0 \qquad \forall u \in \mathcal{U}\,;\ \forall v \in \mathcal{V} \tag{8.19}$$

The condition of consistency is a test for continuity. Indeed, if $B(u,v)$ is continuous, then a constant $M>0$ exists such that

$$|C_h(u,v)| \leqslant M(\|u-\Pi_h u\|_{\mathcal{U}}\|v-P_h v\|_{\mathcal{V}} + \|u\|_{\mathcal{U}}\|v - P_h v\|_{\mathcal{V}} + \|u-\Pi_h u\|_{\mathcal{U}}\|v\|_{\mathcal{V}})$$

Then, in order that the approximations be consistent, it is sufficient that

$$\lim_{h\to 0} \|u-\Pi_h u\|_{\mathcal{U}}=0 \qquad \lim_{h\to 0} \|v-P_h v\|_{\mathcal{V}}=0 \tag{8.20}$$

The conditions (8.20) are then equivalent to the conditions for "completeness of the coordinate functions" described in earlier literature on variational methods of approximation (e.g., [8.2]).

Assume that the approximate problem (8.16) is solvable. Then, for arbitrary $u \in \mathcal{U}$ and $v \in \mathcal{V}$ we may calculate

$$\inf_{\|\Pi_h u\|_{\mathcal{U}}=1} \sup_{\|P_h v\|_{\mathcal{V}} \leqslant 1} |B(\Pi_h u, P_h v)| \geqslant \Gamma(h) > 0 \tag{8.21}$$

$$\sup_{\Pi_h u \in \mathcal{U}_h} |B(\Pi_h u, P_h v)| > A_0(h) > 0 \qquad P_h v \neq 0 \tag{8.22}$$

The Galerkin approximation (8.16) is *stable* whenever (8.21) and (8.22) hold for all members of the families $\{\Pi_h\}$ and $\{P_h\}$, and

$$\lim_{h\to 0} \Gamma(h) = \gamma > 0 \qquad \lim_{h\to 0} A_0(h) \begin{array}{l} >0 \\ <\infty \end{array} \tag{8.23}$$

The stability conditions ensure that the coerciveness of the forms appearing in the approximate problem is carried on to the exact problem in the limit as $h\to 0$. In view of (8.21), the stability of an approximation has to do with the continuity of the *inverse* operator Λ_h^{-1} [or Λ^{-1} of (7.79)]. The solutions of stable approximations depend continuously on the data, as seen in (8.17).

A fundamental equivalence theorem holds for Galerkin approximations of the type in (8.16). It is clear from Theorem 8.1 and the above definitions that consistency plus stability of an approximation imply its convergence. This is therefore a Galerkin analog of the well-known equivalence theorem developed by Lax [8.3] for difference methods for linear differential equations. For additional information on the use of consistency and stability concepts in finite-element analysis, see Aubin [8.4], and the papers of Oden and Reddy [8.5–8.7].

The following theorem is given in [8.8].

THEOREM 8.2. Let the conditions of Theorems 7.8 and 8.1 hold, and let e denote the *approximation error*

$$e = u^* - U^* \tag{8.24}$$

where u^* and U^* are the solutions of (8.1) and (8.16), respectively. Then the following a priori error estimate holds:

$$\|e\|_{\mathcal{U}} \leqslant \left(1+\frac{M}{\Gamma}\right)\|u^*-U\|_{\mathcal{U}} \qquad \forall U \in \mathcal{U}_h \tag{8.25}$$

where M and Γ are the constants in (7.69) and (8.14).

PROOF. (Cf. [8.8]). Solutions u^* and U^* exist and are unique by virtue of Theorems 7.8 and 8.1. Now let Λ denote the operator defined in (7.75); i.e.,

$$(\Lambda u, v)_{\mathcal{V}} = B(u,v) \qquad u \in \mathcal{U}\,;\ v \in \mathcal{V}$$

There must, by similar arguments, exist an analogous operator Λ_h for the approximate problem; i.e.,

$$(\Lambda_h U, V)_{\mathcal{V}} = B(U,V) \qquad U \in \mathcal{U}_h;\ V \in \mathcal{V}_h \tag{8.26}$$

Now let P_h denote an orthogonal projection from $\mathcal{V}$ onto $\mathcal{V}_h$. Since Λ maps $\mathcal{U}$ continuously into $\mathcal{V}$, the operator

$$Q_h = P_h\Lambda$$

maps $\mathcal{U}$ continuously onto $\mathcal{V}_h$. We shall show that Λ_h of (8.26) is the restriction of Q_h to $\mathcal{U}_h$. We first make use of the fact that P_h is symmetric. Then

$$(P_h\Lambda U, V)_{\mathcal{V}} = (\Lambda U, P_h V)_{\mathcal{V}} = (\Lambda U, V)_{\mathcal{V}} = B(U,V)$$

because $P_h V = V$ for $V \in \mathcal{V}_h$. Thus, from (8.26),

$$\Lambda_h = P_h\Lambda|_{\mathcal{U}_h} = Q_h|_{\mathcal{U}_h} \tag{8.27}$$

as asserted.

Let V^* be the unique element in $\mathcal{V}_h$ such that $(V^*, V)_{\mathcal{V}} = l(V)$, $\forall V \in \mathcal{V}_h$. Clearly,

$$B(U^*, V) = (\Lambda_h U^*, V)_{\mathcal{V}} = (V^*, V)_{\mathcal{V}}$$

Thus

$$\Lambda_h U^* = V^* \tag{8.28}$$

But, on restricting v to $\mathcal{V}_h$, we also have

$$B(u^*, V) = l(V) = (P_h\Lambda u^*, V)_{\mathcal{V}} = (V^*, V)_{\mathcal{V}}$$

so that

$$V^* = P_h \Lambda u^* \tag{8.29}$$

Combining (8.28) and (8.29), we see that the approximate solution U^* is related to the exact solution by

$$U^* = \Lambda_h^{-1} P_h \Lambda u^* \tag{8.30}$$

Let U be a fixed but arbitrary element of $\mathfrak{U}_h$ and denote $u_0 = u^* - U$. Then

$$e = u^* - U^* = u_0 + U - \Lambda_h^{-1} P_h \Lambda (u_0 + U) = (1 - \Lambda_h^{-1} P_h \Lambda) u_0 \tag{8.31}$$

because $\Lambda_h^{-1} Q_h |_{\mathfrak{U}_h} = I$, by (8.27). Therefore

$$\begin{aligned} \|e\|_{\mathfrak{U}} &\leqslant \left(1 + \|\Lambda_h^{-1}\|_{\mathcal{L}(\mathcal{V}_h, \mathfrak{U}_h)} \|\Lambda\|_{\mathcal{L}(\mathfrak{U}, \mathcal{V})}\right) \|u_0\|_{\mathfrak{U}} \\ &\leqslant \left(1 + \frac{M}{\Gamma}\right) \|u^* - U\|_{\mathfrak{U}} \end{aligned}$$

■

REMARK. Inequality (8.25) establishes an equivalence of the problem of *approximation* of solutions to elliptic variational boundary-value problems and *interpolation* of functions in Hilbert spaces. Indeed, suppose Π_h is a projection of $\mathfrak{U}$ onto $\mathfrak{U}_h$. Then we can set $U = \Pi_h u$ in (8.26) and immediately bring into the analysis all the interpolation theory developed in Chapter 6. This important observation sets the stage for the next section. ■

8.4 FINITE-ELEMENT APPROXIMATIONS

Galerkin's method cannot be a general and effective method of approximation unless we add to it a systematic and perfectly general method for constructing the subspaces $\mathfrak{U}_h$ and $\mathcal{V}_h$. This is where the finite-element method described in Chapter 6 comes into the picture: In view of the remark made in the previous section, it should provide an exceptionally general and systematic procedure for constructing global polynomial basis functions $\{\phi_k(\mathbf{x})\}_{k=1}^{G}$ which in turn provide a basis for such finite-dimensional subspaces.

There are several special features of this finite-element Galerkin method that deserve some consideration. First among these is the way that finite-element approximations (interpolations) are built up from local interpolation functions by fitting finite elements together to form a connected model. For example, if $\Omega \subset \mathbf{R}^n$, recall that the finite-element method involves decomposing (partitioning) Ω into a finite number E of almost disjoint subdomains Ω_e such that $\bar{\Omega} = \cup_e^E \bar{\Omega}_e$ and $\Omega_e \cap \Omega_f = \varnothing$, $e \neq f$. Over the *connected* model $\tilde{\Omega}$ of $\bar{\Omega}$, we construct a system of G linearly independent global basis functions $\phi_k(\mathbf{x})$. Confining ourselves to Lagrange-type elements for simplicity, we recall from (6.61) that these global functions are generated from local approximations via the formula

$$\phi_k(\mathbf{x}) = \bigcup_{e=1}^{E} \sum_{N=1}^{N_e} \overset{(e)}{\Omega}{}_k^N \psi_N^{(e)}(\mathbf{x}) \tag{8.32}$$

where $\overset{(e)}{\Omega}{}_k^N$ is the Boolean transformation matrix for element Ω_e defined in (6.11), and $\psi_N^{(e)}(\mathbf{x})$ are the local interpolation functions corresponding to element Ω_e.

The way in which this particular choice of basis functions leads to a collection of local Galerkin approximations is a fundamental property of the finite element as a method of approximation. This idea is easier to appreciate when we observe that the *global* approximation of the bilinear form $B(U, V)$ and the linear functional $l(V)$ are themselves put together by adding contributions from each finite element. Indeed, from the simple fact that $\int_\Omega = \sum_{e=1}^E \int_{\Omega_e}$, we have

$$B(U,V) = \sum_{e=1}^{E} B_e(U_e, V_e) \quad \text{and} \quad l(V) = \sum_{e=1}^{E} l_e(V_e) \tag{8.33}$$

where the subscript e denotes a restriction to element Ω_e. For simplicity, let $\mathcal{U} = \mathcal{V}$ and $\mathcal{U}_h = \mathcal{V}_h$ and let

$$U^*(\mathbf{x}) = \sum_{i=1}^{G} A^i \phi_i(\mathbf{x}) \quad \text{and} \quad V(\mathbf{x}) = \sum_{j=1}^{G} B^j \phi_j(\mathbf{x}) \tag{8.34}$$

where the specific coefficients A^i are to be determined and the B^j are arbitrary. On using (8.32), we see that the Galerkin approximation assumes the form

$$B(\phi_i, \phi_j) = B\left[\bigcup_{e}^{E} \sum_{N}^{N_e} \overset{(e)}{\Omega}{}_i^N \psi_N^{(e)}, \quad \bigcup_{f}^{E} \sum_{M}^{N_f} \overset{(f)}{\Omega}{}_j^M \psi_M^{(f)}\right]$$

$$= \sum_{e=1}^{E} \sum_{M,N}^{N_e} \overset{(e)}{\Omega}{}_i^N \overset{(e)}{\Omega}{}_j^M B(\psi_N^{(e)}, \psi_M^{(e)}) \tag{8.35}$$

or, in view of (8.6), the global stiffness matrix is given by

$$K_{ij} = \sum_{e=1}^{E} \sum_{M,N}^{N_e} \overset{(e)}{\Omega}{}_i^N k_{NM}^{(e)} \overset{(e)}{\Omega}{}_j^M \tag{8.36}$$

where $k_{NM}^{(e)}$ is the local stiffness matrix corresponding to element Ω_e:

$$k_{NM}^{(e)} = B\left(\psi_N^{(e)}, \psi_M^{(e)}\right) \tag{8.37}$$

Likewise,

$$f_j = l(\phi_j) = l\left[\bigcup_e^E \sum_N^{N_e} \overset{(e)}{\Omega}{}_j^N \psi_N^{(e)}\right] = \sum_{e=1}^{E} \sum_{N=1}^{N_e} \overset{(e)}{\Omega}{}_j^N f_N^{(e)} \tag{8.38}$$

where $f_N^{(e)}$ is the local generalized force

$$f_N^{(e)} = l\left(\psi_N^{(e)}\right) \tag{8.39}$$

Again, the Galerkin approximation (8.2) of the boundary-value problem (8.1) leads to a system of equations of the form (8.5); i.e.,

$$\sum_i^G K_{ij} A^i = f_j \qquad j = 1, 2, \ldots, G \tag{8.40}$$

However, in the finite-element method, this equation takes on the special form

$$\sum_{i=1}^{G} \sum_{e=1}^{E} \sum_{M,N}^{N_e} \overset{(e)}{\Omega}{}_i^N k_{NM}^{(e)} \overset{(e)}{\Omega}{}_j^M A^i = \sum_{e=1}^{E} \sum_{N}^{N_e} \overset{(e)}{\Omega}{}_j^N f_N^{(e)} \tag{8.41}$$

A remarkable aspect of finite-element approximations comes into view in this equation. If we wish to solve the variational problem *locally*, we would identify local finite-element approximations associated with (8.34) of the forms

$$U_e^*(\mathbf{x}) = \sum_{N=1}^{N_e} a_e^N \psi_N^{(e)}(\mathbf{x}) \qquad \text{and} \qquad V_e(\mathbf{x}) = \sum_{N=1}^{N_e} b_e^N \psi_N^{(e)}(\mathbf{x}) \qquad \mathbf{x} \in \bar{\Omega}_e \tag{8.42}$$

where a_e^N and b_e^N are the values of A^i and B^i viewed locally in element Ω_e:

$$a_e^N = \sum_{i=1}^{G} \overset{(e)}{\Omega}{}_i^N A^i \qquad b_e^N = \sum_{i=1}^{G} \overset{(e)}{\Omega}{}_i^N B^i \tag{8.43}$$

Thus (8.41) is satisfied whenever

$$\sum_{M=1}^{N_e} k_{MN}^{(e)} a_e^M = f_N^{(e)} \qquad 1 \leq e \leq E;\ 1 \leq N \leq N_e \tag{8.44}$$

Hence the structure of the basis functions (8.32) of the finite-element method is such that the Galerkin approximation can be constructed locally, one element at a time, and the discretized problem then assumes the local form (8.44). The final global approximation (8.40) is then obtained by summing up these local contributions in the sense of (8.41).

REMARKS. 1. When Hermite families of finite elements are used, the global and local stiffness matrices assume the forms

$$K_{ij}^{\alpha,\beta} = B\left(\phi_i^\alpha, \phi_j^\beta\right) \tag{8.45}$$

$$k_{NM}^{(e)\alpha,\beta} = B\left(\psi_N^{\alpha(e)}, \psi_M^{\beta(e)}\right) \tag{8.46}$$

where $\phi_i^\alpha(\mathbf{x})$ are the global basis functions of (6.35), and $\psi_N^{\alpha(e)}(\mathbf{x})$ are the corresponding local interpolation functions described in (6.16) and (6.17). Likewise, we denote

$$f_i^\alpha = l(\phi_i^\alpha) \qquad \text{and} \qquad f_N^{\alpha(e)} = l(\psi_N^{\alpha(e)}) \tag{8.47}$$

All other aspects of the formulation remain in tact; so that, for finite-element representations of order q, we are led to equations of the form

$$\sum_{|\alpha| \leq q} \sum_{j=1}^{G} K_{ji}^{\alpha,\beta} A_\alpha^j = f_i^\beta \qquad 1 \leq i \leq G;\ |\boldsymbol{\beta}| \leq q;\ \boldsymbol{\beta} \in \hat{Z}_+^n \tag{8.48}$$

and the corresponding finite-element approximation of the solution of the variational problem is then

$$U^*(\mathbf{x}) = \sum_{|\alpha| \leq q} \sum_{i=1}^{G} A_\alpha^j \phi_j^\alpha(\mathbf{x}) \tag{8.49}$$

with $\boldsymbol{\alpha} \in \hat{Z}_+^n$.

2. The same comments made in Example 6.6 concerning the scheme for generating global Gram matrices of various orders also apply to local and global stiffness matrices. That is, Example 6.6 can also be regarded as an example of the assembly process for stiffness matrices if we merely replace the local Gram matrices of that example by local stiffness matrices of the same order.

3. We also note that, if we construct the *expanded* local stiffness matrices

$$K_{ij}^{(e)} = \sum_{M,N=1}^{N_e} \overset{(e)}{\Omega}{}_i^N k_{NM}^{(e)} \overset{(e)}{\Omega}{}_j^M \tag{8.50}$$

the global stiffness matrix is obtained by the simple summation

$$K_{ij} = \sum_{e=1}^{E} K_{ij}^{(e)} \tag{8.51}$$

This procedure, we recall, is called the *direct stiffness method.* ■

8.5 PROPERTIES OF FINITE-ELEMENT SUBSPACES

The character of the finite-element approximations described in the previous section depends entirely on the properties of the subspaces $\mathcal{U}_h \subset \mathcal{U}$ and $\mathcal{V}_h \subset \mathcal{V}$. In this section we establish a collection of properties of these subspaces when they are generated using finite elements.

An important step toward the establishment of properties of finite-element approximation spaces has already been made. In Chapter 6 we proved [see Theorem 6.8 and also (6.180)] that, if $u \in H^{k+1}(\Omega)$ and if a finite-element projection Π_h: $H^{k+1}(\Omega) \to S_h(\Omega) \subset H^m(\Omega)$ can be constructed such that $\Pi_h u = u$ for every u in the space $\mathcal{P}_k(\Omega)$ of polynomials of degree $\leqslant k$, for regular refinements of a finite-element mesh and Ω with the cone property,

$$\|u - \Pi_h u\|_{H^m(\Omega)} \leqslant K h^{k+1-m} \|u\|_{H^{k+1}(\Omega)} \tag{8.52}$$

where h is the mesh parameter defined in (6.40), and K is a positive constant independent of u and h. If we integrate U (or, equivalently, if we take fewer derivatives) we can obtain higher accuracies, e.g.,

$$\|u - \Pi_h u\|_{H^s(\Omega)} \leqslant K h^{k+1-s} \|u\|_{H^{k+1}(\Omega)} \qquad s \leqslant m \tag{8.53}$$

and if u is *not* too smooth (e.g., if $u \in H^r(\Omega)$, $r \leqslant k+1$), only lower-order terms in the polynomials (e.g., $\leqslant r-1$) are affected in the approximation;

i.e., then

$$\|u-\Pi_h u\|_{H^s(\Omega)} \leqslant Kh^{r-s}\|u\|_{H^r(\Omega)} \tag{8.54}$$

Estimates (8.52) through (8.54) *apply only to the problem of interpolation of a given u, not to the problem of approximation.* They define the interpolation properties of the subspace $S_h(\Omega)$. However, if the conditions of Theorem 8.2 hold, we also see that the *approximation error* satisfies the inequality

$$\|e\|_{H^m(\Omega)} \leqslant Ch^{k+1-m}\|u^*\|_{H^{k+1}(\Omega)} \tag{8.55}$$

provided the space $\mathcal{U}_h \equiv S_h(\Omega)$ has the properties assumed in deriving (8.52), $\mathcal{U} \subset H^m(\Omega)$, and u^* is sufficiently smooth. The problem is that the estimate (8.55) is far too specialized; it does not indicate the error in norms weaker than $H^m(\Omega)$ and, more importantly, it does not account for the regularity (or possible lack of it) of the solution. Moreover, we cannot simply replace the right-hand side of (8.25) with (8.53); again, the latter inequality is an interpolation result, whereas (8.25) deals with approximation.

The Aubin-Nitsche Method

A method for determining the accuracy of finite-element approximations in weaker norms $\|\cdot\|_{H^s(\Omega)}$, $0 \leqslant s \leqslant m$, for sufficiently smooth solutions was developed independently by Aubin [8.9] and Nitsche [8.10]. We summarize the basic ideas by considering a model problem:

$$Au=f \text{ in } \Omega \qquad D^\alpha u=0 \text{ on } \partial\Omega \qquad |\alpha| \leqslant m-1 \tag{8.56}$$

Here A is a properly elliptic differential operator of order $2m$ with smooth coefficients. The associated variational problem is to find $u \in H_0^m(\Omega) \cap H^{k+1}(\Omega)$ such that

$$B(u,v)=\langle f,v\rangle \qquad \forall v \in H_0^m(\Omega) \tag{8.57}$$

where $B(u,v)$ is assumed to be a continuous, symmetric, $H_0^m(\Omega)$-coercive bilinear form on $H_0^m(\Omega)\times H_0^m(\Omega)$. Of course, we assume that f is sufficiently smooth so that $u \in H^{k+1}(\Omega)$ for $k>0$. In addition, suppose that we calculate a Galerkin approximation of (8.57) by constructing a subspace $S_h(\Omega) \subset H_0^m(\Omega)$ such that $\mathcal{P}_k(\Omega) \subset S_h(\Omega)$; i.e., assume (8.55) holds.

The Aubin-Nitsche method is based on duality ideas; specifically, for some $0 \leqslant s \leqslant m$, we know that

$$\|u^*-U^*\|_{H^s(\Omega)} \equiv \|e\|_{H^s(\Omega)} = \sup_{g\in H^{-s}(\Omega)} \frac{|\langle g,e\rangle|}{\|g\|_{H^{-s}(\Omega)}} \qquad (g \neq 0) \tag{8.58}$$

because $e \in H_0^m(\Omega)$ implies that $e \in H_0^s(\Omega)$ for problem (8.56) and $H^{-s}(\Omega) \subset H^{-m}(\Omega)$. Thus we must estimate $\langle g, e \rangle$ for arbitrary $g \in H^{-s}(\Omega)$. Toward this end, we consider an auxiliary problem of finding w such that

$$B(w,v) = \langle g, v \rangle \qquad \forall v \in H_0^m(\Omega) \tag{8.59}$$

Such a w exists, and in fact satisfies the regularity condition

$$\|w\|_{H^{2m-s}(\Omega)} \leqslant C \|g\|_{H^{-s}(\Omega)} \tag{8.60}$$

Setting $v = e$ in (8.59), we obtain

$$\begin{aligned} |\langle g, e \rangle| &= |B(w,e)| \\ &= |B(w - \Pi_h w, e)| \\ &\leqslant M \|w - \Pi_h w\|_{H^m(\Omega)} \|e\|_{H^m(\Omega)} \\ &\leqslant MKh^{k+1-m} \|u^*\|_{H^{k+1}(\Omega)} \|w - \Pi_h w\|_{H^m(\Omega)} \end{aligned} \tag{8.61}$$

Here we have used the orthogonality condition (8.11) [i.e., $B(\Pi_h w, e) = 0$], the continuity of $B(\cdot, \cdot)$, and (8.55).

We next must eliminate the term $\|w - \Pi_h w\|_{H^m(\Omega)}$ in (8.61). From (8.60) it is clear that $w \in H^{2m-s}(\Omega)$; thus (8.52) holds without modification whenever $2m - s \geqslant k+1$. But this is not too interesting, for when $s = m$ then $k+1-m$ would be $\leqslant 0$, which would mean that even $\|e\|_{H^m(\Omega)}$ could not converge to zero as $h \to 0$. However, Strang [8.11] and Strang and Fix [8.12] have pointed out that this situation can arise in higher-order problems for small enough s. What is more likely is that $2m - s - 1 \leqslant k+1$. Then, even though $\mathcal{P}_k(\Omega) \subset S_h(\Omega)$, no terms of degree $> 2m - s - 1$ contribute to the approximation. Indeed, if $\mathcal{P}_k(\Omega) \subset S_h(\Omega)$, certainly $\mathcal{P}_{2m-s}(\Omega) \subset S_h(\Omega)$. Thus we have two possibilities:

$$\|w - \Pi_h w\|_{H^m(\Omega)} \leqslant \begin{cases} Kh^{m-s} \|w\|_{H^{2m-s}(\Omega)} & 2m - s \leqslant k+1 \\ K_1 h^{k+1-m} \|w\|_{H^{2m-s}(\Omega)} & 2m - s \geqslant k+1 \end{cases} \tag{8.62}$$

Thus, using (8.62) and (8.60), inequality (8.61) leads to the bound

$$|\langle g, v \rangle| \leqslant \tilde{C} \left\{ \begin{matrix} h^{k+1-s} \\ h^{2(k+1-m)} \end{matrix} \right\} \|g\|_{H^{-s}(\Omega)} \|u^*\|_{H^{k+1}(\Omega)}$$

which, on substitution into (8.58) gives the final estimate, for sufficiently small h,

$$\|e\|_{H^s(\Omega)} \leqslant Ch^{\mu}\|u^*\|_{H^{k+1}(\Omega)}$$

$$\mu = \min(k+1-s, 2(k+1-m)) \tag{8.63}$$

The $S_h^{k,m}(\Omega)$ Families

The estimates (8.52) and (8.54) can be used to establish some fairly general "approximability" properties of finite-element subspaces $S_h(\Omega)$. We list some obvious ones:

(i) On quasi-uniform refinements of a finite-element mesh, the normalized mesh parameter h, $0<h\leqslant 1$, can be used to identify members of a entire family of finite-dimensional spaces

$$\{S_h(\Omega): 0<h\leqslant 1\}$$

As $h\to 0$, the dimension of $S_h(\Omega)$ increases indefinitely.

(ii) For each $h>0$, the basis functions $\phi_i(\mathbf{x})$ of $S_h(\Omega)$ are generally piecewise polynomials of some degree $k>0$. In particular, we require that each polynomial $p(\mathbf{x})\in \mathcal{P}_k(\Omega)$ be in $S_h(\Omega)$:

$$\mathcal{P}_k(\Omega)\subset S_h(\Omega)$$

(iii) For each $h>0$, $S_h(\Omega)$ is a closed subspace of a Hilbert space $H^m(\Omega)$:

$$S_h(\Omega)\subset H^m(\Omega)$$

(iv) The spaces $S_h(\Omega)$ are used to interpolate a given function u in some space $H^r(\Omega)$, $r\geqslant 0$. Given the functions $\phi_i(\mathbf{x})$ described in (ii), we can always construct a projection Π_h: $H^r(\Omega)\to S_h(\Omega)$ such that $\Pi_h u = u$, $u\in \mathcal{P}_k(\Omega)$. We wish to estimate the interpolation error in some norm $\|\cdot\|_{H^s(\Omega)}$, where $s\leqslant m$ if $r\geqslant m$, and $s\leqslant r$ if $r\leqslant m$. Recall that $k+1>m$.

(v) When u is smooth enough so that $r\geqslant k+1$, $\|u-\Pi_h u\|_{H^s(\Omega)}$ is of order h^{k+1-s}, as can be deduced from (8.53). When u is less regular, i.e., when $r<k+1$, then only polynomials of degree $\leqslant r-1$ contribute to the solution. The interpolation error is then of order h^{r-s}, as implied by (8.54).

All the above properties lead to the notion of an $S_h^{k,m}(\Omega)$ family of finite-element subspaces. A class of finite-dimensional spaces $S_h(\Omega)$, $0<h\leqslant 1$, is referred to as an $S_h^{k,m}(\Omega)$ *family* if and only if the following

conditions are satisfied:

(i) $S_h^{k,m}(\Omega) \subset H^m(\Omega)$, $k+1 > m \geqslant 0$

(ii) $\mathcal{P}_k(\Omega) \subset S_h^{k,m}(\Omega)$ (8.64)

(iii) For each fixed $h \in (0,1)$ and for every $u \in H^r(\Omega)$, $r \geqslant 0$, and $0 \leqslant s \leqslant \min(r,m)$ there exists a $U(\mathbf{x}) \in S_h^{k,m}(\Omega)$ and a constant $C > 0$, independent of u and h, such that

$$\left.\begin{aligned} \|u-U\|_{H^s(\Omega)} &\leqslant Ch^{\sigma}\|u\|_{H^r(\Omega)} \\ \sigma &= \min(k+1-s, r-s) \end{aligned}\right\} \tag{8.65}$$

where

Now if we choose U in (8.65) so that

$$\|u-U\|_{H^s(\Omega)} = \min_{V \in S_h^{k,m}(\Omega)} \|u-V\|_{H^s(\Omega)} \tag{8.66}$$

the function U will depend on s; indeed, according to (6.52), we will have

$$U(\mathbf{x}) = \sum_{i=1}^{G} a^i \phi_i(\mathbf{x}) \qquad a^i = \sum_{j=1}^{G} G_{(s)}^{ij}(u, \phi_j)_{H^s(\Omega)}$$

where $G_{(s)}^{ij}$ is the inverse Gram matrix of category s. The point is this: For $S_h^{k,m}(\Omega)$ families as now defined, the function U in (8.65) defines an s for which the estimate (8.65) holds.

What is more common is to choose U to be the $H^0(\Omega)$ projection of u into $S_h^{k,m}(\Omega)$ and, for each h, to assume that $S_h^{k,m}(\Omega)$ has the property that (8.65) holds for all s, $0 \leqslant s \leqslant \min(m,r)$. When the $S_h^{k,m}(\Omega)$ families have this property [i.e., when U can be chosen independent of s so that (8.65) holds], we follow Babuška and Aziz [8.8] and refer to the collection of spaces $S_h^{k,m}(\Omega)$ as *regular*.

REMARKS. 1. The introduction of such families of finite-element subspaces is entirely based on the interpolation theory of Chapter 6 and is frequently encountered in the literature; see, e.g., Schultz [8.13], Nitsche [8.14], Bramble and Schatz [8.15], Aubin [8.16], Babuška [8.17, 8.18], and Babuška and Aziz [8.8].

2. Property (ii) of the $S_h^{k,m}(\Omega)$ families can of course be eliminated, since nonpolynomial basis functions can be generated which will still lead to (8.65). However, polynomials have a special place in finite-element

methods, and property (ii) leads to some useful interpretations in subsequent developments.

3. Babuška and Aziz [8.8] have used duality arguments and the interpolation properties of Hilbert spaces described in Chapter 4 to show that, if a collection of finite-element subspaces forms on $S_h^{k,m}(\Omega)$ family, (8.65) holds for arbitrary real r and s, even $r, s \leqslant 0$. ■

Boundary Families, $S_h^{k,m}(\partial\Omega)$

We recall from (4.106) that, whenever a function $g \in H^s(\partial\Omega)$ is given $(s \geqslant 0,\ \partial\Omega \in C^\infty)$, there exists a function $u \in H^{s+j+1/2}(\Omega)$ such that $\|u\|_{H^{s+j+1/2}(\Omega)} \leqslant C\|g\|_{H^s(\partial\Omega)}$. Thus for each $g \in H^r(\partial\Omega)$, $r > 0$, we can find a $u \in H^{r+1/2}(\Omega)$ such that

$$\|u\|_{H^{r+1/2}(\Omega)} \leqslant C_1\|\gamma_0 u\|_{H^r(\partial\Omega)} \tag{8.67}$$

where $\gamma_0 u = u|_{\partial\Omega}$. In view of (8.65) and remark 3 above, we can find a function $U \in S_h^{k,m}(\Omega)$ such that

$$\|u - U\|_{H^{s+1/2}(\Omega)} \leqslant Ch^{\bar{\mu}}\|u\|_{H^{r+1/2}(\Omega)}$$
$$\bar{\mu} = \min(k + 1/2 - s, r - s) \tag{8.68}$$

Now using the conventional trace theorem and then (8.68) and (8.67), we have $(s > 0)$

$$\begin{aligned}\|\gamma_0 u - \gamma_0 U\|_{H^s(\partial\Omega)} &\leqslant C_2\|u - U\|_{H^{s+1/2}(\Omega)}\\ &\leqslant C_2 C h^{\bar{\mu}}\|u\|_{H^{r+1/2}(\Omega)}\\ &\leqslant C_2 C C_1 h^{\bar{\mu}}\|\gamma_0 u\|_{H^r(\partial\Omega)}\end{aligned}$$

We have thus proved the following theorem (cf. [8.8, p. 101]).

THEOREM 8.3. Let $\Omega \subset \mathsf{R}^n$ have a smooth boundary $\partial\Omega$. Then the traces of functions U in $S_h^{k,m}(\Omega)$ families with property (8.65) generate families $S_h^{k-1/2,m-1/2}(\partial\Omega)$of functions on the boundary $\partial\Omega$. ■

Several generalizations of Theorem 8.3 have been given by Bramble and Schatz [8.15]. We cite a representative result.

THEOREM 8.4. Let $\{B_j\}_{j=0}^{m-1}$ be a normal system of boundary operators of the type (7.8) of order q_j, $0 \leqslant j \leqslant m-1$, and let there be given a $S_h^{k,m}(\Omega)$ family of finite-element subspaces such that $m > q_{m-1} + 1/2$. Let μ_j and λ_j

denote real numbers such that $0 \leqslant \mu_j \leqslant m-q_j-1/2$ and $0 \leqslant \lambda_j \leqslant k+1/2-q_j-\mu_j$, $j=0,1,\ldots,m-1$. Then, for all $\mathbf{g}=(g_0, g_1,\ldots,g_{m-1}) \in \Pi_{j=0}^{m-1} H^{\mu_j+\lambda_j}(\partial\Omega)$ and sufficiently smooth $\partial\Omega$, we have

$$\inf_{U \in S_h^{k,m}(\Omega)} \left(\sum_{j=0}^{m-1} h^{q_j+\mu_j} \| g_j - B_j U \|_{H^{\mu_j}(\partial\Omega)} \right) \leqslant c \left(\sum_{j=0}^{m-1} h^{q_j+\mu_j+\lambda_j} \| g_j \|_{H^{\mu_j+\lambda_j}(\partial\Omega)} \right) \tag{8.69}$$

where C is a constant independent of h and $\mathbf{g}$. ■

We also have the following corollary to Theorem 8.4 (Cf. [8.15]).

COROLLARY. 8.4.1. The traces of functions $U \in S_h^{k,m}(\Omega)$ generate a boundary family $S_h^{\rho,\alpha}(\partial\Omega)$, where $\alpha \leqslant m-1/2$ and $\alpha \leqslant \rho \leqslant k+1/2$. ■

Replacing g_j by $B_j u$, $u \in H^\alpha(\Omega)$ in Theorem 8.4 we have for $\alpha - q_j - 1/2 \neq$ integer and $\lambda_j = \alpha - q_j - 1/2 - \mu_j$,

$$\inf_{U \in S_h^{k,m}(\Omega)} \sum_{j=0}^{m-1} h^{q_j+\mu_j} \| B_j u - B_j U \|_{H^{\mu_j}(\partial\Omega)} \leqslant C h^{\alpha-1/2} \sum_{j=0}^{m-1} \| B_j u \|_{H^{\alpha-q_j-1/2}(\partial\Omega)}$$

Using (5.51) [or the trace inequalities (4.110)], we obtain

$$\inf_{U \in S_h^{k,m}(\Omega)} \sum_{j=0}^{m-1} h^{q_j+\mu_j} \| B_j u - B_j U \|_{H^{\mu_j}(\partial\Omega)} \leqslant C h^{\alpha-1/2} \| u \|_{H^\alpha(\Omega)} \tag{8.70}$$

We proved the following additional corollary to Theorem 8.4.

COROLLARY. 8.4.2. Let the conditions of Theorem 8.4 hold. Then for $u \in H^\alpha(\Omega)$, $q_{m-1}+1/2 < \alpha < k+1$, and $0 < \mu_j \leqslant m - q_j - 1/2$ the inequality (8.70) holds. ■

For additional results of this type, consult [8.15].

The Inverse Property

Several authors have considered special $S_h^{k,m}(\Omega)$ families of finite-element approximants in which, roughly speaking, increasing negative powers of h are encountered when stronger norms of elements are compared with weaker ones. For example, consider the Lagrange interpolation function

$$\phi(x)=\begin{cases} -\dfrac{x}{h}+2\left(\dfrac{x}{h}\right)^2 & 0\leqslant x\leqslant h \\ \dfrac{x-2h}{h}+2\dfrac{(x-2h)^2}{h^2} & h\leqslant x\leqslant 2h \end{cases}$$

has the property that

$$\|\phi\|_{H^1(\Omega)}\leqslant Ch^{-1}\|\phi\|_{H^0(\Omega)}$$

where $C\geqslant\sqrt{35/2}$. The assumption that such a property holds for all elements in $S_h^{k,m}(\Omega)$ is useful in studying the stability of the finite-element method. See, e.g., Nitsche [8.14], Aubin [8.16], and Babuška and Aziz [8.8].

We make the idea precise by introducing a formal definition: A family of finite-element subspaces $S_h^{k,m}(\Omega)$ has the *inverse property* if and only if there exits an integer $m>0$ such that, for every $s\leqslant m$,

$$\|U\|_{H^m(\Omega)}\leqslant Ch^{-(m-s)}\|U\|_{H^s(\Omega)} \tag{8.71}$$

for every $U\in S_h^{k,m}(\Omega)$, where $C>0$ does not depend on U or h.

Babuška and Aziz [8.8] have shown that, whenever (8.71) holds, the exponent in (8.71) can be replaced by k_1-k_2, where k_1 and k_2 are arbitrary numbers such that $k_1\leqslant k_2\leqslant m$. This follows from the interpolation theory of Chapter 4. Indeed, from part (ii) of Theorem 4.15,

$$\|U\|_{H^{k_2}(\Omega)}\leqslant\|U\|^{\theta}_{H^{k_1}(\Omega)}\|U\|^{1-\theta}_{H^m(\Omega)}$$

where $\theta=(m-k_2)/(m-k_1)$, and from (8.71),

$$\begin{aligned}\|U\|_{H^{k_2}(\Omega)}&\leqslant\|U\|^{\theta}_{H^{k_1}(\Omega)}Ch^{-(m-k_1)(1-\theta)}\|U\|^{1-\theta}_{H^{k_1}(\Omega)}\\ &\leqslant Ch^{-(m-k_1)(1-(m-k_2)/(m-k_1))}\|U\|_{H^{k_1}(\Omega)}\end{aligned}$$

Thus, for any $S_h^{k,m}(\Omega)$ family having the inverse property,

$$\|U\|_{H^{k_2}(\Omega)}\leqslant Ch^{k_1-k_2}\|U\|_{H^{k_1}(\Omega)} \tag{8.72}$$

where $k_1\leqslant k_2\leqslant m$.

8.6 ERROR ESTIMATES

We continue here our investigation of errors in finite-element approximations, which we initiated in the previous section. Now we assume that we have at our disposal finite-element spaces $S_h(\Omega)$ which are members of the $S_h^{k,m}(\Omega)$ families described earlier.

We demonstrate the principal results in connection with a model problem involving the following assumptions:

(i) We consider a regularly elliptic boundary-value problem,

$$\left.\begin{array}{ll} Au=f \text{ in } \Omega & \\ B_j u=g_j \text{ on } \partial\Omega & 0\leqslant j\leqslant m-1 \end{array}\right\} \tag{8.73}$$

where A is the $2m$th-order elliptic partial differential operator with smooth coefficients given in (7.9), and $\{B_j\}_{j=0}^{m-1}$ is a normal system of boundary operators covering A (here $\Omega\subset\mathbb{R}^n$, $\partial\Omega\in C^\infty$). We consider (8.73) a nonhomogeneous Dirichlet problem; i.e., the boundary operators $\{B_j\}_{j=0}^{m-1}$ form a Dirichlet system of order m.

(ii) We assume that the data $(f;g_j)$ are such that a solution $u^*\in H^r(\Omega)$ exists, with $r\geqslant 2m$. We recall from (5.61) that in this case there is a constant $C>0$ such that

$$\|u^*\|_{H^r(\Omega)}\leqslant C\left(\|f\|_{H^{r-2m}(\Omega)}+\sum_{j=0}^{m-1}\|g_j\|_{H^{r-q_j-1/2}(\partial\Omega)}\right) \tag{8.74}$$

(iii) Let $B(u,v)$ be the bilinear form associated with (8.73) such that the variational problem equivalent to (8.73) is to find $u\in H^r(\Omega)\cap H_g^m(\Omega)$ such that

$$B(u,v)=l(v) \qquad \forall v\in H_g^m(\Omega) \tag{8.75}$$

Here $B(u,v)$ is assumed to be a continuous, weakly coercive, bilinear form on $H^m(\Omega)\times H^m(\Omega)$, l is a linear functional on $H^m(\Omega)$ generated by $(f;g_j)$, and

$$H_g^m(\Omega)=\{v\colon v\in H^m(\Omega);\ B_jv=g_j \text{ on } \partial\Omega;\quad 0\leqslant j\leqslant m-1\} \tag{8.76}$$

Of course, we can reduce (8.75) to a homogeneous Dirichlet problem and still include the effects of nonhomogeneous boundary data by introducing a function w such that $B_jw=g_j$, $0\leqslant j\leqslant m-1$. Then, if $u=\tilde{u}-w$, where $\tilde{u}$ is the solution of (8.75), $Au=A\tilde{u}-Aw=f-Aw$ and $B_ju=0$ on $\partial\Omega$. We then have the variational problem of finding $u \in H^r(\Omega) \cap H_0^m(\Omega)$ such that

$$B(u,v) = \tilde{l}(v) \qquad \forall v \in H_0^m(\Omega) \tag{8.77}$$

where $\tilde{l}$ is the linear functional generated by $(f - Aw; 0)$.

(iv) Next we construct a Galerkin approximation to (8.75) or (8.77). For example, the approximation to (8.77) involves finding $U \in S_h(\Omega)$ such that

$$B(U,V) = \tilde{l}(V) \qquad \forall V \in S_h(\Omega) \tag{8.78}$$

where $B(U,V)$ is also a continuous and weakly coercive bilinear form on $S_h(\Omega)$, which is a finite-dimensional subspace of $H_0^m(\Omega)$. We assume that the boundary conditions are exactly satisfied by the Galerkin approximation (we consider boundary errors as a separate problem in Section 8.8).

(v) Finally, we assume that $S_h(\Omega)$ is a member of an $S_h^{k,m}(\Omega)$ family of finite-element approximations such that $k+1>m$.

THEOREM 8.5. Let conditions (i) through (v) above hold, and let e denote the approximation error

$$e = u^* - U^* \tag{8.79}$$

where u^* is the exact solution of (8.77) and U^* is its finite-element approximation; i.e., U^* is the solution of (8.78). Then there exists a constant $C>0$, independent of u^* and h, such that, for sufficiently small h,

$$\|e\|_{H^m(\Omega)} \leqslant C\left(h^{\gamma_1}\|f\|_{H^r(\Omega)} + h^{\gamma_2}\sum_{j=0}^{m-1}\|g_j\|_{H^{p_j}(\partial\Omega)}\right) \tag{8.80}$$

where $r \geqslant 0$, $p_j \geqslant \frac{1}{2}$, and

$$\left.\begin{aligned} \gamma_1 &= \min(k+1-m, r+m) \\ \gamma_2 &= \min\left[k+1-m, \min_{0\leqslant j\leqslant m-1}\left(p_j+q_j+\tfrac{1}{2}-m\right)\right] \end{aligned}\right\} \tag{8.81}$$

PROOF. Conditions (iii) and (iv) guarantee that unique solutions u^* and U^* exist to (8.77) and (8.78), respectively. Thus, in view of (8.25),

$$\|e\|_{H^m(\Omega)} \leqslant C\|u^* - U\|_{H^m(\Omega)}$$

for arbitrary $U \in S_h(\Omega)$. Condition (v) implies that (8.65) holds; hence we choose U so that

$$\|e\|_{H^m(\Omega)} \leqslant Ch^\sigma \|u^*\|_{H^s(\Omega)} \qquad \sigma = \min(k+1-m, s-m) \tag{8.82}$$

Introducing (8.74) gives

$$\|e\|_{H^m(\Omega)} \leqslant Ch^\sigma \left(\|f\|_{H^{s-2m}(\Omega)} + \sum_{j=0}^{m-1} \|g_j\|_{H^{s-q_j-1/2}(\partial\Omega)} \right)$$

Setting $r = s - 2m$ and $p_j = s - q_j - 1/2$ completes the proof. ■

EXAMPLE 8.1. Let us return to the general variational boundary-value problem (7.38), where $B(u,v)$ is the weakly coercive bilinear form (7.36). In view of Theorem 7.7, $\mathcal{U} = \mathcal{V} = H^1(\Omega)$, provided (7.49) holds. In this case, the data is $(f; g_1)$, $q_1 = 1$, and we have the following estimate:

$$\|e\|_{H^1(\Omega)} \leqslant C\left(h^{\gamma_1}\|f\|_{H^r(\Omega)} + h^{\gamma_2}\|g\|_{H^{p_1}(\partial\Omega)}\right)$$

$$\gamma_1 = \min(k, r+1) \qquad \gamma_2 = \min(k, p_1 + \tfrac{1}{2}) \quad ■$$

EXAMPLE 8.2. Consider a finite-element approximation of the boundary-value problem

$$-\Delta u + u = 0 \text{ in } \Omega \subset \mathbf{R}^2$$

$$u = g \text{ on } \partial\Omega$$

where $g \in H^1(\partial\Omega)$. Suppose that a triangular mesh is used, over which U is piecewise linear (i.e., a Lagrange simplex with $k = 1$). Then

$$\|e\|_{H^1(\Omega)} \leqslant Ch^{1/2}\|g\|_{H^1(\partial\Omega)} \qquad ■$$

The question as to what the error bounds look like for lower norms naturally arises. This we resolve in the following theorem.

THEOREM 8.6. Let the conditions of Theorem 8.5 holds, and let $S_h(\Omega) \subset S_h^{k,m}(\Omega)$, where $S_h^{k,m}(\Omega)$ is regular. Moreover, let the finite-element approximation be such that the boundary conditions are exactly satisfied. Then

$$\|e\|_{H^\alpha(\Omega)} \leqslant C\left(h^{\sigma_1}\|f\|_{H^r(\Omega)} + h^{\sigma_2} \sum_{j=0}^{m-1} \|g_j\|_{H^{p_j}(\partial\Omega)} \right) \tag{8.83}$$

where $\alpha \leqslant m$ and

$$\sigma_1 = \gamma_1 + \mu \qquad \sigma_2 = \gamma_2 + \mu \qquad \text{and} \qquad \mu = \min(k+1-m, m-\alpha) \tag{8.84}$$

and γ_1, γ_2 are defined in (8.81).

PROOF. The proof involves an application of the Aubin-Nitsche method described earlier. We solve an auxiliary problem of the type

$$B(w,v) = \langle g, v \rangle \qquad \forall v \in H_0^\alpha(\Omega)$$

$0 \leqslant \alpha \leqslant m$. The key fact here is that the Dirichlet boundary conditions are assumed to be exactly satisfied. Then the error $e \in H_0^\alpha(\Omega)$, and

$$\|e\|_{H^\alpha(\Omega)} = \sup_{g \in H^{-\alpha}(\Omega)} \frac{|B(w,e)|}{\|g\|_{H^{-\alpha}(\Omega)}}$$

$$\leqslant \sup_{g \in H^{-\alpha}(\Omega)} \frac{M}{\|g\|_{H^{-\alpha}(\Omega)}} \|w - \Pi_h w\|_{H^m(\Omega)} \|e\|_{H^m(\Omega)}$$

$(g \neq 0)$ or, in view of (8.61),

$$\|e\|_{H^\alpha(\Omega)} \leqslant \sup_{g \in H^{-\alpha}(\Omega)} \frac{MK}{\|g\|_{H^{-\alpha}(\Omega)}} h^\mu \|w\|_{H^r(\Omega)} \|e\|_{H^m(\Omega)}$$

where $\mu = \min(k+1-m, r-m)$ and, in this case, $r = 2m - \alpha$. Combining these last inequalities, noting that $\|w\|_{H^{2m-\alpha}(\Omega)} \leqslant C\|g\|_{H^{-\alpha}(\Omega)}$, and introducing (8.80) gives (8.83). ■

EXAMPLE 8.3. Theorem 8.6 generalizes (8.63). Consider a homogeneous boundary-value problem of order $2m$ for which the conditions of Theorem 8.6 are satisfied. Then, for $r \geqslant 0$, $\alpha \leqslant m$, (8.82) gives

$$\begin{aligned} &\|e\|_{H^\alpha(\Omega)} \leqslant Ch^\nu \|f\|_{H^r(\Omega)} \\ &\nu = \min(k+1-\alpha, 2(k+1-m), 2m+r-\alpha) \end{aligned} \tag{8.85}$$

Clearly, when f is smooth, $2m+r-\alpha$ is large, and the rate of convergence is governed by $k+1-\alpha$ or, occasionally, $2(k+1-m)$. ■

In certain instances it is possible to obtain estimates of the form in (8.82) for certain $\alpha \geqslant m$. This is true, for example, for regular $S_h^{k,m}(\Omega)$ families whenever they possess the inverse property (8.71). We prove this statement in the following theorem which is adapted from a result obtained by Babuška and Aziz [8.8].

THEOREM 8.7. Let the conditions of Theorem 8.6 hold, and let $S_h(\Omega)$ be a member of a regular $S_h^{k,m}(\Omega)$ family which has the inverse property. Then for $m \leqslant \alpha$, the approximation error e satisfies the inequality

$$\|e\|_{H^\alpha(\Omega)} \leqslant C\left(h^{\mu_1}\|f\|_{H^r(\Omega)} + h^{\mu_2}\sum_{j=0}^{m-1}\|g_j\|_{H^{p_j}(\partial\Omega)}\right) \tag{8.86}$$

$$\left.\begin{aligned} \mu_1 &= \min(k+1-\alpha, r+2m-\alpha) \\ \mu_2 &= \min\left[k+1-\alpha, \min_{0\leqslant j\leqslant m-1}\left(p_j+q_j+\tfrac{1}{2}-\alpha\right)\right] \end{aligned}\right\} \tag{8.87}$$

PROOF. Since the $S_h^{k,m}(\Omega)$ is regular, we can find a $\tilde{U}\in S_h^{k,m}(\Omega)$ such that $\|u^*-\tilde{U}\|_{H^\alpha(\Omega)}$ is bounded above by the right-hand side of inequality (8.65). According to the triangle inequality and the inverse property (8.71),

$$\|e\|_{H^\alpha(\Omega)} \leqslant \|u^*-\tilde{U}\|_{H^\alpha(\Omega)} + Ch^{m-\alpha}\|U^*-\tilde{U}\|_{H^m(\Omega)}$$

By using the triangle inequality, we can show that the term $\|U^*-\tilde{U}\|_{H^m(\Omega)}$ is bounded by terms of the same order as those on the right side of (8.80). Multiplying these by $h^{m-\alpha}$ and collecting terms gives (8.86). ■

EXAMPLE 8.4. In this example, we cite results of some numerical experiments by S. Mochizuki [8.19]. Suppose we are given the simple one-dimensional problem

$$-\frac{d^2u}{dx^2}+u=f \qquad 0<x<1 \qquad u(0)=u(1)=0$$

where f is given by either of the following:

$$\text{(i)}\quad f=\delta_{1/2} \qquad \text{(ii)}\quad f=\begin{cases} 1 & 0<x\leqslant\frac{1}{2} \\ 0 & \frac{1}{2}<x<1 \end{cases}$$

In case (i), $\langle f,v\rangle=\delta_{1/2}(v)=v(\frac{1}{2})$; $f\in H^{-1/2-\varepsilon}(0,1)$, $\varepsilon>0$, and the solution $u\in H^{3/2-\varepsilon}(0,1)$. In case (ii), a simple application of Fourier transforms reveals that $f\in H^{1/2-\varepsilon}(0,1)$, hence $u\in H^{5/2-\varepsilon}(0,1)$. In each case, ε is an arbitrary positive number. We assume that a finite-element approximation of solutions to either problem is obtained for polynomials of degree $k=1$, 3; i.e., we use $S_h^{1,1}(0,1)$ and $S_h^{3,2}(0,1)$ families.

Using (8.82), we have for the respective choices of data and $\varepsilon\approx 0$,

$$\text{(i)}\qquad \|e\|_{H^\varepsilon(0,1)}\leqslant Ch^{3/2-\varepsilon}\|u\|_{H^{3/2}(0,1)} \qquad k=1 \text{ and } 3$$

$$\text{(ii)}\qquad \|e\|_{H^\varepsilon(0,1)}\leqslant\begin{cases} Ch^{2-\varepsilon}\|u\|_{H^{5/2}(0,1)} & k=1 \\ Ch^{5/2-\varepsilon}\|u\|_{H^{5/2}(0,1)} & k=3 \end{cases}$$

Since ε is arbitrarily small, it seems reasonable that these estimates should not depend on it. We have thus set $\varepsilon = 0$. In fact, Scott [8.20] has shown rigorously that ε does not enter such estimates.

The table given below gives rates of convergence actually calculated for this example for the two data choices (i) and (ii) described above. Noticeably different results are obtained when the points of singularity of the data lie on a nodal point. When the singular point $x = \frac{1}{2}$ lies inside an element, the theoretical rates of convergence are exactly those obtained through numerical experiments. When $x = \frac{1}{2}$ is a nodal point, noticeably better results are obtained. These improved results are actually easy to explain. Recall that

$$\|e\|_{H^1(0,1)} \leqslant C\|u - \Pi_h u\|_{H^1(\Omega)} \leqslant \bar{C} \sum_{e=1}^{E} \|u - \Pi_h u\|_{H^1(\Omega_e)}$$

$$\leqslant \tilde{C} \sum_{e=1}^{E} h_e^{\mu} \|u\|_{H^r(\Omega_e)} \leqslant C_0 h^{\mu} \sum_{e=1}^{E} \|u\|_{H^r(\Omega_e)}$$

$\mu = \min(k, r-1)$. Thus, when u is smooth *locally*, within an element, the *global* rate of convergence may be k. This phenomenon is exactly what is observed in the tabulated results.

Notice also that the estimates appear to hold for $\alpha > m = 1$, indicating that the $S_h^{k,1}(\Omega)$ families used in the calculations have an inverse property.

Computed Rates of Convergence: Example 8.4

Data		(i) 		(ii) 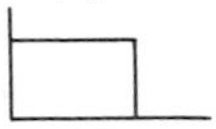	
Node Location		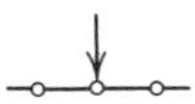		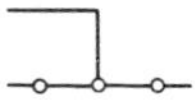	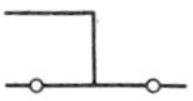
Norm	k				
$H^0(\Omega)$	$k=1$	2.0	1.5	2.0	2.0
	$k=3$	1.5	1.5	4.0	2.5
$H^1(\Omega)$	$k=1$	1.0	0.5	1.0	1.0
	$k=3$	0.5	0.5	2.9	1.5
$H^2(\Omega)$	$k=1$	−0.5	−0.5	0.0	0.0
	$k=3$	−0.3	−0.5	2.0	0.5

■

8.7 POINTWISE AND $L_\infty(\Omega)$ ERROR ESTIMATES

Since the finite-element method is a variational method of approximation, it is natural that the approximate solutions and their errors be given in some Sobolev norm $H^j(\Omega)$, $0 \leqslant j \leqslant m$. However, in many practical situations, the question of the accuracy of finite-element approximations at a point in Ω arises, and whether or not there are points in Ω (such as nodes) at which the approximation is especially accurate or inaccurate. In a similar vein, what is the accuracy of finite elements in the $L_\infty(\Omega)$ norm?

The subject of L_∞ estimates has been studied by several authors; e.g., Ciarlet and Raviart [8.21], Fried [8.22], Wheeler [8.23], Douglas [8.24], Scott [8.25], and Nitsche [8.26]; additional references are given in Scott [8.25]. We give only a brief outline of the basic method for arriving at certain pointwise estimates and cite some representative results on L_∞ estimates. We begin the study by considering the simple one-dimensional problem

$$\left.\begin{aligned} Au \equiv -\frac{d}{dx}\left(a(x)\frac{du}{dx}\right) + b(x)\frac{du}{dx} + c(x)u = f(x) \quad 0<x<1 \\ u(0)=u(1)=0 \end{aligned}\right\} \tag{8.88}$$

where $0 < a_0 \leqslant a(x) \leqslant a_1 \leqslant \infty$, $a(x)$, $b(x)$, $c(x) \in C^\infty(0,1)$. The associated variational problem is

$$B(u,v) \equiv \int_0^1 (au'v' + bu'v + cuv)\,dx = \int_0^1 fv\,dx \equiv f(v) \tag{8.89}$$

$\forall v \in H_0^1(0,1)$ where $B(\cdot,\cdot)$ is continuous and coercive. If we construct a finite-element approximation of (8.89) using piecewise polynomials of degree k, we know from the results of the previous section that [see (8.84) and (8.85)]

$$\|e\|_{H^1(\Omega)} \leqslant Ch^\nu \|f\|_{H^r(\Omega)} \qquad \nu = \min(k, r+1) \tag{8.90}$$

where $e = u^* - U^*$ is the approximation error.

Now let ξ denote an arbitrary point in $(0,1)$ and consider the auxiliary distributional problem

$$B(g,v) = v(\xi) \qquad \forall v \in H_0^1(0,1) \tag{8.91}$$

Here $g(x,\xi)$ is Green's function. Problem (8.91) has meaning because, for $n=1$, $\delta_\xi \in H^{-1}(0,1)$ by the Sobolev embedding theorem. Thus, setting

$v=e$, we have

$$\begin{aligned}
|e(\xi)| &= |B(g,e)| \\
&= |B(g-G,e)| \\
&\leqslant M\|g-G\|_{H^1(\Omega)}\|e\|_{H^1(\Omega)} \\
&\leqslant Ch^\nu\|f\|_{H^r(\Omega)} \inf_{G\in S_h(\Omega)}\|g-G\|_{H^1(\Omega)}
\end{aligned} \tag{8.92}$$

For one-dimensional problems, if we choose ξ to be the coordinate of a nodal point x^N, a simple calculation reveals that

$$\inf_{G\in S_h(\Omega)}\|g-G\|_{H^1(\Omega)} \leqslant C(a,b,c)h^\nu \qquad \nu=\min(k,r+1)$$

where $C(a,b,c)$ is a positive constant, dependent on the functions a, b, and c but independent of h. Thus, if ξ^N is a node,

$$|e(\xi^N)| \leqslant C'(a,b,c)h^{2\nu}\|f\|_{H^r(\Omega)} \tag{8.93}$$

REMARK. For sufficiently smooth data ($r+1>k$), the order of the error at the nodes is seen to be

$$|e(\xi^N)| = O(h^{2k})$$

This rate of convergence is much higher than that which can be obtained globally and is referred to as a *superconvergence* phenomenon (see [8.24]). ■

Essentially the same procedure can be used for the two-dimensional case, with one major exception: The delta distribution δ_ξ is no longer in $H^{-1}(\Omega)$.

Turning now to the question of L_∞ estimates, consider as a model problem the Neuman problem

$$\begin{aligned}
-\Delta u + u &= f \text{ in } \Omega \subset \mathbf{R}^2 \\
\frac{\partial u}{\partial n} &= 0 \text{ on } \partial\Omega
\end{aligned} \tag{8.94}$$

or, equivalently,

$$B(u,v) \equiv \int_\Omega (\nabla u\cdot\nabla v + uv)\,dx = \int_\Omega fv\,dx \tag{8.95}$$

Now we regard $B(u,v)$ as a bilinear form on $W_1^1(\Omega)\times W_\infty^1(\Omega)$, and we approximate (8.91) using a space $S_h(\Omega)$ for which

$$\|u-\tilde{U}\|_{W_\infty^1(\Omega)} \leqslant Kh^k\|u\|_{W_\infty^{k+1}(\Omega)} \tag{8.96}$$

for some $\tilde{U}\in S_h(\Omega)$ and sufficiently smooth u. Thus, repeating the process that yielded (8.92) now gives

$$|e(\xi)| \leqslant Ch^k\|u^*\|_{W_\infty^{k+1}(\Omega)}\|g-G\|_{W_1^1(\Omega)} \tag{8.97}$$

Scott [8.25] has shown that, under the stated assumptions,

$$\|g-G\|_{W_1^1(\Omega)} \leqslant C_h\begin{cases} h|\log h| & \text{if } k=1\\ h & \text{if } k\geqslant 2\end{cases} \tag{8.98}$$

where $C_h>0$ may depend on k, Ω, and $\operatorname{dist}(\xi,\partial\Omega)/h$. Hence

$$\sup_\Omega|e| \leqslant C\left\{\begin{matrix} h^2|\log h| & k=1\\ h^{k+1} & k\geqslant 2\end{matrix}\right\}\|u^*\|_{W_\infty^{k+1}(\Omega)} \tag{8.99}$$

One can then present arguments to the effect that (8.99) also holds for $C>0$ independent of h (see Scott [8.25]). Thus the pointwise error in finite-element approximations of the two-dimensional problem (8.95) satisfies the estimate (8.99). For additional details, see Scott [8.25], and for some results pertaining to fourth-order problems and second-order problems in three dimensions, see Fried [8.22].

8.8 QUADRATURE, BOUNDARY, AND DATA ERRORS

All the error estimates in the previous sections are based on the assumption that the finite-element mesh exactly coincides with the given domain Ω, that the integrals in the definitions of the stiffness and generalized force matrices are evaluated exactly, and that no error is introduced in the definitions of the coefficients in the equations or in the data. In practice, one or all of these assumptions may be frequently violated, and it becomes necessary to evaluate the effects of these formulative errors on the quality of the approximation. In this section, we give a brief account of how this can be done for a model problem. For a more elaborate accounts consult Fix [8.27, 8.28] or Ciarlet and Raviart [8.29].

We consider the Dirichlet problem in two dimensions,

$$\begin{aligned} Au = \sum_{|\alpha|,|\beta| \leqslant m} (-1)^{|\alpha|} D^{\alpha} a_{\alpha\beta}(\mathbf{x}) D^{\beta} u = f \text{ in } \Omega \\ D_n^j u = 0 \qquad j = 0, 1, \ldots, m-1 \text{ on } \partial\Omega \end{aligned} \tag{8.100}$$

where A is strongly elliptic, and the coefficients $a_{\alpha\beta}(\mathbf{x})$ are smooth. The associated variational problem is to find u such that

$$B(u,v) = l(v) \qquad \forall v \in H_0^m(\Omega) \tag{8.101}$$

where $B(u,v)$ is a strongly coercive bilinear form given by

$$B(u,v) = \int_\Omega \sum_{|\alpha|,|\beta| \leqslant m} a_{\alpha\beta} D^{\alpha} v D^{\beta} u \, dx \tag{8.102}$$

and

$$l(v) = \int_\Omega f v \, dx \tag{8.103}$$

We approximate (8.101) by the finite-element method, and the approximate solution is a function $U \in S_h(\Omega)$, where $S_h(\Omega) \in S_h^{k,m}(\Omega)$, $m \geqslant 1$, of the form (8.9) where, for this case, the stiffness matrix and force vectors are

$$K_{ij} = \int_\Omega \left(\sum_{|\alpha|,|\beta| \leqslant m} a_{\alpha\beta} D^{\alpha} \phi_i D^{\beta} \phi_j \right) dx \tag{8.104}$$

$$f_j = \int_\Omega f \phi_j \, dx \tag{8.105}$$

Here $\phi_j(\mathbf{x})$ are the usual finite-element basis functions.

Quadrature Errors

In almost all problems of much importance, the integrals in (8.104) and (8.105) are evaluated numerically via some numerical quadrature formula such as

$$\int_\Omega R(\mathbf{x}) \, dx \approx \sum_{i=1}^{M} W_i R(\xi^i) \equiv Q_\Omega(R) \tag{8.106}$$

where W_i are *weights* and ξ^i, $i=1,2,\ldots,M$, are points in Ω which depend on the size and shape of Ω and also on the quadrature rule. Thus, instead of computing the actual coefficients a^j of (8.8), we obtain

$$\tilde{a}^j = \sum \tilde{K}_{jk}^{-1} \tilde{f}_k \tag{8.107}$$

where

$$\tilde{K}_{jk} = \sum_{e=1}^{E} Q_{\Omega_e}\left[\sum_{|\alpha|,|\beta| \le m} a_{\alpha\beta} D^{\alpha}\phi_j D^{\beta}\phi_k \right] \equiv \tilde{B}(\phi_j, \phi_k) \tag{8.108}$$

$$\tilde{f}_k = \sum_{e=1}^{E} Q_{\Omega_e}(f\phi_k) \tag{8.109}$$

Hence, instead of U in (8.3), we compute $\tilde{U} = \sum_{j=1}^{G} \tilde{a}^j \phi_j$. We now wish to estimate the error introduced by employing quadrature formulas alone; i.e., we wish to find the error $U - \tilde{U}$, measured in an appropriate norm, for a given quadrature rule and subspace $S_h(\Omega)$.

Toward this end, let $S_h^{k,m}(\Omega)$ be the family of finite-element subspaces with the properties described in (8.65). We say that the quadrature rule in (8.106) *is of order* q if the error is $O(h_e^{q+1})$:

$$\left| \int_{\Omega_e} f(\mathbf{x})\,dx - Q_{\Omega_e}(f) \right| \le M_e h_e^{q+1} \sum_{|\alpha| = q+1} \int_{\Omega_e} |D^{\alpha} f(\mathbf{x})|\,dx \tag{8.110}$$

where M_e is a positive number which depends on q, and h_e is the diameter of Ω_e. For example, a Newton–Cotes quadrature formula of order q integrates polynomials of order q exactly. An n-point gaussian quadrature integrates polynomials of order $2n-1$ or less exactly. Hence the n-point gaussian formula is of order $q = 2n-1$ (i.e., the error is of order h_e^{2n}).

We now introduce a parameter that plays an important role in the estimation of quadrature error: we denote by p the largest exponent appearing in the polynomial representing the finite element approximation. For example, for piecewise linear functions defined on triangles $p=1$, whereas for bilinear functions on rectangles $p=2$.

Instead of solving the variational equation (8.101), in practice we solve the approximate problem

$$\tilde{B}(\tilde{U}, V) = \tilde{l}(V) \qquad \forall V \in S_h(\Omega) \subset S_h^{k,m}(\Omega) \tag{8.111}$$

where

$$\tilde{B}(U,V)=\sum_{e=1}^{E} Q_{\Omega_e}\left(\sum_{|\alpha|,|\beta|\leqslant m} a_{\alpha\beta}D^{\alpha}VD^{\beta}U\right)$$
$$\tilde{l}(V)=\sum_{e=1}^{E} Q_{\Omega_e}(fV) \tag{8.112}$$

We next must estimate the differences in $\tilde{B}(U,V)-B(U,V)$ and $\tilde{l}(V)-l(V)$. First consider the latter term. Since f is assumed to be smooth (i.e., sufficiently differentiable), the product fV can be differentiated sufficiently many times. Suppose that Q_{Ω_e} is a quadrature rule of order q. Then the error due to numerical integration (quadrature) is bounded by

$$|\tilde{l}(V)-l(V)|\leqslant\sum_{e=1}^{E} M_e h_e^{q+1}\sum_{|\alpha|=q+1}\int_{\Omega_e}|D^{\alpha}(fV)|\,dx$$

Let $V=U^*-\tilde{U}$, where $U^*\in S_h^{k,m}(\Omega)$ is the finite-element solution of (8.16), $\tilde{U}\in S_h^{k,m}(\Omega)$ is the (finite-element) solution of (8.111), and let $f\in H^s(\Omega)$, $s>q+1-p$. Then

$$|\tilde{l}(U^*-\tilde{U})-l(U^*-\tilde{U})|\leqslant\sum_{e=1}^{E} M_e h_e^{q+1}\sum_{|\alpha|=q+1}\int_{\Omega_e}|D^{\alpha}[f\cdot(U^*-\tilde{U})]|\,dx$$

Since $U^*-\tilde{U}\in S_h^{k,m}(\Omega_e)$, differentiations of order $>p$ must be carried out on f:

$$D^{\alpha}[f(\mathbf{x})\cdot(U^*-\tilde{U})]=\sum_{|\beta|\leqslant|\alpha|}\binom{\alpha}{\beta}D^{\beta}(U^*-\tilde{U})\cdot D^{\alpha-\beta}f(\mathbf{x})$$
$$=\sum_{|\beta|\leqslant q+1}\binom{\alpha}{\beta}D^{\beta}(U^*-\tilde{U})D^{\alpha-\beta}f(\mathbf{x})$$

Noting that $U^*-\tilde{U}$ has nonvanishing derivatives of only order $p\leqslant q+1$, we have

$$D^{\alpha}[f(\mathbf{x})(U^*-\tilde{U})]=\sum_{|\beta|\leqslant p}\binom{\alpha}{\beta}D^{\beta}(U^*-\tilde{U})D^{\alpha-\beta}f(\mathbf{x})$$

Next we assume that the family $S_h^{k,m}(\Omega_e)$ has the inverse property; in other words, wince U^* and U depend on h (e.g., $U^* = \alpha_1 + \alpha_2 x_1/h + \alpha_3 x_2/h + \cdots$), each time we differentiate $U^* - U$ we gain a factor of h^{-1}.

$$\|U\|_{H^p(\Omega_e)} \leq Ch_e^{-(p-m)}\|U\|_{H^m(\Omega_e)} \qquad \forall U \in S_h^{k,m}(\Omega_e)$$

Therefore we have

$$|\tilde{l}(U^* - \tilde{U}) - l(U^* - \tilde{U})|$$

$$\leq \sum_{e=1}^{E} M_e h_e^{q+1} \sum_{|\alpha|=q+1} \int_{\Omega_e} \left| \sum_{|\beta|\leq p} \binom{\alpha}{\beta} D^\beta (U^* - \tilde{U}) D^{\alpha-\beta} f(\mathbf{x}) \right| dx$$

$$\leq M_0 h^{q+1} \sum_{\substack{|\alpha|=q+1 \\ |\beta|\leq p}} \binom{\alpha}{\beta} \int_\Omega |D^\beta (U^* - \tilde{U}) D^{\alpha-\beta} f(x)| \, dx$$

$$\leq \hat{M}_0 h^{q+1} \|f\|_{H^{q+1-p}(\Omega)} \|U^* - \tilde{U}\|_{H^p(\Omega)}$$

$$\leq M h^{q+1-(p-m)} \|U^* - \tilde{U}\|_{H^m(\Omega)} \|f\|_{H^{q+1-p}(\Omega)} \tag{8.113}$$

wherein $h = \max_e h_e$.

Now consider the term $|\tilde{B}(U, V) - B(U, V)|$. We have

$$|\tilde{B}(U^*, U^* - \tilde{U}) - B(U^*, U^* - \tilde{U})| \leq \sum_{e=1}^{E} M_e h_e^{q+1} \sum_{|\gamma|=q+1}$$

$$\cdot \int_{\Omega_e} \left| D^\gamma \left[\sum_{|\alpha|,|\beta|\leq m} a_{\alpha\beta} D^\alpha U^* D^\beta (U^* - \tilde{U}) \right] \right| dx$$

Since U^*, $U^* - \tilde{U} \in S_h^{k,m}(\Omega_e)$, differentiations of order $> p$ are transferred to the coefficients $a_{\alpha\beta}(\mathbf{x})$; i.e.,

$$D^\gamma \left[\sum_{|\alpha|,|\beta|\leq m} a_{\alpha\beta} D^\alpha U^* D^\beta (U^* - \tilde{U}) \right]$$

$$= \sum_{|\eta|\leq|\gamma|} \sum_{|\alpha|,|\beta|\leq m} \binom{\gamma}{\eta} D^{\gamma-\eta} a_{\alpha\beta}(\mathbf{x}) \cdot D^\eta \left[D^\alpha U^* D^\beta (U^* - \tilde{U}) \right]$$

Since $|\eta|+|\alpha|$ are taken, at most, equal to p, we have

$$D^{\gamma}\left[\sum_{|\alpha|,|\beta|\leqslant m} a_{\alpha\beta} D^{\alpha} U^{*} D^{\beta}(U^{*}-\tilde{U})\right]$$

$$=\sum_{|\alpha|,|\beta|\leqslant m}\ \sum_{|\eta|\leqslant p-m}\binom{\gamma}{\eta} D^{\gamma-\eta} a_{\alpha\beta} D^{\eta}\left[D^{\alpha} U^{*} D^{\beta}(U^{*}-\tilde{U})\right]$$

Then

$$|\tilde{B}(U^{*},U^{*}-\tilde{U})-B(U^{*},U^{*}-\tilde{U})|$$

$$\leqslant \sum_{e=1}^{E} M_e h_e^{q+1}\int_{\Omega_e}\sum_{\substack{|\gamma|=q+1\\|\eta|\leqslant p-m\\|\alpha|,|\beta|\leqslant m}}\binom{\gamma}{\eta} D^{\gamma-\eta} a_{\alpha\beta} D^{\eta}\left[D^{\alpha} U^{*} D^{\beta}(U^{*}-\tilde{U})\right] dx$$

$$\leqslant CMh^{q+1}\left(\|U^{*}\|_{H^{m+|\eta|}(\Omega)}\|U^{*}-\tilde{U}\|_{H^{m}(\Omega)}+\|U^{*}\|_{H^{m}(\Omega)}\|U^{*}-\tilde{U}\|_{H^{m+|\eta|}(\Omega)}\right)$$

(8.114)

where C is a positive constant depending upon $a_{\alpha\beta}, q, p$, and m. Now using the inverse property of $S_h^{k,m}(\Omega_e)$, we have

$$|\tilde{B}(U^{*},U^{*}-\tilde{U})-B(U^{*},U^{*}-\tilde{U})|$$

$$\leqslant CMh^{q+1-(p-m)}\|U^{*}\|_{H^{m}(\Omega)}\|U^{*}-\tilde{U}\|_{H^{m}(\Omega)} \qquad (8.115)$$

At this point, we have established sufficient groundwork for the following important result on quadrature errors:

THEOREM 8.8. Suppose that the bilinear form $B(U,V)$ satisfies the conditions

$$\begin{aligned}
&\text{(i)} && \tilde{B}(U,U)\geqslant\tilde{\Gamma}\|U\|^2_{H^{m}(\Omega)} && U\in S_h^{k,m}(\Omega)\\
&\text{(ii)} && |\tilde{B}(U^{*},V)-B(U^{*},V)|+|\tilde{l}(V)-l(V)| &&\\
& && \qquad\leqslant\tilde{M}_0 h^{\mu}\|V\|_{H^{m}(\Omega)} && V\in S_h^{k,m}(\Omega)
\end{aligned} \qquad (8.116)$$

where $S_h^{k,m}(\Omega)$ has the inverse property. Then the error due to quadrature is given by

$$\|U^* - \tilde{U}\|_{H^m(\Omega)} \leqslant \frac{\tilde{M}_0}{\tilde{\Gamma}} h^{\mu} \tag{8.117}$$

PROOF. First note the identity,

$$\begin{aligned}\tilde{B}(U^* - \tilde{U}, U^* - \tilde{U}) &= \tilde{B}(U^*, U^* - \tilde{U}) - \tilde{B}(\tilde{U}, U^* - \tilde{U}) \\ &= \tilde{B}(U^*, U^* - \tilde{U}) - \tilde{l}(U^* - \tilde{U}) \\ &\quad - B(U^*, U^* - \tilde{U}) + l(U^* - \tilde{U})\end{aligned}$$

Then, using condition (i) of (8.116), we have

$$\begin{aligned}\|U^* - \tilde{U}\|^2_{H^m(\Omega)} &\leqslant \frac{1}{\tilde{\Gamma}} \tilde{B}(U^* - \tilde{U}, U^* - \tilde{U}) \\ &= \frac{1}{\tilde{\Gamma}} [\tilde{B}(U^*, U^* - \tilde{U}) - B(U^*, U^* - \tilde{U}) - \tilde{l}(U^* - \tilde{U}) + l(U^* - \tilde{U})] \\ &\leqslant \frac{1}{\tilde{\Gamma}} [|\tilde{B}(U^*, U^* - \tilde{U}) - B(U^*, U^* - \tilde{U})| + |\tilde{l}(U^* - \tilde{U}) - l(U^* - \tilde{U})|]\end{aligned}$$

Now, using condition (ii) of (8.116), we obtain (8.117). ■

REMARKS. 1. The condition (i) of Theorem 8.8 is analogous to the requirement (8.14) on $B(U, V)$. That is, condition (i) of (8.116) requires the bilinear form $\tilde{B}(U, V)$ to be coercive. This is reasonable, since we expect the approximate problem (8.111) to have a solution. Indeed, it is clear that the strong coerciveness of $\tilde{B}(U, V)$ is equivalent to the positive definiteness of the stiffness matrix $\tilde{K}_{ij}$ [see (8.108)]. However, the positive definiteness of $\tilde{K}_{ij}$ depends on the quadrature rule and the finite-element subspace $S_h^{k,m}(\Omega)$; definiteness can be guaranteed by taking a sufficient number of integration points (see Strang and Fix [8.12, p. 189].)

2. Condition (ii) of (8.116) is reminiscent of the consistency requirement (8.20) and involves the choice of the proper degree of the polynomials for which the quadrature rule is exact. Equations (8.113) and (8.115) are useful in arriving at condition (ii) of (8.116). We return to this shortly.

3. The bound in Theorem 8.8 is also valid for the case in which the coefficients of the original differential equation are perturbed; i.e., the result is valid for data errors. The crucial difference between quadrature and data errors lies in establishing inequality (8.113). ■

Now returning to Theorem 8.8, we note that the estimate (8.117) is of no use unless the value of the exponent μ is known. Note that the inequalities (8.113) and (8.115) will lead directly to the estimate (8.116) if the term $\| U^* \|_{H^p(\Omega)}$ in (8.115) can be expressed in terms of the data f.

Note that from the triangle inequality and (8.86) it follows that, as $h \to 0$, the finite element solution $U^* \in S_h^{k,m}(\Omega)$ satisfies the inequality

$$\sum_{e=1}^{E} \sum_{|\alpha| \leqslant m} \int_{\Omega_e} |D^\alpha U^*|^2 \, dx \leqslant \| u^* \|_{H^p(\Omega)}^2 \leqslant c \| f \|_{H^{p-2m}(\Omega)}^2 \tag{8.118}$$

As a result, from (8.114) we obtain, instead of (8.115), the inequality

$$|\tilde{B}(U^*, U^* - \tilde{U}) - B(U^*, U^* - \tilde{U})| \leqslant c_2 M_0 h^{q+1-(p-m)} \| U^* - \tilde{U} \|_{H^m(\Omega)} \| f \|_{H^{p-2m}(\Omega)} \tag{8.119}$$

Combining (8.113) and (8.119), we obtain

$$|\tilde{B}(U^*, U^* - \tilde{U}) - B(U^*, U^* - \tilde{U})| + |\tilde{l}(U^* - \tilde{U}) - l(U^* - \tilde{U})| \leqslant \tilde{M}_0 h^{q+1-(p-m)} \| U^* - \tilde{U} \|_{H^m(\Omega)} \tag{8.120}$$

where $\tilde{M}_0$ now depends on $\| f \|_{H^r(\Omega)}, r = \max(q+1-p, p-2m)$. Comparing (8.117) with (8.120), we see that $\mu = q+1-p+m$. We summarize the results in the following theorem. ■

THEOREM 8.9. Let $S_h^{k,m}(\Omega)$ be a family of finite-element subspaces exhibiting the properties (8.65) and endowed with the inverse property (8.71). Let $f \in H^s(\Omega), s \geqslant 0$. Further, suppose that the approximate bilinear form $\tilde{B}(U,V)$ (corresponding to $2m$th-order elliptic differential operator), obtained by qth-order quadrature rule satisfies the condition (i) of (8.116). Then the error due to quadrature is bounded by

$$\| U^* - \tilde{U} \|_{H^m(\Omega)} \leqslant \frac{\tilde{M}_0}{\tilde{\Gamma}} h^{q+1-p+m} \tag{8.121}$$

Thus, to maintain a rate-of-convergence of order $O(h^{k+1-m})$, we must take

$$q = k + p - 2m \geqslant p - 1 \quad ■$$

EXAMPLE 8.5. Suppose that the problem (8.100) is to be solved on a unit square, and suppose that Ω is divided into a set $\{\Omega_e\}$ of triangles of diameter h. Let $S_h^{k,m}(\Omega)$ be the space of continuous piecewise linear polynomials; i.e., $p=1$, $m=1$, and $S_h^{1,1}(\Omega)\subset H_0^1(\Omega)$. If a quadrature rule of order $q=1$ (such as the midpoint quadrature rule, which is exact for linear functions $(1,x_1,x_2)$) is used, the quadrature error is of the order

$$\|U^* - \tilde{U}\|_{H^1(\Omega)} = O(h^{q+1-p+m}) = O(h^2) \quad \blacksquare$$

EXAMPLE 8.6. Consider again the previous example, but now let $S_h^{k,m}(\Omega)$ be the space of bicubic Hermite functions on a uniform mesh:

$$S_h^{k,m}(\Omega) = \left\{ p : p|_{\Omega_e} = \sum_{i,j=0}^{3} a_{ij}(x_1)^i (x_2)^j, \quad 1 \leqslant e \leqslant E \right\}$$

Here $p=6$, and $m=1$. Consider the one-point gaussian quadrature rule $(n=1, q=1)$,

$$\int_0^1 \int_0^1 R(x_1,x_2)\,dx_1\,dx_2 = R\left(\tfrac{1}{2},\tfrac{1}{2}\right)$$

which is exact for $(1,\ x_1,\ x_2,\ x_1x_2)$. The quadrature error for this rule is

$$\|U^* - \tilde{U}\|_{H^1(\Omega)} = O(h^{1+1-6+1})$$

That is, the method is divergent. However, a five-point gaussian quadrature $(n=5, q=9)$ which is exact for $p(\mathbf{x}) = \sum_{|\boldsymbol{\alpha}|\leqslant 18} a_{\boldsymbol{\alpha}} \mathbf{x}^{\boldsymbol{\alpha}}$, $\boldsymbol{\alpha}=(\alpha_1,\alpha_2)$, $0 \leqslant \alpha_1, \alpha_2 \leqslant 9$,

$$\int_0^1 \int_0^1 R(\mathbf{x})\,dx_1\,dx_2 = \sum_{i,j=1}^{5} W_i W_j R(\xi_i, \xi_j)$$

would lead to an error,

$$\|U^* - \tilde{U}\|_{H^1(\Omega)} = O(h^{q+1-6+1}) = O(h^5) \quad \blacksquare$$

Boundary Errors

We now turn our attention to the estimation of errors introduced in the finite-element solution of the model problem (8.100) due to the approximation of the domain. Approximation of the domain $\bar{\Omega} = \Omega \cup \partial\Omega$ by finite elements inevitably leads to perturbation of its boundary $\partial\Omega$; i.e., the union $\cup_{e=1}^{E}\bar{\Omega}_e=\tilde{\Omega}'$ of finite elements is in general not equal to $\bar{\Omega}$, and the boundary $\partial\tilde{\Omega}$ of the connected model $\tilde{\Omega}'$ is not equal to the original boundary $\partial\Omega$. Consequently, the original problem (8.101) is replaced by the problem of finding $\hat{U}\in S_h(\tilde{\Omega})\subset H_0^m(\tilde{\Omega})$ such that

$$B_h(\hat{U},V)=l_h(V)\qquad \forall V\in S_h(\tilde{\Omega}) \tag{8.122}$$

$$B_h(U,V)=\int_{\tilde{\Omega}}\Bigg(\sum_{|\alpha|,|\beta|\le m}\tilde{a}_{\alpha\beta}D^\alpha V\cdot D^\beta U\Bigg)dx\qquad l_h(V)=\int_{\tilde{\Omega}}\tilde{f}V\,dx \tag{8.123}$$

Here $\tilde{a}_{\alpha\beta}$ and $\tilde{f}$ are, respectively, extensions (by zero) of $a_{\alpha\beta}$ and f to $\tilde{\Omega}$ or restrictions to $\tilde{\Omega}$ depending on whether $\tilde{\Omega}\subset\bar{\Omega}$ or $\bar{\Omega}\subset\tilde{\Omega}$. Thus a change in domain can be interpreted as a change in data. Again taking into account the fact that the integrals in (8.122) are evaluated by quadrature rules, we in effect solve the approximate problem.

$$\hat{B}(\hat{U}^*,V)=\hat{l}(V)\qquad \forall V\in S_h(\tilde{\Omega})\subset H_0^m(\tilde{\Omega}) \tag{8.124}$$

where

$$\begin{aligned}\hat{B}(U,V)&=\sum_{e=1}^{E}Q_{\Omega_e}\Bigg(\sum_{|\alpha|,|\beta|\le m}\tilde{a}_{\alpha\beta}D^\alpha VD^\beta U\Bigg)\\ \hat{l}(V)&=\sum_{e=1}^{E}Q_{\Omega_e}(\tilde{f}V)\end{aligned} \tag{8.125}$$

We now wish to estimate the error $\hat{U}^*-U$ due to change in the domain and due to quadrature (note that, when $\tilde{\Omega}=\bar{\Omega}$, this error is due just to quadrature). The following lemma, due to Ciarlet and Raviart [8.29], is a counterpart of the result in Theorem 8.8.

LEMMA 8.1. Let the bilinear form $\hat{B}$ of (8.125) be coercive in the sense

(i) $\hat{B}(U,U)\ge\hat{\Gamma}\|U\|^2_{H^m(\tilde{\Omega})}$, $U\in S_h(\tilde{\Omega})\subset H_0^m(\tilde{\Omega})$, and suppose that the bilinear form

$$B_h(u,v)=\int_{\tilde{\Omega}}\Bigg(\sum_{|\alpha|,|\beta|\le m}\tilde{a}_{\alpha\beta}D^\alpha vD^\beta u\Bigg)dx\qquad u,v\in H^m(\tilde{\Omega})$$

is continuous,

(ii)

$$|B_h(u,v)|\le\hat{M}\|u\|_{H^m(\tilde{\Omega})}\|v\|_{H^m(\tilde{\Omega})}$$

Then the solution $\hat{U}^*$ of (8.124) satisfies the inequality

$$\|\hat{U}^* - V\|_{H^m(\tilde{\Omega})} \leqslant \frac{\hat{M}}{\hat{\Gamma}}\|u - V\|_{H^m(\tilde{\Omega})} + \frac{1}{\hat{\Gamma}} \sup_{W \in S_h(\tilde{\Omega})} \frac{|B_h(V,W) - \hat{B}(V,W)|}{\|W\|_{H^m(\tilde{\Omega})}} + \frac{1}{\hat{\Gamma}} \sup_{W \in S_h(\tilde{\Omega})} \frac{|B_h(u,W) - \hat{l}(W)|}{\|W\|_{H^m(\tilde{\Omega})}} \quad (W \neq 0)$$

for any $u \in H^m(\hat{\Omega})$, where $\hat{\Omega}$ is a bounded domain containing $\overline{\Omega}$ and $\tilde{\Omega}$.

PROOF. By Theorem 8.1, and in view of assumptions (i) and (ii) above, the approximate problem (8.124) has a unique solution $\hat{U}^*$. We have

$$\|\hat{U}^* - V\|^2_{H^m(\tilde{\Omega})} \leqslant \frac{1}{\hat{\Gamma}}\hat{B}(\hat{U}^* - V, \hat{U}^* - V)$$
$$= \frac{1}{\hat{\Gamma}}[\hat{B}(\hat{U}^*, \hat{U}^* - V) - \hat{B}(V, \hat{U}^* - V)]$$
$$= \frac{1}{\hat{\Gamma}}[\hat{l}(\hat{U}^* - V) - \hat{B}(V, \hat{U}^* - V)]$$

Adding and subtracting the term $B_h(u, \hat{U}^* - V)$, we get

$$\|\hat{U}^* - V\|^2_{H^m(\tilde{\Omega})} \leqslant \frac{1}{\hat{\Gamma}}[\hat{l}(\hat{U}^* - V) - \hat{B}(V, \hat{U}^* - V) + B_h(u - V, \hat{U}^* - V) + B_h(V, \hat{U}^* - V) - B_h(u, \hat{U}^* - V)]$$

Now using the continuity of $B_h(\cdot,\cdot)$,

$$\|\hat{U}^* - V\|^2_{H^m(\tilde{\Omega})} \leqslant \frac{\hat{M}}{\hat{\Gamma}}\|u - V\|_{H^m(\tilde{\Omega})}\|\hat{U}^* - V\|_{H^m(\tilde{\Omega})} + \frac{1}{\hat{\Gamma}}|B_h(V, \hat{U}^* - V) - \hat{B}(V, \hat{U}^* - V)| + \frac{1}{\hat{\Gamma}}|B_h(u, \hat{U}^* - V) - \hat{l}(\hat{U}^* - V)|$$

or

$$\|\hat{U}^* - V\|_{H^m(\tilde{\Omega})} \leqslant \frac{\hat{M}}{\hat{\Gamma}} \|u - V\|_{H^m(\tilde{\Omega})} + \frac{1}{\hat{\Gamma}} \sup_{W \in S_h(\tilde{\Omega})} \frac{|B_h(V,W) - \hat{B}(V,W)|}{\|W\|_{H^m(\tilde{\Omega})}} + \frac{1}{\hat{\Gamma}} \sup_{W \in S_h(\tilde{\Omega})} \frac{|B_h(u,W) - \hat{l}(W)|}{\|W\|_{H^m(\tilde{\Omega})}} \quad (W \neq 0)$$

which was to be proved. ■

Now we are ready to give an important result on boundary (and quadrature) errors.

THEOREM 8.10. Let the conditions of Lemma 8.1 hold. Then

$$\|\hat{U}^* - u\|_{H^m(\tilde{\Omega})} \leqslant c \left\{ \inf_{V \in S_h(\tilde{\Omega})} \left[\|u - V\|_{H^m(\tilde{\Omega})} + \sup_{W \in S_h(\tilde{\Omega})} \frac{|B_h(V,W) - \hat{B}(V,W)|}{\|W\|_{H^m(\tilde{\Omega})}} + \sup_{W \in S_h(\tilde{\Omega})} \frac{|B_h(u,W) - \hat{l}(W)|}{\|W\|_{H^m(\tilde{\Omega})}} \right] \right\} (W \neq 0) \quad (8.126)$$

PROOF. Using the triangle inequality,

$$\|\hat{U}^* - u\|_{H^m(\tilde{\Omega})} \leqslant \inf_{V \in S_h(\tilde{\Omega})} \|\hat{U}^* - V\|_{H^m(\tilde{\Omega})} + \|u - V\|_{H^m(\tilde{\Omega})}$$

and Lemma 8.1, we have

$$\|\hat{U}^* - u\|_{H^m(\tilde{\Omega})} \leqslant \left(1 + \frac{\hat{M}}{\hat{\Gamma}}\right) \inf_{V \in S_h(\tilde{\Omega})} \|u - V\|_{H^m(\tilde{\Omega})} + \frac{1}{\hat{\Gamma}} \inf_{V \in S_h(\tilde{\Omega})} \left[\sup_{W \in S_h(\tilde{\Omega})} \frac{|B_h(V,W) - \hat{B}(V,W)|}{\|W\|_{H^m(\tilde{\Omega})}} + \sup_{W \in S_h(\tilde{\Omega})} \frac{|B_h(u,W) - \hat{l}(W)|}{\|W\|_{H^m(\tilde{\Omega})}} \right]$$

from which the result follows immediately. ■

REMARKS. 1. Note that Theorem 8.10 is a generalization of Theorem 8.8. Indeed, in the special case in which $\tilde{\Omega}=\Omega=\hat{\Omega}$, we have $\hat{U}^*=\tilde{U}$, and setting $u \doteq V = U^*$ and $W=\tilde{U}-U^*$ in (8.124) gives

$$\|\tilde{U}-U^*\|^2_{H^m(\Omega)} \leqslant \frac{1}{\hat{\Gamma}}\big[|B(U^*,\tilde{U}-U^*)-\tilde{B}(U^*,\tilde{U}-U^*)|$$

$$+|l(\tilde{U}-U^*)-\tilde{l}(\tilde{U}-U^*)|\big]$$

which is precisely the inequality in the proof of Theorem 8.8.

In order to estimate the error $\|\hat{U}^*-u\|_{H^m(\tilde{\Omega})}$ in (8.126), the last two terms on the right side of (8.126) must be estimated. Ciarlet and Raviart [8.29, Theorem 9, p. 456] have estimated these terms for simplicial and quadrilateral elements.

2. Suppose that the integrals in (8.122) are computed exactly. Then $\hat{B}(V,W)=B_h(V,W)$, $\hat{l}(W)=l_h(W)$ and $\hat{U}^*=\hat{U}$. Choose $u=u^*$. Then (8.126) yields the error estimate,

$$\|\hat{U}-u^*\|_{H^m(\tilde{\Omega})} \leqslant c \inf_{V\in S_h(\tilde{\Omega})} \|u^*-V\|_{H^m(\tilde{\Omega})} \tag{8.127}$$

which includes, in addition to the approximation error, the error due to approximation of the domain.

3. If the domain is approximated exactly (e.g., in one-dimensional and polygonal domains) and the integrals are evaluated exactly, then (8.127) gives the usual error estimate (8.52).

Returning to the error (8.125) due to approximation of the domain, we must estimate the right side of (8.125) in terms of the mesh parameter h. Suppose that $\overline{\Omega}$ of (8.100) is approximated by a polygonal domain $\tilde{\Omega}\subset\overline{\Omega}$ and that the trial functions are chosen such that they vanish on $\partial\tilde{\Omega}$. We can extend them by zero outside $\partial\tilde{\Omega}$ (i.e., the trial functions are defined to be zero outside $\partial\tilde{\Omega}$). Then $S_h(\tilde{\Omega})$ is a subspace of $H_0^m(\Omega)$. We have

$$\|\hat{U}^*-u^*\|_{H^m(\Omega)} \leqslant c \inf_{V\in S_h(\tilde{\Omega})\subset H_0^m(\Omega)} \|u^*-V\|_{H^m(\Omega)} \quad \blacksquare$$

We now give a few specific examples of the estimate in (8.127). (Cf. [8.12]).

EXAMPLE 8.7. Let $m=1$, and let $S_h(\tilde{\Omega})$ be the space of piecewise linear polynomials on triangles (see Fig. 6.11). The interpolant $\tilde{U}$ of u^* is zero at boundary nodes (since $u^*=0$ on $\partial\Omega$), and therefore $\tilde{U}$ is identically zero on

$\partial\tilde{\Omega}$, and $\tilde{U} \in S_h(\tilde{\Omega}) \subset H_0^1(\tilde{\Omega})$. Hence we can choose V in (8.127) be equal to the interpolant $\tilde{U}$. Then, we have

$$\|\hat{U} - u^*\|_{H^1(\tilde{\Omega})} \leqslant c\|u^* - \tilde{U}\|_{H^1(\tilde{\Omega})} \leqslant ch\|u^*\|_{H^2(\tilde{\Omega})} \quad \blacksquare$$

EXAMPLE 8.8. Again let $m=1$, and let $S_h(\tilde{\Omega})$ be the space of piecewise quadratic polynomials on triangles. Since u^* is zero on the boundary $\partial\Omega$ and nonzero inside Ω, its interpolant $\tilde{U}$ vanishes only at the boundary nodes and assumes nonzero values at the midpoints on edges joining boundary nodes. Therefore $\tilde{U}$ is not zero on $\partial\tilde{\Omega}$, and $\tilde{U} \not\in S_h(\tilde{\Omega}) \subset H_0^1(\tilde{\Omega})$. Hence, V of (8.125) cannot be chosen to be the interpolant $\tilde{U}$. However, we can choose V to be a piecewise quadratic that interpolates u^* at the internal nodes of $\tilde{\Omega}$ and vanishes on $\partial\tilde{\Omega}$. Then, using the triangle inequality,

$$\|u^* - V\|^2_{H^1(\tilde{\Omega})} \leqslant \|u^* - \tilde{U}\|^2_{H^1(\tilde{\Omega})} + \|\tilde{U} - V\|^2_{H^1(\tilde{\Omega})}$$

where $\tilde{U}$ is the interpolant of u^* One way of proceeding further is to make use of the following arguments of Strang and Fix [8.12]: Note that the term $\tilde{U} - V$ is zero at all the interior nodes and is equal to $\tilde{U} = u^*$ at the boundary midpoints. Hence

$$\|\tilde{U} - V\|^2_{H^1(\tilde{\Omega})} = \sum_{e=1}^{\hat{E}} \int_{\Omega_e} |u^*(\mathbf{x}^e)|^2 \left(\sum_{i,j=0}^{2} a_{ij} \frac{\partial\phi_e}{\partial x_i} \frac{\partial\phi_e}{\partial x_j} \right) dx_1\, dx_2$$

where $\hat{E}$ is the number of boundary elements, $\mathbf{x}^e$ is the boundary midpoint, and ϕ_e is the quadratic trial function with value 1 at node $\mathbf{x}^e$ and zero at the other nodes of the triangle e. At the boundary midpoints $u^* = \tilde{U} = O(h^2)$, and the number of boundary triangles is proportionate to $1/h$; we have

$$\|\tilde{U} - V\|^2_{H^1(\tilde{\Omega})} \leqslant c^2 \frac{1}{h} h^4 \int_{\Omega_e} \left(\sum_{i,j=0} a_{ij} \frac{\partial\phi_e}{\partial x_i} \frac{\partial\phi_e}{\partial x_j} \right) dx_1\, dx_2$$

The first derivative of ϕ_e is of order $1/h$ and the area of the triangle Ω_e is of order h^2. Hence

$$\|\tilde{U} - V\|^2_{H^1(\tilde{\Omega})} = O(h^3)$$

We then have

$$\|u^* - V\|^2_{H^1(\tilde{\Omega})} = O(h^4) + O(h^3) = O(h^3)$$

Consequently, the error is governed by

$$\|u^* - \tilde{U}^*\|_{H^1(\tilde{\Omega})} \leqslant c\|u^* - V\|_{H^1(\tilde{\Omega})} \leqslant ch^{3/2} \quad \blacksquare$$

Similar estimates can be obtained for higher-order problems ($m \geqslant 1$). If rectangular elements are used to approximate the domain instead of triangular elements, the error in energy will be of order $h^{1/2}$ (see Strang and Fix [8.12, p. 196]). Error estimates for various types of elements can be found in Ciarlet and Raviart [8.29].

Data Errors

As pointed out earlier, approximation of a domain leads to perturbation of actual data, f and a_{ij}. The data are also perturbed due to finite arithmetics in computers (i.e., due to round-off). Suppose that the data are perturbed by δ,

$$\max_{\mathbf{x}} \left[\max_{|\alpha|,|\beta| \leqslant m} |a_{\alpha\beta}(\mathbf{x}) - \tilde{a}_{\alpha\beta}(\mathbf{x})|, |f(\mathbf{x}) - \tilde{f}(\mathbf{x})| \right] < \delta$$

Then by Theorem 8.8, the actual solution $\tilde{U}^*$ differs from the ideal (or theoretical) finite-element solution U^* by

$$\| U^* - \tilde{U}^* \|_{H^m(\Omega)} \leqslant \frac{\tilde{M}}{\tilde{\Gamma}} \delta$$

A common data error is that introduced by approximating f by piecewise polynomials. Following Fried [8.30], suppose that instead of f in (8.100) we use its interpolant $\Pi_h f = F$ in the space of piecewise polynomials of degree q which have the interpolation property

$$\|f - \Pi_h F\|_{H^{-m}(\Omega)} \leqslant Kh^{q+1+m} \|f\|_{H^{q+1}(\Omega)}$$

The error $e = U^* - \tilde{U}^*$ induced in the finite element solution in $S_h^{k,m}(\Omega)$ clearly satisfies

$$\gamma \|e\|^2_{H^m(\Omega)} \leqslant B(e,e) = \langle f - \Pi_h f, e \rangle \leqslant \|e\|_{H^m(\Omega)} \|f - \Pi_h f\|_{H^{-m}(\Omega)}$$

Hence

$$\|e\|_{H^m(\Omega)} \leqslant Ch^{q+1+m} \|f\|_{H^{q+1}(\Omega)}$$

Thus the full rate of convergence is preserved if we take

$$q = k - 2m$$

8.9 H^{-1} FINITE-ELEMENT METHODS

Rachford and Wheeler [8.31] have proposed an interesting Galerkin method which makes use of discontinuous finite-element approximations as well as different subspaces for the approximation and the weight functions (i.e., $\mathcal{U}_h \neq \mathcal{V}_h$ or the space of trial functions does not coincide with the space of test functions). The whole idea is suggested by distributional differential equations. For example, if u is a distribution and ϕ is a test function, the second-order equation

$$D^2u = f \qquad x \in \mathbb{R}$$

suggests an entire family of distributional differential equations:

$$\left.\begin{aligned} \langle D^2u, \phi\rangle &= \langle f, \phi\rangle \qquad \forall \phi \in \mathcal{D}(\mathbb{R}) \\ -\langle Du, D\phi\rangle &= \langle f, \phi\rangle \qquad \forall \phi \in \mathcal{D}(\mathbb{R}) \\ \langle u, D^2\phi\rangle &= \langle f, \phi\rangle \qquad \forall \phi \in \mathcal{D}(\mathbb{R}) \\ -\langle D^{-1}u, D^3\phi\rangle &= \langle f, \phi\rangle \qquad \forall \phi \in \mathcal{D}(\mathbb{R}) \end{aligned}\right\} \tag{8.128}$$

While all these may be equivalent from a distributional point of view, each suggests a quite different method of approximation. The first member of (8.128) suggests, for example, collocation methods, the second, Galerkin methods, etc. (see Oden [8.32]). The third member of (8.128) suggests the H^{-1} method of [8.31]. We describe the method in connection with the two-point boundary-value problem

$$\left.\begin{aligned} Au \equiv D^2u + \alpha Du + \beta u = f \qquad 0 < x < 1 \\ u(0) = 0 \qquad u(1) = 0 \end{aligned}\right\} \tag{8.129}$$

with α', $\alpha, \beta \in L_\infty(0,1)$ and which, in a distributional setting, is written

$$\langle u, A^*\phi\rangle = \langle f, \phi\rangle \qquad \forall \phi \in \mathcal{D}(0,1) \tag{8.130}$$

where $A^*u = D^2u - D(\alpha u) + \beta u$ is the adjoint of A. We weaken ϕ for practical reasons and consider the problem of finding u such that

$$(u, A^*v) = (f, v) \qquad \forall v \in H^2(I) \cap H_0^1(I) \tag{8.131}$$

where $I = (0,1)$.

Now the H^{-1} finite-element approximation of (8.131) consists of seeking the function U^* in the space

$$\mathcal{U}_h = \{ U: U \in \mathcal{P}_k(\Omega_e); 1 \leqslant e \leqslant E \} \tag{8.132}$$

where $\Omega_e = (x_{e-1}, x_e), 0 = x_0 < x_1 < \cdots < x_E = 1$, such that

$$(U^*, A^*V) = (f, V) \qquad \forall V \in \mathcal{V}_h \tag{8.133}$$

where

$$\mathcal{V}_h = \{ V: V \in \mathcal{P}_{k+2}(\Omega_e) \cap C^1(0,1); 1 \leqslant e \leqslant E; V(0) = V(1) = 0 \} \tag{8.134}$$

Thus $\mathcal{U}_h \neq \mathcal{V}_h$ and the stiffness matrix corresponding to (8.133) is unsymmetric.

To arrive at an estimate of the error of the approximation (8.133) we follow Wheeler [8.23] (and also [8.31]) and introduce a *local projection* $\tilde{U}$ of the solution u defined by

$$(u - \tilde{U}, U)_{H^0(\Omega_e)} = 0 \qquad U \in \mathcal{P}_k(\Omega_e); 1 \leqslant e \leqslant E \tag{8.135}$$

Thus, by confining ourselves to a single element, we easily arrive at the *local* error estimate using the interpolation theory of Chapter 6:

$$\|u - \tilde{U}\|_{H^0(\Omega_e)} \leqslant K_e h^q \|u\|_{H^q(\Omega_e)} \qquad 0 \leqslant q \leqslant k+1; 1 \leqslant e \leqslant E \tag{8.136}$$

As usual, $h = \max_e |x_e - x_{e-1}|$. Consequently,

$$\|u - \tilde{U}\|_{H^0(\Omega)} \leqslant K h^q \|u\|_{H^q(\Omega)} \qquad 0 \leqslant q \leqslant k+1 \tag{8.137}$$

The final error estimate indicates that (8.137) is also a bound for the approximation error $u - U^*$:

THEOREM 8.11. Let U^* denote the solution of the H^{-1} finite-element approximation (8.133) and u the exact solution of (8.131). Then there exists a constant $C > 0$ such that, for sufficiently small h,

$$\|u - U^*\|_{H^0(\Omega)} \leqslant C h^q \|u\|_{H^q(\Omega)} \qquad 0 \leqslant q \leqslant k+1 \tag{8.138}$$

PROOF. The proof involves first solving an auxiliary problem

$$A^*\phi = W \qquad \phi(0) = \phi(1) = 0$$

where $W \in \mathcal{U}_h$. We then choose a function $\Phi \in \mathcal{V}_h$ such that

$$\left(D^2(\phi-\Phi), D^2V\right)_{H^0(\Omega)}=0 \qquad \forall V \in \mathcal{V}_h \tag{8.139}$$

It can be easily shown (e.g., [8.24]) that

$$\|\phi-\Phi\|_{H^1(\Omega)} \leqslant h\|\phi\|_{H^2(\Omega)} \tag{8.140}$$

Now observe that, if $V \in \mathcal{V}_h, D^2V \in \mathcal{U}_h$, so that (8.139) implies that $D^2(\phi-\Phi)$ is orthogonal to $\mathcal{U}_h$. Consequently,

$$\left(U^*, D^2(\phi-\Phi)\right)_{H^0(\Omega)}=\left(\tilde{U}, D^2(\phi-\Phi)\right)_{H^0(\Omega)}=0 \tag{8.141}$$

where $\tilde{U}$ is the local projection defined in (8.135). By subtracting (8.133) from (8.131) we also have the orthogonality condition,

$$(u-U^*, A^*V)_{H^0(\Omega)}=0 \qquad \forall V \in \mathcal{V}_h$$

Thus

$$(u-U^*, A^*\Phi)_{H^0(\Omega)}=0 \tag{8.142}$$

We now use (8.135), (8.141), and (8.142) to show that

$$\begin{aligned}
(U^*-\tilde{U}, W)_{H^0(\Omega)} &= (U^*-u, W)_{H^0(\Omega)} \\
&= (U^*-u, A^*(\phi-\Phi))_{H^0(\Omega)} \\
&= \left(U^*-u, D^2(\phi-\Phi)\right)_{H^0(\Omega)} \\
&\quad + (U^*-u, -(D\alpha(\phi-\Phi))+\beta(\phi-\Phi))_{H^0(\Omega)} \\
&= \left(\tilde{U}-u, D^2(\phi-\Phi)\right)_{H^0(\Omega)} \\
&\quad + (U^*-u, -(\alpha D(\phi-\Phi))+\beta(\phi-\Phi))_{H^0(\Omega)}
\end{aligned}$$

which, with the aid of (8.140), yields

$$|(U^*-\tilde{U}, W)_{H^0(\Omega)}| \leqslant C(\|u-\tilde{U}\|_{H^0(\Omega)}+h\|U^*-u\|_{H^0(\Omega)})\|W\|_{H^0(\Omega)}$$

Now using the triangle inequality on $\|U^*-u\|_{H^0(\Omega)}$, setting $W=U^*-\tilde{U}$, and taking h sufficiently small, we arrive at the inequality

$$\| U^* - \tilde{U} \|_{H^0(\Omega)} \leqslant \bar{C} \| u - \tilde{U} \|_{H^0(\Omega)}$$

which, in view of (8.137), completes the proof of the theorem. ■

REMARK. There are several interesting superconvergence properties of such H^{-1} methods which can also be proved. The methods appear to be well-suited for problems with irregular data. For additional details, see the paper of Rachford and Wheeler [8.31]. For a summary account, see Douglas [8.24]. ■

8.10 HYBRID AND MIXED FINITE-ELEMENT METHODS

We sketch in this section the basic features of an important part of the theory of finite elements: the concept and structure of hybrid and mixed finite-element approximations. We adopt the following definitions:

(i) *Hybrid finite elements* A finite-element method is referred to as *hybrid* whenever independent approximations are used for the dependent variable on the interior of an element and its traces on the boundary. For example, in constructing local hybrid approximations of the Dirichlet problem

$$\left.\begin{aligned} -\Delta u_e &= f_e \quad &&\text{in } \Omega_e \\ \gamma_0 u_e &= g_e \quad &&\text{on } \partial\Omega_e \end{aligned}\right\} \tag{8.143}$$

the boundary condition is viewed as a constraint, and in the variational statement of the problem the normal derivative $\partial u / \partial n$ is identified with a Lagrange multiplier. We then construct independent polynomial approximations for $u_e(\mathbf{x}), \mathbf{x} \in \Omega_e$, and for its normal derivative $\partial u_e(\mathbf{x}) / \partial n_e, \mathbf{x} \in \partial\Omega_e$. In this way, requirements on the continuity of approximations of u_e across interelement boundaries can be relaxed.

(ii) *Mixed finite elements.* A finite-element method is *mixed* whenever independent approximations are used for both the dependent variable and its derivatives. In mixed methods, it is understood that the given operators are to be decomposed to give systems of lower-order equations. For example, instead of (8.143) we consider the equivalent problem

$$\left.\begin{aligned} &\left.\begin{aligned} \nabla u_e &= \boldsymbol{\sigma}_e \\ -\nabla \cdot \boldsymbol{\sigma}_e &= f \end{aligned}\right\} \quad \text{in } \Omega_e \\ &\quad \gamma_0 u_e = g_e \qquad\quad \text{on } \partial\Omega_e \end{aligned}\right\} \tag{8.144}$$

We then construct independent polynomial approximations of both $u_e(\mathbf{x})$ and $\boldsymbol{\sigma}_e(\mathbf{x}) \equiv \operatorname{grad} u_e(\mathbf{x})$.

(iii) *Mixed hybrid finite elements.* A finite-element method is a *mixed hybrid* method if it is both mixed and hybrid; i.e., a decomposition such as (8.144) is used, and independent approximations of u_e, $\boldsymbol{\sigma}_e = \operatorname{grad} u_e$, and $\psi_e = \gamma_1 u_e$ are constructed.

REMARKS. 1. The notion of hybrid finite-element approximations was introduced in 1964 by Jones [8.33], who suggested that reduced continuity requirements on the interpolation functions could be obtained by treating the continuity conditions as constraints and carrying along the Lagrange multipliers as new dependent variables. These ideas were further developed and extended by Pian and Tong (see, e.g., [8.34–8.36] and the references therein). The mathematical theory of hybrid methods has been studied by Babuška, Oden, and Lee [8.37], Oden and Lee [8.38], Raviart [8.39], Thomas [8.40], and Brezzi [8.41].

2. Mixed finite-element models were introduced by Herrmann [8.42], and mathematical properties of conforming mixed elements were investigated by Oden and Reddy (e.g., [8.5–8.7]), Johnson [8.43], Ciarlet and Raviart [8.44], and Oden and Lee [8.38]. Mixed methods and mixed hybrid methods were studied by Babuška, Oden, and Lee [8.37] and Oden and Lee [8.38]. An extensive account of the literature on various hybrid and mixed methods has been compliled by Wolfe [8.45, 8.46]. ■

We present a brief account of the theory of mixed hybrid methods following [8.37]. We confine our attention to the following model Dirichlet boundary-value problem in two-dimensions: Given $f \in L_2(\Omega)$ and $g \in H^{1/2}(\partial\Omega)$, find u such that

$$\begin{aligned} -\Delta u + u &= f \quad \text{in } \Omega \\ u &= g \quad \text{on } \partial\Omega \end{aligned} \tag{8.145}$$

where we assume that Ω is a polyogonal domain in $\mathbb{R}^2$, its boundary being made up of a finite number N straight-line segments meeting at N corners.

As noted earlier [recall (8.143)], the key to mixed finite-element formulations lies in the fact that (8.145) is equivalent to the problem

$$\begin{aligned} \left.\begin{aligned} \nabla u &= \boldsymbol{\sigma} \\ -\nabla \cdot \boldsymbol{\sigma} + u &= f \end{aligned}\right\} &\quad \text{in } \Omega \\ \gamma_0 u = g \qquad &\quad \text{on } \partial\Omega \end{aligned} \tag{8.146}$$

We now set out to construct a special variational principle associated with problem (8.146), which is designed to lead naturally to a class of mixed hybrid finite-element methods. The following steps are used in this construction:

(i) We begin by constructing a partition P of Ω into a finite number $E(P)=E$ of *triangular* subdomains Ω_e; i.e., Ω is viewed as having been constructed by fitting together a collection of simple triangles by connecting them together at their vertices. We select triangular subdomains only for simplicity; more general geometries are dealt with in [8.37]. These subdomains will of course eventually play the role of finite elements, but now we are interested in the formulation of a special "exact" variational statement of problem (8.146) and not its approximation.

(ii) We introduce some special spaces on this partition, as follows:

$$H^1(P)=\{u: u_e=u|_{\Omega_e}\in H^1(\Omega_e);\, 1\leqslant e\leqslant E\}$$

$$\mathbf{L}_2(P)=\{\boldsymbol{\sigma}=(\sigma_1,\sigma_2): \sigma_\alpha|_{\Omega_e}\in L_2(\Omega_e);\, \alpha=1,2;\, 1\leqslant e\leqslant E\} \tag{8.147}$$

Norms on these spaces are

$$\|u\|^2_{H^1(P)}=\sum_{e=1}^{E}\|u_e\|^2_{H^1(\Omega_e)}$$

$$\|\boldsymbol{\sigma}\|^2_{\mathbf{L}_2(P)}=\sum_{e=1}^{E}\left(\|\sigma_1\|^2_{L_2(\Omega_e)}+\|\sigma_2\|^2_{L_2(\Omega_e)}\right) \tag{8.148}$$

In addition, we introduce the spaces $H^{1/2}(\partial\Omega_e)$ of traces which, in the present case, are furnished with the norm

$$\|\phi\|_{H^{1/2}(\partial\Omega_e)}=\inf_{u_e\in H^1(\Omega_e)}\{\|u_e\|_{H^1(\Omega_e)};\, \phi=\gamma_0 u\} \tag{8.149}$$

(iii) Next we note that the boundary $\partial\Omega_e$ of each subdomain consists of three vertices $\{\mathbf{x}_e^N\}_{N=1}^3$ and three straight-line segments $\{\Gamma_e^i\}_{i=1}^3$:

$$\partial\Omega_e=\{\mathbf{x}_e^N\}_{N=1}^3\cup\left[\bigcup_{i=1}^3\Gamma_e^i\right] \qquad 1\leqslant e\leqslant E$$

The total collection of open boundary segments in the partition P is given by

$$\Gamma=\Gamma(P)=\bigcup_{\substack{1\leqslant e\leqslant E\\ 1\leqslant i\leqslant 3}}\Gamma_e^i \tag{8.150}$$

(iv) Now it is clear that families of functions can be constructed whose domain is Γ. Indeed, the space $L_2(\Gamma)$ is well-defined for a given partition P, and if $\psi \in L_2(\Gamma)$, we may define

$$\|\psi\|^2_{L_2(\Gamma)} = \sum_{\substack{1 \leq e \leq E \\ 1 \leq i \leq 3}} \int_{\Gamma_e^i} |\psi_e|^2 \, ds \tag{8.151}$$

where ψ_e is the restriction of ψ to $\partial \Omega_e$.

(v) We now introduce a space $\mathcal{W}(\Gamma)$ of functions on Γ, which is defined as the completion of $L_2(\Gamma)$ in the norm

$$\|\psi\|^2_{\mathcal{W}(\Gamma)} = \sum_{e=1}^{E} \|\psi_e\|^2_{H^{-1/2}(\partial\Omega_e)} \tag{8.152}$$

where $H^{-1/2}(\partial \Omega_e)$ is the completion of $L_2(\partial \Omega_e)$ in the norm

$$\|\psi_e\|_{H^{-1/2}(\partial\Omega_e)} = \sup_{\phi_e \in H^{1/2}(\partial\Omega_e)} \frac{\left|\oint_{\partial\Omega_e} \psi_e \phi_e \, ds\right|}{\|\phi_e\|_{H^{1/2}(\partial\Omega_e)}} \tag{8.153}$$

(vi) Next we introduce the product space

$$\mathfrak{X} = H^1(P) \times \mathbf{L}_2(P) \times \mathcal{W}(\Gamma) \tag{8.154}$$

the elements of which are triples, $\boldsymbol{\lambda} = (u, \boldsymbol{\sigma}, \psi), u \in H^1(P), \boldsymbol{\sigma} \in \mathbf{L}_2(P), \psi \in \mathcal{W}(\Gamma)$. We provide $\mathfrak{X}$ with the norm

$$\|\boldsymbol{\lambda}\|^2_{\mathfrak{X}} = \sum_{e=1}^{E} \|\boldsymbol{\lambda}_e\|^2_{\mathfrak{X}_e} \tag{8.155}$$

where $\boldsymbol{\lambda}_e = (u_e, \boldsymbol{\sigma}_e, \psi_e)$ and

$$\|\boldsymbol{\lambda}_e\|^2_{\mathfrak{X}_e} = \|u_e\|^2_{H^1(\Omega_e)} + \|\boldsymbol{\sigma}_e\|^2_{\mathbf{L}_2(\Omega_e)} + \|\psi_e\|^2_{H^{-1/2}(\partial\Omega_e)} \tag{8.156}$$

(vii) Let $\boldsymbol{\lambda}_e$ and $\bar{\boldsymbol{\lambda}}_e$ denote restrictions to Ω_e of arbitrary elements $\boldsymbol{\lambda}, \bar{\boldsymbol{\lambda}} \in \mathfrak{X}$. We next introduce the special bilinear and linear forms on $\mathfrak{X} \times \mathfrak{X}$ and $\mathfrak{X}$, respectively:

$$B(\boldsymbol{\lambda}, \bar{\boldsymbol{\lambda}}) = \sum_{e=1}^{E} B_e(\boldsymbol{\lambda}_e, \bar{\boldsymbol{\lambda}}_e) \tag{8.157}$$

$$l(\bar{\boldsymbol{\lambda}}) = \sum_{e=1}^{E} l_e(\bar{\boldsymbol{\lambda}}_e) \tag{8.158}$$

where

$$B_e(\boldsymbol{\lambda}_e, \bar{\boldsymbol{\lambda}}_e) = \int_{\Omega_e} \left[\boldsymbol{\sigma}_e \cdot \nabla \bar{u}_e + u_e \bar{u}_e + (\nabla u_e - \boldsymbol{\sigma}_e) \cdot \bar{\boldsymbol{\sigma}}_e\right] dx$$

$$+ \oint_{\partial\Omega_e} \left(\psi_e \gamma_0 \bar{u}_e + \bar{\psi}_e \gamma_0 u_e\right) ds \tag{8.159}$$

$$l_e(\bar{\boldsymbol{\lambda}}_e) = \int_{\Omega_e} f_e \bar{u}_e \, dx + \int_{\partial\Omega_e \cap \partial\Omega} g_e \bar{\psi}_e \, ds \tag{8.160}$$

(viii) Finally, we come to the special (mixed-hybrid) variational boundary-value problem: Find the triple $\boldsymbol{\lambda} \in \mathfrak{X}$ such that

$$B(\boldsymbol{\lambda}, \bar{\boldsymbol{\lambda}}) = l(\bar{\boldsymbol{\lambda}}) \qquad \forall \bar{\boldsymbol{\lambda}} \in \mathfrak{X} \tag{8.161}$$

It remains to be shown that (8.161) has a unique solution and that this solution is equivalent to our original problem (8.146). ∎

THEOREM 8.12. There exists a unique solution $\boldsymbol{\lambda}_0 \in \mathfrak{X}$ to the variational boundary-value problem (8.161). Moreover, there exists a finite positive constant C, independent of the partition P and the solution $\boldsymbol{\lambda}_0$, such that

$$\|\boldsymbol{\lambda}_0\|_{\mathfrak{X}} \leqslant C \|l\|_{\mathfrak{X}'} = C\left[\|f\|^2_{H^0(\Omega)} + \|g\|^2_{H^{1/2}(\partial\Omega)}\right]^{1/2} \tag{8.162}$$

where $\mathfrak{X}'$ is the dual of $\mathfrak{X}$.

PROOF. It is sufficient to prove that $B(\cdot,\cdot)$ satisfies the conditions (7.69) through (7.71) of Theorem 7.8. Toward this end, we first establish a fundamental lemma.

LEMMA 8.2. Let $\psi \in L_2(\partial\Omega)$ be given and let $z \in H^1(\Omega)$ be the weak solution of the Neumann problem

$$\begin{aligned} -\Delta z + z &= 0 \quad \text{in } \Omega \\ \frac{\partial z}{\partial n} &= \psi \quad \text{on } \partial\Omega \end{aligned} \tag{8.163}$$

Then

$$\|\psi\|^2_{H^{-1/2}(\partial\Omega)} = \oint_{\partial\Omega} \psi(\gamma_0 z)\, ds = \|z\|^2_{H^1(\Omega)} \tag{8.164}$$

PROOF. The solution z is uniquely characterized by the equation

$$b(z,v)=\oint_{\partial\Omega}\psi(\gamma_0 v)\,ds \qquad \forall v\in H^1(\Omega)$$

where $b(z,v)$ is a bilinear form on $H^1(\Omega)\times H^1(\Omega)$:

$$b(z,v)=\int_{\partial\Omega}(\nabla z\cdot\nabla v+zv)\,dx=(z,v)_{H^1(\Omega)}$$

By definition,

$$\|\psi\|_{H^{-1/2}(\partial\Omega)}=\sup_{\phi\in H^{1/2}(\partial\Omega)}\frac{\left|\oint_{\partial\Omega}\psi\phi\,ds\right|}{\|\phi\|_{H^{1/2}(\partial\Omega)}}$$

$$\geqslant\frac{\left|\oint_{\partial\Omega}\psi(\gamma_0 z)\,ds\right|}{\|\gamma_0 z\|_{H^{1/2}(\partial\Omega)}}\geqslant\frac{b(z,z)}{\|z\|_{H^1(\Omega)}}=\|z\|_{H^1(\Omega)}$$

We also have from the trace inequality (4.106),

$$\|\psi\|_{H^{-1/2}(\partial\Omega)}=\sup_{\phi\in H^{1/2}(\partial\Omega)}\frac{\left|\oint_{\partial\Omega}\psi\phi\,ds\right|}{\|\phi\|_{H^{1/2}(\partial\Omega)}}$$

$$=\sup_{\phi\in H^{1/2}(\partial\Omega)}\frac{|b(z,\nu\phi)|}{\|\phi\|_{H^{1/2}(\partial\Omega)}}\leqslant\|z\|_{H^1(\Omega)}$$

where ν: $H^{1/2}(\partial\Omega)\rightarrow H^1(\Omega)$. Thus

$$\|\psi\|^2_{H^{-1/2}(\partial\Omega)}=\|z\|^2_{H^1(\Omega)}=\oint_{\partial\Omega}\psi(\gamma_0 z)\,ds$$

which was to be proved. ∎

At this point we introduce the following more compact notation which will be used throughout the remainder of this section:

$$\|\psi\|_{H^{-1/2}(\partial\Omega)}=\|\psi\|_{-1/2,\partial\Omega}$$

$$\|u\|_{H^1(\Omega)}=\|u\|_{1,\Omega}$$

etc.

Returning to the proof of Theorem 8.12, we see that, if z_e is the solution of (8.163) on $\overline{\Omega}_e$, $1 \leqslant e \leqslant E$, then

$$\oint_{\partial\Omega} \psi_e \gamma_0 \bar{u}_e \, ds = b(z_e, \bar{u}_e) \leqslant \|z_e\|_{1,\Omega_e} \|\bar{u}_e\|_{1,\Omega_e}$$

$$= \|\psi_e\|_{-1/2,\partial\Omega_e} \|\bar{u}_e\|_{1,\Omega_e}$$

Similarly,

$$\oint_{\partial\Omega_e} \bar{\psi}_e \gamma_0 u_e \, ds \leqslant \|\bar{\psi}_e\|_{-1/2,\partial\Omega_e} \|u_e\|_{1,\Omega_e}$$

Using these results and the Schwarz inequality, we obtain

$$\begin{aligned}
B_e(\boldsymbol{\lambda}_e, \bar{\boldsymbol{\lambda}}_e) &\leqslant \|\boldsymbol{\sigma}_e\|_{0,\Omega_e} \|\nabla \bar{u}_e\|_{0,\Omega_e} + \|u_e\|_{0,\Omega_e} \|\bar{u}_e\|_{0,\Omega_e} \\
&\quad + \|\nabla u_e\|_{0,\Omega_e} \|\bar{\boldsymbol{\sigma}}_e\|_{0,\Omega_e} + \|\boldsymbol{\sigma}_e\|_{0,\Omega_e} \|\bar{\boldsymbol{\sigma}}_e\|_{0,\Omega_e} \\
&\quad + \|\psi_e\|_{-1/2,\partial\Omega_e} \|\bar{u}_e\|_{1,\Omega_e} \\
&\quad + \|\bar{\psi}_e\|_{-1/2,\partial\Omega_e} \|u_e\|_{1,\Omega_e} \\
&\leqslant \Big(2\|\boldsymbol{\sigma}_e\|_{0,\Omega_e}^2 + \|u_e\|_{0,\Omega_e}^2 + \|\nabla u_e\|_{0,\Omega_e}^2 \\
&\quad + \|u_e\|_{1,\Omega_e}^2 + \|\psi_e\|_{-1/2,\partial\Omega_e}^2\Big)^{1/2} \\
&\quad \cdot \Big(\|\nabla \bar{u}_e\|_{0,\Omega_e}^2 + \|\bar{u}_e\|_{0,\Omega_e}^2 + 2\|\bar{\boldsymbol{\sigma}}_e\|_{0,\Omega_e}^2 \\
&\quad + \|\bar{u}_e\|_{1,\Omega_e}^2 + \|\bar{\psi}_e\|_{-1/2,\Omega_e}^2\Big)^{1/2} \\
&\leqslant \Big(2\|u_e\|_{1,\Omega_e}^2 + 2\|\boldsymbol{\sigma}_e\|_{0,\Omega_e}^2 + \|\psi_e\|_{-1/2,\partial\Omega_e}^2\Big)^{1/2} \\
&\quad \cdot \Big(2\|\bar{u}_e\|_{1,\Omega_e}^2 + 2\|\bar{\boldsymbol{\sigma}}_e\|_{0,\Omega_e}^2 + \|\bar{\psi}_e\|_{-1/2,\partial\Omega_e}^2\Big)^{1/2} \\
&= 2\|\boldsymbol{\lambda}_e\|_{\mathfrak{X}_e} \|\bar{\boldsymbol{\lambda}}\|_{e\,\mathfrak{X}_e}
\end{aligned}$$

Hence

$$\begin{aligned}
B(\boldsymbol{\lambda}, \bar{\boldsymbol{\lambda}}) = \sum_e^E B_e(\boldsymbol{\lambda}_e, \bar{\boldsymbol{\lambda}}_e) &\leqslant 2 \sum_e^E \|\boldsymbol{\lambda}_e\|_{\mathfrak{X}_e} \|\bar{\boldsymbol{\lambda}}\|_{e\,\mathfrak{X}_e} \\
&\leqslant 2\left(\sum_e^E \|\boldsymbol{\lambda}_e\|_{\mathfrak{X}_e}^2\right)^{1/2} \left(\sum_e^E \|\bar{\boldsymbol{\lambda}}\|_{e\,\mathfrak{X}_e}^2\right)^{1/2}
\end{aligned}$$

or

$$B(\lambda,\bar{\lambda}) \leqslant 2\|\lambda\|_{\mathfrak{X}}\|\bar{\lambda}\|_{\mathfrak{X}} \tag{8.165}$$

which shows that $B(\lambda,\lambda)$ satisfies condition (7.69).

Next, we show that $B(\lambda,\bar{\lambda})$ satisfies the conditions (7.70) and (7.71). Toward this end, we introduce, for arbitrary $\lambda=(u,\boldsymbol{\sigma},\psi)$ the *special* triplet $\hat{\lambda}=(\hat{u},\hat{\boldsymbol{\sigma}},\hat{\psi})$, where $\hat{\lambda}|_{\bar{\Omega}_e}=\hat{\lambda}_e$ is given by

$$\hat{\lambda}_e=(\hat{u}_e,\hat{\sigma}_e,\hat{\psi}_e) \qquad \begin{aligned} \hat{u}_e&=2u_e+z_e \\ \hat{\boldsymbol{\sigma}}_e&=\nabla u_e+\nabla z_e-\boldsymbol{\sigma}_e \\ \hat{\psi}_e&=-3\psi_e \end{aligned} \qquad 1\leqslant e\leqslant E \tag{8.166}$$

where z_e is the solution of the local auxiliary problem

$$b(z_e,v_e)=\oint_{\partial\Omega_e}\psi_e\gamma_0 v_e\,ds \qquad \forall v_e\in H^1(\Omega_e)$$

and $b(z_e,v_e)$ is the form defined in the proof of Lemma 8.2. Thus

$$\begin{aligned}\|\hat{\lambda}_e\|^2_{\mathfrak{X}_e}&=\|2u_e+z_e\|^2_{1,\Omega_e}+\|\nabla u_e+\nabla z_e-\boldsymbol{\sigma}_e\|^2_{0,\Omega_e}+9\|\psi_e\|^2_{-1/2,\partial\Omega_e}\\ &\leqslant 15\|\lambda_e\|^2_{\mathfrak{X}_e}\end{aligned} \tag{8.167}$$

Substituting (8.166) into $B_e(\lambda_e,\hat{\lambda}_e)$, we obtain

$$\begin{aligned}B_e(\lambda_e,\hat{\lambda}_e)=&\int_{\Omega_e}(\nabla u_e\cdot\nabla u_e+2u_e^2+\boldsymbol{\sigma}_e\cdot\boldsymbol{\sigma}_e+\nabla z_e\cdot\nabla u_e+z_eu_e)\,dx\\ &-\oint_{\partial\Omega_e}\psi_e\gamma_0u_e\,ds+\int_{\partial\Omega_e}\psi_e\gamma_0z_e\,ds\end{aligned}$$

Therefore

$$\begin{aligned}B_e(\lambda_e,\hat{\lambda}_e)&\geqslant\|u_e\|^2_{1,\Omega_e}+\|\boldsymbol{\sigma}_e\|^2_{0,\Omega_e}+\|z_e\|^2_{1,\Omega_e}\\ &\geqslant\|u_e\|^2_{1,\Omega_e}+\|\boldsymbol{\sigma}_e\|^2_{0,\Omega_e}+\|\psi_e\|^2_{-1/2,\partial\Omega_e}\\ &=\|\lambda_e\|^2_{\mathfrak{X}_e}\end{aligned} \tag{8.168}$$

where we have used Lemma 8.2 in arriving at the last step. It follows that

$$B(\lambda,\hat{\lambda})=\sum_{e=1}^{E}B_e(\lambda_e,\hat{\lambda}_e)\geqslant\sum_{e}^{E}\|\lambda_e\|^2_{\mathfrak{X}_e}=\|\lambda\|^2_{\mathfrak{X}}$$

Now, using (8.167), we write

$$B(\boldsymbol{\lambda},\hat{\boldsymbol{\lambda}}) > \|\boldsymbol{\lambda}\|_{\mathfrak{X}}^2 \geqslant \frac{1}{\sqrt{15}}\|\boldsymbol{\lambda}\|_{\mathfrak{X}}\|\hat{\boldsymbol{\lambda}}\|_{\mathfrak{X}}$$

Thus

$$\inf_{\|\boldsymbol{\lambda}\|_{\mathfrak{X}}=1}\ \sup_{\|\bar{\boldsymbol{\lambda}}\|_{\mathfrak{X}}\leqslant 1} |B(\boldsymbol{\lambda},\bar{\boldsymbol{\lambda}})| \geqslant \inf_{\|\boldsymbol{\lambda}\|_{\mathfrak{X}}=1} \frac{|B(\boldsymbol{\lambda},\hat{\boldsymbol{\lambda}})|}{\|\hat{\boldsymbol{\lambda}}\|_{\mathfrak{X}}}$$

$$\geqslant \inf_{\|\boldsymbol{\lambda}\|_{\mathfrak{X}}=1} \frac{1}{\sqrt{15}}\|\boldsymbol{\lambda}\|_{\mathfrak{X}} = \frac{1}{\sqrt{15}} > 0 \qquad (8.170)$$

which proves that condition (7.70) of Theorem 7.8 holds. By interchanging the roles of $\boldsymbol{\lambda}$ and $\bar{\boldsymbol{\lambda}}$, we can prove that condition (7.71) also holds. Note that the constants $M=2$ and $\gamma=1/\sqrt{15}\approx 0.258$ are independent of the partition P. This completes the proof of Theorem 8.12. ■

THEOREM 8.13. Let $(u^*,\boldsymbol{\sigma}^*)$ denote the solution of (8.146) and let $\psi^*\in\mathcal{W}(\Gamma)$ be such that $\psi^*=-\mathbf{n}_e\cdot\boldsymbol{\sigma}_e^*=0$ on all Γ_e^i, $1\leqslant e\leqslant E, i=1,2,3$, for a given triangulation P of $\tilde{\Omega}$. Denote $(u^*,\boldsymbol{\sigma}^*,\psi^*)=\boldsymbol{\lambda}^*$. In addition, let $\boldsymbol{\lambda}_0\in\mathfrak{X}$ be a solution to (8.161) for the partition P. Then

$$\boldsymbol{\lambda}_0=\boldsymbol{\lambda}^*$$

PROOF. Noting that $\boldsymbol{\lambda}^*\in\mathfrak{X}$, we have, by integrating by parts,

$$B(\boldsymbol{\lambda}^*,\bar{\boldsymbol{\lambda}}) = \sum_{e=1}^{E}\left\{\int_{\Omega_e}\left[(-\nabla\cdot\boldsymbol{\sigma}_e^*)\bar{u}_e+(\nabla u_e^*-\boldsymbol{\sigma}_e^*)\cdot\bar{\boldsymbol{\sigma}}_e\right]dx+\oint_{\partial\Omega_e}\bar{\psi}_e u_e^*\,ds\right\}$$

$$= l(\bar{\boldsymbol{\lambda}}) + \sum_{e=1}^{E}\int_{\partial\Omega_e-\partial\Omega}\bar{\psi}_e u_e^*\,ds$$

$$= \sum_{e=1}^{E}\left\{\int_{\Omega_e} f_e\bar{u}_e\,dx + \int_{\partial\Omega_e\cap\partial\Omega}\bar{\psi}_e u_e^*\,ds\right.$$

$$\left. + \int_{\partial\Omega_e-\partial\Omega}\bar{\psi}_e u_e^*\,ds\right\}$$

However, since $u^*\in H^1(\Omega)$, each segment of $\partial\Omega_e-\partial\Omega$ is shared by two (or more) elements, and the direction of contour integration around one subdomain is opposite that around an adjacent element sharing part of the same boundary, we have

$$\sum_{e=1}^{E} \int_{\partial\Omega_e - \partial\Omega} \bar{\psi}_e u_e^* \, ds = 0$$

Therefore, by noting that $\gamma_0 u^* = g$ on $\partial\Omega$,

$$B(\lambda^*, \bar{\lambda}) = l(\bar{\lambda}) \qquad \forall \bar{\lambda} \in \mathfrak{X}$$

Hence $\lambda^* = \lambda_0$, because of the uniqueness of the solution λ_0. ■

Finite-Element Approximations

Here we describe finite-element approximations of the variational boundary-value problem (8.161). In the mixed hybrid finite-element approximations we introduce, over each element, local approximations of u_e, σ_e, and ψ_e using polynomials of possibly differing degree. We construct a collection of finite-dimensional subspaces over the partition P with the following properties:

1. The diameter h_e of finite element Ω_e is

$$h_e = \max_{\mathbf{x}, \mathbf{y} \in \Omega_e} \left(\sum_{i=1}^{2} |x_i - y_i|^2 \right)^{1/2} \qquad 1 \leqslant e \leqslant E \tag{8.171}$$

and

$$h = \max_{1 \leqslant e \leqslant E} h_e$$

We assume here that all refinements of the mesh (or partition) P are quasi-uniform; i.e., there exists a constant ν such that $h_e/\rho_e \leqslant \nu, 1 \leqslant e \leqslant E$, where ρ_e is the diameter of the largest circle that can be inscribed in $\bar{\Omega}_e$.

2.

$$Q_k^1(P) = \{ U \in H^1(P): U_e \in \mathcal{P}_{k'}(\Omega_e); 1 \leqslant e \leqslant E \} \tag{8.172}$$

where $1 \leqslant k \leqslant k'$, and $\mathcal{P}_{k'}(\Omega_e)$ is the space of polynomials of degree $\leqslant k'$ on Ω_e.

2*a*. $\dim Q_k^1(P) = K < \infty$ (8.173)

2*b*. For any function $u \in H^l(\Omega_e)$, there is a constant $C_1 > 0$, independent of h_e, and a $\tilde{U} \in Q_k^1(P)$ such that

$$\|u - \tilde{U}\|_{s,\Omega_e} \leqslant C_1 h_e^{\eta} \|u\|_{l,\Omega_e} \tag{8.174}$$

where $s = 0, 1, l > 1$, and

$$\eta = \min(k + 1 - s, l - s) \tag{8.175}$$

3. $\mathbf{Q}_r^0(P) = \{\mathbf{\Sigma} \in \mathbf{L}_2(P): \mathbf{\Sigma}_e \in (\mathcal{P}_{r'}(\Omega_e))^2; 1 \leqslant e \leqslant E\} \qquad 0 \leqslant r \leqslant r'.$ (8.176)

3*a*. $\dim \mathbf{Q}_r^0(P) = R < \infty$ (8.177)

3*b*. For any $\boldsymbol{\sigma} \in \mathbf{H}^q(\Omega_e)$, there is a constant $C_2 > 0$, independent of h_e, and a $\tilde{\mathbf{\Sigma}} \in \mathbf{Q}_r^0(P)$ such that

$$\|\boldsymbol{\sigma} - \tilde{\mathbf{\Sigma}}\|_{0,\Omega_e} \leqslant C_2 h_e^{\omega} \|\boldsymbol{\sigma}\|_{q,\Omega_e} \qquad q \geqslant 0 \tag{8.178}$$

where

$$\omega = \min(r + 1, q) \tag{8.179}$$

4. $Q_t^{-1/2}(\Gamma) = \{\Psi \in \mathcal{W}(\Gamma): \ \Psi_{ef} = \Psi|_{\Gamma_{ef}} \in \mathcal{P}_t(\Gamma_{ef}); t \geqslant 0; 0 \leqslant e, f \leqslant E\}$ (8.180)

4*a*. $\dim Q_t^{-1/2}(\Gamma) = T < \infty$ (8.181)

4*b*. For any $\psi \in \mathcal{W}(\Gamma) \cap L_2(\Gamma)$ there exists a constant $C_3 > 0$, independent of h_e, and a $\tilde{\Psi} \in Q_t^{-1/2}(\Gamma)$ such that

$$\|\psi_e - \tilde{\Psi}_e\|_{-1/2,\partial\Omega_e} \leqslant C_3 h_e^{\theta} \|\psi_e\|_{\hat{H}^{m-3/2}(\partial\Omega_e,\Omega_e)} \tag{8.182}$$

where

$$\theta = \min(t + \tfrac{3}{2}, m - 1) \qquad m \geqslant 2 \tag{8.183}$$

$\|\psi_e\|_{\hat{H}^{m-3/2}(\partial\Omega_e,\Omega_e)} = \inf\{\|v\|_{m,\Omega_e}; \psi_e = \partial v/\partial n$ on each side of $\partial\Omega_e\}$

By definition, the product space

$$\mathcal{S} = Q_k^1(P) \times \mathbf{Q}_r^0(P) \times Q_t^{-1/2}(\Gamma) \tag{8.184}$$

is a subspace of $\mathcal{X}$ defined earlier.

The mixed hybrid finite-element method consists of seeking an element

$$\mathbf{\Lambda} = (U, \mathbf{\Sigma}, \Psi) \in \mathcal{S} \tag{8.185}$$

such that

$$B(\Lambda,\overline{\Lambda}) = l(\overline{\Lambda}) \qquad \forall \overline{\Lambda} \in \mathfrak{S} \tag{8.186}$$

where $B(\cdot,\cdot)$ and $l(\cdot)$ are the bilinear and linear forms defined in (8.157) through (8.160). It is natural to inquire whether a unique solution to (8.186) exists. This depends on whether or not $B(\cdot,\cdot)$ in (8.186) satisfies the conditions of Theorem 7.8; i.e., whether or not the approximate problem (8.186) possesses the properties established in Theorem 8.1 for the variational boundary-value problem (8.161). We investigate the conditions under which such properties hold.

Let Π_e^1 and $\mathbf{\Pi}_e^0$ denote orthogonal projections of $H^1(\Omega_e)$ and $\mathbf{L}_e(\Omega_e)$ onto $Q_k^1(\Omega_e)$ and $\mathbf{Q}_r^0(\Omega_e)$, respectively. Again we construct a special element $\hat{\Lambda} \in \mathfrak{S}$ such that

$$\hat{\Lambda}_e = (\hat{U}_e, \hat{\Sigma}_e, \hat{\Psi}_e) \qquad \begin{aligned} \hat{U}_e &= 2U_e + \Pi_e^1 z_e \\ \hat{\mathbf{\Sigma}}_e &= -\mathbf{\Sigma}_e + \mathbf{\Pi}_e^0(\nabla U_e) + \mathbf{\Pi}_e^0(\nabla \Pi_e^1 z_e) \\ \hat{\Psi}_e &= -3\Psi_e \end{aligned} \tag{8.187}$$

where $z_e \in H^1(\Omega_e)$ is the solution of (8.163) on Ω_e for the choice of data, $\psi = \Psi \in Q_t^{-1/2}(\partial\Omega_e)$. Note that

$$b(z_e, U_e) = \oint_{\partial\Omega_e} \Psi_e U_e \, ds = b(\Pi_e^1 z_e, U_e) \qquad \forall U_e \in Q_k^1(\Omega_e) \tag{8.188}$$

Using the continuity of $\mathbf{\Pi}_e^0$ and Π_e^1, we have

$$\|\hat{\Lambda}_e\|_{\mathfrak{X}_e} \leqslant \sqrt{15}\, \|\Lambda_e\|_{\mathfrak{X}_e} \qquad 1 \leqslant e \leqslant E \tag{8.189}$$

Using the following identities (which follow from the orthogonal projection properties),

$$\begin{aligned} &(\mathbf{\Sigma}_e, \nabla U_e - \mathbf{\Pi}_e^0(\nabla U_e))_{0,\Omega_e} = 0 && \forall \mathbf{\Sigma}_e \in \mathbf{Q}_r^0(\Omega_e) \\ &(\mathbf{\Sigma}_e, \nabla(\Pi_e^1 z_e) - \mathbf{\Pi}_e^0(\nabla \Pi_e^1 z_e))_{0,\Omega_e} = 0 && \forall \mathbf{\Sigma}_e \in \mathbf{Q}_r^0(\Omega_e) \\ &(\nabla U_e, \mathbf{\Pi}_e^0 \nabla U_e)_{0,\Omega_e} = \|\mathbf{\Pi}_e^0 \nabla U_e\|_{0,\Omega_e}^2 && \forall U_e \in Q_k^1(\Omega_e) \end{aligned} \tag{8.190}$$

we can show that

$$\begin{aligned} B_e(\Lambda_e, \hat{\Lambda}_e) = {} & 2\|U_e\|_{0,\Omega_e}^2 + \|\mathbf{\Sigma}_e\|_{0,\Omega_e}^2 + \|\Pi_e^1 z_e\|_{1,\Omega_e}^2 \\ & + \|\mathbf{\Pi}_e^0 \nabla U_e\|_{0,\Omega_e}^2 - \int_{\Omega_e} \nabla U_e \cdot (\nabla \Pi_e^1 z_e - \mathbf{\Pi}_e^0 \nabla \Pi_e^1 z_e)\, ds \end{aligned} \tag{8.191}$$

To obtain proper bounds on these terms, we introduce the following three parameters,

$$\mu_e=\mu_e\big(Q_k^1(\Omega_e),Q_t^{-1/2}(\partial\Omega_e)\big)=\inf_{\Psi\in Q_t^{-1/2}(\partial\Omega_e)}\frac{\|\Pi_e^1 z_e\|_{1,\Omega_e}^2}{\|\Psi\|_{-1/2,\partial\Omega_e}^2}\tag{8.192}$$

$$\nu_e=\nu_e\big(Q_k^1(\Omega_e),\mathbf{Q}_r^0(\Omega_e)\big)=\inf_{V_e\in Q_k^1(\Omega_e)}\frac{\|\mathbf{\Pi}_e^0\nabla V_e\|_{0,\Omega_e}^2}{\|\nabla V_e\|_{0,\Omega_e}^2}\tag{8.193}$$

$$\gamma_e=\gamma_e\big(Q_k^1(\Omega_e),\mathbf{Q}_r^0(\Omega_e)\big)=\sup_{V_e\in Q_k^1(\Omega_e)}\frac{\|\nabla V_e-\mathbf{\Pi}_e\nabla V_e\|_{0,\Omega_e}}{\|\nabla V_e\|_{0,\Omega_e}}\tag{8.194}$$

Observe that, by definition

$$0\leqslant \mu_e,\nu_e,\gamma_e\leqslant 1$$

for nonzero denominators.

Making use of (8.192) through (8.194) and the elementary inequality $ab\leqslant\frac{1}{2}(a^2+b^2)$, we obtain from (8.191),

$$\begin{aligned}B_e(\mathbf{\Lambda}_e,\hat{\mathbf{\Lambda}}_e)&\geqslant 2\|U_e\|_{0,\Omega_e}^2+\|\mathbf{\Sigma}_e\|_{0,\Omega_e}^2+\mu_e\|\Psi_e\|_{-1/2,\partial\Omega_e}^2\\&\quad+\nu_e\|\nabla U_e\|_{0,\Omega_e}^2-\|\nabla U_e\|_{0,\Omega_e}\|\nabla\Pi_e^1 z_e-\mathbf{\Pi}_e^0\nabla\Pi_e^1 z_e\|_{0,\Omega_e}\\&\geqslant \nu_e\|U_e\|_{1,\Omega_e}^2+\|\mathbf{\Sigma}_e\|_{0,\Omega_e}^2+\mu_e\|\Psi_e\|_{-1/2,\partial\Omega_e}^2\\&\quad-\frac{\gamma_e}{2}\|\nabla U_e\|_{0,\Omega_e}^2-\frac{\gamma_e}{2}\|\nabla\Pi_e^1 z_e\|_{0,\Omega_e}^2\end{aligned}$$

Since $\|\nabla\Pi_e^1 z_e\|_{0,\Omega_e}\leqslant\|z_e\|_{1,\Omega_e}=\|\Psi_e\|_{-1/2,\partial\Omega_e}$, we have

$$\begin{aligned}B_e(\mathbf{\Lambda}_e,\hat{\mathbf{\Lambda}}_e)&\geqslant\Big(\nu_e-\frac{\gamma_e}{2}\Big)\|U_e\|_{1,\Omega_e}^2+\|\mathbf{\Sigma}_e\|_{0,\Omega_e}^2+\Big(\mu_e-\frac{\gamma_e}{2}\Big)\|\Psi\|_{-1/2,\partial\Omega_e}^2\\&\geqslant\beta_e(\Omega_e)\|\mathbf{\Lambda}_e\|_{\mathfrak{X}_e}^2\end{aligned}\tag{8.195}$$

where

$$\beta_e=\min\Big(\nu_e-\frac{\gamma_e}{2},\mu_e-\frac{\gamma_e}{2}\Big)\tag{8.196}$$

Now summing (8.195) over all E elements, making arguments similar to those used in Theorem 8.13, and recalling Theorem 8.1, we arrive at the following approximation theorem.

THEOREM 8.14. Let $B(\Lambda,\bar{\Lambda})$ denote the bilinear form on $\mathbb{S}\times\mathbb{S}$ in (8.186). Then

$$\inf_{\|\Lambda\|_{\mathfrak{X}}\leqslant 1}\ \sup_{\|\bar{\Lambda}\|_{\mathfrak{X}}=1} |B(\Lambda,\bar{\Lambda})| \geqslant \frac{1}{\sqrt{15}}\,\beta(P) \tag{8.197}$$

where

$$\beta(P) = \min_{1\leqslant e\leqslant E} \beta_e(\Omega_e) \tag{8.198}$$

and $\beta_e(\Omega_e)$ is defined in (8.196). Further, if $\beta(P)>0$ for a given partition P of Ω, there exists a unique solution Λ^0 to the mixed hybrid finite-element approximation problem (8.186), and this solution satisfies the inequality

$$\|\boldsymbol{\lambda}^0-\Lambda^0\|_{\mathfrak{X}} \leqslant \left(1+\frac{M}{\gamma_h}\right)\inf_{\Lambda\in\mathbb{S}} \|\boldsymbol{\lambda}^0-\Lambda\|_{\mathfrak{X}} \tag{8.199}$$

where $\boldsymbol{\lambda}^0$ is the unique solution to (8.161), $M=2$ is the constant appearing in (8.165), and $\gamma_h=\beta(P)/\sqrt{15}$. ■

The condition $\beta(P)>0$ is only a sufficient condition for the existence of a unique solution. A necessary condition is furnished in the following theorem.

THEOREM 8.15. In order that the approximate problem (8.186) have a unique solution $\Lambda^0\in\mathbb{S}\subset\mathfrak{X}$, it is necessary that

$$\sum_e \mu_e>0 \qquad 1\leqslant e\leqslant E \tag{8.200}$$

where μ_e is defined by (8.192).

PROOF. We prove the theorem by contradiction. Suppose that there exists a unique solution to (8.186) for arbitrary $l\in\mathfrak{X}'$, and let $\mu_e=0, 1\leqslant e \leqslant E$. Then from (8.188) and (8.192) it can be shown that there exists a $\Psi^0\in\mathfrak{W}^0(\Gamma)$ such that $\Psi^0\neq 0$ and

$$\oint_{\partial\Omega_e} \Psi_e^0 V_e\,ds=0 \qquad \forall V_e\in Q_k^1(\Omega_e) \qquad \forall e \tag{8.201}$$

Now for the choice of $\tilde{\Lambda}=(0,\mathbf{0},\Psi^0)$, $B(\tilde{\Lambda},\bar{\Lambda})=0$; also, $B((0,0,0),(\bar{\Lambda}))=0$. Hence $\tilde{\Lambda}$ is not a unique solution for $l=0$, which is a contradiction. ■

REMARK. In the special case in which

$$\mathbf{Q}_r^0(P) \supset \nabla\left(Q_k^1(P)\right) \tag{8.202}$$

which occurs whenever $k'-1 \leqslant r'$ [i.e., $\nabla U_e \in \mathbf{Q}_r^0(\Omega_e), \forall U_e \in Q_k^1(\Omega_e)$], then

$$\gamma_e = 0 \qquad \nu_e = 1 \qquad 1 \leqslant e \leqslant E \tag{8.203}$$

and

$$\beta(P) = \min_{1 \leqslant e \leqslant E} \mu_e = \mu \tag{8.204}$$

Therefore $\mu > 0$ is a sufficient condition for the existence of a unique solution in this case. ■

We now give a necessary and sufficient condition for the parameter μ_e to be strictly positive.

THEOREM 8.16. The parameter μ_e defined by (8.192) is strictly positive if and only if the following condition holds: For any $\Psi_e \in Q_t^{-1/2}(\partial\Omega_e)$,

$$\oint_{\partial\Omega_e} \Psi_e U_e \, ds = 0 \qquad \forall U_e \in Q_k^1(\Omega_e) \tag{8.205}$$

implies that $\Psi_e = 0$.

PROOF. Let condition (8.205) hold. Suppose $\mu_e = 0$. Then, in view of (8.192) and (8.188), there is a nonzero $\Psi_e \in Q_t^{-1/2}(\partial\Omega_e)$ such that (8.205) holds, a contradiction. Conversely, let $\mu_e \neq 0$ and $\Psi_e \neq 0$. Then by (8.192) and (8.188),

$$\|\Pi_e^1 z_e\|_{1,\Omega_e}^2 = \oint_{\partial\Omega_e} \Psi_e \Pi_e^1 z_e \, ds \neq 0$$

which is the contrapositive form of (8.205). ■

Condition (8.205) is referred to as the *rank condition*, because its satisfaction depends on the rank of a rectangular matrix obtained by introducing the basis functions of $Q_k^1(\Omega_e)$ and $Q_t^{-1/2}(\partial\Omega_e)$ into the contour integral in (8.205). Condition (8.205) is examined by Raviart [8.39] and Thomas [8.40] globally, i.e., by summing the integral over all elements rather than locally, and is called the *compatibility condition*. We provide more discussions on this condition at the end of this section.

Before deriving error estimates, we now must come to grips with the fact that the fundamental parameters μ_e, ν_e, and γ_e of (8.192), (8.193), and

(8.194), respectively, may depend on the mesh parameter h. The dependence of these parameters on h was studied in [8.37] and, for our purposes here, it suffices to state the basic ideas and results.

We recall from Chapter 6 that the dependence of certain finite-element projections Π_h on h can be ascertained by investigating the form of smooth transformation from some fixed master element to an arbitrary element Ω_e of diameter h_e in a finite-element mesh. Employing the same methodology here, it is not too difficult to show that ν_e and γ_e of (8.193) and (8.194) are independent of h_e. As for the dependence of μ_e of (8.192) on h_e, we have the following theorem, the proof of which is given in [8.37].

THEOREM 8.17. Let $\mu = \min_{1 \leqslant e \leqslant E} \mu_e$ and let the rank condition (8.205) hold. Then there exists a positive number $\mu^* > 0$ such that μ is uniformly bounded below by μ^* as $h \to 0$. ■

A Priori Error Estimate

Using Theorem 8.13 and the properties of subspaces defined in (8.172) through (8.183), we can derive error estimates for the mixed hybrid finite-element approximations. We state the result in the following theorem.

THEOREM 8.18. Let $u^0 \in H^l(\Omega), l \geqslant 2$, be the exact solution of the boundary-value problem (8.145) and suppose that the parameter $\beta(P)$ of (8.198) is strictly positive. If $\Lambda^0 = (U^0, \Sigma^0, \Psi^0) \in \mathfrak{S}$ is the mixed hybrid finite-element solution of (8.186), the error $\mathbf{e} = (u^0 - U^0, \nabla u^0 - \Sigma^0, -\partial u^0/\partial n - \Psi)$ is bounded by

$$\|\mathbf{e}\|_{\mathfrak{X}} \leqslant C h^{\alpha} \|u^0\|_{H^l(\Omega)} \tag{8.206}$$

where

$$\alpha = \min(k, r+1, t+\tfrac{3}{2}, l-1) \tag{8.207}$$

$$C = \left(1 + \frac{2}{\gamma_h}\right) \max(C_1, C_2, C_3) \tag{8.208}$$

C_1, C_2, and C_3 being the constants appearing in (8.174), (8.178), and (8.182), respectively, and γ_h the parameter appearing in (8.199).

PROOF. The proof is simple and follows directly from (8.174), (8.178), (8.182), and (8.199). ■

We remark here that one easy way to guarantee the condition that $\beta(P > 0$ is to choose the polynomial spaces such that (8.202) holds and

$$k \geqslant t+1 \quad \text{if } t \text{ is even}, \qquad k \geqslant t+2 \quad \text{if } t \text{ is odd} \tag{8.209}$$

over a triangular element $\overline{\Omega}_e$. This condition is a direct consequence of the previous remark and forthcoming Theorem 8.19 on the rank condition.

EXAMPLE 8.9 (The Hybrid Displacement Model). Suppose that condition $\boldsymbol{\sigma} = \nabla u$ is identically satisfied. Then we obtain the hybrid displacement model studied by Oden and Lee [8.38] and Raviart [8.39], in which independent approximations of u in Ω and $\partial u / \partial n$ on $\partial \Omega$ are constructed. In this case, we take

$$\mathfrak{X} = H^1(P) \times \mathfrak{W}(\Gamma) \qquad \mathbb{S} = Q_k^1(P) \times Q_t^{-1/2}(\Gamma) \tag{8.210}$$

Following the analogous analysis, it is easy to show that the hybrid displacement finite-element model has a unique solution if and only if the rank condition (8.205) is satisfied. Furthermore, if the rank condition holds, the following estimate holds.

$$\|\mathbf{e}\|_{\mathfrak{X}} \equiv \left\| \left(u^0 - U^0, -\frac{\partial u^0}{\partial n} - \Psi^0 \right) \right\|_{\mathfrak{X}} \leqslant C h^{\alpha} \|u^0\|_{H^l(\Omega)} \tag{8.211}$$

where

$$\begin{aligned} \alpha &= \min\left(k, t + \tfrac{3}{2}, l - 1\right) \qquad l \geqslant 2 \\ C &= \left(1 + \frac{2}{\gamma_h}\right) \max(C_1, C_3) \end{aligned} \tag{8.212}$$

and

$$\gamma_h = \min_{1 \leqslant e \leqslant E} \frac{\mu_e}{\sqrt{12}} \qquad \blacksquare$$

EXAMPLE 8.10. (Mixed Method with Constrained Boundary Conditions). This is a generalization of the Lagrange multiplier method studied by Babuška [8.47]. We set

$$\mathfrak{X} = H^1(\Omega) \times \mathbf{L}_2(\Omega) \times H^{-1/2}(\partial \Omega) \tag{8.213}$$

$$\mathbb{S} = Q_k^1(\Omega) \times \mathbf{Q}_r^0(\Omega) \times Q_t^{-1/2}(\partial \Omega) \tag{8.214}$$

where $Q_k^1(\Omega)$, $\mathbf{Q}_r^0(\Omega)$, and $Q_t^{-1/2}(\partial\Omega)$ are defined as in (8.172) through (8.183), with P replaced by Ω, and $\mathfrak{W}(\Gamma)$ by $H^{-1/2}(\partial\Omega)$. Here only the exterior boundary condition is viewed as a constraint, and the usual continuity requirements must be again imposed on U on the interelement boundaries Γ_e^i. The same analysis also holds with the corresponding changes, and the error in this case is given by

$$\|\mathbf{e}\|_{\mathfrak{X}}=\left\|\left(u^0-U^0,\nabla u^0-\Sigma^0,-\frac{\partial u^0}{\partial n}-\Psi^0\right)\right\|_{\mathfrak{X}}\leqslant Ch^{\alpha}\|u^0\|_{H^l(\Omega)} \quad (8.215)$$

with

$$\alpha=\min(k,r+1,t+\tfrac{3}{2},l-1) \qquad l\geqslant 2$$

$$C=\left(1+\frac{2}{\gamma_h}\right)\max(C_1,C_2,C_3) \quad (8.216)$$

and $\gamma_h=(1/\sqrt{15})\min(\nu-\gamma/2,\mu-\gamma/2)$.

Of course, (8.215) holds if $\gamma_h>0$. ■

EXAMPLE 8.11. (The Mixed Finite-Element Method). If we do not approximate the boundary conditions independently, but approximate u and $\nabla u=\boldsymbol{\sigma}$ independently, we arrive at the usual mixed method. In this case, we set

$$\mathfrak{X}=H^1(\Omega)\times\mathbf{L}_2(\Omega) \qquad \mathfrak{S}=Q_k^1(\Omega)\times\mathbf{Q}_r^0(\Omega) \quad (8.217)$$

Here $Q_k^1(\Omega)$ is the usual C^0 finite-element subspace whose elements satisfy the boundary condition. The error estimate becomes

$$\|\mathbf{e}\|_{\mathfrak{X}}=\|(u^0-U^0,\nabla u^0-\Sigma^0)\|_{\mathfrak{X}}\leqslant Ch^{\alpha}\|u^0\|_{H^l(\Omega)} \quad (8.218)$$

$$\alpha=\min(k,r+1,l-1)$$

$$C=\left(1+\frac{2\sqrt{6}}{\nu}\right)\max(C_3,C_4) \quad (8.219)$$

where $\nu=\min_{1\leqslant e\leqslant E}\nu_e$ and ν_e is given by (8.193). ■

The Rank Condition

The rank condition described in (8.205) deserves a closer look. Suppose that the domain under consideration is triangulated; we denote the reference or parent element by Ω_M and its boundary by $\partial\Omega_M$ (Ω_M is now the unit isosceles triangle). For simplicity we take k and k', defined in (8.172), equal. If we denote the ith ($i=1,2,3$) side of the triangle by $\hat{\Gamma}^i$, we have

$$\hat{\Gamma}=\bigcup_{i=1}^{3}\hat{\Gamma}^i$$

and an element $\Psi \in Q_t^{-1/2}(\hat{\Gamma})$ can be expressed as

$$\Psi = \bigcup_{i=1}^{3} \Psi_i \qquad \Psi_i \in \mathcal{P}_t(\hat{\Gamma}^i)$$

Here

$$\Psi_i = \sum_{\eta=1}^{t+1} b_i^{\eta} \phi_{\eta}(s) \tag{8.220}$$

where b_i^{η} are the nodal values of Ψ_i, and $\phi_{\eta}(s)$ are interpolation functions defined on the one-dimensional domains $\hat{\Gamma}^i$. It should be observed that the nodal points are not restricted to vertices of the triangle. In fact, any function U in $Q_k^1(\Omega_M)$ can be written as

$$U = \sum_{0 \leqslant \alpha+\beta \leqslant k} a_{\alpha\beta} x_1^{\alpha} x_2^{\beta} \tag{8.221}$$

Now note that the rank condition can be interpreted as follows. Given a $\Psi \in Q_t^{-1/2}(\hat{\Gamma})$, the condition

$$\sum_{i=1}^{3} \int_{\hat{\Gamma}^i} \Psi_i \gamma_0^i U \, d\hat{s} = 0 \qquad \forall U \in Q_k^1(\Omega_M) \tag{8.222}$$

implies that $\Psi = 0$, where $\gamma_0^i U$ is the restriction of U to $\hat{\Gamma}^i$. In view of (8.220) and (8.221), (8.222) leads to equations of the form

$$\mathbf{b}^T \mathbf{B} \mathbf{a} = 0 \qquad \forall \mathbf{a} \in \mathsf{R}^K \tag{8.223}$$

where, because of our assumption that complete polynomials are used,

$$K = \tfrac{1}{2}(k+1)(k+2)$$

Here $\mathbf{b}$ is a column vector of $3(t+1)$-nodal values of Ψ, $\mathbf{B}$ is a $3(t+1) \times K$ matrix, and $\mathbf{a}$ is a column vector of order K. Condition (8.223) [hence (8.222)] is satisfied if and only if the rank of the matrix $\mathbf{B}$ is

$$\rho(\mathbf{B}) = 3(t+1) \tag{8.224}$$

The following theorem establishes the relation between k and t in order that the rank condition (8.222) be satisfied. For its proof see [8.48].

THEOREM 8.19. Suppose that the domain Ω is triangulated and that the nodal points are not placed at vertices. Then the rank condition (8.222) is satisfied if and only if

$$k \geqslant t+1 \quad \text{if } t \text{ is even,} \qquad k \geqslant t+2 \quad \text{if } t \text{ is odd} \tag{8.225}$$

where $k=k'$, and t, are as defined in (8.172) and (8.180), respectively. ■

For the mixed method described in Example 8.10, the rank condition takes an alternate form: given a $\Psi \in Q_t^{-1/2}(\partial\Omega_M)$ the condition

$$\int_{\partial\Omega} \Psi U\, ds = 0 \qquad \forall U \in Q_k^1(\Omega) \text{ implies that } \Psi = 0 \tag{8.226}$$

which is satisfied if and only if $k \geqslant t$, and no more than two sides of $\partial\Omega_e$ coincide with $\partial\Omega$. For additional details, see [8.37] and [8.38].

REFERENCES

8.1. Galerkin, B. G., "Rods and Plates. Series on Some Problems of Elastic Equilibrium of Rods and Plates," *Vestn. Inzh. Tech.*, Vol. 19, pp. 897–908, 1915. In Russian.

8.2. Mikhlin, S. G., *Variational Methods in Mathematical Physics*, Translated from the 1957 Russian edition by T. Boddington, Pergamon Press, Oxford, 1964.

8.3. Lax, P. D., "On the Stability of Difference Approximations to Solutions of Hyperbolic Equations with Variable Coefficients," *Commun. Pure Appl. Math.*, Vol. 14, p. 497, 1961.

8.4. Aubin, P., *Approximation of Elliptic Boundary Value Problems*, Wiley-Interscience, New York, 1972.

8.5. Oden, J. T., "Some Contributions to the Mathematical Theory of Mixed Finite Element Approximations," *Theory and Practice in Finite Element Structural Analysis*, University of Tokyo Press, Tokyo, 1973, pp. 3–23.

8.6. Reddy, J. N. and Oden, J. T., "Mixed Finite Element Approximations of Linear Boundary-Value Problems," *Quart. Appl. Math.*, Vol. 33, No. 3, pp. 255–280, 1975.

8.7. Oden, J. T. and Reddy, J. N., "On Mixed Finite-Element Approximations," *SIAM J. Numer. Analy.*, Vol. 13, No.3, 1976.

8.8. Babuška, I. and Aziz, A. K., "Survey Lectures on the Mathematical Foundations of the Finite Element Method," in Aziz, A. K., Ed., *The Mathematical Foundations of the Finite Element Method with Applications to Partial Differential Equations*, Academic Press, New York, 1972, pp. 5–359.

8.9. Aubin, J. P., "Behavior of the Error of the Approximate Solutions of Boundary-Value Problems for Linear Elliptic Operators by Galerkin's and Finite Difference Methods," *Annali della Scuola Normale di Pisa*, Series 3, Vol. 21, pp. 599–637, 1967.

8.10. Nitsche, J., "Ein Kriterium fur die Quasi-Optimalitat des Ritzchen Verfahrens," *Numer. Math.*, Vol. 11, pp. 346–348, 1968.

8.11. Strang, G., "Approximation in the Finite Element Method," *Numer. Math.*, Vol. 19, pp. 31–98, 1972.

8.12. Strang, G. and Fix, G., *An Analysis of the Finite Element Method*, Prentice-Hall, Englewood Cliffs, N.J., 1973.

8.13. Schultz, M. H., "L^2-Error Bounds for the Rayleigh-Ritz-Galerkin Method," *SIAM J. Numer. Anal.*, Vol. 8, No. 4, pp. 737–748, 1972.

8.14. Nitsche, J., "On Dirichlet Problems Using Subspaces with Nearly Zero Boundary Conditions," in Aziz, A. K., Ed., *The Mathematical Foundations of the Finite Element Method with Applications to Partial Differential Equations*, Academic Press, New York, 1972, pp. 603–627.

8.15. Bramble, J. H. and Schatz, A., "Least Squares Method for $2m$th Order Elliptic Boundary-Value Problems," *Math. Comput.*, Vol. 25, pp. 1–32, 1971.

8.16. Aubin, J. P., "Evaluation des Erreurs de Troncature des Approximations des Expaces de Soboleve," *J. Math. Anal. Appl.*, Vol. 21, pp. 356–368, 1968.

8.17. Babuška, I., "Error-Bounds for Finite Element Method," *Numer. Math.*, Vol. 16, pp. 322–333, 1971.

8.18. Babuška, I., "Approximation by Hill Functions," *Comment Math., Univ. Carolinae*, Vol. 11, No. 4, pp. 787–811, 1970.

8.19. Mochizuki, S., "Rate of Convergence of the Finite Element Methods for Singular Data," *Proceedings, Sixth Southwestern Graduate Research Conference on Applied Mechanics*, Arlington, Texas, March 1975.

8.20. Scott, R., "The Finite Element Method with Singular Data," *Numer. Math.*, Vol. 21, pp. 317–327, 1973.

8.21. Ciarlet, P. G. and Raviart, P. A., "Maximum Principles and Uniform Convergence for the Finite Element Method," *Comput. Meth. Appl. Mech. Eng.*, Vol. 2, pp. 17–31, 1971.

8.22. Fried, I., "Finite-Element Method: Accuracy at a Point," *Quart. Appl. Math.*, pp. 149–161, July 1974.

8.23. Wheeler, M. F., "L_∞ Estimates of Optimal Order for Galerkin Methods for One-Dimensional Second-Order Parabolic and Hyberbolic Equations," *SIAM J. Numer. Anal.*, Vol. 10, pp. 908–913, 1973.

8.24. Douglas, J., *Lectures on Finite Element Methods for Elliptic and Parabolic Equations*, unpublished manuscript, 1974.

8.25. Scott, R., "Optimal L^∞ Estimates for Finite Element Method on Irregular Meshes," *Math Comp.*, (to appear).

8.26. Nitsche, J., "L_∞-Convergence of Finite Element Approximation," *Second Conference on Finite Elements*, Rennes, France, May 1975.

8.27. Fix, G., "Effects of Quadrature in the Finite Element Method," in J. T. Oden, R. W. Clough, and Y. Yamamoto, Eds., *Advances in Computational Methods in Structural Analysis and Design*, U.A.H. Press, Huntsville, Ala., 1972, pp. 55–68.

8.28. Fix, G., "Effects of Quadrature Errors in Finite Element Approximation of Steady State, Eigenvalue, and Parabolic Problems," in Aziz, A. K., Ed., *The Mathematical Foundations of the Finite Element Method with Applications to Partial Differential Equations*, Academic Press, New York, 1972, pp. 525–556.

8.29. Ciarlet, P. G. and Raviart, P. A., "The Combined Effect of Curved Boundaries and Numerical Integration in Isoparametric Finite Element Methods," in Aziz, A. K., Ed., *The Mathematical Foundations of the Finite Element Method with Applications to Partial Differential Equations,* Academic Press, New York, 1972, pp. 409–474.

8.30. Fried, I., "Boundary and Interior Approximation Errors in the Finite-Element Method," *J. Appl. Mech.,* Vol. 40, No. 4, 1973, pp. 1113–1117.

8.31. Rachford, H. H. and Wheeler, M. F., "An H^{-1} Galerkin Procedure for the Two-Point Boundary-Value Problem," in de Boor, C., Ed., *Mathematical Aspects of Finite Element Methods in Partial Differential Equations*, Academic Press, New York, 1975, pp. 353–382.

8.32. Oden, J. T., "Mathematical Aspects of Finite Element Approximations in Theoretical Mechanics," in Nemat-Nasser, Ed., *Mechanics Today*, Vol. II, Pergamon Press, Oxford, 1975, pp. 159–250.

8.33. Jones, E., "A Generalization of the Direct-Stiffness Method of Structural Analysis," *AIAA J.*, Vol. 2, No. 5, pp. 821–826, 1964.

8.34. Pian, T. H. H., "Element Stiffness Matrices for Boundary Compatibility and for Prescribed Boundary Stresses," *Proceedings, Conference on Matrix Methods in Structural Mechanics*, AFFDL-TR-66-80, Wright Patterson Air Force Base, Ohio, 1966, pp. 457–477.

8.35. Pian, T. H. H. and Tong, P., "Basis of Finite Element Methods for Solid Continua," *Int. J. Numer. Meth. Eng.*, Vol. 1, pp. 3–85, 1969.

8.36. Tong, P., "New Displacement Hybrid Finite Element Models for Solid Continua," *Int. J. Numer. Meth. Eng.*, Vol. 2, pp. 73–85, 1970.

8.37. Babuška, I., Oden, J. T., and Lee, J. K., "Mixed Hybrid Finite Elements for Second-Order Elliptic Boundary-Value Problems," in press.

8.38. Oden, J. T. and Lee, J. K., "Theory of Mixed and Hybrid Finite-Element Approximations in Linear Elasticity," (IUTAM/IUM Symposium on Applications of Methods of Functional Analysis to Problems of Mechanics), *Lecture Notes in Mathematics*, Springer-Verlag, Berlin, 1975.

8.39. Raviart, P. A., "Hybrid Finite Element Methods for Solving 2nd Order Elliptic Equations," *Conference on Numerical Analysis*, Royal Irish Academy, Dublin, 1974.

8.40. Thomas, J. M., "Methodes des Elements Finis Hybrides Duaux pour les Problems Elliptiques du Second-Order," *Report 189*, Universite Paris IV et Centre National de la Recherche Scientifique, 1975.

8.41. Brezzi, F., "Sur la Methode des Elements Finis Hybrides pour le Probleme Biharmonique," *Numer. Math.*, Vol. 24, pp. 103–131, 1975.

8.42. Herrmann, L. R., "Finite Element Bending Analysis for Plates," *J. Mech. Div., ASCE*, Vol. 93, No. EM5, pp. 13–26, 1967.

8.43. Johnson, C., "On the Convergence of a Mixed Finite Element Method for Plate Bending Problems," *Numer. Math.*, Vol. 21, pp. 43–62, 1973.

8.44. Ciarlet, P. G. and Raviart, P. A., "A Mixed Finite Element Method for the Biharmonic Equation," in de Boor, C., Ed., *Mathematical Aspects of Finite Elements in Partial Differential Equations*, Academic Press, New York, pp. 125–145, 1974.

8.45. Wolf, J. P., "Generalized Hybrid Stress Finite Element Models," *AIAA J.*, Vol. 11, No. 3, pp. 386–388, 1973.

8.46. Wolf, J. P., "Generalized Stress Models for Finite Element Analysis," Ph.D. Thesis, Swiss Federal Institute of Technology (ETH), Dissertation No. 5263, Zurich, 1974.

8.47. Babuška, I., "The Finite Element Method with Lagrange Multipliers," *Numer. Math.*, Vol. 20, pp. 172–192, 1973.

8.48. Raviart, P. G. and Thomas, J. M., "Primal Hybrid Finite Element Methods for 2nd order Elliptic Equations," Report No. 75021, Universite Paris VI, Laboratoire Analyse Numerique.

9

TIME-DEPENDENT PROBLEMS

9.1 INTRODUCTION

This chapter is devoted to an introduction to the theory of finite-element approximations of time-dependent problems, with emphasis on techniques for establishing error estimates of certain finite-element-finite-difference methods. To keep the scope within reasonable limits, we emphasize linear parabolic equations. The underlying theory is more completely developed for this class of problems. However, we also consider finite-element approximations of certain linear hyperbolic problems which, to date, have not been treated in depth in the literature. Moreover, many of the techniques we consider can be applied to finite-element approximations of many other types of linear boundary-initial-value problems.

We describe several methods that have been developed for studying the accuracy and convergence of approximations to time-dependent problems. There are many techniques for arriving at error estimates, and these include the L_2 methods, energy methods, parabolic regularization for hyperbolic problems, semigroup theoretic methods, and others. The L_2 theory, together with a variety of other Galerkin-type methods for parabolic equations, has been largely developed by Douglas and Dupont (e.g., [9.1–9.7]), with important contributions by Wheeler [9.8] and others (see, e.g., Price and Varga [9.9], Thomee [9.10], or the summary accounts in Varga [9.11], Oden [9.12], Strang and Fix [9.13]); what we refer to as energy methods have been used by Fujii [9.14], Oden and Fost [9.15]; parabolic regularization schemes for finite-element approximations of both linear and nonlinear hyperbolic problems have been discussed by Wellford and Oden [9.16]. For semigroup methods for finite-element approximations we follow Wellford and Oden [9.17, 9.18]. We are content in this chapter only to outline some representative methods for studying certain linear parabolic and hyperbolic problems. For additional references and details, the specific sources cited here should be consulted.

9.2 FINITE-ELEMENT MODELS OF THE DIFFUSION EQUATION

We begin by considering a class of time-dependent problems characterized by equations of the form

$$\left.\begin{aligned} &\frac{\partial u(\mathbf{x},t)}{\partial t} + A(\mathbf{x})u(\mathbf{x},t) = f(\mathbf{x},t) \qquad \mathbf{x}\in\Omega;\ t\in(0,T] \\ &\gamma_j u(\mathbf{x},t) = 0 \qquad \mathbf{x}\in\partial\Omega;\ t\in[0,T];\ 0<j\leqslant m-1 \\ &u(\mathbf{x},0) = u_0(\mathbf{x}) \qquad \mathbf{x}\in\Omega \end{aligned}\right\} \tag{9.1}$$

where Ω is, as usual, a smooth, bounded, open domain in $\mathbf{R}^n$ with a smooth boundary $\partial\Omega$, $A(\mathbf{x})$ is a strongly elliptic $2m$th-order partial differential operator with C^∞ coefficients of the form

$$A(\mathbf{x}) = \sum_{|\alpha|,|\beta|\leqslant m} (-1)^{|\alpha|} D^\alpha a_{\alpha\beta}(\mathbf{x}) D^\beta \tag{9.2}$$

T is a real number >0, and $\gamma_j u(\mathbf{x},t) \equiv \partial^j u/\partial n^j|_{\mathbf{x}\in\partial\Omega}$. As usual, we associate with $A(\mathbf{x})$ a bilinear form $B(u,v)$ of the form

$$B(u,v) = \int_\Omega \sum_{|\alpha|,|\beta|\leqslant m} a_{\alpha\beta}(\mathbf{x}) D^\alpha v D^\beta u\, dx \tag{9.3}$$

which, for simplicity, is assumed to be continuous and strongly coercive on $H_0^m(\Omega)$ for every $t\in[0,\infty)$; i.e., there exist constants M and γ, >0, such that

$$B(u,v) \leqslant M\|u\|_{H^m(\Omega)}\|v\|_{H^m(\Omega)} \qquad \text{and} \qquad B(u,u) \geqslant \gamma\|u\|^2_{H^m(\Omega)} \tag{9.4}$$

for every $u,v\in H_0^m(\Omega)$. We can then replace (9.1) by the equivalent variational problem of finding u such that

$$\begin{aligned} &\left(\frac{\partial u(t)}{\partial t}, v\right) + B(u(t),v) = (f(t),v) \qquad \forall v\in H_0^m(\Omega);\ t\in(0,T] \\ &(u(0),v) = (u_0,v) \qquad \forall v\in H_0^m(\Omega) \end{aligned} \tag{9.5}$$

where $(\cdot,\cdot)$ denotes the $L_2(\Omega)$ inner product, and we use the notation $(u(\cdot,t),v)\equiv(u(t),v)$.

The Galerkin method for the approximate solution of (9.5) follows lines similar to that of the stationary problem. We identify, for each $t\geqslant 0$, a subspace $S_h(\Omega)$ of $H_0^m(\Omega)$ spanned by a system of G linearly independent functions $\{\phi_i(\mathbf{x})\}_{i=1}^G$ generated using finite-element techniques. We assume that $S_h(\Omega)$ belongs to a regular $S_h^{k,m}(\Omega)$ family, as described in the previous chapter. Each element in $S_h(\Omega)$ is therefore of the form

$$V(\mathbf{x})=\sum_{i=1}^{G} c^i\phi_i(\mathbf{x}) \tag{9.6}$$

where the c^i are constants. The Galerkin approximation to (9.5) is, then, the function

$$U(\mathbf{x},t)=\sum_{i=1}^{G} A^i(t)\phi_i(\mathbf{x}) \tag{9.7}$$

assumedly unique, which satisfies

$$\begin{aligned}\left(\frac{\partial U(t)}{\partial t},V\right)+B(U(t),V)&=(f(t),V) \qquad t\in(0,T]\\ (U(0),V)&=(u_0,V) \qquad \forall V\in S_h(\Omega)\end{aligned} \tag{9.8}$$

Since a continuous dependence on t is still assumed, $U(\mathbf{x},t)$ is referred to as a *semidiscrete* Galerkin approximation. Note that the coefficients A^i in (9.7) are functions of time t. Thus, on introducing (9.7) into (9.8), we obtain a system of first-order differential equations for the specific coefficients $A^i(t)$ corresponding to the finite-element approximation

$$\left.\begin{aligned}\sum_{j=1}^{G}\left[G_{ij}\dot{A}^j(t)+K_{ij}A^j(t)\right]&=f_i(t)\\ \sum_{j=1}^{G}G_{ij}A^j(0)&=g_i\end{aligned}\right\} \tag{9.9}$$

where

$$\left.\begin{aligned}G_{ij}&=(\phi_i,\phi_j) \qquad K_{ij}=B(\phi_i,\phi_j)\\ f_i&=(f,\phi_i) \qquad g_i=(u_0,\phi_i)\end{aligned}\right\} \tag{9.10}$$

and $dA^j(t)/dt\equiv\dot{A}^j(t)$.

In practical calculations, we introduce the partition P of $[0,T]$ composed of the set $\{t_0,t_1,\ldots,t_R\}$, where $0=t_0<t_1<\cdots<t_R=T$, with $t_{n+1}-t_n=\Delta t$; and we introduce the sequence $\{U^n\}_{n=0}^R$ to denote the value of the function $U(\mathbf{x},t)\in S_h(\Omega)\times C^1(0,T)$ at the time points of partition P. Then we construct a family of *finite-difference Galerkin* approximations associated with parameter θ $(0\leqslant\theta\leqslant 1)$ which represent solutions to the following equation:

$$\left.\begin{aligned}(\delta_t U^n,V)+(1-\theta)B(U^{n+1},V)+\theta B(U^n,V)&=(f(t),V) \quad t\in(0,T]\\ (U^0,V)&=(u_0,V) \quad \forall V\in S_h(\Omega)\end{aligned}\right\} \tag{9.11}$$

where δ_t denotes the forward difference operator; i.e., $\delta_t U^n=(U^{n+1}-U^n)/\Delta t$. Expansion of (9.11) with the finite-element approximation (9.9) leads to a system of algebraic equations for the coefficients $A_j^n\approx A^j(n\Delta t)$:

$$\sum_{j=1}^{G}\left[G_{ij}+\Delta t(1-\theta)\delta_{ik}K_{kj}\right]A_j^{n+1}=\sum_{j=1}^{G}(G_{ij}-\Delta t\theta\,\delta_{ik}K_{kj})A_j^n+\Delta t f_i$$

$$\sum_{j=1}^{G}G_{ij}A_j^0=g_i \tag{9.12}$$

For a given R, we solve this set of equations for $\{A_j^n\}_{\substack{0\leqslant n\leqslant R\\0\leqslant j\leqslant G}}$ and, with the aid of (9.7), thereby determine the fully discrete approximation of (9.5).

9.3 SEMIDISCRETE L_2 GALERKIN APPROXIMATIONS

We outline briefly in this section some techniques that have become quite popular for determining a priori L_2 estimates for semidiscrete Galerkin approximations of the diffusion equation. The basic idea is to solve an *auxiliary* steady-state Galerkin problem: Find $\tilde{U}\in S_h(\Omega)$ such that

$$B(\tilde{U},V)=B(u,V) \qquad \forall V\in S_h(\Omega);\ t\geqslant 0 \tag{9.13}$$

where u is the exact solution of (9.5). Since $S_h(\Omega)\in S_h^{k,m}(\Omega)$, this Galerkin approximation $\tilde{U}$ has several familiar properties for sufficiently smooth u; e.g.,

$$\left.\begin{aligned}\|u-\tilde{U}\|_{H^m(\Omega)}&\leqslant Kh^{k+1-m}\|u(t)\|_{H^{k+1}(\Omega)}\\ \left\|\frac{\partial}{\partial t}(u-\tilde{U})\right\|_{H^m(\Omega)}&\leqslant Kh^{k+1-m}\left\|\frac{\partial u(t)}{\partial t}\right\|_{H^{k+1}(\Omega)}\end{aligned}\right\} \tag{9.14}$$

where $\|u(\cdot,t)\|_{H^{k+1}(\Omega)} \equiv \|u(t)\|_{H^{k+1}(\Omega)}$, etc. In addition, we have from (9.5) and (9.8) that

$$\left(\frac{\partial}{\partial t}(u-\tilde{U}),V\right)=\left(-\frac{\partial}{\partial t}\tilde{U},V\right)+\left(\frac{\partial U}{\partial t},V\right)+B(U,V)-B(u,V)$$

where U is the semidiscrete Galerkin solution defined in (9.8). Introducing (9.13), we obtain

$$\left(\frac{\partial}{\partial t}(u-\tilde{U}),V\right)=\left(\frac{\partial}{\partial t}(U-\tilde{U}),V\right)+B(U-\tilde{U},V)$$

or, setting $V=U-\tilde{U}$,

$$\left(\frac{\partial}{\partial t}(u-\tilde{U}),U-\tilde{U}\right)=\left(\frac{\partial}{\partial t}(U-\tilde{U}),U-\tilde{U}\right)+B(U-\tilde{U},U-\tilde{U})$$

We note that for any $\phi(x,t)\in L_2(\Omega)$, $\forall t\geqslant 0$, we have the identity

$$\frac{1}{2}\frac{d}{dt}\|\phi(\cdot,t)\|^2=\left(\frac{\partial}{\partial t}\phi(\cdot,t),\phi(\cdot,t)\right)=\|\phi(\cdot,t)\|\frac{d}{dt}\|\phi(\cdot,t)\|$$

where $\|\cdot\|=(\cdot,\cdot)^{1/2}$ is the $L_2(\Omega)$ norm. Thus, we are led to the inequality,

$$\left(\frac{\partial\tilde{e}}{\partial t},\tilde{E}\right)=\left(\frac{\partial\tilde{E}}{\partial t},\tilde{E}\right)+B(\tilde{E},\tilde{E})\geqslant\|\tilde{E}\|\frac{d}{dt}\|\tilde{E}\|+\gamma\|\tilde{E}\|^2 \tag{9.15}$$

where, for simplicity in notation, we have denoted

$$\tilde{e}=u-\tilde{U} \qquad \text{and} \qquad \tilde{E}=U-\tilde{U} \tag{9.16}$$

and the dependence on t is, for the moment, understood. Since

$$\left(\frac{\partial\tilde{e}}{\partial t},\tilde{E}\right)\leqslant\left\|\frac{\partial\tilde{e}}{\partial t}\right\|\|\tilde{E}\|$$

inequality (9.15) leads to the first-order differential inequality

$$\frac{d}{dt}\|\tilde{E}\|+\gamma\|\tilde{E}\|\leqslant\left\|\frac{\partial\tilde{e}}{\partial t}\right\| \tag{9.17}$$

the solution of which is

$$\|\tilde{E}(t)\|\leqslant\exp(-\gamma t)\left[\|\tilde{E}(0)\|+\int_0^t\exp(\gamma s)\left\|\frac{\partial\tilde{e}}{\partial t}\right\|ds\right] \tag{9.18}$$

However, from the triangle inequality we have

$$
\begin{aligned}
\|u(t)-U(t)\| &\equiv \|e(t)\| \\
&= \|u(t)-\tilde{U}(t)+\tilde{U}(t)-U(t)\| \\
&\leqslant \|\tilde{e}(t)\| + \|\tilde{E}(t)\| \\
&\leqslant Ch^{k+1}\|u(t)\|_{H^{k+1}(\Omega)} \\
&\quad + \exp(-\gamma t)\left[\|\tilde{E}(0)\| + \int_0^t \exp(\gamma s)\left\|\frac{\partial \tilde{e}}{\partial t}\right\| ds\right]
\end{aligned}
$$

Collecting all these results, and using (9.14), we arrive at the following approximation theorem.

THEOREM 9.1. Let $U(\mathbf{x},t)$ denote the semidiscrete finite-element approximation in (9.8) of the weak diffusion equation (9.5) and suppose that the finite-element subspace $S_h(\Omega)\subset H_0^m(\Omega)$ has the interpolation property (8.52) for smooth $u(\mathbf{x},t)$. Then approximation error $e(t)=u(\cdot,t)-U(\cdot,t)$ is such that $\forall t\geqslant 0$,

$$
\begin{aligned}
\|e(t)\| \leqslant \exp(-\gamma t)\|e(0)\| + Ch^{k+1}\Bigg[&\|u(t)\|_{H^{k+1}(\Omega)} + \exp(-\gamma t)\|u(0)\|_{H^{k+1}(\Omega)} \\
&+ \int_0^t \exp[\gamma(s-t)]\left\|\frac{\partial u(s)}{\partial t}\right\|_{H^{k+1}(\Omega)} ds\Bigg]
\end{aligned}
\tag{9.19}
$$

where $\|\cdot\| = \|\cdot\|_{L_2(\Omega)}$ and C is a positive constant. ■

Thus, for the semidiscrete approximation, the accuracy is of the same order as the comparable elliptic problem.

9.4 ELEMENTS OF SEMIGROUP THEORY

There are several fairly elegant techniques for studying evolution equations of all types using the concept of a semigroup. The foundations of semigroup theory were laid down in Hille and Phillips [9.19], and readable accounts can be found in Butzer and Berens [9.20] and Goldstein [9.21]; for an introduction see Schechter [9.22]. Semigroup techniques often prove to be very powerful and to allow us to go further than the L_2 semidiscrete theory of the previous section. We now record some of the basic features of semigroup theory, and we will put these to use in developing an associated theory of approximation in Section 9.5.

The whole idea of semigroups springs from properties of solutions of the elementary equation of evolution:

$$\left.\begin{aligned} \dot{u}(t) &= Au \qquad t>0 \\ u(0) &= u_0 \end{aligned}\right\} \tag{9.20}$$

where A is a constant. The solution of course is

$$u(t)=E(t)u_0 \tag{9.21}$$

where

$$E(t)u_0 = u_0 e^{tA} \tag{9.22}$$

The operator $E(t)$ has some obvious properties:

$$\left.\begin{aligned} &1.\quad \lim_{t\to 0} E(t) = I(=1) \\ &2.\quad E(t_1)E(t_2) = E(t_1+t_2) \\ &3.\quad E(t_1)(E(t_2)E(t_3)) = (E(t_1)E(t_2))E(t_3) \end{aligned}\right\} \tag{9.23}$$

Effectively, $E(t)$ maps the initial data u_0 into the current value of the solution $u(t)$. A collection of operators, closed under the operation of composition, which satisfy properties (9.23) is called a *semigroup*.

The importance of semigroups is that relations such as (9.20) through (9.23) can be shown to hold for operator equations on Banach spaces. Indeed, now suppose A is the elliptic operator of (9.2) and u is, for each $t \geqslant 0$, an element of a Banach space $\mathscr{U}$. Then consider the *abstract evolution problem*:

$$\left.\begin{aligned} \dot{u}(t) &= Au(t) \qquad t>0 \\ u(0) &= u_0 \in \mathscr{U} \end{aligned}\right\} \tag{9.24}$$

where $\dot{u}(t_0)$ is the strong derivative of $u(t)$ at t_0; i.e., $\forall \varepsilon>0$, there is a $\delta>0$ such that $|t-t_0|<\delta$ implies

$$\left\| \frac{u(t)-u(t_0)}{t-t_0} - \dot{u}(t_0) \right\|_{\mathscr{U}} < \varepsilon \tag{9.25}$$

for $u(t)$, $u(t_0) \in \mathscr{U}$.

We assume that A is a bounded operator on $\mathscr{U}$, and we make use of the spectral theory of linear operators [recall (4.87)]. We define for this problem,

$$E(t)=e^{tA}=\sum_{k=0}^{\infty}\frac{1}{k!}t^kA^k \tag{9.26}$$

Then $E(t)$ is a bounded operator, and we may write down the following properties:

$$\lim_{m,n\to\infty}\left\|\sum_{k=m}^{n}\frac{1}{k!}t^kA^k\right\| \leqslant \lim_{m,n\to\infty}\sum_{k=m}^{n}\frac{1}{k!}|t|^k\|A\|^k=0 \tag{9.27}$$

where $\|\cdot\|$ is the operator norm on $\mathcal{L}(\mathcal{U},\mathcal{U})$, and

$$\|E(t)\|=\|e^{tA}\|\leqslant e^{|t|\,\|A\|} \tag{9.28}$$

We can easily verify that

$$u(t)=u_0e^{tA}=E(t)u_0 \tag{9.29}$$

is a solution of (9.24):

$$\frac{u(t+\Delta t)-u(t)}{\Delta t}=\frac{e^{(t+\Delta t)A}-e^{tA}}{\Delta t}u_0$$

$$=\frac{e^{\Delta tA}-I}{\Delta t}e^{tA}u_0$$

where we have assumed that $e^{A+B}=e^Ae^B$. Thus

$$\frac{u(t+\Delta t)-u(t)}{\Delta t}-Au=\left[\sum_{k=2}^{\infty}\frac{1}{k!}(\Delta t)^{k-1}A^k\right]u(t)$$

and therefore

$$\left\|\frac{u(t+\Delta t)-u(t)}{\Delta t}-Au(t)\right\|_{\mathcal{U}}\leqslant\sum_{k=2}^{\infty}\frac{1}{k!}|\Delta t|^{k-1}\|A\|^k\|u\|_{\mathcal{U}}$$

Clearly, the right-hand side of this inequality $\to 0$ as $\Delta t\to 0$, so that (9.29) is in fact a solution of (9.24).

We also may verify that this solution is unique. Assume otherwise:

$$\dot{u}_1=Au_1 \qquad u_1(0)=u_0 \qquad \dot{u}_2=Au_2 \qquad u_2(0)=u_0$$

Then, if $w = u_1 - u_2$,

$$\dot{w} = Aw \qquad \text{and} \qquad w(0) = 0$$

We must show that $w = 0$. Set $v = e^{-tA}w$. Then

$$\frac{v(t+\Delta t) - v(t)}{\Delta t} = e^{-(t+\Delta t)A}\left[\frac{w(t+\Delta t) - w(t)}{\Delta t}\right] + \frac{e^{-\Delta t A} - I}{\Delta t} v(t)$$

Noting that $\lim_{\Delta t \to 0}(1/\Delta t)(e^{-\Delta t A} - I) = -A$, we see that, as $\Delta t \to 0$, $\dot{v} = e^{-tA}(\dot{w} - Aw)$. Thus

$$\overline{(e^{-tA}w)}^{\,\cdot} = \dot{v} = 0 \tag{9.30}$$

Now let l denote a continuous linear functional on $\mathcal{U}$, and set

$$f(t) = l(e^{-tA}w)$$

It is easily verified that $f(t)$ is differentiable if $t > 0$ and continuous if $t \geqslant 0$, and that $f(t) = 0$, $t > 0$, and $f(0) = 0$. Thus $f \equiv 0$. Therefore (9.30) does indeed hold. Finally, we set

$$w(t) = e^{tA}(e^{-tA}0) = 0$$

so that the solution is unique.

EXAMPLE 9.1. Let $\mathbf{q} = \{q_1, q_2, \dots, q_n\}^T$ be an n-vector, let $\mathbf{A}$ be a $n \times n$ invertible matrix of constants, and consider the linear dynamical system

$$\dot{\mathbf{q}} - \mathbf{A}\mathbf{q} = \mathbf{f} \qquad \mathbf{q}(0) = \mathbf{q}_0$$

the solution of which is

$$\mathbf{q} = e^{\mathbf{A}t}\mathbf{q}_0 + \int_0^t e^{\mathbf{A}(t-s)}\mathbf{f}(s)\, ds$$

For simplicity, let us assume that $\mathbf{f}(s) = 0$. Then the operator $\mathbf{E}(t)$ is a $n \times n$ matrix given by

$$\mathbf{E}(t) = e^{\mathbf{A}t}$$

and

$$\mathbf{q}(t) = \mathbf{E}(t)\mathbf{q}_0$$

How does one construct the fundamental solution operator $\mathbf{E}(t)$ from a matrix $\mathbf{A}$? Let $\mathcal{L}(\mathbf{q}(t)) = \bar{\mathbf{q}}$ denote the Laplace transform of $\mathbf{q}(t)$ [i.e., $\mathcal{L}(\mathbf{q}(t)) = \int_0^\infty e^{-st}\mathbf{q}(t)\,dt$, where s is the complex transformed variable]. Then

$$s\bar{\mathbf{q}} - \mathbf{q}(0) = \mathbf{A}\bar{\mathbf{q}}$$

Thus, if s is not an eigenvalue of $\mathbf{A}$,

$$\bar{\mathbf{q}} = (s\mathbf{I} - \mathbf{A})^{-1}\mathbf{q}_0$$

and

$$\mathbf{E}(t) = \mathcal{L}^{-1}(s\mathbf{I} - \mathbf{A})^{-1} \quad \blacksquare$$

EXAMPLE 9.2. Now a similar situation is encountered in the use of the weak formulation of partial differential equations of the type (9.5). In this case we define a fundamental solution operator $E(\mathbf{x}, t)$ such that

$$u(\cdot, t) = E(\cdot, t)u_0(\cdot)$$

Then it is clear from the previous examples and the definition of the resolvent operator $R(\mathbf{x}, s) = (sI - A)^{-1}$ that

$$R(\mathbf{x}, s) = \int_0^\infty e^{-st}E(\mathbf{x}, t)\,dt.$$

Hence, if $\bar{u} = \mathcal{L}(u)$, then

$$\bar{u}(\mathbf{0}, s) = R(\mathbf{x}, s)u_0(\mathbf{0})$$

and, of course,

$$u(t) = \frac{1}{2\pi i}\int_\Gamma R(\mathbf{x}, s)e^{st}\,ds\,u_0$$

where Γ is a contour in the complex plane.

Now we take the Laplace transform of the weak parabolic partial differential equation (9.5) and set $f(t) = 0$, for convenience. Then

$$\mathcal{L}\left[\left(\frac{\partial u(t)}{\partial t}, v\right)\right] + \mathcal{L}[B(u(t), v)] = 0 \qquad \forall v \in H_0^m(\Omega) \tag{9.31}$$

Now

$$\mathfrak{L}\left[\left(\frac{\partial u(t)}{\partial t},v\right)\right]=\int_\Omega\int_0^\infty e^{-st}\frac{\partial u(\mathbf{x},t)}{\partial t}v(\mathbf{x})\,dt\,dx$$

$$=s\int_\Omega \bar{u}(\mathbf{x},s)v(\mathbf{x})\,dx-\int_\Omega u_0(\mathbf{x})v(\mathbf{x})\,dx$$

$$=s(\bar{u}(s),v)-(u_0,v) \tag{9.32}$$

and

$$\mathfrak{L}[B(u(t),v)]=B(\bar{u}(s),v) \tag{9.33}$$

Hence, introducing (9.32) and (9.33) into (9.31), we obtain the *boundary-value problem*

$$s(\bar{u}(s),v)+B(\bar{u}(s),v)=(u_0,v) \qquad \forall v\in H_0^m(\Omega) \tag{9.34}$$

■

The above examples lead us directly to the following theorem:

THEOREM 9.2. Let A be a closed linear operator on a Banach space $\mathfrak{U}$ having the positive real axis of the complex plane in its resolvent set $\rho(A)$ and let A be such that there is a constant a satisfying

$$\frac{1}{a+\lambda}\geqslant\|(\lambda I-A)^{-1}\| \qquad \lambda>\max(0,-a) \tag{9.35}$$

Then there is a family $\{E(t)\}$ of bounded operators in $\mathfrak{L}(\mathfrak{U},\mathfrak{U})$, $t\geqslant 0$, such that

(i) $$E(s)E(t)=E(t+s) \qquad s,t\geqslant 0 \tag{9.36}$$

(ii) $$E(0)=I \tag{9.37}$$

(iii) $E(t)u(t)$ is continuous in t, $\forall t\geqslant 0$ and $u\in\mathfrak{U}$.

(iv) $E(t)u(t)$ is differentiable in t, $\forall t\geqslant 0$ and

$\forall u\in\mathfrak{D}(A)$.

(v) $$\overline{(E(t)u(t))^{\cdot}}=AE(t)u(t) \tag{9.38}$$

(vi) $$E(t)(\lambda I-A)^{-1}=(\lambda\cdot I-A)^{-1}E(t) \tag{9.39}$$

PROOF. A clear and detailed proof of this theorem is given in Schechter [9.22]. ■

REMARKS. As noted earlier, the one-parameter family $E(t)$ described in Theorem 9.2 is called a semigroup. The operator A is called an *infinitesimal generator* of this semigroup; such infinitesimal generators are unique. Notice that, if A is an infinitesimal generator, its domain is dense in $\mathfrak{U}$; if $u \in \mathfrak{D}(A)$, $\overline{\dot{E(t)u(t)}}$ exists for $t \geqslant 0$ and $\overline{\dot{E(t)u(t)}} = AE(t)u(t)$. Conversely, it can be shown that for every one-parameter semigroup $\{E(t)\}$ in $\mathfrak{L}(\mathfrak{U}, \mathfrak{U})$ and continuous for $t \geqslant 0$, there exists an infinitesimal generator A. ■

9.5 SEMIGROUP METHODS FOR GALERKIN APPROXIMATIONS

We now describe how the semigroup ideas can be used to develop error estimates for fully discrete Galerkin approximations of evolution equations. Our starting point is Example 9.2, and in fact a repetition of the procedure described there for semidiscrete Galerkin approximations. For additional information on applications in numerical analysis, see [9.17, 9.18, 9.23–9.25].

Initially, we must define the components of the error of the approximation scheme: $u(\mathbf{x},t)$ is the exact solution to (9.5); $U(\mathbf{x},t)$ is the solution to the semidiscrete Galerkin approximation (9.8); and $U^n(\mathbf{x})$ is the solution to the finite-difference Galerkin approximation (9.11) at time $t = n\Delta t$. We introduce the definitions,

$$\left.\begin{aligned} e(\mathbf{x}, n\Delta t) &= u(\mathbf{x}, n\Delta t) - U^n(\mathbf{x}) = \text{approximation error} \\ \sigma(\mathbf{x}, n\Delta t) &= u(\mathbf{x}, n\Delta t) - U(\mathbf{x}, n\Delta t) = \begin{array}{l}\text{semidiscrete}\\ \text{approximation error}\end{array} \\ \tau(\mathbf{x}, n\Delta t) &= U(\mathbf{x}, n\Delta t) - U^n(\mathbf{x}) = \begin{array}{l}\text{temporal approximation}\\ \text{error}\end{array} \end{aligned}\right\} \tag{9.40}$$

Then, for any choice of norm,

$$\begin{aligned} \|e(t)\| &= \|u(\cdot, n\Delta t) - U^n(\cdot) + U(\cdot, n\Delta t) - U(\cdot, n\Delta t)\| \\ &= \|\sigma + \tau\| \leqslant \|\sigma\| + \|\tau\| \end{aligned} \tag{9.41}$$

To use semigroup theory for the semidiscrete approximation, we introduce an *approximate fundamental solution* operator $\mathbf{E}_h(\mathbf{x}, t)$ such that

$$U(\mathbf{x}, t) = \mathbf{E}_h(\mathbf{x}, t) u_0(\mathbf{x}) \tag{9.42}$$

If we introduce the approximation $\mathbf{A}_h$ to matrix $\mathbf{A}$, an approximate resolvent operator $\mathbf{R}_h(\mathbf{x},s)$ is defined such that

$$\mathbf{R}_h(\mathbf{x},s)=(s\mathbf{I}-\mathbf{A}_h)^{-1}$$

Then the approximate resolvent is again related to the approximate fundamental solution operator through

$$\mathbf{E}_h(\mathbf{x},t)=\mathfrak{L}^{-1}\left[(s\mathbf{I}-\mathbf{A}_h)^{-1}\right]=\mathfrak{L}^{-1}\left[\mathbf{R}_h(\mathbf{x},s)\right]$$

Thus

$$\mathbf{R}_h(\mathbf{x},s)=\int_0^\infty e^{-st}\mathbf{E}_h(\mathbf{x},t)\,dt \tag{9.43}$$

Hence, if $\bar{U}=\mathfrak{L}[U]$, the transformed version of (9.42) is

$$\bar{U}(\mathbf{x},s)=\mathbf{R}_h(\mathbf{x},s)u_0(\mathbf{x})$$

and of course

$$U(\mathbf{x},t)=\frac{1}{2\pi i}\int_\Gamma \mathbf{R}_h(\mathbf{x},s)e^{st}\,ds\,u_0(\mathbf{x}) \tag{9.44}$$

Now, if we take the Laplace transform of the semidiscrete Galerkin equation (9.8), setting $f(t)=0$ for convenience, we obtain an approximate boundary-value problem

$$s\big(\bar{U}(s),V\big)+B\big(\bar{U}(s),V\big)=(u_0,V)\qquad \forall V\in S_h(\Omega) \tag{9.45}$$

Subtracting (9.45) from (9.34) and defining the transformed approximation error $\bar{\sigma}(s)$ by $\bar{u}(s)-\bar{U}(s)=\mathfrak{L}[\sigma(t)]=\mathfrak{L}[u(t)-U(t)]$, we obtain

$$s(\bar{\sigma}(s),V)+B(\bar{\sigma}(s),V)=0\qquad \forall V\in S_h(\Omega) \tag{9.46}$$

Thus it is clear that, if we can estimate the approximation error $\bar{\sigma}(s)$ for the boundary-value problem (9.34), we can deduce the approximation error for problem (9.8) by using the inverse Laplace transform:

$$\begin{aligned}\sigma(t)&=\mathfrak{L}^{-1}\left[\bar{\sigma}(s)\right]=\mathfrak{L}^{-1}\left[\bar{u}(s)-\bar{U}(s)\right]\\&=\frac{1}{2\pi i}\int_\Gamma\left[\mathbf{R}(\mathbf{x},s)-\mathbf{R}_h(\mathbf{x},s)\right]e^{st}\,ds\,u_0(\mathbf{x})\end{aligned}$$

To ensure that the inverse Laplace transform exists, we assume that the operator $\mathbf{A}$ is the infinitesimal generator of a strongly continuous semigroup. The Hille-Yosida-Phillips theorem (e.g., [9.19]) guarantees that, if the operator **A** is the infinitesimal generator of a strongly continuous semigroup, there exist real numbers M and ω such that, for every $\operatorname{Re} s > \omega$, s is in the resolvent set of **A** (i.e., $(s\mathbf{I}-\mathbf{A})^{-1}$ exists), and $\|\mathbf{R}(\mathbf{x},s)^n\| \leqslant M/(\operatorname{Re} s-\omega)^n$, $n=1,2,3,\ldots$. Thus, if we select the contour Γ in the complex plane so that, if $s\in\Gamma$, s is in the resolvent set, $\operatorname{Re} s>\beta$, and $\operatorname{Re} s=\beta$ as $s\to\infty$, where β is a small negative constant, then the integrand in (9.44) is bounded as s increases, and the inverse Laplace transform exists. We choose β so that $\gamma+\beta\equiv\gamma_1>0$.

The operator **A** satisfies the $H_0^m(\Omega)$-elliptic condition (9.4). Thus the operator $\mathbf{A}+\operatorname{Re} s\mathbf{I}$ also satisfies the $H_0^m(\Omega)$-elliptic condition, since

$$\begin{aligned}((\mathbf{A}+\operatorname{Re} s\mathbf{I})v,v)&=(\mathbf{A}v,v)+\operatorname{Re} s(v,v)\\ &=B(v,v)+\operatorname{Re} s(v,v)\\ &\geqslant B(v,v)+\beta\|v\|^2_{H^m(\Omega)}\\ &\geqslant \gamma_1\|v\|^2_{H^m(\Omega)} \qquad \forall v\in H_0^m(\Omega)\end{aligned} \tag{9.47}$$

Thus the techniques developed earlier for elliptic problems are now applicable to the transformed problem.

THEOREM 9.3. If $\overline{U}(s)$ is the Galerkin approximation in $S_h(\Omega)$ to the solution of problem (9.34),

$$\|\bar{u}(s)-\overline{U}(s)\|_{H^m(\Omega)}\leqslant \overline{C}_1 h^{k+1-m}\|\bar{u}(s)\|_{H^{k+1}(\Omega)} \tag{9.48}$$

PROOF. From the $H_0^m(\Omega)$-elliptic property of operator $\mathbf{A}+\operatorname{Re} s\mathbf{I}$, we have that, if $U^*(s)$ is an arbitrary element of $S_h(\Omega)$,

$$\begin{aligned}\gamma_1\|\bar{u}(s)-\overline{U}(s)\|^2_{H^m(\Omega)}\leqslant&\operatorname{Re} s\big(\bar{u}(s)-\overline{U}(s),\bar{u}(s)-\overline{U}(s)\big)\\ &+B\big(\bar{u}(s)-\overline{U}(s),\bar{u}(s)-\overline{U}(s)\big)\\ =&\operatorname{Re} s\big(\bar{u}(s)-\overline{U}(s),\bar{u}(s)-\overline{U}^*(s)+\overline{U}^*(s)-\overline{U}(s)\big)\\ &+B\big(\bar{u}(s)-\overline{U}(s),\bar{u}(s)-\overline{U}^*(s)+\overline{U}^*(s)-\overline{U}(s)\big)\\ =&\operatorname{Re} s\big(\bar{u}(s)-\overline{U}(s),\bar{u}(s)-\overline{U}^*(s)\big)\\ &+B\big(\bar{u}(s)-\overline{U}(s),\bar{u}(s)-\overline{U}^*(s)\big)\end{aligned}$$

To obtain the last result, we have used the identity

$$\operatorname{Re} s\big(\bar{u}(s)-\bar{U}(s),\bar{U}^*(s)-\bar{U}(s)\big)+B\big(\bar{u}(s)-\bar{U}(s),\bar{U}^*(s)-\bar{U}(s)\big)=0$$

obtained by taking the real part of (9.46) and setting $V=\bar{U}^*(s)-\bar{U}(s)$. Now using (9.4), we obtain

$$\begin{aligned}\gamma_1\|\bar{u}(s)-\bar{U}(s)\|^2_{H^m(\Omega)}&\leqslant \operatorname{Re} s\|\bar{u}(s)-\bar{U}(s)\|_{H^m(\Omega)}\|\bar{U}^*(s)-\bar{u}(s)\|_{H^m(\Omega)}\\&\quad+M\|\bar{u}(s)-\bar{U}(s)\|_{H^m(\Omega)}\|\bar{U}^*(s)-\bar{u}(s)\|_{H^m(\Omega)}\\&=(\operatorname{Re} s+M)\|\bar{u}(s)-\bar{U}(s)\|_{H^m(\Omega)}\|\bar{U}^*(s)-\bar{u}(s)\|_{H^m(\Omega)}\end{aligned}$$

Thus

$$\|\bar{u}(s)-\bar{U}(s)\|_{H^m(\Omega)}\leqslant\frac{\operatorname{Re} s+M}{\gamma_1}\|\bar{U}^*(s)-\bar{u}(s)\|_{H^m(\Omega)}\tag{9.49}$$

Now let $\bar{U}^*(s)$ be the arbitrary element of $S_h(\Omega)$ which interpolates $\bar{u}(s)$. Then we see from the usual interpolation results that

$$\begin{aligned}\|\bar{u}(s)-\bar{U}(s)\|_{H^m(\Omega)}&\leqslant\frac{\operatorname{Re} s+M}{\gamma_1}Ch^{k+1-m}\|\bar{u}(s)\|_{H^{k+1}(\Omega)}\\&=\bar{C}_1h^{k+1-m}\|\bar{u}(s)\|_{H^{k+1}(\Omega)}\end{aligned}$$

This completes the proof. ■

Thus we have determined the error estimate for the semidiscrete approximation (9.8):

THEOREM 9.4. If $u(\cdot,n\Delta t)$ is the solution to (9.5) at time $t=n\Delta t$, and $U(\mathbf{x},n\Delta t)$ is the solution to (9.8) at time $t=n\Delta t$, then there exists a constant $C_2>0$ such that

$$\|\sigma(\cdot,n\Delta t)\|_{H^m(\Omega)}\leqslant C_2h^{k+1-m}\|u(\cdot,n\Delta t)\|_{H^{k+1}(\Omega)}\tag{9.50}$$

PROOF.

$$\begin{aligned}\|\sigma(\cdot,n\Delta t)\|_{H^m(\Omega)}&=\left\|\frac{1}{2\pi i}\int_\Gamma\big[\bar{u}(s)-\bar{U}(s)\big]e^{st}\,ds\right\|_{H^m(\Omega)}\\&\leqslant\left|\frac{\bar{C}_1}{2\pi i}\right|h^{k+1-m}\int_\Gamma\|\bar{u}(s)\|_{H^{k+1}(\Omega)}|e^{st}|\,ds\\&=C_2h^{k+1-m}\|u(\cdot,n\Delta t)\|_{H^{k+1}(\Omega)}\end{aligned}$$ ■

Now let $E_h(\mathbf{x};t,\tau)$ be the *fundamental solution operator* associated with the semidiscrete Galerkin approximation (9.8), and $E_h^{\Delta t,\theta}(\mathbf{x})$ be the fundamental solution operator (often called the *amplification* matrix) associated with the finite-difference Galerkin approximation (9.11):

$$U(\mathbf{x},(\nu+1)\Delta t)=E_h(\mathbf{x};(\nu+1)\Delta t,\nu\Delta t)U(\mathbf{x},\nu\Delta t) \tag{9.51}$$

where $\nu=0,1,2,\cdots$ and

$$U^n(\mathbf{x})=\left(E_h^{\Delta t,\theta}(\mathbf{x})\right)^n U^0(\mathbf{x}) \tag{9.52}$$

The operator E_h is assumed to have the semigroup property. Now the semidiscrete problem (9.8) is assumed to be well posed, and the finite-difference Galerkin problem is assumed to be stable in the sense that the solution depends continuously on the initial data. Thus the fundamental solution operator and the amplification matrix are bounded; i.e., we assume that there exist constants $C_3, C_4>0$ such that

$$\|E_h(\mathbf{x};(\nu-1)\Delta t,0)U(\mathbf{x})\|_{H^m(\Omega)}\leqslant C_3\|U(\mathbf{x})\|_{H^m(\Omega)}$$
$$\nu=1,2,\ldots,N;\quad \forall U\in S_h(\Omega) \tag{9.53}$$

and

$$\|E_h^{\Delta t,\theta}(\mathbf{x})V(\mathbf{x})\|_{H^m(\Omega)}\leqslant C_4\|V(\mathbf{x})\|_{H^m(\Omega)}\qquad \forall V\in S_h(\Omega) \tag{9.54}$$

Effectively, $E_h^{\Delta t,\theta}(\mathbf{x})$ is assumed to be uniformly bounded: there exists a $\tau>0$ such that for $0<\Delta t<\tau$ and $0\leqslant n\Delta t\leqslant T$, the family of maps

$$\left(E_h^{\Delta t,\theta}(\mathbf{x})\right)^n,\qquad 0<\Delta t<\tau,\qquad 0\leqslant n\Delta t\leqslant T \tag{9.55}$$

is uniformly bounded; i.e.

$$\|E_h^{\Delta t,\theta}(\mathbf{x})\|=\sup_{\substack{V\in S_h(\Omega)\\(V\neq 0)}}\frac{\|E_h^{\Delta t,\theta}(\mathbf{x})V(\mathbf{x})\|_{H^m(\Omega)}}{\|V\|_{H^m(\Omega)}}\leqslant C_4 \tag{9.56}$$

Consider again the semidiscrete finite element-Galerkin equation, and solve for the value of the solution vector $A^M(\nu\Delta t)$ based on the "initial value" of $A^M((\nu-1)\Delta t)$ (setting $f_i=0$). We find

$$A^i(\nu\Delta t)=\sum_{j=1}^{G}E_j^iA^j((\nu-1)\Delta t) \tag{9.57}$$

where

$$E^i_{\cdot j} = e^{-\Delta t B^i_{\cdot j}} \qquad \text{and} \qquad B^i_{\cdot j} = \sum_{k=1}^{G} G^{-1}_{ik} K_{kj} \tag{9.58}$$

The matrix $E^i_{\cdot j}$ is a semidiscrete approximation for the fundamental solution operator E. Similarly, solving the finite-difference Galerkin equation (9.12) for the values of A^ν_j in terms of $A^{\nu-1}_j$ and setting $f_j = 0$, we obtain

$$A^\nu_i = \sum_{j=1}^{G} F^\theta_{ij} A^{\nu-1}_j \tag{9.59}$$

where

$$F^\theta_{ij} = \sum_{k=1}^{G} \left[\delta_{ik} + \Delta t(1-\theta) B^i_{\cdot k} \right]^{-1} \left(\delta_{kj} - \Delta t \theta B^k_{\cdot j} \right) \tag{9.60}$$

The operator F^θ_{ij} obviously defines $E^{\Delta t, \theta}_h$.

Padé Approximations

A useful approximation for the term $E^i_{\cdot j}$ in (9.57) can be obtained through the Padé approximations. The Padé approximations $R_{p,q}(\Delta t B^i_{\cdot j})$ are a rational matrix approximation for $e^{-\Delta t B^i_{\cdot j}}$ defined by

$$\begin{aligned} R_{p,q}\left(\Delta t B^i_{\cdot j}\right) &= n_{p,q}\left(\Delta t B^i_{\cdot j}\right)\left[d_{p,q}\left(\Delta t B^i_{\cdot j}\right)\right]^{-1} \\ &= e^{-\Delta t B^i_{\cdot j}} + O\left(\Delta t B^i_{\cdot j}\right)^r \end{aligned} \tag{9.61}$$

where

$$\left.\begin{aligned} n_{p,q}\left(\Delta t B^i_{\cdot j}\right) &= \sum_{k=0}^{q} (p+q-k)!\left[(p+q)!k!(q-k)!\right]^{-1}\left(-\Delta t B^i_{\cdot j}\right)^k \\ d_{p,q}\left(\Delta t B^i_{\cdot j}\right) &= \sum_{k=0}^{p} (p+q-k)!\left[(p+q)!k!(p-k)!\right]^{-1}\left(\Delta t B^i_{\cdot j}\right)^k \end{aligned}\right\} \tag{9.62}$$

$$r = p + q + 1$$

From (9.58), (9.60), and (9.61) it can be shown that the following relationships hold.

$$\left.\begin{aligned} R_{0,1} &= F_{ij}^{1} = E_{\cdot j}^{i} + O(\Delta t^2) \\ R_{1,0} &= F_{ij}^{0} = E_{\cdot j}^{i} + O(\Delta t^2) \\ R_{1,1} &= F_{ij}^{1/2} = E_{\cdot j}^{i} + O(\Delta t^3) \end{aligned}\right\} \tag{9.63}$$

Here the choice of θ corresponds to the forward difference, backward difference, and Crank-Nicholson schemes, respectively. The notation indicates that each term of the matrix F_{ij}^{θ} can be expressed as the sum of the corresponding term in the matrix $E_{\cdot j}^{i}$ plus a term of order of magnitude Δt^r. According to (9.63),

$$E_{\cdot j}^{i} - F_{ij}^{1} = O(\Delta t^2)$$

$$E_{\cdot j}^{i} - F_{ij}^{0} = O(\Delta t^2)$$

$$E_{\cdot j}^{i} - F_{ij}^{1/2} = O(\Delta t^3)$$

Now for some $V(\mathbf{x},t) = \sum_{i=1}^{G} C^i(t)\phi_i(\mathbf{x})$, we can define

$$(E_h - E_h^{\Delta t,\theta})V = \sum_{i,j=1}^{G} (E_{\cdot j}^{i} - F_{ij}^{\theta})C^j(\nu\Delta t)\phi_i(\mathbf{x})$$

so that $\| E_h - E_h^{\Delta t,\theta} \|$ is equivalent to a matrix norm on $E_{\cdot j}^{i} - F_{ij}^{\theta}$. Thus, using (9.63), we have

$$\left.\begin{aligned} \| E_h(\mathbf{x};\nu\Delta t,(\nu-1)\Delta t) - (E_h^{\Delta t,1}(\mathbf{x}))^{\nu} \| &\leqslant C_5\Delta t^2 \\ \| E_h(\mathbf{x};\nu\Delta t,(\nu-1)\Delta t) - (E_h^{\Delta t,0}(\mathbf{x}))^{\nu} \| &\leqslant C_6\Delta t^2 \\ \| E_h(\mathbf{x};\nu\Delta t,(\nu-1)\Delta t) - (E_h^{\Delta t,1/2}(\mathbf{x}))^{\nu} \| &\leqslant C_7\Delta t^3 \end{aligned}\right\} \tag{9.64}$$

The above results can be used to obtain temporal error estimates. We derive such temporal error estimates only for the forward difference approximation ($\theta = 1$). Error estimates for other temporal operators can be derived in similar fashion.

THEOREM 9.5. Let $U(\mathbf{x}, n\Delta t)$ be the solution to the semidiscrete Galerkin equation (9.8) at time $t = n\Delta t$, and $U^n(\mathbf{x})$ be the solution to the forward difference Galerkin approximation (9.11), with $\theta=1$, at time $t=n\Delta t$. In addition, let (9.54), (9.56), and (9.64) hold. Then the temporal approximation error τ is such that

$$\|\tau(\mathbf{x},n\Delta t)\|_{H^m(\Omega)}=\|U(\mathbf{x},n\Delta t)-U^n(\mathbf{x})\|_{H^m(\Omega)}\leqslant C_8\Delta t\|U^0(\mathbf{x})\|_{H^m(\Omega)} \tag{9.65}$$

PROOF. Using the semigroup properties (9.51) and (9.52), we have

$$\begin{aligned}
\|\tau(\mathbf{x},n\Delta t)\|_{H^m(\Omega)}&=\|U(\mathbf{x},n\Delta t)-U^n(\mathbf{x})\|_{H^m(\Omega)}\\
&=\left\|\left(E_h(\mathbf{x};n\Delta t,0)-\left(E_h^{\Delta t,1}(\mathbf{x})\right)^n\right)U^0(\mathbf{x})\right\|_{H^m(\Omega)}\\
&=\left\|\sum_{\nu=1}^{n}\left(E_h^{\Delta t,1}(\mathbf{x})\right)^{n-\nu}\left[E_h^{\Delta t,1}(\mathbf{x})-E_h(\mathbf{x};\nu\Delta t,(\nu-1)\Delta t)\right]\right.\\
&\qquad\left.\times E_h(\mathbf{x};(\nu-1)\Delta t,0)\,U^0(\mathbf{x})\right\|_{H^m(\Omega)}\\
&\leqslant\sum_{\nu=1}^{n}\left\|\left(E_h^{\Delta t,1}(\mathbf{x})\right)^{n-\nu}\right\|\left\|E_h^{\Delta t,1}(\mathbf{x})-E_h(\mathbf{x};\nu\Delta t,(\nu-1)\Delta t)\right\|\\
&\qquad\times\left\|E_h(\mathbf{x};(\nu-1)\Delta t,0)\,U^0(\mathbf{x})\right\|_{H^m(\Delta)}
\end{aligned}$$

Now using (9.53), (9.56), and (9.64), we obtain

$$\begin{aligned}
\|\tau(\mathbf{x},n\Delta t)\|_{H^m(\Omega)}&\leqslant nC_4C_5\Delta t^2C_3\|U^0(\mathbf{x})\|_{H^m(\Omega)}\\
&\leqslant C_8\Delta t\|U^0(\mathbf{x})\|_{H^m(\Omega)} \quad \blacksquare
\end{aligned}$$

The total error estimate of the approximation is given in the following theorem.

THEOREM 9.6. If the components of the error of the approximation are given by (9.40),

$$\|e(n\Delta t)\|_{H^m(\Omega)}\leqslant C_2h^{k+1-m}\|u(n\Delta t)\|_{H^{k+1}(\Omega)}+C_8\Delta t\|U^0\|_{H^m(\Omega)} \tag{9.66}$$

PROOF. The theorem follows immediately from (9.41), (9.50), and (9.65). ■

9.6 HYPERBOLIC EQUATIONS OF SECOND ORDER

We outline briefly in this section a study of the stability and convergence of a finite-element central difference approximation of the model hyperbolic problem

$$\left.\begin{aligned} &\left(\frac{\partial^2 u(t)}{\partial t^2}, v\right) + B(u(t), v) = (f(t), v) \qquad t \in (0, T] \\ &\left(\frac{\partial u(0)}{\partial t}, v\right) = 0 \\ &(u(0), v) = 0 \qquad \forall v \in H_0^1(\Omega) \end{aligned}\right\} \tag{9.67}$$

where

$$B(u, v) = \int_\Omega \nabla u \cdot \nabla v \, dx \tag{9.68}$$

We employ a fully discrete approximation to (9.67) using finite elements in **x** and central differences in t such that the following hold:

(i) The interval $[0, T]$ is partitioned into R equal subintervals, $0 = t_0 < t_1 < \cdots < t_R = T$, $t_{n+1} - t_n = \Delta t$.

(ii) For each $t \in [0, T]$ and each $0 < h \leqslant 1$, the approximate solution U of (9.67) lies in a subspace $S_h(\Omega) \subset H_0^1(\Omega)$ such that

(ii.1) $S_h(\Omega) \in S_h^{k,1}(\Omega)$

(ii.2) The $S_h^{k,1}(\Omega)$ family is regular and has the inverse property (8.71).

(iii) The difference quotients $\delta_t^{1/2}$ and δ_t^2 are given by the forward and central difference formulas,

$$\delta_t^{1/2} u^n = \frac{u^{n+1} - u^n}{\Delta t} \qquad \delta_t^2 u^n = \frac{u^{n+1} - 2u^n + u^{n-1}}{\Delta t^2} \tag{9.69}$$

where $u^n = u(n\Delta t)$.

Thus, our approximation to (9.67) assumes the form

$$\begin{aligned} &(\delta_t^2 U^n, V) + B(U^n, V) = (f^n, V) \qquad \forall V \in S_h(\Omega) \\ &\qquad\qquad\qquad\qquad\qquad\qquad\qquad\qquad 0 < n \leqslant R \\ &(U^0, V) = 0, \ (\delta_t^{1/2} U^0, V) = 0 \qquad \forall V \in S_h(\Omega) \end{aligned} \tag{9.70}$$

We also use the following standard notation:

$$\|u\|^2_{L_2(H(\Omega))} = \int_0^T \|u(t)\|^2_{H(\Omega)}\,dt \tag{9.71}$$

and, following Dupont [9.25], we assume that the solution u of (9.67) is such that $\partial^4 u/\partial t^4 \in L_2(L_2(\Omega))$. In this case it can be shown (see [9.25]) that

$$\|\varepsilon_n\|^2_{L_2(\Omega)} \leqslant C\Delta t^3 \int_{t_{n-1}}^{t_{n+1}} \left\| \frac{\partial^4 u(\tau)}{\partial \tau^4} \right\|^2_{L_2(\Omega)} d\tau \tag{9.72}$$

where

$$\varepsilon_n \equiv \delta_t^2 u^n - \left(\frac{\partial^2 u}{\partial t^2} \right)\bigg|_{t = n\Delta t} \tag{9.73}$$

With these preliminaries now behind us, we proceed to the study of errors in (9.70). As a first step, we evaluate (9.67) at $t = n\Delta t$, set $v = V \in S_h(\Omega)$, and add $(\delta_t^2 u^n, V)$ to both sides of the resulting equation. Then subtracting (9.70) from this result, we obtain

$$(\delta_t^2 e_n, V) + B(e_n, V) = (\varepsilon_n, V) \qquad \forall V \in S_h(\Omega) \tag{9.74}$$

As in the case of the semidiscrete parabolic problem discussed in Section 9.3, we introduce an auxiliary problem. Let $W^n \in S_h(\Omega)$ be such that

$$B(W^n, V) = B(u^n, V) \qquad \forall V \in S_h(\Omega);\ 0 \leqslant n \leqslant R \tag{9.75}$$

Also, we introduce the notation

$$e_n = E_n + \mathcal{E}_n$$

where

$$E_n = u^n - W^n \qquad \text{and} \qquad \mathcal{E}_n = W^n - U^n$$

Here of course u^n and U^n are the exact solution and its approximation, respectively, evaluated at $t = n\Delta t$. We also use the abbreviated notations

$$u_{n+1/2} = \tfrac{1}{2}(u^{n+1} + u^n)$$

$$\delta_t u_{n+1/2} = \frac{u^{n+1} - u^n}{\Delta t}$$

$$\delta_{t_{n+1/2}}(X) = \frac{1}{\Delta t}\left(X|_{t=(n+1)\Delta t} - X|_{t=n\Delta t}\right)$$

Collecting all these results, we are led to the following theorem concerning the behavior of the error component $\mathcal{E}_n$.

THEOREM 9.7. Let the solution u of (9.67) be such that u exists and $\partial^4 u/\partial t^4 \in L_2(L_2(\Omega))$. Further, let conditions (i), (ii), and (iii) stated above hold for problem (9.70) and let Δt and h be chosen so that the following stability criterion is satisfied:

$$\frac{\Delta t^2}{h^2} \leqslant \hat{C} \tag{9.76}$$

where $\hat{C}$ is a specific positive constant independent of Δt and h but dependent on the choice of a basis for $S_h(\Omega)$. Then there exist positive constants C_1, C_2, not depending on the discretization parameters, such that

$$\begin{aligned}\|\delta_t \mathcal{E}\|_{\tilde{L}_\infty(L_2(\Omega))} &+ C_1 \|\mathcal{E}\|_{\tilde{L}_\infty(H^1(\Omega))}\\ &\leqslant C_2 \Bigg(\|\mathcal{E}_0\|_{H^1(\Omega)} + \|\mathcal{E}_1\|_{H^1(\Omega)} + \|\delta_t \mathcal{E}_{1/2}\| \\ &\qquad + \left\| \frac{\partial^2 E}{\partial t^2} \right\|_{L_2(L_2(\Omega))} + \Delta t^2 \left\| \frac{\partial^4 u}{\partial t^4} \right\|_{L_2(L_2(\Omega))} \Bigg)\end{aligned} \tag{9.77}$$

where

$$\|\delta_t \mathcal{E}\|_{\tilde{L}_\infty(L_2(\Omega))} = \sup_{0 \leqslant n \leqslant R} \|\delta_t \mathcal{E}_{n-1/2}\| \tag{9.78}$$

PROOF. It follows from the decomposition of e_n and (9.74) that

$$(\delta_t^2 \mathcal{E}_n, V) + B(\mathcal{E}_n, V) = -(\delta_t^2 E_n, V) - B(E_n, V) + (\varepsilon_n, V) \qquad \forall V \in S_h(\Omega)$$

Now let $V = \delta_t \mathcal{E}_{n+1/2} + \delta_t \mathcal{E}_{n-1/2}$; then

$$\begin{aligned}(\delta_t^2 \mathcal{E}_n, \delta_t \mathcal{E}_{n+1/2} + \delta_t \mathcal{E}_{n-1/2}) &+ B(\mathcal{E}_n, \delta_t \mathcal{E}_{n+1/2} + \delta_t \mathcal{E}_{n-1/2})\\ &= -(\delta_t^2 E_n, \delta_t \mathcal{E}_{n+1/2} + \delta_t \mathcal{E}_{n-1/2})\\ &\quad - B(E_n, \delta_t \mathcal{E}_{n+1/2} + \delta_t \mathcal{E}_{n-1/2}) + (\varepsilon_n, \delta_t \mathcal{E}_{n+1/2} + \delta_t \mathcal{E}_{n-1/2})\end{aligned}$$

But

$$B(\mathcal{E}_n, \delta_t \mathcal{E}_{n+1/2}) = -\frac{\Delta t}{2} B(\delta_t \mathcal{E}_{n+1/2}, \delta_t \mathcal{E}_{n+1/2}) + \tfrac{1}{2} \delta_{t_{n+1/2}} B(\mathcal{E}, \mathcal{E}) \tag{9.79}$$

Similarly,

$$B(\mathcal{E}_n, \delta_t \mathcal{E}_{n-1/2}) = \frac{\Delta t}{2} B(\delta_t \mathcal{E}_{n-1/2}, \delta_t \mathcal{E}_{n-1/2}) + \tfrac{1}{2}\delta_{t_{n-1/2}} B(\mathcal{E}, \mathcal{E})$$

and

$$\left(\delta_t^2 \mathcal{E}_n, \delta_t \mathcal{E}_{n+1/2} + \delta_t \mathcal{E}_{n-1/2}\right) = \frac{1}{\Delta t}\left[\|\delta_t \mathcal{E}_{n+1/2}\|^2 - \|\delta_t \mathcal{E}_{n-1/2}\|^2\right]$$

Collecting these results, we obtain

$$\begin{aligned}
&\frac{1}{\Delta t}\left[\|\delta_t \mathcal{E}_{n+1/2}\|^2 - \|\delta_t \mathcal{E}_{n-1/2}\|^2\right] - \frac{\Delta t}{2} B(\delta_t \mathcal{E}_{n+1/2}, \delta_t \mathcal{E}_{n+1/2}) \\
&\qquad + \frac{\Delta t}{2} B(\delta_t \mathcal{E}_{n-1/2}, \delta_t \mathcal{E}_{n-1/2}) + \tfrac{1}{2}\delta_{t_{n+1/2}} B(\mathcal{E}, \mathcal{E}) + \tfrac{1}{2}\delta_{t_{n-1/2}} B(\mathcal{E}, \mathcal{E}) \\
&\quad = -\left(\delta_t^2 E_n, \delta_t \mathcal{E}_{n+1/2} + \delta_t \mathcal{E}_{n-1/2}\right) - B(E_n, \delta_t \mathcal{E}_{n+1/2} + \delta_t \mathcal{E}_{n-1/2}) \\
&\qquad + (\varepsilon_n, \delta_t \mathcal{E}_{n+1/2} + \delta_t \mathcal{E}_{n-1/2})
\end{aligned} \tag{9.80}$$

We next perform the following operations: (1) Eliminate the second term on the right-hand side using (9.75); (2) estimate the remaining terms on the right-hand side using the Schwarz inequality and the elementary inequality $ab \leqslant (a^2/4\varepsilon) + \varepsilon b^2, \varepsilon > 0$; (3) multiply by Δt; and (4) sum from 1 to $R-1$. This gives

$$\begin{aligned}
&\|\delta_t \mathcal{E}_{R-1/2}\|^2 - \|\delta_t \mathcal{E}_{1/2}\|^2 - \frac{\Delta t^2}{2} B(\delta_t \mathcal{E}_{R-1/2}, \delta_t \mathcal{E}_{R-1/2}) \\
&\qquad + \frac{\Delta t^2}{2} B(\delta_t \mathcal{E}_{1/2}, \delta_t \mathcal{E}_{1/2}) + \frac{1}{2} B(\mathcal{E}_R, \mathcal{E}_R) + \frac{1}{2} B(\mathcal{E}_{R-1}, \mathcal{E}_{R-1}) \\
&\qquad - \frac{1}{2} B(\mathcal{E}_1, \mathcal{E}_1) - \frac{1}{2} B(\mathcal{E}_0, \mathcal{E}_0) \leqslant \Delta t \sum_{n=1}^{R-1} \left(\kappa \|\delta_t^2 E_n\|^2 + \xi \|\varepsilon_n\|^2\right) \\
&\qquad + \Delta t \alpha \sum_{n=1}^{R-1} \left(\|\delta_t \mathcal{E}_{n+1/2}\|^2 + \|\delta_t \mathcal{E}_{n-1/2}\|^2\right)
\end{aligned} \tag{9.81}$$

where κ, ξ, and α are positive constants. Then using inequality (9.4), we conclude that there exists a positive constant M such that

$$B(\delta_t \mathcal{E}_{R-1/2}, \delta_t \mathcal{E}_{R-1/2}) \leqslant M \|\delta_t \mathcal{E}_{R-1/2}\|^2_{H^1(\Omega)}$$

We next make use of the inverse property of the subspace $S_h(\Omega)$. From it, there follows the inequality,

$$B\left(\delta_t \mathcal{E}_{R-1/2}, \delta_t \mathcal{E}_{R-1/2}\right) \leqslant \frac{M\overline{C}}{h^2} \|\delta_t \mathcal{E}_{R-1/2}\|^2 \tag{9.82}$$

Introducing (9.82) into (9.81) and using properties (9.4), we obtain

$$\begin{aligned}\left(1-\frac{M\overline{C}}{2}\frac{\Delta t^2}{h^2}\right)\|\delta_t \mathcal{E}_{R-1/2}\|^2 + \frac{\mu}{2}\|\mathcal{E}_R\|^2_{H^1(\Omega)} \\ +\frac{\mu}{2}\|\mathcal{E}_{R-1}\|^2_{H^1(\Omega)} \leqslant \frac{M}{2}\|\mathcal{E}_0\|^2_{H^1(\Omega)} + \frac{M}{2}\|\mathcal{E}_1\|^2_{H^1(\Omega)} + \|\delta_t \mathcal{E}_{1/2}\|^2 \\ +\Delta t \sum_{n=1}^{R-1}\left(\kappa\|\delta_t^2 E_n\|^2 + \xi\|\varepsilon_n\|^2\right) \\ +\Delta t \alpha \sum_{n=1}^{R-1}\left(\|\delta_t \mathcal{E}_{n+1/2}\|^2 + \|\delta_t \mathcal{E}_{n-1/2}\|^2\right)\end{aligned} \tag{9.83}$$

As a *condition of stability* we require that

$$1-\frac{M\overline{C}}{2}\frac{\Delta t^2}{h^2} = C'' \geqslant 0 \tag{9.84}$$

which, as expected, places a constraint on the permissible values of the discretization parameters.

We next need a discrete version of the Gronwall inequality (cf. Lees [9.27]).

LEMMA 9.1. If $\phi(t)$ and $\psi(t)$ are nonnegative functions with $\psi(t)$ nondecreasing and

$$\phi(R\Delta t) \leqslant \psi(R\Delta t) + C\Delta t \sum_{n=0}^{R-1} \phi(n\Delta t) \tag{9.85}$$

then

$$\phi(R\Delta t) \leqslant \psi(R\Delta t) e^{CR\Delta t} \tag{9.86}$$

where $C \geqslant 0$. ■

This is merely a discretization of the classical Gronwall inequality, which asserts that, if a and c are positive constants and if $u(t)$ is a function such that

$$|u(t)| \leqslant a + c\int_0^t |u(s)|\, ds \tag{9.87}$$

then

$$|u(t)| \leqslant ae^{Ct} \tag{9.88}$$

See, e.g., Bellman [9.28]. ■

Returning to (9.83), we now apply (9.86) and the inequality

$$\Delta t \sum_{n=1}^{R-1} \|\delta_t^2 V_n\| \leqslant \left\| \frac{\partial^2 V}{\partial t^2} \right\|_{L_2(L_2(\Omega))}$$

and we obtain (9.77). ■

The stage is now set for developing a final estimate of the approximation error. We need only quote a lemma given by Wheeler [9.8], which is established using essentially the same manipulations used in the proof of Theorem 9.7.

LEMMA 9.2. Let u, $\partial u/\partial t \in L_\infty(H^{k+1}(\Omega))$ and $\partial^2 u/\partial t^2 \in L_2(H^{k+1}(\Omega))$. Then there exists a constant C, independent of the discretization parameters, such that

$$\|\mathcal{E}\|_{L_\infty(L_2(\Omega))} + \left\|\frac{\partial \mathcal{E}}{\partial t}\right\|_{L_\infty(L_2(\Omega))} + \left\|\frac{\partial^2 \mathcal{E}}{\partial t^2}\right\|_{L_2(L_2(\Omega))}$$

$$\leqslant C\left(h^{k+1}\|u\|_{L_\infty(H^{k+1}(\Omega))} + h^{k+1}\left\|\frac{\partial u}{\partial t}\right\|_{L_\infty(H^{k+1}(\Omega))} + h^{k+1}\|u\|_{L_2(H^{k+1}(\Omega))}\right) \tag{9.89}$$

■

Summing up, we now need only apply the results of Theorem 9.7, Lemma 9.2, and the triangle inequality to arrive at the final estimate.

THEOREM 9.8. Let the conditions of Theorem 9.7 and Lemma 9.2. hold. Then there exist positive constants C_3 and C_4 which are independent of h and Δt, such that

$$\|\delta_t e\|_{\tilde{L}_\infty(L_2(\Omega))} + C_3\|e\|_{\tilde{L}_\infty(L_2(\Omega))}$$

$$\leqslant C_4\Bigg(\|e_0\|_{H^1(\Omega)} + \|e_1\|_{H^1(\Omega)} + \|\delta_t e_{1/2}\| + h^{k+1}\|u\|_{L_\infty(H^{k+1}(\Omega))} + h^{k+1}\left\|\frac{\partial u}{\partial t}\right\|_{L_\infty(H^{k+1}(\Omega))} + h^{k+1}\|u\|_{L_2(H^{k+1}(\Omega))} + \Delta t^2\left\|\frac{\partial^4 u}{\partial t^4}\right\|_{L_2(L_2(\Omega))}\Bigg) \tag{9.90}$$

■

REMARKS. 1. The procedures outlined in the lengthy proof of (9.90) can, with minor modifications, be applied to problems in which temporal discretizations other than elementary central differences are used. In this regard, see Wellford and Oden [9.16]. Of course, the character of the error estimates depends on which scheme is used.

The techniques used here represent adaptations of those of Dupont [9.25] to an explicit fully discrete scheme and, in some respects, are similar to those proposed by Fujii [9.14]. See also [9.26].

2. The numerical stability of the scheme described here is governed by (9.76), and the stability constant $\hat{C}$ depends on the continuity constant M and on the finite-element subspaces $S_h(\Omega)$. For a discussion of this point, see Fujii [9.14]. ■

9.7 FIRST-ORDER HYPERBOLIC EQUATIONS

As a final example, we consider a first-order problem of the form

$$\begin{aligned}\left(\frac{\partial u(t)}{\partial t},v\right)+(Lu(t),v)&=(f(t),v) \qquad \forall v\in H_0^1(\Omega);\ t\in(0,T]\\ (u(0),v)&=(u_0,v) \qquad \forall v\in H_0^1(\Omega)\end{aligned} \tag{9.91}$$

where $L(t)$ is the first-order linear operator

$$L(t)=\sum_{i=1}^{n} a_i(\mathbf{x},t)D^i \qquad D^i=\partial/\partial x_i \tag{9.92}$$

Now we construct a semidiscrete Galerkin approximation of (9.91), which involves seeking a function $U(\mathbf{x},\mathrm{t})\in C^1[0,T;S_h(\Omega)]$ such that

$$\begin{aligned}\left(\frac{\partial U(t)}{\partial t},V\right)+(LU(t),V)&=(f(t),V) \qquad \forall V\in S_h(\Omega)\\ (U(\cdot,0),V)&=(u_0,V) \qquad \forall t\in(0,T]\end{aligned} \tag{9.93}$$

Introducing the finite-element approximations (9.6) and (9.7) into (9.93) we obtain a system of first-order differential equations for the coefficients $A^j(t)$:

$$\left.\begin{aligned}\sum_{j=1}^{G} G_{ij}\dot{A}^j(t)+\sum_{j=1}^{G} D_{ij}A^j(t)&=f_i(t)\\ \sum_{j=1}^{G} G_{ij}A^j(0)&=l_i\end{aligned}\right\} \tag{9.94}$$

where

$$\left.\begin{aligned} G_{ij} = (\phi_i, \phi_j) \qquad D_{ij} = (L\phi_i, \phi_j) \qquad f_i = (f, \phi_i) \\ l_i = (u_0, \phi_i) \quad \text{and} \quad \dot{A}^i(t) = \frac{dA^i(t)}{dt} \end{aligned}\right\} \tag{9.95}$$

We may also construct a Galerkin approximation in which the temporal operator is replaced by a finite difference operator in time through a procedure similar to those described earlier. In this way we construct a family of finite-difference Galerkin approximations associated with the parameter θ $(0 \leqslant \theta \leqslant 1)$ which represent solutions to the following equation:

$$\begin{gathered} (\delta_t U^n, V) + (1-\theta)(LU^{n+1}, V) + \theta(LU^n, V) = (f(t), V) \\ t \in (0, T]; \ \forall V \in S_h(\Omega) \\ (U^0, V) = (u_0, V) \end{gathered} \tag{9.96}$$

where δ_t denotes the forward difference operator in time; i.e., $\delta_t U^n = (U^{n+1} - U^n)/\Delta t$. This process leads to a system of algebraic equations for the coefficients $A_i^n \approx A^i(n\Delta t)$:

$$\left.\begin{aligned} \sum_{j=1}^{G} \left[G_{ij} + \Delta t(1-\theta) D_{ij} \right] A_j^{n+1} = \sum_{j=1}^{G} \left[G_{ij} - \Delta t \theta D_{ij} \right] A_j^n + \Delta t f_i \\ \sum_{j=1}^{G} G_{ij} A_j^0 = l_i; \ i = 1, 2, \ldots, G \end{aligned}\right\} \tag{9.97}$$

We briefly describe a modification of the semigroup methods developed in Section 9.5 to (9.96) following the plan in [9.18]. Again, we decompose the error into the three components given in (9.40) and make use of (9.41) after determining error bounds for each component. To establish the semidiscrete approximation error, we use a technique essentially the same as adopted in Section 9.5; i.e., we take the Laplace transform of the weak hyperbolic equation (9.91), with $f(t) = 0$, and denote the transformed solution by $\bar{u}(\mathbf{x}, s) = \mathfrak{L}[u(\mathbf{x}, t)]$. This leads to the problem

$$s(\bar{u}(s), v) + (L\bar{u}(s), v) = (u_0, v) \qquad \forall v \in H_0^1(\Omega) \tag{9.98}$$

Similarly, taking the Laplace transform of the semidiscrete Galerkin approximation (9.93) and denoting the transformed solution by $\bar{U}(\mathbf{x}, s)$

$= \mathcal{L}[U(\mathbf{x},t)]$, we have

$$s(\overline{U}(s),V)+(L\overline{U}(s),V)=(u_0,V) \qquad \forall V\in S_h(\Omega) \tag{9.99}$$

Now subtracting (9.99) from (9.98) and defining the transformed approximation error by $\bar{u}(s)-\overline{U}(s)=\bar{\sigma}(s)=\mathcal{L}[\sigma(t)]=\mathcal{L}[u(t)-U(t)]$, we obtain

$$s(\bar{\sigma}(s),V)+(L\bar{\sigma}(s),V)=0 \qquad \forall V\in S_h(\Omega) \tag{9.100}$$

We can now introduce a theorem establishing the magnitude of $\bar{\sigma}$, the transformed semidiscrete approximation error.

THEOREM 9.9. Let $u(\mathbf{x},n\Delta t)$ be the solution to (9.93), $U(\mathbf{x},n\Delta t)$ be the solution to (9.96) with $\theta=1$, and $\sigma(x,n\Delta t)$ be the semidiscrete approximation error. Then there exists a constant C_1 such that

$$\|\bar{\sigma}\| \leqslant 2\|\overline{E}\| + \frac{C_1}{s}\|\overline{E}\|_{H^1(\Omega)} \tag{9.101}$$

PROOF. Let $\overline{U}^*$ be the element of $S_h(\Omega)$ that interpolates $\bar{u}$, and let $\overline{E}=\bar{u}-\overline{U}^*$ and $\overline{\mathcal{E}}=\overline{U}^*-\overline{U}$. Then $\bar{\sigma}=\overline{E}+\overline{\mathcal{E}}$. Thus (9.100), with $V=\overline{\mathcal{E}}$, leads directly to

$$(\overline{\mathcal{E}},\overline{\mathcal{E}})=-(\overline{E},\overline{\mathcal{E}})-\frac{1}{s}\left[(L\overline{E},\overline{\mathcal{E}})+(L\overline{\mathcal{E}},\overline{\mathcal{E}})\right]$$

Note that $(L\overline{\mathcal{E}},\overline{\mathcal{E}})$ can be expressed as an integral over the boundary. Thus, since the boundary conditions are homogeneous, $(L\overline{\mathcal{E}},\overline{\mathcal{E}})=0$. Using this result and the Schwarz inequality, we obtain.

$$\|\overline{\mathcal{E}}\| \leqslant \|\overline{E}\| + \frac{1}{s}\|L\overline{E}\| \leqslant \|\overline{E}\| + \frac{C_1}{s}\|\overline{E}\|_{H^1(\Omega)}$$

Now using the triangle inequality, we obtain (9.101). ■

The semidiscrete approximation error is established by taking the inverse Laplace transform in (9.101):

THEOREM 9.10. Let the hypotheses of Theorem 9.9 hold. Then

$$\begin{aligned}\|\sigma(\cdot,n\Delta t)\| &\leqslant 2\|E(\cdot,n\Delta t)\| + C_1\int_0^{n\Delta t}\|E(\cdot,t)\|_{H^1(\Omega)}dt \\ &\leqslant C_2h^{k+1}\|u\|_{H^{k+1}(\Omega)} + C_3h^k\int_0^{n\Delta t}\|u(t)\|_{H^{k+1}(\Omega)}dt \end{aligned} \tag{9.102}$$

■

Note that the rate of convergence is (h^k); iee., the rate of convergence is one power lower than the rate of convergence for the finite-element approximation of the second-order diffusion equation.

The temporal approximation error is established by exactly the same procedure as was used in Section 9.5 and is given in Theorem 9.5. The total error for the finite-element Galerkin approximation is summarized in the following theorem.

THEOREM 9.11. If the components of the error of the approximation (9.96), $\theta=1$, of (9.91) are given by (9.40), there exist constants C_2, C_3, and C_4 such that

$$\|e(\cdot,n\Delta t)\| \leqslant C_2 h^{k+1}\|u\|_{H^{k+1}(\Omega)} + C_3 h^k \int_0^{n\Delta t} \|u\|_{H^{k+1}(\Omega)}\,dt \quad (9.103)$$

■

REFERENCES

9.1. Douglas, J. and Dupont, T., "Galerkin Methods for Parabolic Equations," *SIAM J. Numer. Anal.* Vol. 7, No. 4, pp. 575–626, 1970.

9.2. Dupont, T., "Some L^2 Error Estimates for Parabolic Galerkin Methods," in Aziz, A. K., Ed., *The Mathematical Foundations of the Finite Element Method with Applications to Partial Differential Equations*, Academic Press, New York, 1972, pp. 291–504.

9.3. Douglas, J. and Dupont, T., "Galerkin Methods for Parabolic Equations with Nonlinear Boundary Conditions," *Numer. Math.*, Vol. 20, pp. 213–237, 1973.

9.4. Douglas, J. and Dupont, T., *Collocation Methods for Parabolic Equations in a Single Space Variable*, Lecture Notes in Mathematics, No. 385, Springer-Verlag, Berlin, 1974.

9.5. Dupont, T., "Galerkin Methods for First-Order Hyperbolics: An Example," *SIAM J. Numer. Anal.* Vol. 10, No. 5, pp. 890–899, 1973.

9.6. Douglas, J., *Lectures on Finite Element Methods for Elliptic and Parabolic Problems*, unpublished manuscript, University of Chicago, 1974.

9.7. Douglas, J. and Dupont, T., "A Finite Element Collocation Method for the Heat Equation," *Symposia Mathematica*, Vol. X, Istituto Nazionale di Alta Matematica, Bologna, 1972.

9.8. Wheeler, M. F., "A Priori L_2 Error Estimates for Galerkin Approximations to Parabolic Partial Differential Equations," *SIAM J. Numer. Anal.*, Vol. 10, No. 4, pp. 723–759, 1973.

9.9. Price, H. S. and Varga, R. S., "Error Bounds for Semidiscrete Galerkin Approximations of Parabolic Problems with Applications to Petroleum Reservoir Mechanics," in Birkhoff, G. and Varga, R. S., Eds., *Numerical Solution to Field Problems in Continuum Mechanics*, Vol. II, SIAM-AMS Proceedings, Providence, R. I., 1970, pp. 74–94.

9.10. Thomee, V., "Spline Approximation and Difference Schemes for the Heat Equation," in Aziz, A. K., Ed., *The Mathematical Foundations of the Finite Element Method with Application to Partial Differential Equations*, Academic Press, New York, 1972, pp. 711–746.

9.11. Varga, R. S., *Functional Analysis and Approximation Theory in Numerical Analysis*, Society for Industrial and Applied Mathematics, Philadelphia, 1971.

9.12. Oden, J. T., "Mathematical Aspects of Finite Element Approximations in Theoretical Mechanics," in *Mechanics Today*, Nemat-Nasser, S., Ed., Vol. 2, Pergamon Press, Oxford, 1975 pp. 159–250.

9.13. Strang, G. and Fix, G., *An Analysis of the Finite Element Method*, Prentice-Hall, Englewood Cliffs, N. J., 1973.

9.14. Fujii, H., "Finite Element Schemes: Stability and Convergence," in Oden, J. T., Yamamoto, Y., and Clough, R. W., Eds., *Advances in Computational Methods in Structural Mechanics and Design*, UAH Press, Huntsville, Ala., 1972, pp. 201–218.

9.15. Oden, J. T. and Fost, R. B., "Convergence, Accuracy, and Stability of Finite Element Approximations of a Class of Non-Linear Hyperbolic Equations," *Int. J. Numer. Meth. Eng.*, Vol. 6, pp. 357–365, 1973.

9.16. Wellford, L. C., Jr. and Oden, J. T., "On Some Finite Element Methods for Certain Nonlinear Second-Order Hyperbolic Equations," *TICOM Report*, No. 74–7, 1974.

9.17. Wellford, L. C., Jr. and Oden, J. T., "Accuracy and Convergence of Finite Element/Galerkin Approximations of Time Dependent Problems with Emphasis on Diffusion," in Gallagher, R. H., Oden, J. T., Taylor, C., and Zienkiewicz, O. C., Eds., *Finite Element Analyses of Flow Problems*, John Wiley, London, 1975, pp. 31–54.

9.18. Wellford, L. C., Jr. and Oden, J. T., "A Note on the Accuracy and Convergence of Finite Element Approximations of the Convection Equations," *Comput. Meth. in Appl. Mech. Eng.*, Vol. 5, 1, pp. 83–96, 1975.

9.19. Hille, E. and Phillips, R., *Functional Analysis and Semigroups*, American Mathematical Society Colloquium Publications, Vol. 31, 1957.

9.20. Butzer, P. L. and Berens, H., *Semigroups of Operators and Approximations*, Springer-Verlag, New York, 1967.

9.21. Goldstein, J., *Semigroups of Operators and Abstract Cauchy Problems*, Tulane University, New Orleans, La., 1970.

9.22. Schechter, M., *Principles of Functional Analysis*, Academic Press, New York, 1971.

9.23. Widlund, O. B., "On the Rate of Convergence for Parabolic Difference Schemes, II," *Commun. Pure Appl. Math.*, Vol. 23, pp. 79–96, 1970.

9.24. Richtmyer, R. and Morton, K., *Difference Methods for Initial-Value Problems*, Interscience, New York, 1967.

9.25. Dupont, T., "L^2-Estimates for Galerkin Methods for Second-Order Hyperbolic Equations," *SIAM J. Numer. Anal.*, Vol. 10, pp. 880–889, 1973.

9.26. Peetre, J. and Thomee, V., "On the Rate of Convergence for Discrete Initial-Value Problems," *Math. Scand.*, Vol. 21, pp. 159–176, 1967.

9.27. Lees, M., "A Priori Estimates for the Solution of Difference Approximations to Parabolic Partial Differential Equations," *Duke Math. J.*, Vol. 27, pp. 297–311, 1960.

9.28. Bellman, R., *Stability Theory of Differential Equations*, McGraw-Hill, New York, 1952.

AUTHOR INDEX

SUBJECT INDEX